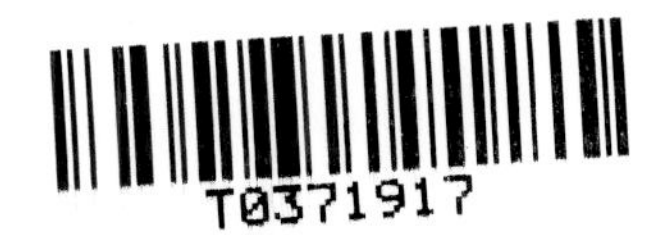

Routledge Handbook of Human Rights and Climate Governance

Over the last decade, the world has increasingly grappled with the complex linkages emerging between efforts to combat climate change and to protect human rights around the world. The Paris Climate Agreement adopted in December 2015 recognized the necessity for governments to take into consideration their human rights obligations when taking climate action. However, important gaps remain in understanding how human rights can be used in practice to develop and implement effective and equitable solutions to climate change at multiple levels of governance.

This book brings together leading scholars and practitioners to offer a timely and comprehensive analysis of the opportunities and challenges for integrating human rights in diverse areas and forms of global climate governance. The first half of the book explores how human rights principles and obligations can be used to reconceive climate governance and shape responses to particular aspects of climate change. The second half of the book identifies lessons in the integration of human rights in climate advocacy and governance and sets out future directions in this burgeoning domain.

Featuring a diverse range of contributors and case studies, this Handbook will be an essential resource for students, scholars, practitioners and policy makers with an interest in climate law and governance, human rights and international environmental law.

Sébastien Duyck is a Senior Attorney at the Center for International Environmental Law, USA, and an affiliated researcher at the Northern Institute for Environmental and Minority Law/ Arctic Centre, University of Lapland, Finland.

Sébastien Jodoin is an Assistant Professor in the Faculty of Law at McGill University, Canada, and an Associate Member of the McGill School of Environment.

Alyssa Johl is a founding member of the Climate Rights Collective. She previously served as a Senior Attorney at the Center for International Environmental Law, USA.

Routledge Handbook of Human Rights and Climate Governance

Edited by Sébastien Duyck, Sébastien Jodoin and Alyssa Johl

First published 2018
by Routledge

2 Park Square, Milton Park, Abingdon, Oxfordshire OX14 4RN
52 Vanderbilt Avenue, New York, NY 10017

Routledge is an imprint of the Taylor & Francis Group, an informa business

First issued in paperback 2020

British Library Cataloguing-in-Publication Data
A catalogue record for this book is available from the British Library

Library of Congress Cataloging-in-Publication Data
A catalog record for this book has been requested

ISBN: 978-1-138-23245-7 (hbk)
ISBN: 978-0-367-51876-9 (pbk)

Typeset in Bembo
by Apex CoVantage, LLC

Contents

PART VI
Future directions 357

Figures

Contributors

Sébastien Duyck is a Senior Attorney at the Center for International Environmental Law (CIEL). As coordinator of the Human Rights & Climate Change Working Group, Sébastien contributes to advocacy for the promotion and protection of human rights in the context of the Paris Agreement as well as through human rights–related forums. Sébastien is also an affiliated researcher at the Northern Institute for Environmental and Minority Law/Arctic Centre (University of Lapland) and a research fellow at the Institute of European and International Economic Law (University of Bern). He holds an LL.M. on Public International Law from the University of Helsinki and an LL.M. in International Environmental Law and the Law of Natural Resources from the University of Iceland. His main areas of research relate to human rights in the context of environmental policy and to UN climate governance. Sébastien has taught courses related to climate governance, environmental law and human rights law at the University of Lapland, University of Bern, University of Münster and Wake Forest University. Prior to joining CIEL, Sébastien contributed to several projects related to European forest governance, the role of the EU in the Arctic, the promotion of human rights in climate action, the relevance of agriculture in international climate governance, the promotion of safeguards in climate finance mechanisms, the implementation of the Paris Agreement and the Agenda 2030 through the work of the Arctic Council, and the responsibility of the European Union with regards to human rights impacts of climate-related actions.

Sébastien Jodoin is an Assistant Professor in the Faculty of Law of McGill University, where he directs the Law, Governance & Society Lab. He is also an Associate Member of the McGill School of Environment, a member of the McGill Centre for Human Rights & Legal Pluralism, and a Faculty Associate of the Governance, Environment & Markets Initiative at Yale University. Prior to his appointment at McGill, Sébastien worked for the Centre for International Sustainable Development Law, the Canadian Centre for International Justice, Amnesty International Canada and the United Nations. He continues to work with a range of international and non-governmental organizations on issues that relate to his research on transnational law, climate change, environmental governance, human rights and sustainable development. Sébastien holds a Ph.D. in environmental studies from Yale University, an M.Phil. in international relations from the University of Cambridge, an LL.M. in international law from the London School of Economics, and a B.C.L. and LL.B. from McGill University. He has received numerous awards and honors, including the 2012 Public Scholar Award from the Yale Graduate School of Arts & Sciences and a Doctoral Scholarship from the Pierre Elliott Trudeau Foundation, and currently serves as a principal investigator for grants awarded by the Social Sciences and Humanities and Research Council of Canada and the Fonds québécois de recherche société et culture.

Alyssa Johl is founder of the Climate Rights Collective, a collaborative network of individuals developing strategic and practical solutions to the climate crisis through a rights and justice lens. Prior to establishing the Climate Rights Collective, she worked as a Senior Attorney at the Center for International Environmental Law, advocating to protect the rights of those most vulnerable to climate change impacts and developing new legal strategies to accelerate action on climate change. She previously worked as a Staff Attorney at the Climate Law and Policy Project, where she advocated for stronger emission reduction targets, adaptation assistance, and inclusion in policy processes at all levels of governments. Alyssa also serves on the Steering Committee of Many Strong Voices, a network of individuals and communities in the Arctic and small island developing states (SIDS) taking strategic action on climate mitigation and adaptation. Alyssa received her J.D., as well as a certification of completion in Environmental and Natural Resources Law, from the University of Oregon School of Law. She also holds a B.A. in Development Studies from the University of California at Berkeley.

Sumudu Anopama Atapattu is the Director of Research Centers and Senior Lecturer at the University of Wisconsin Law School. She teaches seminar classes on 'International Environmental Law' and 'Climate Change, Human Rights and the Environment'. She is affiliated with UW-Madison's Nelson Institute for Environmental Studies and the Center for South Asia and is the Executive Director of the campus-wide interdisciplinary Human Rights Program. She was a visiting professor at Doshisha University Law School, Japan, in summer 2014 and at the Giessen University Law School in summer 2016. She serves as the Lead Counsel for Human Rights at the Center for International Sustainable Development Law based in Montreal, Canada, and is on the advisory board of the *McGill Journal of Sustainable Development Law*. She has participated in several consultations organized by the UN Independent Expert on Human Rights and the Environment, including a consultation on Climate Change and Human Rights with the UN Office of the High Commissioner for Human Rights. She has published widely in the fields of international environmental law, climate change, environmental rights and sustainable development and her book *Human Rights Approaches to Climate Change: Challenges and Opportunities* was published by Routledge in 2016. She was a co-editor of *International Environmental Law and the Global South* published by Cambridge University Press in 2015 and is currently working on a textbook on human rights and environment for Routledge. She holds an LL.M. (Public International Law) and a Ph.D. (International Environmental Law) from the University of Cambridge, UK, and is an Attorney-at-Law of the Supreme Court of Sri Lanka.

Anne Barre is the Gender and Climate Policy Coordinator at WECF – Women Engage for a Common Future, a member of the Women and Gender Constituency of the UNFCCC. Anne has worked with WECF since 2000 on different gender and sustainable development programmes. She founded WECF France in 2008, today a member of the French Climate Action Network. Anne Barre co-coordinates the civil society activities for the Women and Gender Constituency, including the Gender Just Climate Solutions Award.

Lisa Benjamin is an Assistant Professor at The University of The Bahamas, a member of the Compliance Committee (Facilitative Branch) of the UNFCCC, and co-founder of the Climate Change Initiative. She is a director of the Bahamas Protected Areas Fund and has been an advisor to the Bahamian national delegation to the UNFCCC. Her current research looks at climate change and SIDS, companies and investors and climate change, as well as trade and environmental law.

Eleanor Blomstrom is Co-Director and Head of Office at WEDO where she manages relationships and partnerships and works closely with program staff in strategic development, implementation and monitoring of WEDO programs and projects. Her research, capacity building and advocacy focuses on sustainable development, climate change, disaster risk reduction and urbanization. She represents WEDO as organizing partner of the Women's Major Group, facilitating space for women's rights action and collaboration around the SDGs. She holds a master's degree in international affairs.

Joanna Bourke Martignoni has a Doctor of Laws from the University of Fribourg, Switzerland, as well as degrees in law and South East Asian history from the University of New South Wales, Australia. She has carried out research and published on a range of human rights issues, including non-discrimination guarantees in international law, the rights to food and education, children's rights, labour law, violence against women, the protection of cultural heritage and the rights of Indigenous Peoples. Prior to taking up her academic role at the Geneva Academy of International Humanitarian Law and Human Rights, she worked with the International Committee of the Red Cross, the World Organisation Against Torture and the Office of the UN High Commissioner for Human Rights.

Donald A. Brown is a Scholar in Residence for Sustainability Ethics and Law at the Widener University's Environmental Law and Sustainability Center. Donald directed the Pennsylvania Environmental Research Consortium, an organization of 56 Pennsylvania universities, and the Pennsylvania Departments of Environmental Protection and Conservation and Natural Resources. He previously worked as program manager for United Nations Organizations at the US Environmental Protection Agency's Office of International Environmental Policy. He represented the agency on US delegations to the United Nations, negotiating climate change, biodiversity and sustainable development issues. At Widener Law, he researches and presents public presentations on climate change, ethics and sustainability.

Bridget Burns is the Co-Director of the Women's Environment and Development Organization (WEDO). In her international policy work, Bridget has facilitated travel support and capacity building for over 250 women from Least Developed Countries to participate as part of their national climate delegations. Bridget also serves as co-Focal Point of the Women and Gender Constituency, supporting political participation of women's rights advocates into the climate process. Bridget has a degree in Gender, Development and Globalization from the London School of Economics (LSE).

Isabell Büschel (Ph.D.) is specialized in EU law and policies. As Senior Policy Consultant for the European Federation for Transport and Environment (T&E), she advises on climate-friendly mobility solutions in Spain. Among over 20 publications, she authored a book chapter titled 'Les politiques européennes d'adaptation et les droits de l'homme' (Bruylant, Bruxelles, 2013) and Chapters 4 and 11 of 'European Union Adaptation Policies and Human Rights' of the European Parliament's report *Human Rights and Climate Change: EU Policy Options* (2012).

Nathalie Chalifour is an Associate Professor at the Faculty of Law and Co-Director of the Centre for Environmental Law and Global Sustainability at the University of Ottawa. Her research focuses on the intersection between environmental law, the economy and social justice. Her most recent articles focus on environmental rights, climate justice, carbon pricing and sustainable food. She is the co-editor of three international books, including *Energy, Governance and Sustainability* (Edward Elgar 2017).

Lynda M. Collins is a Professor in the Centre for Environmental Law & Global Sustainability at the University of Ottawa. She is an expert in environmental human rights, including constitutional environmental rights, indigenous environmental rights and environmental rights in private law. She is past co-chair of Ontario's Toxic Reduction Scientific Expert Panel and co-author of *The Canadian Law of Toxic Torts*, with Heather McLeod-Kilmurray.

Irene Dankelman is Director of IRDANA advice on gender and sustainable development and Lecturer at the Radboud University in Nijmegen (Netherlands). During her almost 40 years professional life, she has worked for UNIFEM and Oxfam-Novib, IUCN-Netherlands and several universities. She is actively involved in international, national and local NGOs and advises the United Nations and governmental and academic organizations. She co-authored the book *Women and Environment in the Third World* (Earthscan, 1988; with Joan Davidson), was editor of *Gender and Climate Change: An Introduction* (Earthscan, 2010) and lead author of UNEP's *Global Gender and Environment Outlook* (2016).

Deborah Delgado Pugley is Assistant Professor at the Pontificia Universidad Catolica del Peru. She has conducted research on climate change policies, indigenous social movements, human rights and natural resources management. She has a Ph.D. in Development Studies and Sociology from the Université Catholique de Louvain and the Ecole des hautes études en sciences sociales of Paris.

Kate Dooley holds a Masters in Environmental Technology from Imperial College London and is currently a Ph.D. candidate at the University of Melbourne, where her research focuses on the environmental integrity of terrestrial carbon accounting and the justice and equity implications of land-based climate mitigation. Kate also advises a range of non-governmental organisations on issues related to land use, carbon trading, equity and human rights at the UNFCCC.

Rosine Faucher is a Research Assistant with the Law, Governance & Sustainability Lab, where she is working on projects on human rights and climate governance, REDD+ and carbon rights. She is currently completing her B.C.L. and LL.B. degrees at the McGill Faculty of Law. She also holds a B.A. honours in political science and environment from McGill University.

Sven Harmeling has been working as CARE International's Climate Change Advocacy Coordinator since 2013 and was previously Team Leader for International Climate Policy with Germanwatch (2005–2013). He has been engaged in the international policy debate on climate change impacts, adaptation, and loss and damage for almost 10 years, with various publications.

Rueanna Haynes was former First Secretary at the Trinidad and Tobago Permanent Mission to the United Nations, New York, and possesses extensive experience in international negotiations on climate change and sustainable development. Currently, she is a Legal Advisor with Climate Analytics gGmbH and provides support to island nations in climate change negotiations. She is pursuing a Masters in International Affairs and International Law with the Paris School of International Affairs and the Georgetown University Law Centre in Washington, DC.

Heta Heiskanen (M.Soc.Sci., M.Adm.Sci.) is a human rights researcher. She is at the final stage of preparing her Ph.D. thesis on the green jurisprudence of the European Court of Human Rights. Heiskanen has also researched, for example, the Finnish constitutional environmental right and rights of Indigenous Peoples. Currently she works as an Instructor in the Faculty of Management, University of Tampere, Finland.

Felix Horne is the Senior Ethiopia and Eritrea researcher for Human Rights Watch. Based in Ottawa, his research publications have included the human rights dimensions of Ethiopia's development programs, telecom surveillance, media freedoms, and abuses by security forces. Felix has also worked on a variety of indigenous rights and land issues in northern Canada and internationally, including research into the impacts of agricultural investment in several African countries. He holds a Masters in Resource and Environmental Management from Dalhousie University.

Eniko Horvath is Senior Researcher at Business & Human Rights Resource Centre, where she covers climate justice, community-driven initiatives and the UN Guiding Principles on Business and Human Rights. She has experience working on human rights in Argentina, Hungary, Spain and the UK. Eniko holds a B.A. from Harvard University and an M.Phil. from University of Cambridge.

Ademola Oluborode Jegede is a Senior Lecturer at the Department of Public and International Law, School of Law, University of Venda, Thohoyandou, South Africa. He obtained a doctoral degree and an LL.M. (Human Rights & Democratisation in Africa) from the University of Pretoria. Also, he possesses an LL.B. with honors from the Obafemi Awolowo University, Ile-Ife, a Master of Public Health (M.P.H.) from the University of Ibadan in Nigeria. His general areas of research interest are the interface of climate change and environment with the human rights of vulnerable populations, negotiation, conflicts and constitutionalism. He has published in peer-reviewed journals on these areas. Ademola is a research visitor to the Centre for International Environmental Law, USA, a Fellow of the Institute for Human Rights, Abo Akademi, Finland, and the Salzburg Global Seminar, Austria.

Sabaa A. Khan is a Senior Researcher and Lecturer at the Center for Climate Change, Energy and Environmental Law based at the School of Law, University of Eastern Finland. In 2017, she was appointed by the Government of Canada to serve on the Joint Public Advisory Committee of the Commission for Environmental Cooperation. Sabaa is a Member of the Barreau du Québec and holds a doctorate from McGill University's Faculty of Law, with specialization in international trade and environmental law.

Felix Kirchmeier manages the conceptualization, development and implementation of the Geneva Academy Policy Studies Section's research projects, fundraising and outreach activities, including the publication of research results, and activities related to the UN Human Rights Council, other human rights mechanisms and the disarmament and security debates. Before joining the Academy, he worked at the Geneva office of the Friedrich-Ebert-Stiftung, coordinating projects on security policy, development and human rights, with a special focus on the links between human rights and climate change.

Yves Lador works as a Consultant and represents Earthjustice to the UN in Geneva. He carries out advocacy at the Human Rights Council (and previously at the Commission on Human

Rights) with regard to issues such as human rights and the environment, climate change or the right to water and sanitation. He was actively involved, with the NGO coalition, in the negotiation of the Aarhus Convention on access to information, public participation and access to justice in environmental matters. He has been monitoring the communications brought by the public to the Aarhus Convention Compliance Committee since its creation.

Katherine Lofts is a Research Associate with McGill University's Law, Governance and Sustainability Lab, as well as a Legal Research Fellow with the Centre for International Sustainable Development Law's Climate Change Programme. She recently completed an LL.M. at McGill University, focused on rights-based discourses in the international climate regime. Katherine also holds a B.C.L. and LL.B. (McGill), an M.A. in English literature (McGill), and a B.A. in English literature (University of Victoria), and has studied law at the Université Paris 1 (Panthéon-Sorbonne).

Kasumi Maeda was formerly a Researcher at the Business & Human Rights Resource Centre, where she focused on climate change and the UN Guiding Principles. She currently works at the Institute for War & Peace Reporting. She has worked on various human rights issues, including labour rights, genocides, child marriages, the right to education in India, and child disappearances in El Salvador. Kasumi holds an M.Sc. Human Rights from the London School of Economics and Political Science.

Heather McLeod-Kilmurray is an Associate at the Faculty of Law and a Co-Director of the Centre for Environmental Law and Global Sustainability at the University of Ottawa (CELGS). Her recent research deals with toxic torts, environmental justice and food law, including GMOs and industrial factory farming. She is co-author of *The Canadian Law of Toxic Torts* (Canada law book), with Prof. Lynda Collins, and co-editor of several books in the IUCN Academy Environmental Law Series (Edward Elgar), such as *Climate Law and Developing Countries, Biodiversity and Climate Change* and *The Law and Policy of Biofuels*.

Edouard Morena lectures in French and European politics at the University of London Institute in Paris (ULIP) and New York University Paris. His research focuses on non-state actors' engagement in international environmental and development processes. Recent publications include *The Price of Climate Action: Philanthropic Foundations in the International Climate Debate* (Palgrave, 2016) and *Globalising the Climate: COP21 and the Climatisation of Global Debates* (Routledge, 2017).

Gita Parihar is an Independent Legal Consultant, who was Head of Legal at Friends of the Earth England, Wales and Northern Ireland for 10 years. She is an expert on international climate and environmental rights issues and an experienced litigator.

Joni Pegram joined Unicef UK as their Climate Change Policy & Advocacy Adviser in 2014, having previously held positions at the UN Environment Programme, European Union External Action Service and UK Foreign and Commonwealth Office. She has wide-ranging experience in shaping and delivering policy in the field of sustainable development at the national and multilateral levels, with a particular focus on environmental and human rights issues. Her current focus is on raising awareness of the impact of climate change on child rights and advocating on this basis within the UNFCCC.

Ben Powless is of Mohawk and Ojibwe nationhood and is based in Ottawa, Canada. He has a degree in Human Rights, Indigenous and Environmental Studies from Carleton University. He has worked with the Indigenous Environmental Network on issues of climate justice and Indigenous rights, and was a co-founder of the Canadian Youth Climate Coalition and the Defenders of the Land network.

Katharina Rall is a Researcher with the Environment and Human Rights Division at Human Rights Watch, where her work focuses on human rights violations in the context of climate change and environmental health. Katharina also researches economic, social and cultural rights more broadly with the Center for Human Rights and Global Justice at New York University School of Law. She holds a law degree from the University of Goettingen and graduated from New York University School of Law with an LL.M. in International Legal Studies.

Andrea Rodriguez is an expert in climate finance. She is an Environmental Lawyer from Bolivia, with Masters degrees in sustainable development and environmental law from Uppsala University and Stockholm University respectively. Andrea led the Climate Change Program at AIDA in Latin America, in which she helped the design of the Green Climate Fund's operational rules. She also served as an active civil society observer at the fund. Andrea currently works at Fundación Avina, supporting Latin American governments in accessing climate financing from the GCF.

Annalisa Savaresi is a Lecturer in Environmental Law at the University of Stirling, UK, where she co-directs the Masters in Environmental Policy and Governance. She has several years' experience researching, teaching and working with environmental law. Her research focuses on climate change, biodiversity, forestry, environmental liability, renewable energy and the interplay between environmental and human rights law. Her work has been published in numerous peer-reviewed outlets and has been widely cited, including by the United Nations Environment Programme and the United Nations Permanent Forum on Indigenous Issues. Annalisa is a member of the IUCN Commission on Environmental Law and of Women's Energy and Climate Law Network. She is associate editor of the *Review of European, Comparative and International Law* and has reported on the negotiations of the Paris Agreement for the Earth Negotiation Bulletin, published by the International Institute for Sustainable Development.

Andrea Schapper (Dipl.-Sozialwiss., Ph.D.) is a Lecturer in International Politics at the University of Stirling, UK. Prior to this, she was an Assistant Professor in International Relations at the Technische Universität Darmstadt in Germany (2012–2015). Her Ph.D. is from the Bremen International Graduate School of Social Sciences (Universität Bremen, 2011) and she has previously also studied at Cornell University (USA), Leibniz Universität Hannover (Germany) and the United Nations Office at Geneva (United Nations Graduate Study Program, Switzerland).

Allison Silverman is a Vice President of the SC Group, a nonprofit philanthropic collective, responsible for environmental, energy and social justice issues. Previously, Allie was a Solar Ombudsman for New York, where she led project-based change strategies to advance equitable access to solar energy and to encourage urban resiliency. She has been working on climate and energy justice issues for over a decade, including implementing a solar rural electrification and sustainable development project in Panama as a Fulbright Scholar, coordinating a campaign with the Natural Resources Defense Council to promote energy alternatives to an ill-conceived massive hydroelectric scheme in Chile, and advocating a rights-based approach to international climate change policies that safeguard communities and ecosystems as an attorney for the Center for International Environmental Law.

Doreen Stabinsky is Professor of Global Environmental Politics at College of the Atlantic in Bar Harbor, Maine, USA. Her research focuses on the impacts of climate change on agriculture and food security; adaptation and adaptation institutions under the UN Framework Convention on Climate Change (UNFCCC); and the emerging issue of loss and damage from slow onset impacts of climate change. In 2015–2016, she held the first Zennström visiting professorship in Climate Change Leadership at Uppsala University, Sweden.

Anke Stock is an Expert in gender issues and sustainable development. She worked for human rights organisations and since 2005 has been employed at WECF in Munich, a network of NGOs focusing on integrating gender issues into environmentally sound and sustainable development work. Her work experience contains consultancies for various UN entities, e.g. UNECE and UNDP. Her special focus areas are gender mainstreaming into environmental projects and policies as well as public participation.

Alice Thomas is the Climate Displacement Program Manager at Refugees International. She previously worked as a staff attorney in the international program at Earthjustice, where she implemented programs designed to defend and seek redress for communities from the impacts of pollution and climate change. She has a law degree from the University of Wisconsin Law School and a Bachelor's degree in History from Princeton University.

Basil Ugochukwu holds a Bachelor of Laws (LL.B.), Abia State University, Uturu, Nigeria; a Master of Laws (LL.M.), Central European University, Budapest, Hungary; and a Doctor of Philosophy (Ph.D.), Osgoode Hall Law School, York University, Canada. He is currently a Post-Doctoral Fellow at the Centre for International Governance Innovation (CIGI) in Waterloo, Ontario, Canada. He is an Open Society Justice Initiative Fellow and winner of the Osgoode Hall Law School Walter Williston Essay Prize on Civil Liberties in 2010. He was Lead Editor of the *Osgoode Hall Review of Law and Policy*, Contributing Editor to the Dissent and Democracy Network and is currently a co-Managing Editor of the *Transnational Human Rights Review*. He is also a 2016 Fellow of the Transnational Law Summer Institute at Kings College, London.

María José Veramendi Villa is a lawyer with law degrees from the Universidad de Los Andes in Bogota, Colombia and the Pontificia Universidad Católica del Perú, and an LL.M. (Master of Laws) in International Legal Studies from American University Washington College of Law. Since December 2016, she has been a Researcher for South America at the Americas Regional Office of Amnesty International. Previously, she worked as a senior attorney for the Human Rights and Environment Program at the Interamerican Association for Environmental Defense (AIDA); attorney at the Legal Defense Institute (IDL); human rights lawyer for the Inter-American Commission on Human Rights (IACHR); and researcher for Iberoamericana University in Mexico City. The chapter authored by María José in this volume was written before joining Amnesty International and therefore does not reflect the views of the organization.

Abby Rubinson Vollmer is a Consultant on international, environmental, and human rights law, with a focus on the intersection of human rights and climate change. She has taught International Environmental Law as an adjunct professor at the University of San Francisco School of Law; litigated and advocated for strong environmental and human rights protections before domestic, regional, and international bodies, including the Conference of the Parties to the United Nations Framework Convention on Climate Change and the United Nations Human Rights Council, as an attorney in the International Program of Earthjustice; and represented plaintiffs in three international human rights cases brought under the US Alien Tort Statute. She received her B.A. from Duke University and J.D. from University of Michigan Law School.

Linda Wallbott is Research Fellow at the Institute of Political Science at Technische Universität Darmstadt, Germany. In her research she focuses on the conceptual, empirical and normative linkages between sustainability transformations, safeguards, human rights and historical responsibility, especially in climate and biodiversity politics.

Margaretha Wewerinke-Singh is a Lecturer in Environmental Law at the University of the South Pacific's School of Law (Vanuatu) and an Affiliate of the Centre for Climate Change Mitigation Research at the University of Cambridge (UK).

Vositha Wijenayake is an Attorney-at-Law specialising in international environmental law and UN human rights law. She has an LL.M. in pubic international law from University of London, and an LL.B. from Queen Mary, University of London. Her areas of focus include climate change and human rights, climate change adaptation, loss and damage, gender and the UNFCCC process. She works as a country support consultant of the UNDP NAP Global Support Programme, regional facilitator for Asia for Southern Voices on Adaptation, and executive director of SLYCAN Trust.

Foreword

To me, and to all Indigenous Peoples, and to all those affected by the harmful impacts of climate change, the issues I speak to are all connected – our rights as Indigenous Peoples are one and the same as our environmental and cultural rights. Indeed, everything is connected.

As one of the world's most vulnerable regions, the Arctic is undergoing an historic environmental and social change. For decades, the North and its peoples have been subjected to the most dramatic environmental effects of globalization. Most recently, dramatic climate change caused by greenhouse gases has left virtually no feature of our landscape or our way of life untouched, and it now threatens our very culture.

The latest reports of climate change coming in from all of our communities are starker than ever. Virtually every community across the North is now struggling to cope with extreme coastal erosion, melting permafrost, and rapid, destructive runoff that threatens to erode away whole towns, especially in Alaska and the western part of our own country. Despite our last particularly cold winter, our sea ice remains in rapid decline. Glacial melt long relied on for drinking water is now unpredictable, and invasive species travel much further north than ever before. While the size and type of each change varies across the North, the trends are consistent. The change is not just coming – it is already here.

These last decades, however, have seen more than just dramatic environmental change – they have also witnessed a remarkable awakening of a global environmental consciousness, a realization that we are all connected by a common atmosphere and oceans. As Inuit, we have been and remain a hunting people of the land, ice, and snow. The process of the hunt teaches our young people to be patient, courageous, bold under pressure, and reflective and gives them a sense of identity and self-worth. The international community has learned from us as well. International agencies, national governments, civil society, and the media have begun to see that the Inuit hunter, falling through the melting ice, is connected to the cars we drive, the policies we create, and the disposable world we have become.

As this consciousness has emerged, so too have new and innovative partnerships and solutions to address these problems. Global environmental challenges have been successfully addressed when the international community has come together to acknowledge the connections between far-off sources of pollution and the local impacts on health, environment, and human rights.

While you would never know it today, the links between climate change and human rights were virtually unknown in the broader world just a few years ago before we submitted a climate change petition to the Inter-American Commission on Human Rights on behalf of Inuit in Alaska and Canada (myself included). The purpose of the petition was to educate and encourage the United States to join the community of nations in a global campaign to combat climate change. This petition was a "gift" from Inuit hunters and elders to the world. It was an act of generosity from an ancient culture deeply tied to the natural environment and still in tune with its wisdom to an urban, industrial, and "modern" culture that has largely lost its sense of place and position in the natural world.

We successfully translated the human rights dimensions of climate change and its impacts into legal arguments and brought our message to an international human rights tribunal. It was the first case in which the links between human rights and climate change were made clear, but not the last as evidenced by the lessons learned and stories told throughout this handbook.

This handbook is essential and timely as we reflect on our successes and failures and chart a path forward for human rights and climate governance. As elders, youth, scholars, policymakers, activists, and the public, we must come together as a collective to address the greatest human rights challenge of our time. This understanding of our "collective and interconnectedness" as a shared humanity is what is needed to spur decision-makers to act urgently and ambitiously to protect our right to a safe climate.

Sheila Watt-Cloutier

Preface

For more than a decade, we have been privileged to be part of an unprecedented coalition including Indigenous Peoples; labor and trade unions; and women's, gender-based, youth, faith-based, human rights, environmental, development, and climate justice groups and scholars, calling for human rights to be respected, protected, and fulfilled in the context of changing climate. The Paris Agreement in 2015 and its references to human rights obligations has confirmed not only the relevance and importance of these efforts, but also highlighted the work that remains to be done to protect all human rights from climate disruption and the responses thereto.

The development of this book specifically originates in a report published by the Center for International Environmental Law and CARE International titled *Climate change: tackling the greatest human rights challenge of our time: Recommendations for effective action on climate change and human rights*. This report identified key policy recommendations for taking human rights seriously in the field of climate change. Its release underscored the necessity to generate and provide additional knowledge to help inform the development and implementation of climate governance around the world. This volume – as a cooperation between scholars and practitioners – seeks to respond to this need.

This volume would not have been possible without the support and assistance of many individuals. To begin with, we are indebted to the authors who accepted our invitations to contribute chapters to this volume.

Thanks are also due to Annabelle Harris, Margaret Farrelly, and the whole team at Routledge for their enthusiasm for this project and their patience during the production process. Kit Vaughan helped shape the initial direction that this project followed.

We are also very grateful to the team of McGill law students who helped format the text and references for the chapters in the book: Rosine Faucher, Katherine Hansen, and Jinnie Liu. In this regard, we wish to acknowledge the financial support provided by the Social Sciences and Humanities Research Council of Canada for the completion of this volume.

Last, but not least, we owe a debt of gratitude to our families for their patience and support.

We dedicate this book to activists, lawyers, and policy-makers working on human rights and climate change.

Sébastien Duyck, Sébastien Jodoin, and Alyssa Johl

Part I
Conceptual foundations

1

Integrating human rights in global climate governance

An introduction

Sébastien Duyck, Sébastien Jodoin, and Alyssa Johl

The intersections of human rights and climate change

Over the last decade, the world has increasingly grappled with the complex linkages emerging between efforts to combat climate change and to protect human rights around the world. As is well recognised by several international experts and bodies, climate change has a range of negative implications for the rights to life, food, health, housing, and self-determination, among others, and its consequences will be felt most acutely by groups in vulnerable situations such as Indigenous Peoples, the world's poor, women, youth, communities located in low-lying regions, and small-island states.[1] At the same time, on-the-ground experience demonstrates that responses to climate change may themselves have negative repercussions for human rights by failing to abide by the participatory rights of marginalised peoples and communities and restricting their access to the lands, food, energy, and resources on which their livelihoods depend.[2] While the Paris Agreement adopted in December 2015 recognises the necessity for governments to "respect, promote, and consider" their human rights obligations when taking climate action,[3] important gaps remain in understanding how human rights can and should be used in practice to develop and implement effective and equitable solutions to climate change.

This book seeks to fill this gap between needs and action by offering a timely and comprehensive analysis of the opportunities and challenges for integrating human rights with climate governance. It builds on nearly a decade of experience in thinking about and working at the intersections of human rights and climate change. The book considers important recent developments in the United Nations Framework Convention on Climate Change (UNFCCC) (Part I) and the United Nations Human Rights System (Part II), documents emerging lessons in the development and implementation of human rights obligations and principles with respect to different issues raised by climate change (Part III), and provides regional case studies of experiences on the ground (Part V).

One of the unique features of this volume is that it brings together scholars and practitioners that have played key roles in studying the relationship between human rights and climate change or in framing climate change as a human rights issue in a range of venues, including multilateral institutions, non-governmental organisations (NGOs), and the private sector. Indeed, Part IV of the book features contributions from civil society experts, advocates, and stakeholders who are on the leading edge of building linkages between the fields of climate change and human rights. As a result, the chapters in this book employ and reflect multiple disciplines, approaches, and perspectives. Several chapters are written by social scientists or draw on interdisciplinary research and methods.[4] Other chapters are written by civil society activists who can speak to the lived realities of fighting for the protection of rights in the context of climate change.[5] These varied contributions provide the reader with unique and nuanced insights into past, ongoing, and future efforts to enhance the protection of human rights in a changing climate and are of great relevance to the work of scholars, policy-makers, lawyers, activists, and other practitioners working on this dynamic and timely topic.

The remainder of this introductory chapter proceeds as follows. We begin by providing a historical overview of the emergence of human rights in the field of global climate governance. We then identify some of the key themes about the relationship between human rights and climate change that recur throughout the book. We conclude by identifying outstanding questions for scholars writing about human rights and climate change and summarizing key issues and implications for advocacy and policy-making in the future.

The emergence of human rights in the field of global climate governance

The emergence of human rights in the field of climate change is the result of international negotiations and advocacy efforts that extend back to the very origins of the UNFCCC.[6] During the final stages of drafting the UNFCCC in 1992, the G-77 proposed that it recognise in its article 2 that "the right to development is an inalienable human right. All peoples have an equal right in matters relating to reasonable living standards."[7] This language was rejected by the United States, however, as it did not accept the right to development as a human right.[8] Consequently, this language was abandoned and the UNFCCC instead affirms that "the parties have a right to, and should, promote sustainable development."[9] Following the adoption of the UNFCCC, references to human rights all but disappeared from discussions of climate governance, with a few exceptions in the related academic literature continuing to link human rights and equity.[10]

In the late 1990s, the negotiations for the operationalisation of the Kyoto Protocol provided the context for the consideration of a second dimension of the interlinkages between climate governance and human rights.[11] Indigenous Peoples and their allies opposed the establishment of the clean development mechanism (CDM) to incentivise carbon mitigation projects in developing countries, highlighting among other arguments that such a scheme would support activities that could result in violations of the participatory and substantive rights of Indigenous Peoples.[12] Scholars specifically highlighted that the inclusion of land use and afforestation projects within the scope of the CDM could lead to adverse social implications for Indigenous and forest-dependent communities, if adequate safeguards and accountability measures were not developed.[13] Unfortunately, the rules adopted by the UNFCCC conference of the parties (COP) for the CDM did not adequately consider these important matters, and many CDM projects have infringed on the human rights of affected populations.[14] Similar concerns have been raised in the context of the voluntary carbon market, where carbon projects have been developed and implemented with little guidance and scrutiny, often with adverse social implications for local communities.[15] Concerns over the negative consequences of climate mitigation projects also resulted in the development of non-governmental certification programmes that set more stringent rules and processes to enhance the social performance of carbon mitigation activities.[16] One such programme, launched in 2003, was the Gold Standard, which included safeguards providing that CDM projects should respect international human rights and labour standards and not be involved or complicit in human rights abuses, involuntary resettlement, or the alteration, damage, or removal of any critical cultural heritage.[17]

Indigenous Peoples were again at the forefront of the first campaign demanding that governments consider the human rights implications of climate change.[18] In 2005, Inuit in the Canadian and Alaskan Arctic filed a complaint before the Inter-American Commission on Human Rights (IACHR) seeking relief from the United States for alleged violations of their human rights resulting from climate change.[19] Although the IACHR deemed the complaint inadmissible,[20] this petition, in combination with the Inuit Circumpolar Council's advocacy efforts in the UN climate negotiations,[21] is generally considered to have played a critical role in raising awareness of the human rights dimensions of climate change.[22] This landmark case was followed by the 2007 Malé Declaration on the Human Dimension of Global Climate Change adopted by small-island developing states,[23] which called for assistance from the UN Human Rights Council (UNHCR) and the Office of the High Commissioner on Human Rights (OHCHR) in recognising and assessing the human rights implications of climate change. This marked the beginning of efforts led by small-island developing states and their allies in international organisations and civil society to introduce climate change as a human rights matter in the context of the UN Human Rights System.[24]

The launch of a new cycle of international negotiations towards a comprehensive climate agreement in 2007 offered an important opportunity to build on the work of UN human rights bodies and experts to introduce human rights considerations, principles, and obligations to the UNFCCC.[25] In addition, negative experiences with several CDM projects and efforts to develop a new mechanism to reduce carbon emissions from deforestation and forest degradation in developing countries (known as REDD+) in turn spurred Indigenous Peoples and non-governmental organisations to advocate for the need to recognise and prevent the adverse human rights implications of climate mitigation activities.[26] In the wake of the collapse of the Copenhagen climate talks in 2009, civil society advocacy efforts to frame climate change as a human rights issue intensified in the UNFCCC and other multilateral venues.[27] One year later, in the context of the Cancun Agreements, state parties to the UNFCCC recognised, for the first time, several linkages between human rights and climate change. They emphasised that governments should fully respect human rights when taking climate action,[28] referenced a resolution from the UNHCR to the effect that climate change has adverse impacts for a range of human rights,[29] and developed safeguards for REDD+ that refer to human rights and note the relevance of the UN Declaration on the Rights of Indigenous Peoples.[30] These references demonstrated that governments recognised both the adverse human rights impacts of climate change as well as the potentially harmful nature of responses to climate change.[31]

Beyond the UNFCCC, the human rights implications of climate policies and programmes were also increasingly recognised in the extensive array of mechanisms that were established to govern the development and implementation of carbon mitigation programmes and projects in developing countries. In particular, the multilateral, bilateral, and non-governmental initiatives that emerged in the late 2000s to support the pursuit of REDD+ in developing countries has, in various ways and with different levels of effectiveness, integrated several human rights standards and principles.[32] As well, building on its experience with REDD+, the UN Development Programme (UNDP) developed social and environmental standards[33] that apply to all UNDP programmes and projects, including climate mitigation and adaptation activities – which have been recognised as best practices in terms of their application of human rights standards and principles.

To further guarantee that human rights obligations inform climate policies more effectively, supportive governments and organisations proposed the inclusion of relevant references during the drafting process leading to the adoption of the Paris Agreement in 2015. This mobilisation led to a strengthening of the linkages with human rights–related institutions and actors, for instance, with a greater involvement of UNHCR Special Procedures mandate holders.[34] The involvement of human rights experts also led to the launch of a process known as the 'Geneva Pledge for Human Rights in Climate Action' through which individual governments commit to better integrate human rights and climate change expertise into their international policies.[35] As a result of these combined efforts, the Paris Agreement provides that states "should, when taking action to address climate change, respect, promote and consider their respective obligations on human rights".[36] This provision was described as a breakthrough in the integration of human rights concerns in environmental policy-making, the Paris Agreement being the first international environmental instrument to include such a reference.[37]

Despite these important advances, there continues to be a significant gap between the articulation of human rights standards and principles in international and transnational climate initiatives and their implementation in practice. While state parties recognised in the Cancun Agreements that they should fully respect human rights when taking climate action, there is a still a lack of human rights guidance and oversight in the approval and monitoring of CDM projects pursued under the auspices of the UNFCCC.[38] Likewise, the effectiveness of certification programmes seeking to operationalise rights-based safeguards and emphasising social benefits in the pursuit of REDD+ programmes and projects has also been bedevilled by numerous challenges on the ground.[39] Finally, notwithstanding the growing international awareness of the human rights implications of climate change, the recognition and consideration of human rights has yet to be reflected into domestic climate policies, as demonstrated by the limited references to these principles in the context of states reporting to the UNFCCC.[40]

Key themes in the relationship between human rights and climate change

In what follows, we identify some of the most important themes about the relationship between human rights and climate change that recur throughout this book. All the chapters in this book underscore the importance of understanding climate change as a human rights issue in different ways. The direct adverse impacts of climate change on the enjoyment of human rights constitute a central theme for several chapters of this volume.[41] In particular, the evidence provided in the chapters composing Part V echo two key findings of the scientific community in terms of climate impacts: while climate change has already had negative repercussions for human systems on all continents, the nature and intensity of these impacts diverge widely across regions and communities.[42] Consequently, the duty to protect rights in the context of climate change applies to all governments and requires a differentiated response depending on physical and social circumstances.

Wewerinke-Singh and Brown most notably argue that states bear the responsibility of reducing carbon emissions in order to prevent the adverse human rights repercussions of climate change.[43] While the focus of climate governance remains primarily on curbing carbon emissions, Khan also points out that the responsibility of states to mitigate climate change also applies to the emission of other pollutants contributing to climate change.[44] The consideration of states' responsibility to mitigate climate change in order to meet their human rights obligations echoes calls to define thresholds for adequate action on the basis of these obligations.[45] At the same time, several chapters offer more critical perspectives that highlight the challenges involved in bringing human rights norms to bear on climate change. In particular, Lofts notes the limited transformative potential of anthropocentric discourses that focus on human rights, in comparison to ecocentric discourses that emphasise the rights of nature and which have received much less formal recognition within the UNFCCC.[46] As well, Jodoin, Lofts, and Faucher address the extent to which the mechanisms of international human rights law may be useful in influencing domestic climate policies and argue that civil society activists should focus their efforts on domestic litigation and mobilisation in the years to come.[47]

Many contributors in this book also grapple with the human rights implications of *responses to* climate change. These chapters offer an important perspective that has often been lacking in the existing literature on the intersections of human rights and climate change.[48] Indeed, human rights arguments and principles have been most often deployed to serve the objective of combating climate change rather than protecting human rights as such.[49] Accordingly, as Dudai points out,

> With the justified concerns regarding the effects of climate change, and enthusiasm regarding the potential of human rights and the human rights movement in assisting the campaign against it, it seems that some potential points of friction and tension between human rights defenders and environmentalists are being tacitly ignored.[50]

Several chapters in this book thus highlight the ways in which a human rights framework may be most relevant to the climate change regime by championing its ability to ensure that laws, policies, programmes, and projects adopted to mitigate or adapt to climate change respect, protect, and fulfil human rights protected in national, regional, and international law.[51] Although the necessity of responding to climate change is urgent and serious, this does not provide a legal justification for violating binding human rights obligations as a matter of international law. While treaties do set out conditions under which limitations can be imposed by states on the exercise of some human rights in exceptional circumstances, the long-term, cumulative effects of climate change are unlikely to fall within these exceptional circumstances (although particular climatic-related natural disasters might well qualify).[52]

Thinking about the human rights implications of climate actions also raises important issues of climate justice focusing on the potential for the global response to climate change to result in the inequitable distribution of burdens and the further marginalisation of certain communities. Similarly, vulnerability to climate impacts is exacerbated by existing social inequalities resulting from intersecting social processes leading to marginalisation, discrimination, and increased exposure.[53] Discrimination based on factors such as gender, age, sexual orientation, class, nationality, and ethnicity – in particular when such factors intersect – consequently increase the risk that people's rights be adversely impacted by climate impacts or that climate policies fail to respect and protect their rights effectively.[54] This context requires that climate-related policies and projects take into consideration inequalities and populations in vulnerable situations in order to remedy impacts that might threaten the rights of a specific group of the population more acutely. The international community has repeatedly recognised the importance of addressing these factors to guarantee the protection of rights in the context of climate change.[55] States have mandated relevant bodies to consider how to ensure that the protection of the rights of women, children, or migrants be protected in the context of climate change.[56] In light of the above, several contributions to this volume address the importance of considering the protection and promotion of the rights of specific social groups such as Indigenous Peoples, women, children, and workers impacted by the decarbonisation of the economy, while referring to relevant legal and policy frameworks.[57] These chapters propose specific approaches that can ensure a just and equitable protection guaranteeing the rights of all, including of those most at risk to see their rights infringed. Several of these chapters also emphasise that guaranteeing these rights and empowering all segments of societies can strengthen the effectiveness of climate responses.

Another key theme that is found in many chapters relates to an interest in exploring the role of human rights in the broader domain of *climate governance*. In addition to chapters examining how human rights have been reflected in decisions adopted within the UNFCCC, this theme is also explored in chapters that consider the practices of a broader variety of multilateral, bilateral, and non-governmental entities and forms of authority that have emerged to govern climate change as a domain.[58] In particular, three chapters in this handbook address various dimensions of corporate accountability for the human rights abuses associated with climate change and climate policies, focusing on the responsibility of the renewable energy industry, opportunities to further leverage private law remedies in the context of climate change, and the role of investors.[59] At the domestic level, an increasing number of court cases are challenging the adequacy of national mitigation or adaptation policies based wholly or partly on human rights arguments.[60] This broader focus on climate governance not only enhances its relevance to practitioners working in the field of climate change, but also builds important bridges with the burgeoning scholarship examining new forms of transnational climate law and governance.[61]

Conclusion: developing an agenda for scholars and practitioners

The chapters in this book suggest several key topics of interest for scholars and practitioners committed to studying or enhancing the protection of human rights in the context of a changing climate. Additional research on some of the themes discussed below would deepen our collective understanding of the interplay between human rights norms and practical aspects of climate governance and empower practitioners in effectively employing different institutional and judicial mechanisms dedicated to the protection of human rights of communities affected by climate change and responses thereto.

First, legal scholars could support the role of judicial institutions dealing with climate-related human rights by developing theories determining thresholds for actions below which the responsibility of states with regards to their human rights obligations can be engaged.[62] While Simon Caney already made a strong case in 2010 for the use of human rights as 'moral thresholds' in the context of climate action,[63] judicial bodies would benefit from practical analysis translating this ethical argument into a legal one supported by scientific evidence. In relation to the global increase of average temperatures, the special procedures of the Human Rights Council have already stated that an increase of the temperatures by even 2°C would already have drastic consequences for the full enjoyment of human rights.[64] Additionally, human rights bodies often draw from environmental agreements that states have committed to do in order to define the threshold engaging their responsibility.[65] In this context, the Paris Agreement's objective to keep temperature 'well below' 2 degrees and to pursue efforts to limit the temperature increase to 1.5 degrees is significant. But to review the responsibility of individual states, judicial bodies, and other institutions, it would need to be determined whether national mitigation – or energy policies in some cases[66] – are compatible with these global objectives.[67] Defining legal thresholds engaging the responsibility of states will be crucial to turn slogans into legal arguments actionable through domestic and international human rights institutions.

Second, further efforts are needed to tie a rights-based approach to climate change to the broader framework of international policies that support the pursuit of sustainable development.

Shortly prior to the adoption of the Paris Agreement, the UN General Assembly had adopted the 2015 Agenda and its 17 Sustainable Development Goals.[68] In 2015, governments also adopted the Sendai Framework for Disaster Risk Reduction and the Addis Ababa Action Agenda on financing for development, and as this book goes to press, governments are finalising two new global compacts for safe, orderly, and regular migration, and on refugees.[69] The effective implementation of these instruments and their ability to promote human rights will depend to a significant extent on the ability of states and other actors to identify synergies and adopt a coherent approach to sustainable development governance. The stronger focus of some of these other frameworks on the protection of human rights and good governance could contribute to place human rights-related principles and obligations at the core of sustainable and climate-related policies.[70] The research community will have an important role to play to identify synergies and co-benefits across these new policy frameworks.

Third, policy-makers, scholars, and practitioners should continue to work to develop guidance and good practices on how human rights principles and standards can be used to inform climate policies at the national level. In doing so, we can build on and replicate the many rights-based tools and approaches that have emerged over several decades to integrate and mainstream human rights in the design, planning, implementation, evaluation, and oversight of laws, policies, and programmes in such fields as civil liberties, international development, education, and health.[71] Three items that will be particularly important in this regard relate to the need to develop and employ guidance and training materials, indicators and criteria to ensure proper monitoring and follow-up, and approaches for ensuring the full and effective participation of multiple communities and stakeholders, especially among those groups that are marginalised by society or are likely to be most affected by climate change and responses thereto. As many countries seek to adapt their legislation in order to implement the Paris Agreement, guidance regarding what the promotion and respect of human rights implies for national climate policy – including good practices – has a significant role to play to ensure that climate responses and human rights are more effectively integrated at the national level.[72] Global initiatives and research related to the role of national legislation and parliaments to implement the Paris Agreement have yet to include rights-related approaches in the support that they offer to national governments.[73]

Fourth, further efforts are needed to develop legal theories and strategies to hold states and private actors accountable for their contributions to climate change and the resulting harms. While climate litigation is gaining traction in courts and tribunals around the world, there is significant need and opportunity for scholars and practitioners to develop strategic cases that will effectively curb carbon emissions on the scale and timeline needed and hold those responsible to account for the damages. With respect to private actors, the first generation of cases, including the petition filed with the Commission for Human Rights of the Philippines concerning the responsibilities of the major carbon producers for climate damages,[74] provides a useful model for bringing cases before other human rights bodies, including national human rights institutions and regional human rights bodies. The outcomes and lessons learned from these cases, as well as recent developments in climate attribution science and a growing body of evidence demonstrating the knowledge that fossil fuel producers had of the causes and consequences of climate change,[75] will help to inform the next generation of cases seeking to establish liability of these actors.[76] In addition to efforts to seek judicial remedies, further research and advocacy is needed on the use of treaty-monitoring bodies, which review the compliance of human rights obligations within their purview, as a means to consider how such obligations must be upheld in the context of climate change.[77] Other processes seeking to further define the scope of corporate responsibilities in the context of climate change – for instance, at the national level with the adoption of National Action Plans and at the international level with the opening of negotiations for a treaty on the human rights obligations of transnational corporations – present opportunities for further research and analysis.[78]

Fifth, the negotiations leading to the inclusion of rights-related language in the Paris Agreement have highlighted that many actors perceive the integration of human rights in climate policy as a risk to delay climate responses.[79] Additional evidence-based analysis of the benefits of a human rights–based approach for the effectiveness of climate policies would thus strengthen the case for policy coherence – particularly among policy actors who do not perceive the respect and promotion of human rights as their own mandate. Recent research has demonstrated the existence of a correlation in the majority of countries between the level of protection of land tenure rights and the conservation of forests carbon stocks.[80] Additional empirical research regarding the mitigation and adaptation benefits of respect for human rights, public participation in decision-making, the rights of Indigenous Peoples, or gender equality would strengthen the case for policy coherence. Of course, such evidence should in no way be presented or understood as undermining the fundamental principles that underlie a rights-based approach to climate governance – the primacy accorded to human dignity and autonomy in international human rights law as well as the primary responsibility of governments to respect, protect, and fulfil all basic human rights, including in the context of a changing climate.

Notes

1 See UNHCR Res 10/4, *41st Meeting* (25 March 2009), UN Doc. A/HRC/10/L.11 (UNHCR Resolution 10/4); OHCHR, *Report of the Office of the United Nations High Commissioner for Human Rights on the Relationship Between Climate Change and Human Rights* (2009), U.N. Doc. A/HRC/10/61; UNFCCC COP, *The Cancun Agreements: Outcome of the Work of the Ad Hoc Working Group on Long-Term Cooperative Action Under the Convention* (15 March 2011), Decision 1/CP.16, *referred in* Report of the Conference of the Parties on its sixteenth session, Addendum, Part Two: Action taken by the Conference of the Parties, FCCC/CP/2010/7/Add.1, at preamble; OHCHR, *A New Climate Change Agreement Must Include Human Rights Protections for All: An Open Letter From Special Procedures Mandate-Holders of the Human Rights Council to the State Parties to the UN Framework Convention on Climate Change on the Occasion of the Meeting of the Ad Hoc Working Group on the Durban Platform for Enhanced Action in Bonn (20–25 October 2014)* (17 October 2014), available online at www.ohchr.org/Documents/HRBodies/SP/SP_To_UNFCCC.pdf; *IACHR Expresses Concern Regarding Effects of Climate Change on Human Rights*, Inter-American Commission on Human Rights (2 December 2015), available online at www.oas.org/en/iachr/media_center/preleases/2015/140.asp.

2 N. Roht-Arriaza, '"First, Do No Harm": Human Rights and Efforts to Combat Climate Change', 38 *Georgia Journal of International & Comparative Law* (2010) 593; S. Jodoin, *Forest Preservation in a Changing Climate: REDD+ and Indigenous and Community Rights in Indonesia and Tanzania* (2017).

3 UNFCCC COP, *Adoption of the Paris Agreement* (29 January 2016), Decision 1/CP.21, *referred in* Report of the Conference of the Parties on its twenty-first session, held in Paris from 30 November to 13 December 2, UN Doc. FCCC/CP/2015/10/Add.1, at preamble.

4 See in this volume: K. Lofts, 'Analysing Rights Discourses in the International Climate Regime'; A. Schapper, 'Local Rights Claims in International Negotiations: Transnational Human Rights Networks at the Climate Conferences'; L. Wallbott, 'Rights, Representation, and Recognition: Practicing Advocacy for Women and Indigenous Peoples in UN Climate Negotiations'; S. Jodoin, R. Faucher and K. Lofts, 'Look Before You Jump: Assessing the Potential Influence of the Human Rights Bandwagon on Domestic Climate Policy'; E. Morena, 'Securing Workers' Rights in the Transition to a Low-Carbon World: The Just Transition Concept and Its Evolution'.

5 See in this volume: B. Powless, 'The Indigenous Rights Framework and Climate Change'; A. Barre *et al.*, 'From Marrakesh to Marrakesh: The Rise of Gender Equality in the Global Climate Governance and Climate Action'; A. Silverman, 'Energy Justice: The Intersection of Human Rights and Climate Justice'; J. Pegram, 'Overlooked and Undermined: Child Rights and Climate Change'; G. Parihar, 'Human Rights, Differentiated Responsibilities? Advancing Equity and Human Rights in the Climate Change Regime'; D. Stabinsky, 'Climate Justice and Human Rights'; K. Rall and F. Horne, '"There Is No Time

Left"' – Climate Change, Environmental Threats, and Human Rights in Turkana County, Kenya'; V. Wijenayake, 'Human Rights and Climate Change: Focusing on South Asia'; A. Rodriguez and M. J. Veramendi Villa, 'Integrating a Human Rights–Based Approach to Address Climate Change Impacts in Latin America: Case Studies From Bolivia and Peru'; E. Horvath and K. Maeda, 'Towards Responsible Renewable Energy: Assessing 50 Wind and Hydropower Companies' Human Rights Policies in the Context of Rising Allegations of Abuse'.

6 For an in-depth chronological account of the successive campaigns seeking consideration of human rights under the UNFCCC, see S. Duyck , 'The Paris Climate Agreement and the Protection of Human Rights in a Changing Climate', 26 *Yearbook of International Environmental Law* (2017).

7 UNFCCC, *Report of the Intergovernmental Negotiating Committee for a Framework Convention on Climate Change on the Work of Its Fourth Session, Held at Geneva From 9 to 20 December 1991* (29 January 1992), UN Doc. A/AC.237/15.

8 D. Bodansky, 'The United Nations Framework Convention on Climate Change: A Commentary', 18 *Yale Journal of International Law* (1993) 451, at 504.

9 *United Nation Framework Convention on Climate Change*, 9 May 1992, UN Doc. FCCC/INFORMAL/84 (entry into force 21 March 1994), at Art 3.

10 See Bodansky, *supra* note 8; J. Gupta, 'North-South Aspects of the Climate Change Issue: Towards a Constructive Negotiating Package for Developing Countries', 8 *Review of European, Comparative & International Environmental Law* 198, at 202.

11 *Kyoto Protocol to the United Nations Framework Convention on Climate Change*, 10 December 1997, 37 ILM 22 (1998), UN Doc. FCCC/CP/1997/7/Add.1 (entered into force 16 February 2005).

12 *Position Paper Presented to the 13th Session of the Subsidiary Bodies to the UNFCCC*, Forum of Indigenous Peoples and Local Communities on Climate Change (September 2000), available online at http://wrm.org.uy/oldsite/actors/CCC/IPpaperLyon.html. See also K. Fogel, 'The Local, the Global and the Kyoto Protocol', in S. Jasanoff and M. Long Martello (eds.), *Earthly Politics: Local and Global in Environmental Governance* (2004) 117.

13 E. L. Vine, J. A. Sathaye and W. R. Makundi, 'Forestry Projects for Climate Change Mitigation: An Overview of Guidelines and Issues for Monitoring, Evaluation, Reporting, Verification, and Certification', 3 *Environmental Science & Policy* (2000) 99, at 110; J. Smith, 'Afforestation and Reforestation in the Clean Development Mechanism of the Kyoto Protocol: Implications for Forests and Forest People', 2 *International Journal of Global Environmental Issues* (2002) 322, at 335; L. Aukland *et al.*, *Laying the Foundations for Clean Development: Preparing the Land Use Sector: A Quick Guide to the Clean Development Mechanism*, IIED Natural Resource Issues Paper (2002), available online at http://pubs.iied.org/pdfs/9136IIED.pdf, at 12.

14 Eva Filzmoser *et al.*, *The Need for a Rights-Based Approach to the Clean Development Mechanism*, Public Participation and Climate Governance Working Paper Series (2015), available online at http://cisdl.org/public/docs/FILZMOSER.pdf; J-P. Brasier, *How Climate Projects Can Lead to Human Rights Violations: The Case of the Barro Blanco Hydroelectric Dam*, Carbon Market Watch (14 July 2017), available online at www.germanclimatefinance.de/2017/07/14/climate-projects-can-lead-human-rights-violations-case-barro-blanco-hydroelectric-dam/.

15 N. Roht-Arriaza, 'Human Rights in the Climate Change Regime', 1 *Journal of Human Rights and the Environment* (2010) 211, at 213–222. See also K. Dooley, 'Human Rights and Land-Based Carbon Mitigation' (this volume).

16 Roht-Arriaza, *supra* note 2, at 607–608.

17 See *Gold Standard Requirements, Version 2.1*, available online at www.goldstandard.org/sites/default/files/gsv2.1_requirements-11.pdf.

18 See B. Powless, 'The Indigenous Rights Framework and Climate Change' (this volume).

19 Inuit Circumpolar Conference, *Petition to the Inter-American Commission on Human Rights Seeking Relief From Violations Resulting From Global Warming Caused by Acts and Omissions of the United States, Submitted by Sheila Watt-Cloutier* (7 December 2005), with the support of the Inuit Circumpolar Conference, on behalf of all Inuit of the Arctic Regions of the United States and Canada, available online at http://inuitcircumpolar.com/files/uploads/icc-files/FINALPetitionICC.pdf.

20 In a letter dated November 16, 2006, the IACHR informed the petitioners that it would not consider the petition because the information it provided was not sufficient for making a determination. In March 2007, however, the IACHR did hold hearings with the petitioners to address matters raised by the petition without revisiting the issue of its admissibility.

21 S. Watt-Cloutier, *Presentation at the Tenth Conference of Parties to the UN Framework Convention on Climate Change, Buenos Aires, Argentina* (17 December 2004), available online at http://unfccc.int/resource/docs/2004/cop10/stmt/ngo/005.pdf.

22 T. Koivurova, 'International Legal Avenues to Address the Plight of Victims of Climate Change: Problems and Prospects', 22 *Journal of Environmental Law & Litigation* (2007) 267; S. Kravchenko, 'Right to Carbon or Right to Life: Human Rights Approaches to Climate Change', 9 *Vermont Journal of Environmental Law* (2007) 513; H. M. Osofsky, 'The Inuit Petition as a Bridge? Beyond Dialectics of Climate Change and Indigenous Peoples' Rights', 31 *American Indian Law Review* (2006) 675.
23 *Male' Declaration on the Human Dimension of Global Climate Change* (14 November 2007), available online at www.ciel.org/Publications/Male_Declaration_Nov07.pdf.
24 See F. Kirchmeier and Y. Lador, 'From Copenhagen to Paris at the UN Human Rights Council' (this volume).
25 M. Limon, 'Human Rights and Climate Change: Constructing a Case for Political Action', 33 *Harvard Environmental Law Review* (2009) 439; J. Knox, 'Linking Human Rights and Climate Change at the United Nations', 33 *Harvard Environmental Law Review* (2009) 476; L. Rajamani, 'The Increasing Currency and Relevance of Rights-Based Perspectives in the International Negotiations on Climate Change', 22 *Journal of Environmental Law* (2010) 391.
26 See, *e.g.*, V. Tauli-Corpuz and A. Lynge, *Impact of Climate Change Mitigation Measures on Indigenous Peoples and on Their Territories and Lands*, Permanent Forum on Indigenous Issues, 7th session, E/C.19/2008/10 (19 March 2008); *Indigenous Peoples' Statement on Shared Vision Under AWG-LCA* (9 December 2009), available online at www.stakeholderforum.org/mediafiles/Outreach-UNFCCC%20COP%2015%20Issue%203.pdf; A. Johl and M. Wagner, *Recognizing and Protecting Human Rights in the Copenhagen Agreement* (14 December 2009), available online at www.stakeholderforum.org/sf/mediafiles/Outreach-UNFCCC%20COP%2015%20Issue%206.pdf.
27 See Schapper, *supra* note 4.
28 UNFCCC, *Cancun Agreements*, at preamble.
29 *Ibid.*, at preamble.
30 *Ibid.*
31 E. Cameron and M. Limon, 'Restoring the Climate by Realizing Rights: The Role of the International Human Rights System', 21 *Review of European, Comparative & International Environmental Law* (2012) 204.
32 See Jodoin, *supra* note 2.
33 UNDP, *Social and Environmental Standards* (2014), available online at www.undp.org/content/dam/undp/library/corporate/Social-and-Environmental-Policies-and-Procedures/UNDPs-Social-and-Environmental-Standards-ENGLISH.pdf.
34 OHCHR, *Statement of the United Nations Special Procedures Mandate Holders on the Occasion of the Human Rights Day* (10 December 2014), available online at www.ohchr.org/EN/NewsEvents/Pages/DisplayNews.aspx?NewsID=15393&.
35 *Geneva Pledge for Human Rights in Climate Action* (2015), available online at http://carbonmarketwatch.org/wp-content/uploads/2015/02/The-Geneva-Pledge-13FEB2015.pdf.
36 *Paris Agreement*, 12 December 2015, No. 54113 (entry into force 4 November 2016), at preamble, para. 11.
37 See, for instance, J. H. Knox, *Report of the Independent Expert on the Issue of Human Rights Obligations Relating to the Enjoyment of a Safe, Clean, Healthy and Sustainable Environment* (21 March 2014), Wake Forest University Legal Studies Paper No. 2510948, UN Doc. A/HRC/28/61, at para 22.
38 J. Schade and W. Obergassel, 'Human Rights and the Clean Development Mechanism', 27 *Cambridge Review of International Affairs* (2014) 717, at 725.
39 T. D. L. Fuente *et al.*, 'Do Current Forest Carbon Standards Include Adequate Requirements to Ensure Indigenous Peoples' Rights in REDD Projects?' 15(4) *International Forestry Review* (2013) 427; M. Bayrak and L. Marafa, 'Ten Years of REDD+: A Critical Review of the Impact of REDD+ on Forest-Dependent Communities', 8 *Sustainability* (2016) 620.
40 Mary Robinson Foundation for Climate Justice, *Incorporating Human Rights Into Climate Action* (2014), available online at www.mrfcj.org/pdf/2014-10-20-Incorporating-Human-Rights-into-Climate-Action.pdf. See also Jodoin, Faucher and Lofts, 'Look Before You Jump', *supra* note 4.
41 See in this volume: S. Harmeling, 'Climate Change Impacts: Human Rights in Climate Adaptation and Loss and Damage'; Rall and Horne, 'There Is No Time Left', *supra* note 5; Wijenayake, 'Human Rights and Climate Change: Focusing on South Asia', *supra* note 5; I. Büschel, 'Are Europeans Equal With Regard to the Health Impact of Climate Change?'; Rodriguez and Veramendi Villa, 'Integrating a Human Rights–Based Approach', *supra* note 5; S. Khan, 'Connecting Human Rights and Short-Lived Climate Pollutants: The Arctic Angle'; L. Benjamin and R. Haynes, 'Climate Change and Human Rights in the Commonwealth Caribbean: Case Studies of The Bahamas and Trinidad & Tobago'.

42 See Working Group 2, 'IPCC: Summary for Policy Makers', in M. L. Parry, O. F. Canziani, J. P. Palutikof, P. J. van der Linden and C. E. Hanson (eds.), *Climate Change 2007: Impacts, Adaptation and Vulnerability: Contribution of Working Group II to the Fourth Assessment Report of the Intergovernmental Panel on Climate Change* (2013), available online at www.ipcc.ch/pdf/assessment-report/ar4/wg2/ar4-wg2-spm.pdf, at 4–7.
43 See in this volume: M. Wewerinke-Singh, 'State Responsibility for Human Rights Violations Associated With Climate Change'; D. Brown, 'Using the Paris Agreement's Ambition Ratcheting Mechanisms to Expose Insufficient Protection of Human Rights in Formulating National Climate Policies'.
44 See in this volume: S. Khan, 'Connecting Human Rights and Short-Lived Climate Pollutants', *supra* note 41.
45 S. Caney, 'Climate Change, Human Rights and Moral Thresholds', in Gardner *et al.* (eds.), *Climate Ethics* (2010).
46 Lofts, 'Analysing Rights Discourses in the International Climate Regime', *supra* note 4.
47 Jodoin, 'Look Before You Jump', *supra* note 4.
48 For notable exceptions, see Roht-Arriaza, *supra* note 2; Jodoin, *supra* note 2.
49 S. Humphreys, 'Introduction', in S. Humphreys (ed.), *Human Rights and Climate Change* (2009) 16.
50 R. Dudai, 'Climate Change and Human Rights Practice: Observations On and Around the Report of the Office of the High Commissioner for Human Rights on the Relationship Between Climate Change and Human Rights', 1 *Journal of Human Rights Practice* (2009) 294, at 295.
51 See in this volume: D. D. Pugley, 'Rights, Justice, and REDD+: Lessons From Climate Advocacy and Early Implementation in the Amazon Basin'; N. Chalifour, H. Mcleod-Kilmurray and L. Collins, 'Climate Change: Human Rights and Private Remedies'; Horvath and Maeda, 'Towards Responsible Renewable Energy', *supra* note 5.
52 Generally, such limitations will be allowed provided that they are determined by law and necessary in a democratic society to ensure respect for the rights and freedoms of others or to meet the just requirements of public order, public health or morals, national security, or public safety (see, *e.g.*, *International Covenant on Civil and Political Rights* (19 December 1966) (ICCPR), No. 146668, Article 4). In such cases, these limitations must be prescribed by law; address a specific legitimate purpose allowed by international law; and be demonstrably necessary and proportionate (*United Nations Economic and Social Council UN Sub-Commission on Prevention of Discrimination and Protection of Minorities: The Siracusa Principles on the Limitation and Derogation Provisions in the International Covenant on Civil and Political Rights, Annex* (1985), UN Doc. E/CN.4/1985/4). Human rights treaties also provide that derogation (temporary suspension) from human rights is allowed "in time of public emergency of which threatens the life of the nation and the existence of which is officially proclaimed". (ICCPR, Article 4(1)). Such derogations must be strictly required by the exigencies of the situation, must not be inconsistent with other obligations under international law, and may not be discriminatory. In addition, there are a number of rights from which derogation is never allowed: the rights to be free from arbitrary deprivation of life; torture and other ill-treatment; slavery; imprisonment for debt; retroactive penalty; non-recognition of the law; and infringement of freedom of thought, conscience, and religion (see, e.g., ICCPR, Article 4(2). See *Economic, Social and Cultural Rights and Climate Change*, Yale School of Forestry and Environmental Studies, available online at https://environment.yale.edu/gem/publications/ESCR-CC-guide/#gsc.tab=0, Human Rights Committee, General Comment no. 29 (2001), UN Doc. CCPR/C/21/Rev.1/Add.11).
53 See IPCC, *supra* note 42.
54 For a discussion of the implications of the intersection of several of these factors and relevant recommendations from UN human rights bodies, see J. B. Martignoni, 'Intersectionalities, Human Rights, and Climate Change: Emerging Linkages in the Practice of the UN Human Rights Monitoring System' (this volume).
55 UNHCR Resolution 10/4, *supra* note 1, preambular recital 8.
56 For a short overview of specific processes and work mandated under the Human Rights Council to address the rights of children or migrants in the context of climate change, see Kirchmeier and Lador, 'From Copenhagen to Paris at the UN Human Rights Council', *supra* note 24.
57 See in this volume: A. Jegede, 'Protecting Indigenous Peoples' Land Rights in Global Climate Governance'; Powless, 'The Indigenous Rights Framework and Climate Change', *supra* note 5; Barre *et al.*, 'From Marrakesh to Marrakesh', *supra* note 5; Pegram, 'Overlooked and Undermined', *supra* note 5; Morena, 'Securing Workers' Rights', *supra* note 4.
58 See in this volume: A. Savaresi, 'Climate Change and Human Rights: Fragmentation, Interplay, and Institutional Linkages'; Schapper, 'Local Rights Claims in International Negotiations', *supra* note 4;

Wallbott, 'Rights, Representation, and Recognition', *supra* note 4; A. Thomas, 'Human Rights and Climate Displacement and Migration'; Brown, 'Using the Paris Agreement's Ambition Ratcheting Mechanisms to Expose Insufficient Protection of Human Rights in Formulating National Climate Policies', *supra* note 43; H. Heiskanen, 'Climate Change and the European Court of Human Rights: Future Potentials'.

59 See in this volume: Chalifour, Mcleod-Kilmurray and Collins, 'Climate Change: Human Rights and Private Remedies', *supra* note 51; Horvath and Maeda, 'Towards Responsible Renewable Energy', *supra* note 5; B. Ugochukwu, 'Climate Change, Human Rights, and Divestment'.

60 For a discussion of some of these cases and of their impacts on domestic action, see, in this volume: Jodoin, Faucher and Lofts, 'Look Before You Jump', *supra* note 4; A. Rubinson Vollmer, 'Mobilizing Human Rights to Combat Climate Change Through Litigation'.

61 L. B. Andonova, M. M. Betsill and H. Bulkeley, 'Transnational Climate Governance', 9 *Global Environmental Politics* (2009) 52; C. Okereke, H. Bulkeley and H. Schroeder, 'Conceptualizing Climate Governance Beyond the International Regime', 9 *Global Environmental Politics* (2009) 58.

62 See D. Bodansky, 'Introduction: Climate Change and Human Rights: Unpacking the Issues', 38 *Georgia. Journal of International & Comparative Law* (2010) 512, at 515.

63 See Caney, *supra* note 45.

64 See the report by the UN Special Rapporteurs on the rights of persons with disabilities; the issue of human rights obligations relating to the enjoyment of a safe, clean, healthy, and sustainable environment; extreme poverty and human rights; the human right to safe drinking water and sanitation; and the Independent Expert on human rights and international solidarity: C. D. Aguilar, J. H. Knox, P. Alston, L. Heller and V. Dandan, *The Effects of Climate Change on the Full Enjoyment of Human Rights* (2015), available online at http://www4.unfccc.int/Submissions/Lists/OSPSubmissionUpload/202_109_130758775867568762-CVF%20submission%20Annex%201_Human%20Rights.pdf.

65 See for instance the case law of the ECrtHRs, *Borysiewicz v. Poland*, Application no. 71146/01, Judgment 1 July 2008. para. 53; ECrtHRs, *Tătar v. Romania*, Application no. 67021/01, Judgment 27 January 2009, para. 95 and 120.

66 See *Concluding Observation by Committee on Economic, Social and Cultural Rights to Australia Suggesting That the State Review Its Coal Extraction and Export Policies Considering Their Climate Implications* (11 July 2017), UN Doc. E/C.12/AUS/CO/5, para. 12.

67 See for instance *Urgenda*, C/09/456689, 13–1396, 24 May 2015, Rechtbank Den Haag, The District Court of the Hague, para. 4.31–4.35; R. H. J. Cox, 'The Liability of European States for Climate Change', 30(78) *Utrecht Journal of International and European Law* (2014) 125, at 128–129.

68 UN General Assembly, *Transforming Our World: The 2030 Agenda for Sustainable Development* (21 October 2015), A/RES/70/1.

69 United Nations Office for Disaster Risk Reduction (UNISDR), *Sendai Framework for Disaster Risk Reduction 2015–2030* (2015), available at www.unisdr.org/files/43291_sendaiframeworkfordrren.pdf (last visited August 2017); United Nations Department of Economic and Social Affairs, *Addis Ababa Action Agenda of the Third International Conference on Financing for Development* (2015), available online at www.un.org/esa/ffd/wp-content/uploads/2015/08/AAAA_Outcome.pdf; UN General Assembly, *New York Declaration for Refugees and Migrants*, UN GA Res. 71/1 (3 October 2016), UN Doc. A/RES/71/1.

70 See, for instance, UN General Assembly, Transforming our world: the 2030 Agenda for Sustainable Development, 21 October 2015, A/RES/701/, Sustainable Development Goal 16 and targets 16.3, 16.7, 16.10, 16.b; UN General Assembly, New York Declaration for Refugees and Migrants: resolution/adopted by the General Assembly, 3 October 2016, A/RES/71/1, Annex II: Towards a global compact for safe, orderly and regular migration.

71 See, e.g., S. Gruskin, D. Bogecho and L. Ferguson, '"Rights-Based Approaches" to Health Policies and Programs: Articulations, Ambiguities, and Assessment', 31 *Journal of Public Health Policy* (2010) 129; P. J. Nelson and E. Dorsey, 'At the Nexus of Human Rights and Development: New Methods and Strategies of Global NGOs', 31 *World Development* (2003) 2013.

72 For an early discussion of the importance of considering the interplay between human rights and climate change in the context of national legislation, see R. Bratpsies, 'The Intersection of International Human Rights and Domestic Environmental Regulation', 38 *Georgia Journal of International & Comparative Law* (2010) 649.

73 M. Nachmany, S. Fankhauser, T. Townshend, M. Collins, T. Landesman, A. Matthews, C. Pavese, K. Rietig, P. Schleifer and J. Setzer, *The GLOBE Climate Legislation Study: A Review of Climate Change Legislation*

in 66 Countries, 4th ed. (London: GLOBE International and the Grantham Research Institute, London School of Economics, 2014).

74 For more information on the petition, see https://business-humanrights.org/en/philippines-commission-on-human-rights-investigation-of-47-fossil-fuel-companies-contribution-to-climate-human-rights-impacts.

75 G. Supran and N. Oreskes, 'Assessing ExxonMobil's Climate Change Communications (1977–2014)', 12 *Environmental Research Letters* (2017), 1–18.

76 M. Olszynski, S. Mascher and M. Doelle, 'From Smokes to Smokestacks: Lessons From Tobacco for the Future of Climate Change Liability', *Georgetown Environmental Law Review* (2017).

77 On the role of the first cycle of the Universal Periodic Review, see E. Cameron and M. Limon, 'Restoring the Climate by Realizing Rights: The Role of the International Human Rights System', 21 *Review of European, Comparative & International Environmental Law* (2012) 204. See also, in this volume: Kirchmeier and Lador, 'From Copenhagen to Paris at the UN Human Rights Council', *supra* note 24; Martignoni, 'Intersectionalities, Human Rights, and Climate Change', *supra* note 54.

78 See the ongoing work of the UN open-ended intergovernmental working group on transnational corporations and other business enterprises with respect to human rights based on a mandate provided by the Human Rights Council. UNHCR Res 26/9, *Twenty-Sixth Session* (14 July 2014), UN Doc. A/HRC/RES/26/9.

79 B. Mayer, 'Human Rights in the Paris Agreement', 6 *Climate Law* (2016) 109–117.

80 For an example of such empiric research demonstrating this point, see C. Stevens *et al.*, *Securing Rights, Combating Climate Change* (World Resource Institute, 2014), available online at www.wri.org/sites/default/files/securingrights-full-report-english.pdf.

2

Analyzing rights discourses in the international climate regime

Katherine Lofts

Introduction

To date, the 21st century has been a time of environmental firsts. Reports predict that 2016 will be the hottest year on record,[1] and scientists indicate that the earth has crossed four of nine "planetary boundaries" – the limits within which humanity can live sustainably.[2] The earth's atmosphere has now passed the threshold of 400 ppm of carbon dioxide,[3] and the first species of mammal has gone extinct due to human-induced climate change.[4] Meanwhile, the global population continues to soar, projected to reach 9.7 billion by the year 2050.[5] We have undeniably reached the Anthropocene – the "Age of the Human" – with devastating implications for the sustainability of life on earth.

Against this backdrop, a number of actors have begun to conceptualize the linkages between environmental governance and rights, increasingly framing environmental harms in terms of rights violations. This conceptual linkage is not inevitable. Just as the idea of "the environment" itself had to be crystallized in politics and policy making through a series of social, historical and political developments,[6] so too has the formulation of environmental impacts or effects on human populations in terms of rights violations been a process rather than a foregone conclusion, slowly taking root as part of the more general ascendance of rights beginning in the 1970s. Today, as Nedelsky notes:

> [T]he language of rights has become a worldwide phenomenon. People use "rights" to identify serious harms, to make claims from and against governments, to make claims for international intervention and assistance. The battle over the use of the term has been decidedly won in its favor.[7]

The language of rights has also inevitably seeped into the realm of climate change – believed by many to be the "defining issue" of our time, but until relatively recently, articulated almost exclusively in environmental and economic terms.[8] Because of its potential to impact nearly every facet of human civilization and the earth's systems upon which we rely, the changing climate will affect an array of rights, variously defined. The complexities of climate change nevertheless pose a number of challenges in terms of how we conceptualize rights and how we seek to operationalize them in search of a more sustainable relationship with the earth. Climate change, by its very nature, crosses borders, involves long timeframes, and implicates numerous actors, making more conventional rights claims difficult. It is a "hyperobject" – a thing "massively distributed in time and space relative to humans,"[9] too big to fit comfortably within news or election cycles or quarterly financial reporting. It is a "super wicked problem" – one "that defies resolution because of the enormous interdependencies, uncertainties, circularities, and conflicting stakeholders implicated by any effort to develop a solution."[10]

In addition to the enormity and complexity of climate change, those with the greatest capacity to address the problem largely lack the impetus to do so, instead seeking to preserve a global economic system that has incentivized overconsumption and the overexploitation of natural resources. The most powerful, industrialized countries will not feel the effects of climate change as severely as less powerful, less developed countries or the most vulnerable segments of society within those countries. On the other hand, those with the least historic responsibility for greenhouse gas emissions will be most acutely impacted.

The extent and seriousness of climate change, and the difficulty in arriving at effective solutions, also raises fundamental questions about the nature of the relationship between human beings and the environment. Indeed, the very survival of human civilization is now at stake due to the way in which the environment and our relationship to it has been imagined, defined and acted upon to date. As Purdy points out:

> What we become conscious of, how we see it, and what we believe it means – and everything we leave out – are keys to navigating the world . . . Imagination also enables us to do things together politically: a new way of seeing the world can be a way of valuing it – a map of things worth saving, or of a future worth creating.[11]

To this end, this chapter adopts a discursive analytic approach to interrogate our ways of seeing and imagining the environment, and the moral and intellectual commitments that underpin them. Discourse can be broadly defined "as an ensemble of ideas, concepts and categories through which meaning is given to social and physical phenomena, and which is produced and reproduced through an identifiable set of practices."[12] By analyzing the discourses of human rights and the rights of nature, this chapter explores how climate change is articulated as a social product of discursive struggles, as well as how rights structure our relationship with nature and with "relations of power, trust, responsibility, and care."[13] How do the ways in which we frame rights in the realm of the environment and climate change replicate or seek to disrupt hierarchical relationships that lead to unsustainable development? What do these discourses reveal and what do they obscure in mediating the relationship between humans and nature? And what are the possibilities for redefining this relationship in more sustainable ways?

Discourse analysis

The methodology used in this chapter draws primarily on social constructivist discourse analysis. In particular, it examines how the "storylines" within a discourse – narratives "that [allow] actors to draw upon various discursive categories to give meaning to specific physical or social phenomena"[14] – create meaning and produce narrative coherence from otherwise disparate and value-neutral phenomena.[15] The language we use does not neutrally reflect or describe the world "out there"; rather, our access to reality is mediated through and constructed by discourse.[16] In describing the world around us, we are always already creating it discursively through the historical rules determining the conditions of possibility for the truth, meaning, and validity of statements within a particular discourse.[17] In this context, "[c]alamities only become a political issue if they are constructed as such in environmental discourse, if story-lines are created around them that indicate the significance of the physical events."[18] The melting of a glacier or ocean acidification are only disasters to the extent that they are framed in terms of their effects within the matrix defined by the storyline itself.

A social-constructivist discursive analysis reveals that there is no coherent, *a priori* idea of what we call "the environment." Instead, this understanding has developed through particular socio-historical circumstances. The terms "environment" and "nature" are themselves freighted with meanings and associations that in turn shape the discourses of which they are a part. Policy making thus requires the construction of social phenomena in ways that render them intelligible and amenable to resolution,[19] and the types of solutions envisioned depend upon the ways in which phenomena are formulated through discourse. As discourse differs, so will the framing of issues, determining the range of possible political consequences. This discursive process of constructing social reality necessarily structures relations of power between actors.[20] Yet, the truths and subjectivities created through discourse are contingent and contestable.[21]

Flowing from these observations on power and the construction of knowledge is the insight that knowledge and social action are intimately connected.[22] Different social understandings of human societies' relationship to the natural world, constituted by and expressed through discourse, will produce tangible consequences in the form of policies, laws and regulations determining the protection or exploitation of natural resources and the environment. The environmental realm in particular becomes an "interesting site to interrogate the exercise of power," as nature foregrounds "the messy politics of representation, articulation, essentialism and discursive construction."[23] The very idea that nature is *natural* – an original or base state from which varying degrees of "civilization" emerge and develop – is inextricably linked to systems of power, including the continuing legacies of colonialism, imperialism and modern forms of market capitalism.

While there are many ways to approach discourse analysis in methodological terms, this chapter draws on the model outlined by Dryzek, which sets out key questions for the analysis of discourses.[24] According to Dryzek, discourses are comprised of four principal elements: (1) the basic entities whose existence is recognized or constructed; (2) assumptions about natural relationships; (3) agents and their motives; and (4) key metaphors and other rhetorical devices. Taking these elements into account, Dryzek contends that the effects of a discourse can be measured by examining the politics associated with that discourse, its effect on the policies of governments and institutions, the arguments of critics, and the flaws that such arguments reveal.[25]

The discourses of human rights and the rights of nature in the international climate change regime

The analysis in the following sections focuses on rights discourses within the international climate change regime, broadly defined. Transnational climate governance is fragmented and decentralized in nature, involving multiple organizations and entities, across multiple sites of authority, with little centralized coordination.[26] While the United Nations Framework Convention on Climate Change (UNFCCC) remains the primary site of multilateral intergovernmental cooperation, non-governmental organizations, corporations, subnational governments, and governmental and intergovernmental agencies are also actively involved, resulting in a governance regime that is multi-faceted and multi-level.[27]

Human rights discourse in the realm of climate change

Since the mid-2000s, the discourse of human rights has become increasingly prominent within the international climate change regime. It is possible to distinguish between its weaker and stronger variants. The former largely seeks to maintain the broader socio-economic status quo, with certain provisions and modifications to ensure greater rights protections. This variant is perhaps most closely aligned with the discourse of sustainable development, which emphasizes the betterment of living conditions for human populations – including "greener" societies – while still prioritizing continual economic growth. In this weaker form of human rights discourse, mitigation to limit the adverse impacts of climate change on the enjoyment of rights and human rights protections in climate change response measures are part of an ensemble of responses linked to "sustainability," leading to better outcomes for environmental protection and a growing economy.

The weaker variant of human rights discourse in the climate regime also shares some features with the discourse of ecological modernization, which emphasizes "the compatibility of economic growth and environmental protection."[28] Ecological modernization contends that development and capitalism can be greened in a win-win scenario of continued economic growth within a liberal market order, alongside increased environmental protection; "environmental problems can be solved in accordance with the workings of the main institutional arrangements of society,"[29] which merely require refinement and reform, rather than a wholesale re-imagination. In a similar way, the weaker variant of human rights discourse in the climate regime espouses a deep faith in human resilience in the face of challenges and an understanding of humans as rational decision makers acting in their own best social and economic interests. By this logic, empowered communities and individuals able to fully exercise their rights within existing socio-economic structures will be better able to meet the challenges posed by climate change, as well as enjoy a number of co-benefits that will further improve their lives.

In contrast, the stronger version of human rights discourse in the climate regime seeks more transformative change and is closely aligned with concepts such as climate justice and equity – emphasizing the need for deeper, more abiding shifts in the world socio-economic order to ensure dignity, human rights and justice for all. For example, the Delhi Climate Justice Declaration, which emerged from the Climate Justice Summit in New Delhi in October 2002, cites unsustainable consumption, industrialized nations and the practices of transnational corporations as amongst the key causes of climate change, while proposing a human rights–based approach to addressing the climate crisis, grounded in the principles of social justice.[30]

Basic entities whose existence is recognized or constructed

The basic entities stressed in human rights discourse are individual human beings – viewed as the primary unit of moral concern, regardless of their geographic location, nationality or other characteristics. According to this discourse, individuals are the possessors of inherent rights by virtue of their humanity. Moreover, "international human rights are not designed as a form of collective power or vehicle of popular governance, but as individual shields against power."[31] Indeed, the conceptual project of human rights has been critiqued for espousing a Eurocentric form of individualism that does not necessarily translate across cultures.[32]

In recognizing individual human beings as the basic entities in this discursive field, the human rights discourse also emphasizes the ontological separateness of humans from other non-human entities. For the most part, ecosystems, elements of the natural world and non-human animals are recognized less as entities with agency than as factors that impact human beings.

Nevertheless, the individualism of human rights discourse has been tempered somewhat in the climate change arena by the scale of impacts around the world and the large numbers of people affected – factors that have highlighted the plight of entire communities, regions and, in some cases, nation-states whose very existence is threatened by sea level rise. "Peoples" and "local communities" are often referred to, emphasizing the collective dimensions of certain rights impacts, as well as their potential scale and pervasiveness. This collective recognition is also prevalent in aspects of the human rights discourse that overlap with the discourse on the rights of Indigenous Peoples, which are more communal in nature than civil and political rights, and some economic, social and cultural rights.

In human rights discourse relating to climate change, the State also looms large – as a potential protector and promoter of rights, as well as that which threatens to interfere with the enjoyment of an individual's rights and freedoms. As Mutua points out: "The state is the guarantor of human rights; it is also the target and *raison d'être* of human rights law."[33] As States are the primary duty bearers with respect to human rights, the system of State sovereignty is reinforced, with an emphasis on the duties States owe to their citizens.

Nevertheless, the private sector and entities such as corporations are also increasingly recognized within the discourse, if not as primary duty holders, then as nonetheless powerful players. In the strong variant of human rights discourse, this recognition of the power wielded by corporate actors manifests as a rejection of capitalist, market-based solutions to climate change, which are viewed as inherently unjust and inequitable. Instead, this discourse recognizes "that unsustainable production and consumption practices are at the root of this and other global environmental problems,"[34] and that corporate power must be restricted, rather than further enabled through the green economy, carbon markets and other capitalist solutions. Corporate actors are thus regarded with suspicion, exerting an undue influence within the climate regime and subverting processes intended to protect communities from climate change and related rights violations.

On the other hand, the weaker variant of human rights discourse takes the capitalist market economy more or less for granted, viewing corporate concern for the human rights implications of climate change as essential to tackling the problem. Indeed, corporations tend to be viewed as agents who might be persuaded – by means of carrot or stick – to respect and promote rights. They are thus seen as potential allies, and numerous business and human rights initiatives seek to harness the influence and power of corporations in trying to secure greater human rights safeguards in the service of "sustainable development."

Assumptions about natural relationships

While human beings are viewed as separate, autonomous entities, the environment serves largely as the medium in which human beings survive, and ideally, thrive. The human being is the figure, and the environment – conceptually and terminologically – is the ground. Although the wellbeing of humans is inextricably tied to the environment, the relationship is hierarchical, with the environment serving human needs. In this way, and perhaps unsurprisingly, the human rights discourse relating to climate change remains anthropocentric. As with sustainable development discourse, "[i]t is the sustainability of human populations and their wellbeing which is at issue, rather than that of nature."[35]

Ecosystems are thus in a subservient relationship to humans, providing vital "services" or amenities, such as water filtration, pollination or flood control. The disruption of these services due to climate change is measurable in relation to their decreased utility to humans. When anthropogenic climate change impacts natural processes, jeopardizing humans' ability to meet their needs as they have in the past, these changes are framed as infringements on the enjoyment of human rights. Furthermore, to the extent that the components of ecosystems are themselves viewed as being in an interconnected relationship with one another, this interconnection is "premised on an Apollonian assessment of connectedness, and a call to efficient management of resources in a closed system."[36] Ecosystems – if permitted to function as they should – will do so in a predictable and "productive" way.

As between individuals, the human rights–based discourse emphasizes equality as a basic starting point for all human rights. In the stronger variant of human rights discourse relating to climate change, this notion of equality is also closely tied to equity, emphasizing the injustice of climate change's unequal impacts on more vulnerable populations. In his recent Papal Encyclical on climate change, for example, Pope Francis cites material inequality between individuals on a global scale as a root cause of environmental degradation. He notes:

> Inequity affects not only individuals but entire countries; it compels us to consider an ethics of international relations. A true "ecological debt" exists, particularly between the global north and south, connected to commercial imbalances with effects on the environment, and the disproportionate use of natural resources by certain countries over long periods of time.[37]

In this conceptualization, the relationships between individuals within countries, as well as between the populations of different countries, are out of balance and exploitative, rather than equal and harmonious as they should be.

Agents and their motives

According to human rights discourse, climate change is primarily an issue of justice and equity. In this context, it should come as no surprise that the primary agents of human rights discourse are individuals. As conventionally conceived, human rights discourse presents "a model presupposing atomistic individuals with equal potential for rationality."[38] Indeed, the concept of human rights is underpinned by "a set of assumptions about individual autonomy."[39] In order to become the subject of human rights, "people had to be perceived as separate individuals who were capable of exercising independent moral judgment."[40] This notion of autonomy is also linked to narratives of empowerment. As Brown points out, "to the extent that human rights are understood as the ability to protect oneself against injustice and define one's own ends in life, this is a form of 'empowerment' that fully equates empowerment with liberal individualism."[41]

Nevertheless, the concept of agency is complicated to some extent in the human rights discourse relating to climate change. There is a tendency on the part of some (primarily Northern) non-governmental organizations and others to emphasize the vulnerability of populations as "victims" of climate change. While "[n]o-one doubts that climate change has victims – specific individuals who undergo suffering"[42] – this framing in terms of victimhood and vulnerability diminishes the agency of affected individuals. The implicit conclusion of such framing is that corporate actors and the governments of industrialized countries – those who are responsible for emissions and have failed to act – are the ones who truly possess agency, while affected populations suffer passively.[43]

On the other hand, strains of human rights discourse that espouse a more radically participatory view of climate governance tend to promote the agency of these vulnerable populations through their active engagement and involvement in decision-making. For example, People's Climate Forums and People's Summits – events organized by grassroots and other non-governmental organizations – recognize the agency of the public by promoting engagement on climate change and opening discussion to a broader range of participants.

Key metaphors and other rhetorical devices

Mutua has argued that the "damning metaphor" of human rights is a tripartite division "pitting savages, on the one hand, against victims and saviors, on the other."[44] In this metaphor – or complex of metaphors – the State is presented as the savage, or the "operational instrument of savagery" which must be guarded against.[45] The victims constructed by rights discourse are those whose dignity has been violated by the State, and the savior – the one "who protects, vindicates, civilizes, restrains, and safeguards" – comes in the form of the United Nations, charities and NGOs, and Western governments.[46]

In the weaker variant of human rights discourse relating to climate change, it is possible to see this dynamic at play. As already mentioned, human rights discourse tends to emphasize the victimization of individuals and communities. In this dynamic, we can also see the "naming and shaming" of certain States, private actors, or other entities that have violated human rights (for example, in relation to the dispossession of communities from their traditional lands due to a project under the Clean Development Mechanism), along with the championing of other States' "best practices" relating to human rights. Connected to this metaphor is the rhetoric of violation and responsibility, which is particularly prominent in the weaker variant of human rights discourse. Climate change is framed as a violation of human rights, and those who have caused climate change are responsible for the violation.

In contrast, the stronger variant of human rights discourse tends to focus more heavily on metaphors of colonization and corporate imperialism. For example, a report by the non-profit research group Corporate Europe Observatory entitled "Corporate Conquistadors: The Many Ways Multinationals Both Drive and Benefit from Climate Destruction"[47] explicitly compares foreign-owned corporate actors to conquering nations exploiting local populations.

The rhetoric of "false solutions" is also prevalent in the strong form of human rights discourse. This rhetoric identifies a number of the solutions to climate change espoused and promoted within the discourses of ecological modernization and green governmentality as being ineffective and perpetuating injustices, such as market-based mechanisms and technological quick fixes. Such false solutions are framed as dangerous – lulling the population into thinking that the problems of climate change are being addressed, while in fact continuing to perpetuate the systems and power relations that caused the problem in the first place.

Rights of nature discourse in the realm of climate change

Compared to human rights discourse, the discourse of the rights of nature in the climate change arena has tended to be more marginal and politicized – linked to a more radical agenda that also decries systems of domination such as capitalism. It is associated primarily with a handful of States that have enacted rights of nature legislation domestically (i.e. Bolivia and Ecuador), as well as with some Indigenous groups.

While at first blush there may appear to be "weaker" variations of this discourse that have taken a more prominent role in the international climate regime, on closer examination, these differ significantly from a true rights-of-nature approach. For example, a number of countries' Intended Nationally Determined Contributions under the Paris Agreement refer to so-called "nature-based solutions" for climate change mitigation, adaptation and risk management. The discourse of nature-based solutions is in many ways antithetical to the rights of nature, however, as it emphasizes the utilitarian aspects of nature, including "natural capital" and ecosystem services. In contrast to this functional view of nature, in which "the notion of commons is strongly associated with that of a particular type of 'resources,'" the rights of nature discourse instead frames the commons as "a systemic entity and a relational field in which 'things' qua resources are only part of."[48]

Basic entities whose existence is recognized or constructed

In this context, it is perhaps obvious to say that one of the primary entities recognized by this discourse is nature itself. In this conceptualization, nature is discursively constructed with specific attributes, composed of systemic and environmental aspects (ecosystems and natural landscapes), as well as their component parts. These component parts are both animate and inanimate – including plants, animals, rocks and streams. For example, Article 4(1) of the Universal Declaration of the Rights of Mother Earth (UDRME) states: "The term 'being' includes ecosystems, natural communities, species and all other natural entities which exist as part of Mother Earth."[49] Taken together, these systems and entities form Mother Earth, who is a living being.

In addition to physical entities and natural systems, nature or Mother Earth is simultaneously discursively constructed as a "relational field and a set a processes at a scale that comprises and binds pretty much everything."[50] As all things arise from Mother Earth and will ultimately return to her, she is the source and facilitator of all relation – "a unique, indivisible, self-regulating community of interrelated beings that sustains, contains and reproduces all beings."[51] To this end, the discourse also recognizes humans as merely one component of Mother Earth. Human beings are not granted any special ontological status, as they are in the discourse of human rights. Rather, the inherent rights of human beings arise from the same source as, and are therefore equal to, the rights of all other entities. As a result, the field of moral concern in the discourse of the rights of nature is greatly expanded.

Assumptions about natural relationships

The discourse of the rights of nature posits familial relationships between human and non-human entities. It speaks of an anthropomorphized "Mother Earth," who cares for the systems and beings that she has created and of which she is composed. Human and non-human entities alike are both a part of and sustained by this mother figure. For example, Article 1(3) of the UDRME states: "Each being is defined by its relationships as an integral part of Mother Earth." These relationships are by their nature nurturing, mutually supportive and harmonious. The preamble of the UDRME proclaims: "[W]e are all part of Mother Earth, an indivisible, living community of interrelated and interdependent beings with a common destiny."[52] While such interdependency and harmony are framed as the default state of relationships amongst systems and entities on earth (and indeed, in the cosmos more broadly), the discourse recognizes that many of our actions as humans have disrupted these relationships. For example, the UDRME recognizes that in "an interdependent living community it is not possible to recognize the rights of only human beings without causing an imbalance within Mother Earth."[53] Indeed, this discourse posits such imbalance as one of the root causes of the climate crisis. It is only by recognizing the roles and responsibilities of all beings that Mother Earth can maintain her harmonious functioning. The flipside of this observation is the idea that the protection of the rights of nature is essential for the protection of human rights; more often than not, violations of the rights of nature are accompanied by violations of human rights and the rights of Indigenous Peoples.[54]

The discourse also links destructive relationships that involve the domination of human beings over nature to other systems of violence and domination that exist amongst and between humans, including patriarchy, militarism, and capitalism. For example, commentators have pointed out that the Paris Agreement includes the words "economic" and "economy" dozens of times, but only includes the word "earth" once, and does not mention the word "nature" – a situation indicative of the way in which the international climate regime has destructively framed the earth as an exploitable resource in a capitalist system.[55] Similarly, the preamble of the UDRME "recogniz[es] that the capitalist system and all forms of depredation, exploitation, abuse and contamination have caused great destruction, degradation and disruption of Mother Earth, putting life as we know it today at risk through phenomena such as climate change."[56]

Agents and their motives

The rights of nature discourse posits that nature has agency, as do the animate and inanimate entities and systems of which nature is composed. As a result, non-human elements such as rivers, trees and animals are not considered to be *objects* or property – as they are by the majority of the world's legal and economic systems – but rather *subjects* with legal standing in their own right. In this regard, the rights of nature include, as per Article 2(1)(c) of the UDRME, the right of the earth "to regenerate its bio-capacity and to continue its vital cycles and processes free from human disruptions," and as per Article 2(1)(j), "the right to full and prompt restoration for violation of the rights recognized in this Declaration caused by human activities."[57] The agency of nature is thus derived from and also subtends the rights of nature.

The "motives" of nature or Mother Earth, as an entity with agency, are perhaps less clear in the discourse. For the most part, nature is anthropomorphized and framed as a benevolent, nurturing mother who will love and provide for her children – human and otherwise – if they respect and love her in return. In this way, the rights of nature discourse posits that earth is driven by a desire for balance, and will seek to maintain equilibrium and ecological homeostasis.

People – particularly communities – also hold tremendous power and agency. While governments are largely beholden to powerful and destructive corporate interests, proponents of the rights of nature view grassroots human collectivities as points of resistance and agents of change. On the other hand, corporate entities are recognized as being powerful agents, whose profit-driven bottom line and exploitative, instrumental view of nature are antithetical to the rights of Mother Earth.

Key metaphors and other rhetorical devices

One of the primary metaphors espoused by the rights of nature discourse is that of balance. The earth is seen to be a self-stabilizing and interconnected system, which is in turn part of a larger cosmic order. If each component fulfills its proper role, the system functions harmoniously, to the mutual benefit of all. If, on the other hand, aspects of the system become unbalanced, the harmony is destroyed. In the context of climate change, many of the modern structures and systems put into place by humans have caused this imbalance, leading to the current climate crisis. For example, capitalism, as an economic system of "boundless accumulation,"[58] has caused human populations to exceed the planetary boundaries within which human beings must operate in order to maintain balance and harmony.

In this sense, the rights of nature discourse tends to place an emphasis on the root causes of climate change, whereas the discourse of human rights in the climate regime focuses more on its effects. This causal focus is connected to an emphasis on the need for radical transformation, rather than "false solutions." The kind of abiding change necessary to correct and redress the broken relationship between human beings and Mother Earth cannot come from ineffective, symptom-based responses to the effects of climate change, but rather must come from the transformation of how human individuals and collectivities navigate their place in the world and their relationship to nature. Indeed, in this view, climate change is a symptom of a more fundamental failure of humans to understand their proper role in the interconnected and interdependent web of existence.

Finally, the discourse of the rights of nature espouses the idea that to move forward into the future, societies must return to traditional ways, which are viewed as more compatible with the rights of nature. The so-called modern world has lost its way and has forsaken the interdependent relationship with the natural environment that it once had. The discourse recognizes that many Indigenous communities have managed to maintain a harmonious relationship with the land and with nature; however, the preservation of their traditional ways is under extreme threat and the goal is to return to a state of harmony.

Conclusion: rights discourses in the age of the Anthropocene

The foregoing analysis reveals a range of ideological and moral commitments that both underpin and are constructed by the discourses of human rights and the rights of nature. In their strongest forms, these rights-based discourses call for transformative change, including the dismantling of existing power structures and the destructive systems and relations of exploitation that perpetuate the climate crisis. They represent utopian visions, "draw[ing] on the image of a place that has not yet been called into being."[59] But each discourse is also susceptible to co-option in support of the socio-economic status quo, advocating for individual protections or for greater recognition of nature, while simultaneously propping up existing dominant interests.

The variations within and between these discourses also reveal the different forms of social and environmental imagination espoused by each – the different ways in which they conceive of and construct the "natural" world and humans' place within it, and the essence of relationships amongst entities and systems on the planet and with the earth itself. If we accept in turn that "[l]aw is used by a society as a means of creating and defining itself in accordance with its worldview,"[60] then our environmental imagination will shape our laws governing the natural world as much as these laws will shape the natural world itself.

Of course, it would be disingenuous to compare the discourses of human rights and the rights of nature within the climate regime as though they were somehow on opposite ends of a spectrum of possible human responses to the social, environmental and indeed existential challenges posed by climate change – one end anthropocentric, the other ecocentric. Operating within the socially constructed framework of rights already situates both discourses firmly within the realm of the human. This framing is perhaps in contradistinction to the recent turn in a number of disciplines – including anthropology, art and philosophy – towards object-oriented ontology, which rejects the privileging of human existence over non-human existence. Proponents of the rights of nature have thus made efforts to

> 'widen the circle of the human' and to thereby include, as active agents with a kind of personhood, history, voice, freedom, and responsibility of their own, 'those subaltern members of the collective, things, that have been silenced and "othered" by the imperialist social and humanist discourses.'[61]

Yet even this call to "re-think the personhood, and particularly the potential juridical personhood, of trees, rivers, and mountains,"[62] is a far cry from true object-oriented ontology, which acknowledges that other objects – be they animals, forest ecosystems or specks of dust in the atmosphere – are ultimately unknown and unknowable to us, inaccessible to our human understanding. Post-human egalitarianism of the sort discussed in relation to object-oriented ontology would presumably be at odds with attempts to bestow upon rivers and other "objects" the entirely human construct of juridical personhood.

At the same moment that we are witnessing an "object turn" and post-*human* shift in many disciplines, we have also seen the widespread embrace of the concept of the Anthropocene – the notion that we now live in a post-*natural* world, in which all the major earth system processes – atmospheric, biospheric, geologic and hydrologic – have been altered by humans. As Purdy states:

> The Anthropocene finds its most radical expression in our acknowledgement that the familiar divide between people and the natural world is no longer useful or accurate. Because we shape everything, from the upper atmosphere to the deep seas, there is no more nature that stands apart from human beings.[63]

How, then, can we at once acknowledge that human beings have altered the world to such a great extent that nature is no longer natural, while also taking account of the post-human – the ontology of objects that exceed our human knowing? For Purdy, these positions are not irreconcilable. In fact, the very acknowledgement that we are now living in the Anthropocene entails the corollary acknowledgement of the otherness of all that surrounds us:

> Once we recognize that the "meaning" of nature has always been a way of talking about human life and purposes, all our relations to the nonhuman world must be touched by the uncanny. We simply do not know what is behind another pair of eyes, and what is projection from behind our own.[64]

In this way, human-induced climate change, as the Anthropocenic crisis *par excellence*, is both an existential crisis – a very real threat to the continued survival of humans, nonhumans and planetary systems as we know them – as well as an ontological one, challenging our fundamental notions of humanness, naturalness and artifice. As Purdy points out, "[a]s greenhouse-gas levels rise and the earth's systems shift, climate change has also begun to overwhelm the very idea that there is a 'nature' to be saved or preserved."[65] Climate change – articulated as the product of discursive struggles and rendered "governable" through discourse – thus requires us to fundamentally rethink our relationship with an environment that has already irrevocably changed, and in so doing, to reconceive of its ontological status and our own. This precludes any simple reliance on the old binaries that have underpinned Western thought, and have consequently structured Western legal orders[66] – human/non-human, nature/artifice, subject/object, animate/inanimate. At the same moment that the illusion of our ontological separateness as humans and our hubris as masters of our earthly domain can no longer be sustained, we are also confronted with the radical otherness of that which is not us.

In this context, then, the consideration of rights-based discourses within the climate change regime brings the challenges and paradoxes of environmental governance in our era into sharp focus. Can the discourses of human rights and the rights of nature come to grips with the post-human, post-natural state of the Anthropocene? Can they help facilitate the transformations necessary to ensure modes of environmental governance suited to such a world?

Rights are a social construct, but they are nonetheless morally persuasive and rhetorically powerful. They "represent reasonable minimum demands upon society that are rooted in moral values and thus place compelling principles on the side of the person [or entity] asserting a right."[67] When enshrined in constitutions at the State level, "systems of constitutional rights as limits to the power of governments have been important institutional means for articulating a society's core values and for holding governments accountable to those values."[68] Incorporating rights language in international agreements, or deploying it in public campaigning, can have a similar effect. Rights-based discourse can therefore articulate the "moral case" for action on climate change, linking it to broader issues of global justice and equity and creating rallying points around which legal tools for the furtherance of rights and protections can be honed.

Yet in order to remain useful and progressive, our understanding of rights must be reconceptualized in the age of climate change, as old categories no longer hold and unexamined rights-based approaches may risk perpetuating the kinds of relationships that have led to the current climate crisis. In this respect, Nedelsky argues cogently that the traditional liberal conception that views rights as boundaries around a freestanding individualist self has failed to account for the fundamentally constitutive nature of relationships vis-à-vis the self.[69] This includes not only intimate relationships, but also relationships with strangers, broader economic and societal relationships, and relationships between the human and non-human world.

This "rights as trumps" or "rights as shield" conception has underpinned much of the Western liberal tradition and continues to define it. As Nedelsky points out:

> rights serve to mark and protect the bounded self and, thus, the legitimate scope of the state. But neither the 'bounded self' nor the 'boundaries of state power' are optimal concepts for articulating and protecting core values, such as autonomy or equality.[70]

Nor, I would argue, is this conception of rights well suited to the kind of re-imagining necessary in a post-nature Anthropocene era. In reality, if "human beings are both constituted by, and contribute to, changing or reinforcing the intersecting relationships of which they are a part," then "the earth itself is both condition and effect of these relationships."[71] Indeed, as Nedelsky notes "[t]he very concept of ecology is relational. It is about fundamental interdependence."[72]

We must therefore expand our notion of the types of entities with whom we have relationships – recognizing and protecting certain types of relationships between and amongst humans, other organisms and ecosystems – rather than zealously erecting rights as further barriers to relationality. The challenge lies in how to recognize and attend to such interdependence in a way that does not also assimilate the Other to our understanding or exploit it only for human purposes. Absent a radical appreciation of relationality and the position of the Other, the discourse of human rights in the realm of climate change risks the continued totalizing ontology cautioned against by Lévinas – subsuming all that is nonhuman within our own subjectivity, as raw material for human interests and uses. Endowing non-human entities with rights runs a similar risk and, moreover, is a difficult concept to reconcile with the increasing "unnaturalness" of the Anthropocene.

In particular, the discourses of human rights and the rights of nature run into trouble when confronted with complex situations extending beyond the realm of the human. The need to "balance" competing rights and interests of different actors has been a mainstay of human rights jurisprudence – weighing, for example, freedom of expression against the prohibition of the incitement of hatred. But expanding the field of relationality beyond the human requires new tools; the infinite incommensurability of the Other cannot be apprehended by any straightforward balancing of rights as trumps.

The ascendancy of the concept of rights is not likely to abate anytime soon. But in order for rights discourse to lead us in the direction of better conditions on planet earth, contributing to greater environmental and social justice, the content and context of rights, and the impact of framing situations in terms of rights, must be better understood. Much work remains in interrogating the foundations and manifestations of rights discourses in the environmental realm, particularly considering their rapid proliferation and expansion in recent years, and in the face of the shifting challenges of the Anthropocene. As rights discourses continue to be adopted and mobilized by activists, advocates, scholars and others involved in climate and environmental governance, a parallel questioning and ongoing re-evaluation must occur – one which takes up the call for a more relational understanding of rights and expands it to encompass the non-human. In this way, we may perhaps bridge some of the paradoxes and challenges of reconciling the rights of humans and non-humans, and of living sustainably and ethically in both a post-human and post-natural world.

Notes

1 A. Thompson, *99 Percent Chance 2016 Will Be the Hottest Year on Record* (18 May 2016), available online at www.scientificamerican.com/article/99-percent-chance-2016-will-be-the-hottest-year-on-record/.
2 RT News, *Earth Closer to 'Irreversible Changes' as Humanity Crosses 4 of 9 Planetary Boundaries* (18 January 2015), available online at http://rt.com/news/223835-earth-planetary-boundaries-humanity/.
3 B. Kahn, *Antarctic CO_2 Hits 400ppm for First Time in 4m Years* (16 June 2016), available online at www.theguardian.com/environment/2016/jun/16/antarctic-co2-hits-400ppm-for-first-time-in-4m-years.
4 B. C. Howard, *First Mammal Species Goes Extinct Due to Climate Change* (14 June 2016), available online at http://news.nationalgeographic.com/2016/06/first-mammal-extinct-climate-change-bramble-cay-melomys/.
5 United Nations Department of Economic and Social Affairs, *World Population Projected to Reach 9.7 Billion by 2050* (29 July 2015), available online at www.un.org/en/development/desa/news/population/2015-report.html.
6 J. S. Dryzek, *The Politics of the Earth: Environmental Discourses* (1997) 4.
7 J. Nedelsky, *Law's Relations: A Relational Theory of Self, Autonomy, and Law* (2011) 73.
8 J. Camilleri and J. Falk, *Worlds in Transition: Evolving Governance Across a Stressed Planet* (2009) 273.
9 T. Morton, *Hyperobjects: Philosophy and Ecology After the End of the World* (2013).
10 R. J. Lazarus, 'Super Wicked Problems and Climate Change: Restraining the Present to Liberate the Future', 94 *Cornell Law Review* (2008) 1153, at 1159.
11 J. Purdy, *After Nature: A Politics for the Anthropocene* (2015) 7.
12 M. Hajer and W. Versteeg, 'A Decade of Discourse Analysis of Environmental Politics: Achievements, Challenges, Perspectives', 7 *Journal of Environmental Policy & Planning* (2005) 175, at 175.
13 Nedelsky, *supra* note 7, at 74.
14 M. A. Hajer, *The Politics of Environmental Discourse Ecological Modernization and the Policy Process* (1995) 56.
15 *Ibid.*
16 *Ibid.*, at 1.
17 See, in particular, M. Foucault, *The Archaeology of Knowledge* (2012).
18 Hajer, *supra* note 14, at 20–21.
19 Hajer, *supra* note 14, at 2.
20 D. Howarth and Y. Stavrakakis, 'Introduction', in D. R. Howarth, A. J. Norval and Y. Stavrakakis (eds.), *Discourse Theory and Political Analysis: Identities, Hegemonies and Social Change* (2000) 1, at 4.
21 M. W. Jørgensen and L. J. Phillips, *Discourse Analysis as Theory and Method* (2002) 9.
22 *Ibid.*, at 6.
23 S. Rutherford, 'Green Governmentality: Insights and Opportunities in the Study of Nature's Rule', 31 *Progress in Human Geography* (2007) 3, at 294–295.
24 Dryzek, *supra* note 6, at 15.
25 Dryzek, *supra* note 6, at 20.
26 K. W. Abbott, 'The Transnational Regime Complex for Climate Change', 30 *Environmental and Planning C Government and Policy* (2012) 4, at 571.
27 L. B. Andonova, M. M. Betsill and H. Bulkeley, 'Transnational Climate Governance', 9 *Global Environmental Politic* (2009) 2, at 52.

28 K. Bäckstrand and E. Lövbrand, 'Planting Trees to Mitigate Climate Change: Contested Discourses of Ecological Modernization, Green Governmentality and Civic Environmentalism', 6 *Global Environmental Politic* (2006) 1, at 52.
29 Hajer, *supra* note 14, at 3.
30 India Climate Justice Forum, *Delhi Climate Justice Declaration* (2002), available online at www.indiaresource.org/issues/energycc/2003/delhicjdeclare.html.
31 W. Brown, '"The Most We Can Hope for. . . ": Human Rights and the Politics of Fatalism', in A. S. Rathore and A. Cistelecan (eds.), *Wronging Rights: Philosophical Challenges for Human Rights* (2011) 132, at 144.
32 F. Bragato, 'Human Rights and Eurocentrism: An Analysis From the Decolonial Studies Perspective', 5 *Global Studies Journal* (2013) 3.
33 M. Mutua, 'Savages, Victims, and Saviors: The Metaphor of Human Rights', 42 *Harvard International Law Journal* (2001) 201, at 203.
34 India Climate Justice Forum, *supra* note 30.
35 Dryzek, *supra* note 6, at 130.
36 S. Cubitt, 'Affect and Environment in Two Artists' Film and a Video', in A. W. von Mossner (ed.), *Moving Environments: Affect, Emotion, Ecology and Film* (2014) 249, at 249.
37 Pope Francis, *Laudato Si' – Encyclical Letter of the Holy Father Francis* (2015), available online at http://w2.vatican.va/content/francesco/en/encyclicals/documents/papa-francesco_20150524_enciclica-laudato-si.html, at para. 51.
38 V. S. Peterson, 'Whose Rights? A Critique of the "Givens" in Human Rights Discourse', 15 *Alternatives* (1990) 3, at 304.
39 L. Hunt, *Inventing Human Rights: A History* (2007) 27.
40 *Ibid.*
41 Brown, *supra* note 31, at 137.
42 International Council on Human Rights Policy, *Climate Change and Human Rights: A Rough Guide* (2008), available online at www.ohchr.org/Documents/Issues/ClimateChange/Submissions/136_report.pdf, at 65.
43 Climate change litigation founded on human rights grounds also involves framing in terms of victims and victimization, requiring – as a condition of justiciability – the identification of specific victims, along with the articulation of specific injuries caused by specific perpetrators.
44 Mutua, *supra* note 33, at 201.
45 *Ibid.*, at 202.
46 *Ibid.*, at 204.
47 Corporate Europe Observatory, *Corporate Conquistadors: The Many Ways Multinationals Both Drive and Profit From Climate Destruction*, available online at http://corporateeurope.org/sites/default/files/corporate_conquistadors-en-web-0912.pdf, at 328.
48 M. De Angelis, 'Climate Change, Mother Earth and the Commons: Reflections on El Cumbre', 54 *Development* (2011) 2, at 184.
49 World People's Conference on Climate Change and the Rights of Mother Earth, *Rights of Mother Earth*, available online at https://pwccc.wordpress.com/programa/, at Art 1.
50 De Angelis, *supra* note 48, at 185.
51 World People's Conference on Climate Change and the Rights of Mother Earth, *supra* note 49, at Art. 1(2).
52 *Ibid.*
53 World People's Conference on Climate Change and the Rights of Mother Earth, *supra* note 49, at Preamble.
54 L. Sheehan and G. Wilson, *Fighting for Our Shared Future: Protecting Both Human Rights and Nature's Rights* (2015), available online at http://bit.ly/1Ng3VyQ, at 8–11.
55 See L. Sheehan, *Economics for Earth's Rights* (4 January 2016), available online at http://wordpress.vermontlaw.edu/nelc/2016/01/04/economics-for-earths-rights/#_edn2.
56 World People's Conference on Climate Change and the Rights of Mother Earth, *supra* note 49.
57 *Ibid.*
58 De Angelis, *supra* note 48, at 184.
59 S. Moyn, *The Last Utopia: Human Rights in History* (2010) 1.
60 C. Cullinan, *Wild Law: A Manifesto for Earth Justice* (2011) 57.
61 E. Fitz-Henry, 'The Natural Contract: From Lévi-Strauss to the Ecuadorian Constitutional Court', 82 *Oceania* (2012) 3, at 267.

62 *Ibid.*
63 Purdy, *supra* note 11, at 2–3.
64 *Ibid.*, at 244.
65 Purdy, *supra* note 11, at 249.
66 Of course, it is important to acknowledge that for millennia, many non-Western cultures and their legal orders have held similar conceptions to those currently coming into fashion in Western academic thought.
67 D. R. Boyd, *The Environmental Rights Revolution: A Global Study of Constitutions, Human Rights, and the Environment* (2012) 8.
68 Nedelsky, *supra* note 7, at 231.
69 Nedelsky, *supra* note 7.
70 Nedelsky, *supra* note 7, at 91.
71 *Ibid.*, at 22.
72 *Ibid.*, at 12.

3

Climate change and human rights

Fragmentation, interplay, and institutional linkages

Annalisa Savaresi

Introduction: the interplay between human rights and climate change law

The adverse effects of climate change threaten the enjoyment of a range of human rights, such as the right to life, adequate housing, food and the highest attainable standard of health.[1] Qualifying the effects of climate change as human rights violations, however, poses a series of technical difficulties, including disentangling complex causal relationships and projections about future impacts.[2] Conversely, measures adopted to tackle climate change may themselves have (and indeed have reportedly already had) negative impacts on the enjoyment of human rights.[3] This is especially the case for activities affecting access to and the use of natural resources, such as land, water and forests. Adaptation and mitigation action can interfere with the enjoyment of human rights, such as that to culture, the respect for family life, access to safe drinking water and sanitation, Indigenous Peoples' self-determination, as well as the gamut of procedural rights concerning access to information, justice and participation in decision-making.[4] While climate change response measures can engender perverse outcomes for the protection of human rights, human rights protection can also engender problematic outcomes when it is pursued without factoring in climate change concerns.[5]

This complex relationship between climate change and human rights obligations has increasingly been recognized in the literature,[6] as well as by human rights bodies. Starting with 2008, the Human Rights Council (HRC) has adopted a string of resolutions emphasising the potential of human rights obligations, standards and principles to 'inform and strengthen' climate change law- and policy-making, by 'promoting policy coherence, legitimacy and sustainable outcomes'.[7] The HRC also encouraged its special procedures mandate holders to consider the issue of climate change and human rights within their respective mandates.[8] As a result, the special procedures mandate holders unprecedentedly engaged with the making of the Paris Agreement, suggesting, amongst others, the inclusion of human rights language in the treaty and that Parties refrain from viewing their human rights responsibilities as stopping at their borders.[9]

This impassionate plea was laden with potentially significant legal implications. As not all Parties to climate treaties have ratified the same human rights treaties, adhesion to the Paris Agreement could have become a means to impose upon state obligations enshrined in treaties they have not ratified.[10] Furthermore, states commonly interpret their human rights instruments as jurisdictionally limited to individuals or entities within their effective control.[11] It is, in other words, difficult to argue that states have specific obligations to undertake positive action to secure the protection of human rights associated with climate change impacts beyond their territorial boundaries. Finally, states' discretion in choosing the means for implementing their international obligations renders striking a balance between competing societal interests and needs a rather context-specific matter, on which legislators enjoy a great deal of leeway.[12]

These arguments were forcefully made in the lead-up to the adoption of the Paris Agreement.[13] As a result, the Paris Agreement only partially follows the suggestions made by the special procedures mandate holders. The preamble specifies that Parties 'should, when taking action to address climate change, respect, promote and consider their respective obligations on human rights', citing 'the right to health, the rights of Indigenous Peoples, local communities, migrants, children, persons with disabilities and people in vulnerable situations and the right to development, as well as gender equality, empowerment of women and intergenerational equity'.[14] The operative part of the Paris Agreement also makes specific references to the need to be responsive to gender concerns as well as to the rights of Indigenous Peoples.[15] Albeit timid, these textual references break new ground and may have significant implications for the interpretation and further development of Parties' obligations under the climate regime, especially in the context of the newly established platform on indigenous and local community climate action.[16]

This chapter analyses these recent developments, placing them in the context of the scholarly debate on the fragmentation of international law. This scholarship has investigated at length questions of coherence within and interplay between areas of the international legal order. It is used here as a conceptual lens to better understand interactions between international human rights and climate change law, as well as the means available to manage the interplay between the two. The chapter is structured as follows. The second section introduces the debate on the fragmentation of international law, as well as tools that have been devised to tackle it. The third section analyses how the tools identified in the scholarship on fragmentation have been deployed to address the interplay between the climate change and the human rights regimes. The conclusion offers some reflections on future interrelations between these two regimes.

The fragmentation of international law

The debate on the fragmentation of international law emerges from concerns that international law-making and institution-building increasingly tend to take place 'with relative ignorance of legislative and institutional activities in the adjoining fields and of the general principles and practices of international law.'[17] The perceived compartmentalization of the law into highly specialized branches that develop in relative autonomy from each other is not only *inter-specific*, i.e. between different areas of the law (e.g. environmental law, human rights law, trade law, etc.), but also *intra-specific*, thus affecting instruments belonging to the same area of international law (e.g. climate change law, biodiversity law, etc.). This state of affairs is arguably the result of institutional deficiencies of the international legal system, which is inherently devoid of a clear normative and institutional hierarchy and a comprehensive judicial jurisdiction,[18] as well as of the progressive transposition of governance functions from the national to the international plane.[19]

After two decades of debate, fragmentation is widely accepted as an intrinsic characteristic of the international legal order. So while early scholarship focused on problematizing fragmentation,[20] more recent scholarship acknowledges fragmentation as part of the natural state of things, and rather focuses on ways to manage it.[21] How, in other words, is it possible to enhance the coherence between elements in the international legal architecture, avoiding the threat of 'antagonistic developments'?[22] And what is the relationship between rules embedded in separate but overlapping international regimes?

The International Law Commission (ILC) addressed these matters in one of its reports.[23] The report points out that the increased specialization of international law poses challenges associated with the collision of norms and regimes, deploying the term 'conflict' to refer to a situation whereby 'two rules or principles suggest different ways of dealing with a problem'.[24] This definition encompasses not only mere 'logical incompatibility' between norms, but also 'policy conflicts', i.e. when a treaty frustrates the goals of another without there being any strict incompatibility between their provisions.[25] These conflicts have been also described in the literature as 'implementation conflicts' – i.e. conflicts that are engendered by implementation of perfectly compatible treaty obligations[26] – or 'functional conflicts' – i.e. interference in the operation of concurrent norms, that takes place for example when a norm reinforces the behaviour another seeks to discourage.[27] The notion of policy conflict draws attention to the fact that interaction between international law regimes in not a one-off phenomenon, but concerns the 'day-to-day working' of legal instruments, starting from their very making and continuing with their interpretation and implementation.[28] So, rather than simply focus on establishing the applicable legal regime and related set of rules in a given context, scholars are increasingly presuming that multiple international regimes interact with one another in an iterative manner and searching for constructive ways of making them work together.[29]

The ILC report suggests three possible avenues to address conflict: conflict avoidance; resolution through the application of interpretative principles; and institutional cooperation and coordination.[30] On conflict avoidance, the report suggests including in treaties guidance on how to deal with subsequent or prior conflicting treaties, distinguishing different typologies of so-called conflict clauses.[31] These clauses typically specify that a treaty 'is subject to', or that 'it is not to be considered as incompatible with, an earlier or later treaty', or that 'the provisions of that other treaty prevail'. The report nevertheless recognizes the limits to such clauses, which oftentimes merely 'push' the resolution of problems to the future.[32]

On treaty interpretation, the ILC points to rules in the Vienna Convention on the Law of Treaties (VCLT)[33] as the 'tool-box' for dealing with fragmentation.[34] The VCLT codifies treaty interpretation rules that are commonly regarded as an embodiment of customary international law.[35] Whilst recognising that VCLT rules give insufficient recognition to special types of treaties and rules concerning their interpretation and implementation, the ILC points to the continued relevance of these rules to dealing with conflict of norms.[36]

In particular, the report underscores the role of systemic integration as an aid to deal specifically with policy conflicts.[37] Systemic integration suggests that when creating new obligations, states are assumed not to derogate from their obligations, as embodied in any rules of international law that are both 'relevant' and 'applicable' in the relations between the Parties.[38] The rationale behind this interpretation tenet is quite simple: rights and obligations established by treaty provisions exist alongside rights and obligations enshrined in other treaties. As none of these rights or obligations has any intrinsic priority against the others, the ILC suggests that their relationship be approached through a process of reasoning that 'makes them appear as parts of some coherent and meaningful whole'.[39] Such systemic thinking is arguably part of the very essence of legal reasoning and perhaps the only possible solution to the 'clustered' nature in which legal rules and principles appear.[40]

In times of increasing fragmentation of the international legal order, systemic integration has unsurprisingly been the subject of much scholarly attention.[41] Some scholars have cautioned against the dangers of conflating treaty 'interpretation' with treaty 'modification',[42] pointing out that the presumption of coherence is to be 'handled with care' and assessed on a 'case by case basis', as states may indeed have adopted the new instrument with the specific purpose to do away with their extant international commitments.[43] Admittedly, systemic integration may only resolve 'apparent' conflicts and not instances of actual incompatibility.[44] Systemic integration therefore tends to operate before an irreconcilable conflict of norms has arisen and provides a tool to engender coherence in international law,[45] by urging the interpretation of state obligations as much as possible in an integrated fashion.

Finally, the ILC report sets aside the question of institutional interaction, expressing the conviction that 'the issue of institutional competencies is best dealt with by the institutions themselves'.[46] This matter has nevertheless subsequently been addressed in a literature strand that builds upon international relations theories on interplay management.[47] One of the most important studies on the issue distinguishes between various levels of coordination and institutionalisation, ranging from the macro level, where the interplay between regimes is managed by an overarching institution, to the micro level, where interplay management is left to the autonomous efforts of national governments.[48] The next section considers how the techniques reviewed here may be used to address the interplay between climate change and human rights instruments.

Addressing conflicts between the climate change and the human rights regime

By virtue of its subject matter, the climate regime is particularly likely to overlap with other international regimes, and, as such, it is particularly prone to policy conflicts.[49] As acknowledged above, this is especially the case in relation to human rights law.[50] So whereas in principle there is no incompatibility between these two sets of international norms, in practice policy conflicts between the two may well emerge. This section considers how tools to manage the fragmentation of international law have been deployed and may be deployed in future as an aid to address the interplay between the human rights and climate change regimes. The role of conflict avoidance techniques, such as conflict clauses and systemic integration, is considered first, to then look at institutional cooperation as a means to better integrate human rights concerns into climate change action.

Conflict clauses

Neither the UNFCCC nor the Kyoto Protocol includes a conflict clause. Nevertheless, the Cancun Agreements say that Parties 'should, in all climate change related actions, fully respect human rights'.[51] Because of its hortatory tone and the fact that it was included in a COP decision, rather than in treaty text, the all-embracing reference to human rights in the Cancun Agreements was not particularly contentious. Conversely, the inclusion of a textual reference to human rights in the Paris Agreement was hotly debated.[52] On the one hand, some Parties supported a blanket reference to human rights – e.g. 'All Parties . . . shall ensure respect for human rights and gender equality in the implementation of the provisions of this Agreement'.[53] On the other, some Parties expressed reservations, based on the fact that not all states have ratified international or regional human rights treaties.[54]

The possibility to address this tension by means of a conflict clause was put forward in a report by the International Law Association (ILA), which suggested the following formulation: 'states shall formulate, elaborate and implement international law relating to climate change in a mutually supportive manner with other relevant international law.'[55] Such a conflict clause formulation would not create new obligations for states that are not Parties to human rights treaties already. Instead, it would merely underscore states' existing obligations in relation, *inter alia*, to human rights, signalling to Parties that these too should be taken into account when implementing the Paris Agreement. The importance of this interpretative guidance would be limited, but not insignificant. As by its own nature the climate regime is prone to policy conflicts with other international instruments, reminding states of the need to align also with these when implementing climate treaties seems important. Not only would a conflict clause have the effect to emphasise that Parties should take their existing human rights commitments into account when they implement the Paris Agreement. It would also provide an important signal for institutions, within and without the climate regime, on the need to consider human rights in their guidance and standards concerning climate change response measures.

No conflict clause was eventually included in the Paris Agreement or even made it in the negotiating text. Nevertheless, the Paris Agreement's preamble points to Parties' 'respective human rights obligations'. This reference draws attention to Parties' obligations under treaties they have ratified already, or may ratify in future, rather than foreshadowing new ones. Even with this limited remit, the reference to human rights in the Paris Agreement is not devoid of legal consequence. Preambular text carries political and moral weight. By forging an explicit link with human rights instruments, the Paris Agreement's preamble engenders an expectation that Parties will take into account their existing human rights obligations concerning matters such as, for example, public participation or the rights of women and Indigenous Peoples when they adopt climate change response measures. So in spite of its limited legal force and the lack of a conflict clause, the reference to human rights in the Paris Agreement is in many connections groundbreaking, especially in relation to the interpretation of Parties' obligations. The next section looks at this issue in detail.

Treaty interpretation and systemic integration

The ILC report points out the fact that conflict resolution maxims, such as *lex specialis derogat generali*, have clear limitations in addressing policy conflicts such as those engendered by overlaps between the climate and human rights regimes.[56] Identifying what is to be regarded as *lex specialis*, for example, may be difficult with regimes that tend to be all-encompassing in scope.[57] Similarly, the *lex posterior* rule is of limited utility when dealing with 'living instruments' that are constantly kept under review by their treaty bodies, such as those on climate change.[58] As a result, the ILC rather suggests relying upon systemic integration to deal with policy conflicts. Consequently, state obligations under climate treaties should be interpreted in a way that is mutually supportive, rather than conflicting with, obligations under other treaties, including human rights ones.

In the lead-up to the adoption of the Paris Agreement, a string of HRC resolutions drew attention to potential conflicts, overlaps and synergies between the climate change and human rights regimes.[59] These decisions underscore the need for policy coherence, thus implicitly making reference to systemic integration in the interpretation of states' obligations concerning human rights and climate change. Most saliently in 2014 the (then) Independent Expert on Human Rights and the Environment issued a report on the human rights threatened by climate change and the human rights obligations relating to climate change.[60] This report was the first comprehensive effort to systematically map the human rights affected by climate change, as well as relevant guidance adopted by human rights bodies, thus providing an important *vademecum* for systemic integration.

The Paris Agreement's preambular reference may be read as an invitation to practice systemic integration in the interpretation of Parties' obligations, at least insofar as human rights are concerned. Well ahead of the adoption of the Paris Agreement, such an approach has already been experimented with in some areas of the climate regime, where potential conflicts with human rights obligations are particularly evident. As other contributions in this volume show,[61] matters like REDD+[62] and climate finance[63] have already confronted states and international agencies with challenging questions over the interplay between climate change and human rights law. COP decisions on REDD+ make reference to systemic integration.[64] The need to ensure compatibility with human rights has instead been emphasised by one of the international agencies facilitating REDD+, which has adopted a human rights–based approach to its work,[65] including free prior consent guidelines elaborated in partnership with human rights bodies.[66] Equally, standards adopted by some climate finance institutions specifically refer to human rights.[67] In both connections, therefore, while not all countries seeking climate/REDD+ finance may have ratified human rights treaties, human rights protection has been elected as one of the criteria they should satisfy to obtain such finance.[68]

Experience accrued thus far with REDD+ and climate finance standards is an important term of reference to understand how obligations under the Paris Agreement may be interpreted in light of human rights law and practice, as well as challenges that can emerge in this process. This experience is likely to be particularly useful in relation to inter-state collaboration through the so-called Sustainable Development Mechanism (SDM).[69] The Special Rapporteur on human rights and the environment has already drawn attention to the need to ensure that the latter mechanism incorporates strong social safeguards that accord with international human rights obligations.[70] Institutional cooperation could be an important means to streamline human rights considerations into such safeguards. The next section looks at this matter.

Institutional cooperation

Ensuring that obligations under the climate regime are interpreted and implemented in line with states' human rights obligations has long been left to the autonomous efforts of national decision-makers and single institutions, in what Sebastian Oberthür has aptly described as 'autonomous' or 'unilateral interplay management'[71] – as opposed to forms of interplay management, where such coordination endeavours are carried out by a set of institutions together or by an overarching international institution. The risk that autonomous and unilateral interplay management end in incoherence is already palpable when one considers that standards already vary greatly, for example, in relation to climate finance.

In the lead-up to the adoption of the Paris Agreement, human rights bodies have become more and more proactive in their efforts to engage with legal developments in the climate regime. HRC resolutions set the premises for increased institutional cooperation, by encouraging the OHCHR and the HRC special procedures mandate holders to engage with the climate regime. Ensuing initiatives include the special procedures mandate holders' open letter to climate negotiators issued in the lead-up to the adoption of the Paris Agreement,[72] as well as OHCHR's submissions on various matters under considerations at climate negotiations, such as gender, adaptation, the local and indigenous communities platform and the SDM,[73] and the elaboration of expert recommendations on climate change and human rights.[74] Especially notable in this context are the activities undertaken by the Special Rapporteur on human rights and the environment, who through his reports, statements and letters has made significant efforts to engage with the climate change regime and provide recommendations on how to better factor in human rights in climate change law and policy.[75]

Arguably, the very inclusion of a reference to human rights in the Paris Agreement is the result of advocacy by key epistemic actors, including the Special Rapporteur and former United Nations High Commissioner for Human Rights Mary Robinson. Aside from this milestone achievement, however, it is hard to say how receptive the climate regime has been to human rights bodies' institutional cooperation efforts. Historically, Parties to the climate regime have been reluctant to establish inter-institutional linkages. Even when they have done so, as, for example, in the context of the Joint Liaison Group to enhance coordination between the UNFCCC, the Convention on Biological Diversity and the United Nations Convention to Combat Desertification, very limited results have been obtained, based on the argument that the Rio Conventions have a 'distinct legal character, mandate and membership'.[76] Institutional cooperation with human rights bodies could be even more problematic, as treaty membership is more heterogeneous than that of the Rio Conventions. Yet again, the inclusion of a reference to human rights in the Paris Agreement may be a game-changer in this connection.

The HRC special procedures mandate holders have, for example, invited Parties to the climate regime to launch a work program to ensure that human rights are integrated into all aspects of climate actions.[77] The creation of a work program would constitute an institutional space for Parties to consider whether and how to better integrate human rights in the climate regime. The work program could also become a forum for Parties to exchange information on experience with integrating human rights into climate action and share good practices. Finally, a work program could discuss institutional linkages between climate change bodies and international and regional bodies with a specific mandate on the protection of human rights. For example, Parties could entrust the UNFCCC Secretariat to collaborate with the OHCHR to integrate human rights consideration into climate action. Moving forward, a dedicated grievance mechanism for those complaining for human rights violations specifically associated with the implementation of climate change response measures could be established, such as a Special Rapporteur on climate change and human rights.[78] Alternatively, human rights considerations may be specifically factored into the mandate of existing grievance mechanisms, such as, for example, the Independent Redress Mechanism established under the Green Climate Fund.

Inter-institutional cooperation could furthermore galvanise the use of extant human rights bodies as means to seek redress for human rights breaches associated with the implementation of climate change measures and/or impacts. There are already precedents of this happening in practice[79] and more may be in the pipeline, due to imaginative climate change litigations strategies emerging around the globe.[80] The UN Universal Periodic Review (UPR) could be tasked to specifically highlight human rights concerns associated with climate change,[81] and thus become a means to see how Parties to the climate regime address human rights concerns associated with climate change and a way to disseminate best practices. A dedicated institution to support the consideration of human rights issues could also be established. This institution could be entrusted with the task to promote the sharing between Parties of information and of good practices on the integration of human rights in climate action. To support this task, Parties could be required to report efforts to integrate human rights into climate actions and policies in their national communications, or as part of their reporting obligations under human rights instruments.

Conclusion: where next?

As other chapters in this volume will show, the interplay between human rights and climate change law is far from unproblematic.[82] As not all Parties to the climate regime have ratified human rights treaties, adherence to the Paris Agreement cannot be a means to impose upon state obligations enshrined in treaties they have not ratified. Furthermore, even for those states that do have human rights obligations, the conventional interpretation of the jurisdictional limitations of these obligations presently undermines arguments concerning the protection of human rights beyond state territorial boundaries. States' margin of appreciation in implementing their obligations under both climate and human rights treaties is a considerable obstacle. Finally, thus far fundamental limitations in the climate regime have constrained synergies between human rights and climate change law, due to the primacy accorded to economic concerns and development in climate politics; the construction of climate change as a technocratic and scientific policy problem, rather than a human-centred one; and differences between mechanisms to assess compliance with climate obligations versus those used in the human rights regime.

Even bearing these complexities in mind, states' human rights obligations in relation to both the impacts of climate change and response measures have pervasive legal ramifications. With the adoption of the Paris Agreement, these ramifications have been put in the spotlight. The agreement is potentially a game changer, opening up new avenues to improve coordination and address synergies between distinct international legal regimes. This chapter has shown that techniques devised to address the fragmentation of international law have already been deployed in this connection. Whilst not a conflict clause, the Paris Agreement's reference to human rights draws attention to systemic integration, at least for those Parties that have ratified human rights treaties already. Human rights bodies have underscored the potential for systemic integration, and there is evidence that, at least in some cases, institutions in the climate regime have attempted to address human rights concerns in the standards they adopted. At the institutional level, human rights bodies have increasingly engaged with the making of international climate change law, from the drafting of the Paris Agreement to the nitty-gritty decision-making of climate treaty bodies.

Moving ahead, much more could be done to address the limitations of the climate regime: institutional cooperation could be systematised and become instrumental to the streamlining of human rights considerations into the climate regime. Human rights bodies may even provide institutionalized pathways to monitor and sanction human rights violations associated with climate change impacts and the implementation of climate change response measures. The Paris Agreement could thus become the foundation for unprecedented cross-fertilisation between international human rights and environmental law. Indeed, when an issue has over-arching implications for a range of different international regimes, it seems wise to emphasise and vigorously explore avenues for coordination.[83] Yet, how far states will be willing to go down this route largely remains to be seen.

Notes

1 OHCHR, *Report on the Relationship between Climate Change and Human Rights*, UN Doc A/HRC/10/61 (15 January 2009) 16.

2 *Ibid.*, at 70.

3 *Ibid.*, at 65–68.

4 OHCHR, *Report of the Special Rapporteur on the Issue of Human Rights Obligations Relating to the Enjoyment of a Safe, Clean, Healthy and Sustainable Environment*, UN Doc A/HRC/31/52 (1 February 2016) 50–64.

5 See for example O.W. Pedersen, 'The Janus-Head of Human Rights and Climate Change: Adaptation and Mitigation', 80(4) *Nordic Journal of International Law* (2011) 403 and B. Lewis, 'Balancing Human Rights in Climate Policies', in O. Quirico and M. Boumghar (eds.), *Climate Change and Human Rights: An International and Comparative Law Perspective* (2016) 39.
6 See for example S. Humphreys (ed.), *Human Rights and Climate Change* (2009); 'Special Issue', 38 *Georgia Journal of International and Comparative Law* (2009); Rajamani, 'The Increasing Currency and Relevance of Rights-Based Perspectives in the International Negotiations on Climate Change', 22(3) *Journal of Environmental Law* (2010) 391; S. McInerney-Lankford, M. Darrow and L. Rajamani, *Human Rights and Climate Change: A Review of the International Legal Dimensions* (2011); O. Quirico and M. Boumghar, *supra* note 5; and 'Special Issue', 34 *Journal of Energy & Natural Resources Law* (2016).
7 HRC, *Human Rights and Climate Change*, UN Doc A/HRC/7/78 (29 March 2008); HRC, *Human Rights and Climate Change*, UN Doc A/HRC/RES/10/4 (25 March 2009), at preamble; HRC, *Human Rights and Climate Change*, UN Doc A/HRC/RES/18/22 (17 October 2011), at preamble; and HRC, *Human Rights and Climate Change*, UN Doc A/HRC/RES/26/27 (15 July 2014), at preamble.
8 HRC Res. 26/27, *supra* note 7, at 7.
9 Open Letter from special procedures mandate holders of the *Human Rights Council to the State Parties to the UN Framework Convention on Climate Change on the Occasion of the Meeting of the Ad Hoc Working Group on the Durban Platform for Enhanced Action in Bonn (20–25 October 2014)* (17 October 2014), available online at http://newsroom.unfccc.int/media/127348/human-rights-open-letter.pdf.
10 While 197 states have ratified the United Nations Framework Convention on Climate Change 1771 UNTS 163 (adopted 9 May 1992; entered into force 21 March 1994) ('UNFCCC'), the International Covenant on Civil and Political Rights 999 UNTS 171 (adopted 16 December 1966, entered into force 23 March 1976) ('ICCPR') has 169 parties; the International Covenant on Economic, Social and Cultural Rights, 993 UNTS 3 (adopted 16 December 1966, entered into force 3 January 1976) ('ICESCR') has 165 parties; the UN Convention on the Elimination of All Forms of Racial Discrimination, 660 UNTS 195 (adopted 7 March 1966, entered into force 4 January 1969) ('CERD') has 178 parties; the Convention on the Rights of the Child 1577 UNTS 3 (adopted 20 November 1989, entered into force 2 September 1990) ('CRC') has 196 parties; the Convention on the Elimination of All Forms of Discrimination against Women, 1249 UNTS 18 (adopted 18 December 1979, entered into force 3 September 1981) ('CEDAW') has 189 parties; and Convention on the Rights of Persons with Disabilities, 2515 UNTS 3 (adopted 13 December 2006, entered into force 3 May 2008) ('CRPD') has 172 parties. Treaty membership data up to date as of 26 February 2017.
11 As reiterated for example in OHCHR, *supra* note 4, at 41.
12 As argued also in O. Quirico, J. Bröhmer and M. Szabó, 'States, Climate Change and Tripartite Human Rights: The Missing Link', in O. Quirico and M. Boumghar (eds.), *Climate Change and Human Rights: An International and Comparative Law Perspective* (2016) 7, at 21.
13 As reported for example in Human Rights Watch, *Human Rights in Climate Pact Under Fire* (7 December 2015), available online at www.hrw.org/news/2015/12/07/human-rights-climate-pact-under-fire.
14 United Nations Framework Convention on Climate Change, *Adoption of the Paris Agreement*, U.N. Doc. FCCC/CP/2015/L.9/Rev. 1 (12 December 2015), at preamble [*Paris Agreement*].
15 Art. 7.5, Paris Agreement, *supra* note 14 and 11.2; B. Powless, 'The Indigenous Rights Framework and Climate Change' (this volume) and A. Barre *et al.*, 'From Marrakesh to Marrakesh: The Rise of Gender Equality in the Global Climate Governance and Climate Action' (this volume).
16 Decision 1/CP.21, Adoption of the Paris Agreement, FCCC/CP/2015/10/Add.1 (2015), at 135–136; and Report of the Conference of the Parties on its twenty-second session, held in Marrakech from 7 to 18 November 2016, FCCC/CP/2016/10 (2016), at 163–167.
17 ILC, *Fragmentation of International Law: Difficulties Arising From the Diversification and Expansion of International Law. Report of the Study Group of the International Law Commission*, UN Doc A/CN.4/L.682 (13 April 2006), at 11.
18 *Ibid.*, at 493.
19 B. Simma, 'Universality of International Law From the Perspective of a Practitioner', 20(2) *European Journal of International Law* (2009) 265, at 270.
20 G. Teubner, *Global Law Without a State* (1997); and A. Fischer-Lescano and G. Teubner, 'Regime-Collisions: The Vain Search for Legal Unity in the Fragmentation of Global Law', 25 *Michigan Journal of International Law* (2003) 999.
21 See for example M. Young (ed.), *Regime Interaction in International Law: Facing Fragmentation* (2012); R. Michaels and J. Pauwelyn, 'Conflict of Norms or Conflict of Laws? Different Techniques in the

Fragmentation of Public International Law', 22 *Duke Journal of Comparative & International Law* (2012) 349; and H. van Asselt, *The Fragmentation of Global Climate Governance: Consequences and Management of Regime Interactions* (2014).

22 For the use of this term, see C.C. Pollack and M.A. Shaffer, 'The Interaction of Formal and Informal Lawmaking', in J. Pauwelyn, R.A. Wessel and J. Wouters, *Informal International Lawmaking* (2012) 241, at 250–254.

23 *Ibid.*

24 ILC, *supra* note 17, at 25.

25 *Ibid.*, at 24.

26 R. Wolfrum and N. Matz, *Conflicts in International Environmental Law* (2003) 24.

27 M. Prost, *The Concept of Unity in Public International Law* (2012) 63.

28 See M. A. Young, 'Regime Interaction in Creating, Implementing and Enforcing International Law', in M. A. Young, *supra* note 21, at 89 and 91.

29 J. L. Dunoff, 'A New Approach to Regime Interaction', in M. A. Young, *supra* note 21, at 137 and 157.

30 ILC, *supra* note 17, at 13–19.

31 *Ibid.*, at 278–281.

32 *Ibid.*, at 276.

33 United Nations, *Vienna Convention on the Law of Treaties ('VCLT')*, 23 May 1969 (entered into force: 27 January 1980), United Nations, Treaty Series, vol. 1155, p. 331.

34 ILC, *supra* note 17, at 250.

35 Cf. e.g. *Territorial Dispute (Libyan Arab Jamahiriya/Chad)*, Judgment, 3 February 1994, ICJ Reports (1994), at 41; *Kasikilil/Sedudu Island (Botswana/Namibia)*, Judgement, 13 December 1999, ICJ Reports (1999), at 18; *LaGrand (Germany v. United States of America)*, Judgement, 27 June 2001, ICJ Reports (2001), at 99.

36 ILC, *supra* note 17, at 251.

37 *Ibid.*, at 410–480.

38 VCLT, *supra* note 33, Art. 31.3(c) and ILC, *supra* note 17, at 38.

39 ILC, *Ibid.*, at 414.

40 *Ibid.*, at 35.

41 For example, C. McLachlan, 'The Principle of Systemic Integration and Article 31(3)(c) of the Vienna Convention', 54 *International and Comparative Law Quarterly* (2005) 279; J. Klabbers, 'Reluctant Grundnormen: Articles 31(3)(c) and 42 of the Vienna Convention on the Law of Treaties and the Fragmentation of International Law', in M. C. R. Craven, M. Fitzmaurice and M. Vogiatzi (eds.), *Time, History and International Law* (2007) 141; U. Linderfalk, 'Who Are "The Parties"? Article 31, Paragraph 3(c) of the 1969 Vienna Convention and the "Principle of Systemic Integration" Revisited', 55(3) *Netherlands International Law Review* (2008) 343; R. Pavoni, 'Mutual Supportiveness as a Principle of Interpretation and Law-Making: A Watershed for the "WTO-and-Competing-Regimes" Debate?', 21(3) *European Journal of International Law* (2010) 649; M. Samson, 'High Hopes, Scant Resources: A Word of Scepticism About the Anti-Fragmentation Function of Article 31(3)(c) of the Vienna Convention on the Law of Treaties', 24(3) *Leiden Journal of International Law* (2011) 701.

42 B. Simma and T. Kill, 'Harmonizing Investment Protection and International Human Rights: First Steps Towards a Methodology', in C. Binder, U. Kriebaum, A. Reinisch and S. Wittich (eds.), *International Investment Law for the 21st Century: Essays in Honour of Christoph Schreuer* (2009) 678, at 692–4; The same view is expressed in R. K. Gardiner, *Treaty Interpretation* (2008) 266.

43 B. Simma, 'Foreign Investment Arbitration: A Place for Human Rights', 60 *International and Comparative Law Quarterly* (2011) 573.

44 ILC, *supra* note 17, at 42.

45 McLachlan, *supra* note 41, at 318. Similarly, B. Chambers, *Interlinkages and the Effectiveness of Multilateral Environmental Agreements* (2008) 248.

46 ILC, *supra* note 17, at 13.

47 See for example S. Oberthür and O. S. Stokke, *Managing Institutional Complexity: Regime Interplay and Global Environmental Change* (2011); Young, *supra* note 28, at 89; Dunoff, *supra* note 29, at 157; and van Asselt, *supra* note 21.

48 S. Oberthür, 'Interplay Management: Enhancing Environmental Policy Integration Among International Institutions', 9(4) *International Environmental Agreements: Politics, Law and Economics* (2009) 371, at 375–6.

49 On the issue, see for example H. van Asselt, F. Sindico and M. A. Mehling, 'Global Climate Change and the Fragmentation of International Law', 30 *Law & Policy* (2008) 424; C. P. Carlarne, 'Good Climate

Governance: Only a Fragmented System of International Law Away?', 30(4) *Law & Policy* (2008) 423, at 450; F. Haines and N. Reichman, 'The Problem That Is Global Warming: Introduction', 30(4) *Law & Policy* (2008) 385; and J. A. McNeely, 'Applying the Diversity of International Conventions to Address the Challenges of Climate Change', 17 *Michigan State University College of Law Journal of International Law* (2008) 123.

50 See HRC, *supra* note 7 and corresponding text.

51 *The Cancun Agreements: Outcome of the Work of the Ad Hoc Working Group on Long-Term Cooperative Action Under the Convention*, Decision 1/CP.16, UN Doc. FCCC/CP/2010/7/Add.1 (15 March 2011), at 8.

52 For an overview of references to human rights in the Paris Agreement negotiating text, see A. Savaresi and J. Hartmann, *Human Rights in the 2015 Agreement*, available online at http://legalresponseinitiative.org/wp-content/uploads/2015/05/LRI_human-rights_2015-Agreement.pdf.

53 UNFCCC, *Negotiating Text* (12 February 2015), available online at https://unfccc.int/files/bodies/awg/application/pdf/negotiating_text_12022015@2200.pdf, at 12bis.

54 Human Rights Watch, *supra* note 13.

55 International Law Association, *Legal Principles Relating to Climate Change* (2014), available online at www.ila-hq.org/en/committees/index.cfm/cid/1029. Draft Article 10.1. Draft Article 10.3(b) included a specific reference to human rights, according to which: 'States and competent international organizations shall respect international human rights when developing and implementing policies and actions at international, national, and subnational levels regarding climate change'.

56 ILC, *supra* note 17, at 22.

57 As argued also in Wolfrum and Matz, *supra* note 26, at 170–173; Chambers, *supra* note 45, at 50–60; and H. van Asselt, 'Managing the Fragmentation of International Environmental Law: Forests at the Intersection of the Climate and Biodiversity Regimes', 44(5) *New York University Journal of International Law and Politics* (2012) 1205, at 1250–1252.

58 As argued also in Michaels and Pauwelyn, *supra* note 21, at 378; and van Asselt, *supra* note 57, at 1250–1252.

59 HRC, *supra* note 7, and corresponding text.

60 Independent Expert on the issue of human rights obligations relating to the enjoyment of a safe, clean, healthy, and sustainable environment, 'Mapping Human Rights Obligations Relating to the Enjoyment of a Safe, Clean, Healthy and Sustainable Environment', *Focus Report on Human Rights and Climate Change* (June 2014), available online at http://srenvironment.org/mapping-report-2014-2/.

61 D. D. Pugley, 'Rights, Justice, and REDD+: Lessons From Climate Advocacy and Early Implementation in the Amazon Basin' (this volume).

62 As argued for example in A. Savaresi, 'The Human Rights Dimension of REDD', 21 *Review of European Comparative & International Environmental Law* (2012) 102; A. Savaresi, 'REDD+ and Human Rights: Addressing Synergies Between International Regimes', 18(3) *Ecology and Society* (2013) 5; A. Savaresi, 'The Role of REDD in Harmonising Overlapping International Obligations', in E. Hollo, K. Kulovesi and M. Mehling (eds.), *Climate Change and the Law: A Global Perspective* (2013) 391.

63 See A. Johl and Y. Lador, *A Human-Rights Based Approach to Climate Finance* (2012), available online at http://library.fes.de/pdf-files/iez/global/08933.pdf; and D. S. Olawi, *The Human Rights-Based Approach to Carbon Finance* (2016).

64 Decision 1/CP.16, *supra* note 51, at Appendix I, at 2 (a), where specific reference is made to the fact that REDD+ actions 'complement or are consistent with the objectives of national forest programmes and relevant international Conventions and agreements'. For a commentary, see A. Savaresi, 'The Legal Status and Role of Safeguards', in Christina Voigt (ed.), *Research Handbook on REDD+ and International Law* (Edward Elgar Publishing, 2016) 126.

65 See e.g. UN-REDD Programme, *UN-REDD Programme Social and Environmental Principles and Criteria* (2012), UNREDD/PB8/2012/V/1, at 2.

66 UN-REDD Programme, *Guidelines on Free, Prior and Informed Consent* (2012), available online at www.un-redd.org/Launch_of_FPIC_Guidlines/tabid/105976/Default.aspx; and UN-REDD Programme, *Legal Companion to the UN-REDD Programme Guidelines on FPIC* (2012), available online at www.unredd.net/index.php?view=document&alias=8792-legal-companion-to-the-un-redd-programme-guidelines-on-fpic-8792&category_slug=legal-companion-to-fpic-guidelines-2655&layout=default&option=com_docman&Itemid=134.

67 See e.g. Adaptation Fund, *Environmental and Social Policy* (2013) 15: 'Projects/programmes supported by the Fund shall respect and where applicable promote international human rights'. GCF Environmental and Social Safeguards explicitly mention ensuring full respect of the human rights of indigenous peoples, and their FPIC, at least in certain circumstances. Compare: GCF, *Guiding Framework and Procedures for Accrediting National, Regional and International Implementing Entities and Intermediaries, Including*

the Fund's Fiduciary Principles and Standards and Environmental and Social Safeguards, GCF/B.07/02 (7 May 2014), at 1.7.

68 These examples are further discussed in: A. Savaresi, 'The Legal Status and Role of Safeguards', in C. Voigt (ed.), *Research Handbook on REDD+ and International Law* (2016) 126.

69 Paris Agreement, *supra* note 14, Art. 6.4.

70 J. H. Knox, *Letter From the Special Rapporteur on Human Rights and the Environment to Climate Negotiators* (4 May 2016), available online at http://srenvironment.org/wp-content/uploads/2016/06/Letter-to-SBSTA-UNFCCC-final.pdf.

71 Oberthür and Stokke, *supra* note 47, at 376.

72 Open Letter from special procedures mandate holders, *supra* note 9.

73 See: OHCHR response to UNFCCC Secretariat request for submissions on the Nairobi Work Programme: impacts, vulnerability and adaptation to climate change: Health impacts, including occupational health, safety and social protection, FCCC/SBSTA/2016/2, para 15(a)(i), 2016; OHCHR response to the UNFCCC Secretariat request for submissions on the Lima Work Programme on Gender: Views on possible elements and guiding principles for continuing and enhancing the work programme (SBI), FCCC/SBI/2016/L.16, paragraph 5, 2016; OHCHR response to the UNFCCC Secretariat request for submissions on the Paris committee on Capacity-Building: Views on the annual focus area or theme for the Paris Committee on Capacity-Building for 2017 (SBI), FCCC/SBI/2016/L.24, 2016; OHCHR response to UNFCCC Secretariat request for submissions on the Paris Agreement (APA): Views and guidance related to intended nationally determined contributions, adaption communications, the transparency framework, and the global stocktake, and for information, views and proposals on any work of the APA, FCCC/APA/2016/2, 2016; and OHCHR response to UNFCCC Secretariat request for submissions on the future UNFCCC Sustainable Development Mechanism: Regarding the rules, modalities and procedures for the mechanism established by Article 6, paragraph 4, of the Paris Agreement, *supra* note 14, para 100, 2016. All submissions are available online at www.ohchr.org/EN/Issues/HRAndClimateChange/Pages/UNFCCC.aspx.

74 The OHCHR hosted an expert meeting on climate change and human rights on 6–7 October 2016 in Geneva. The Draft Recommendations elaborated at the meeting are available online at www.ohchr.org/EN/Issues/HRAndClimateChange/Pages/ClimateChange.aspx.

75 Most notably: Report of the Special Rapporteur on human rights and the environment, *supra* note 4, and Letter from the Special Rapporteur on human rights and the environment to climate negotiators, *supra* note 70. The OHCHR has also been mandated to organise an expert meeting providing guidance on the same issue: Expert Meeting on Climate Change and Human Rights 6–7 October 2016, Draft Recommendations, available online at www.ohchr.org/EN/Issues/HRAndClimateChange/Pages/ClimateChange.aspx.

76 See for example the position by the US in Views on the Paper on Options for Enhanced Cooperation Among the Three Rio Conventions, *Submissions From Parties*, UN Doc. FCCC/SBSTA/ 2006/MISC.4 (23 March 2006), at 16. The same point was made by Australia, *ibid.*, at 5, as observed also in van Asselt, *supra* note 57, at 41–24.

77 Open Letter from Special Procedures Mandate-Holders, *supra* note 9.

78 Cf. the petition launched by Environmental Justice Foundation, available online at http://ejfoundation.org/petition/special_rapporteur.

79 For an example of how the UN Convention on the Elimination of All Forms of Racial Discrimination has been applied in connection with REDD+ in Indonesia, see Savaresi, *supra* note 62; and N. Johnstone, 'Indonesia in the "REDD": Climate Change, Indigenous Peoples and Global Legal Pluralism', 12(1) *Asian-Pacific Law & Policy Journal* (2011) 93.

80 As suggested e.g. in N. Chia, M. Mueller and J. Warland, *Roundtable Summary – Human Rights & Climate Change: Connecting the Dots* (2016), available online at www.ucl.ac.uk/global-governance/ggi-publications/john-knox-climatechange-publication.

81 This suggestion has been made in e.g. International Bar Association, *Achieving Justice and Human Rights in an Era of Climate Disruption* (2014), available online at www.ibanet.org/PresidentialTaskForceClimateChangeJustice2014Report.aspx.

82 S. Jodoin *et al.*, 'Look Before You Jump: Assessing the Potential Influence of the Human Rights Bandwagon on Domestic Climate Policy' (this volume).

83 As suggested also in A. Boyle, 'Climate Change and International Law – a Post-Kyoto Perspective', 42 *Environmental Policy and Law* (2012) 6, at 342.

4

Local rights claims in international negotiations

Transnational human rights networks at the climate conferences

Andrea Schapper

Introduction

Every year, member states of the United Nations Framework Convention on Climate Change (UNFCCC) negotiate at the Conference of the Parties (COP) to review the convention's implementation, to adopt legal instruments or to make additional institutional arrangements. During the negotiation process, governments organize themselves in blocs and coalitions in order to combine strengths, increase their influence and push forward their common agendas.[1] Civil society organizations (CSOs) can participate in UNFCCC negotiations as accredited observers representing the interests of particular societal groups. To be able to make oral interventions at the negotiations, CSOs need to be recognized as an official constituency by the UNFCCC's Secretariat. Whereas the UNFCCC has initially been characterized by strong engagement of business stakeholders and environmental organizations, other actors have now entered the scene, among them Indigenous Peoples, faith-based groups, gender advocates and human rights activists often organized in transnational advocacy networks (TANs).[2]

TANs can be understood as communicative structures in which a range of activists guided by principled ideas and values interact. These ideas, values and norms are central to the networks' activities and they determine criteria for evaluating whether particular actions and their outcomes are just or unjust. TANs create new linkages, multiply access channels to the international system, make resources available to new actors and help to transform practices of national sovereignty. Within these networks, international and local CSOs, foundations, the media, churches, trade unions, academics and even members of regional or international state organizations as well as single representatives of a state government can collaborate. TANs' overall objective is to change the policies of states and international organizations (IOs).[3]

Strong civil society and TAN participation can make state negotiations more complex and difficult – tabling new themes and issues, such as the adverse human rights impacts of climate change and climate politics. Human rights concerns, in particular, give negotiations a more intricate character; they cannot be easily ignored and usually require immediate political action.[4] Concerns voiced in relation to human rights at the climate conferences usually take two directions. On the one hand, climate change impacts hamper the full realization of economic, social and cultural rights, such as the right to health, water, food and adequate housing. In certain cases, like extreme weather events, the right to life can also be affected. On the other hand, climate policy implementation can also infringe on human rights; large-scale mitigation projects like hydroelectric dams, for instance, have previously led to limited access to land, water or cultural sites for local populations.[5] Therefore, civil society actors lobby for governmental commitment to human rights in climate agreements. States, however, have for a considerable period of time refrained from applying a human rights language in climate negotiation processes. Developed states – that are usually financing climate policies – feared even more costly obligations if an observance of certain standards is required. Developing countries emphasized their sovereignty and were afraid that deficiencies in the rights situation on their own territory could become unveiled.[6]

Nevertheless, human rights entered the perambulatory and operative clauses of the climate agreements made at COP 16 in Cancun, Mexico, in 2010 and have again been institutionalized[7] in the preamble of the 2015 Paris Agreement at COP 21 in France. Moreover, in 2010, procedural human rights have been installed as a requirement for forest protection and management programs called Reducing Emissions from Deforestation and Forest Degradation (REDD+).[8] Their introduction for projects under the Clean Development Mechanism (CDM) has been debated at COP 19 in Warsaw in 2013 and at the following negotiations with a view to revising the modalities and procedures of the CDM. Procedural human rights are of particular importance in environmental law. They establish a link between the state and civil society by fostering transparency and participation in environmental decision-making.[9] The most important procedural rights are the right to information, the right to participation in decision-making and the right to justice, the latter usually meaning access to judicial and administrative recourse procedures. All of these rights are anchored in the International Covenant on Civil and Political Rights.[10] Of far more influence in environmental matters, however, is the United Nations Economic Commission for Europe's (UNECE) Convention on Access to Information, Public Participation in Decision-Making and Access to Justice in Environmental Matters, known as the Aarhus Convention from 1998. Although it is only binding for the states that ratified it, the Aarhus Convention has turned out to become the main reference document when it comes to procedural rights in climate change–related matters.[11]

In this chapter, I argue that the activities of TANs explain why and how human rights have entered climate agreements. TANs use local experiences and case studies at the international climate negotiations to demonstrate that both climate change and climate policies have adverse implications for the human rights of populations.[12] Simultaneously, they emphasize the need for observing existing rights obligations in all climate-related actions and for strengthening procedural rights in climate policies to protect local societal groups, especially Indigenous Peoples.[13] By interacting very closely with state representatives, international non-state partners within TANs transmit local claims to the states' negotiation table. They persuade state actors that are receptive to human rights arguments to introduce rights language into the draft texts being negotiated. States, in which rights infringements in climate policy implementation occur and that have an active domestic civil society, can experience pressure from below (through civil society) and from above (through TANs). In this way, pressure exerted by state governments on their local population in the course of policy implementation can come back to them at the international climate conferences – like a boomerang.[14]

To elaborate my argument, I have structured my chapter in the following way. First, I briefly review the literature on civil society participation and the activities of TANs at the international climate conferences. Second, I will present a case study on the activities and influence of the

Human Rights and Climate Change Working Group (HRCCWG). I selected the HRCCWG because it constitutes the only TAN active at the COPs that is exclusively focusing on institutionalizing human rights in the climate regime and that was established for exactly that reason. The HRCCWG was the initiator and coordinator of a broader inter-constituency alliance at COP 21 promoting human rights in the Paris Agreement. My empirical assessment builds on a content analysis of primary documents, including observer submissions, press releases, social media, as well as reports from IOs and CSOs. Additionally, I draw on semi-structured interviews conducted with activists of the HRCCWG at the international climate negotiations in Warsaw (Poland) in 2013 and in Paris (France) in 2015. Third, I will reflect on the relevance of information delivery and the boomerang pattern as the main mechanisms explaining institutional interaction between the human rights and the climate regime, before I conclude.

Transnational human rights advocacy at the international climate conferences

In 1992, the Rio Conference on Environment and Development has marked, among other substantive themes, the strengthening of civil society participation in decision-making processes within the UN system.[15] Civil society participants are understood to bring expertise and credibility to IOs and negotiations within their fora.[16] In open and transparent negotiation processes, states increasingly take up non-state demands.[17] CSO participation is considered to improve the democratic development of IOs and their decisions.[18] Environmental and climate politics can be regarded as unique policy fields facilitating advanced institutional mechanisms for access and participation of civil society actors.[19]

In climate politics, the focus of CSOs in negotiations is on addressing justice concerns or employing a climate justice frame.[20] Climate justice is a fluid framework that is diversely utilized,[21] but it broadly embraces the observation that those people who have contributed the least to climate change are those who are affected the most[22] – and often have the fewest resources to adapt.[23] The climate justice movement is characterized by antagonism between a moderate wing accepting capitalism and lobbying for change within established institutions and a radical wing viewing capitalism as a root cause for climate challenges that needs to be questioned.[24] This results in cooperation and conflict between TANs and states within and outside of the UNFCCC process.[25]

Moderate-wing CSOs seek to shape the development of climate policies by actively engaging in international climate conferences.[26] They can participate in the official UNFCCC process by acting as accredited organizations,[27] which grants them the opportunity to raise awareness for certain issues, lobby with governmental representatives and create networks of influence. Those organizations with strong ties to state delegations have the most advanced access, and in some cases, civil society actors can even become members of national delegations and are "formally granted a 'seat at the table'".[28] This increases their opportunities to influence governmental decisions since it provides them with access to closed sessions, official state documents and the possibility to present their own proposals.[29] Governmental delegations are interested in including CSOs because they provide expertise[30] and can enhance the legitimacy of their decisions.[31] CSOs, in contrast, use their close interaction with governments to exert pressure for negotiating, ratifying and complying with international environmental agreements.[32] Hence, relations between IOs and CSOs are changing; state and society are not entirely separate entities anymore, but they are in flux and their specific interaction has to be analytically captured.

There are some particulars about TANs active at the international climate conferences: their networks are hybrid – actors may join for a short period of time and then leave again – and participating organizations can be quite diverse.[33] Although groups of the global South are usually underrepresented in these networks,[34] local CSOs from developing countries are increasingly funded by foundations and international CSOs as part of TANs to voice the concerns of local people adversely affected by climate change and climate policies at this international venue. TANs often have the expertise and experience needed to strategically use the information provided by local actors for lobbying purposes within the UNFCCC. Thus, TANs help explain how justice and human rights claims are transported from civil society to state negotiators.

This means actors work at various scales; they differ in their degree of institutionalization and in their positioning toward the UNFCCC process (inside/outside). Newer (and still evolving) network structures engage very closely with international institutions and sometimes even invite governmental delegates to participate, in particular if they bring in information from closed (intergovernmental) sessions or if they are open to introduce text passages prepared by CSOs in these sessions.[35]

Case study: the human rights and climate change working group

Network building and initial successes

One important example of a TAN working closely with governmental delegations is the Human Rights and Climate Change Working Group (HRCCWG). The HRCCWG has been particularly active and successful in their attempt to promote human rights within the climate regime. The idea to launch this group was born in 2008 at COP 14 in Poznan, Poland, and the network became operative during the following COP in 2009 in Copenhagen, Denmark.[36] The HRCCWG can be described as a hybrid network of predominantly civil society and single government representatives (mostly from IOs) operating at various scales – from the local to the global[37] – with the common objective of promoting and institutionalizing human rights in the climate regime. Among the networks' members are prominent international CSOs, such as the Center for International Environmental Law, Earthjustice, Friends of the Earth, Carbon Market Watch, Human Rights Watch, Amnesty International, but also local CSOs from developing countries, gender advocates, Indigenous Peoples' representatives, academics, representatives from IOs, like the Office of the High Commissioner for Human Rights (OHCHR) and UNICEF, as well as single actors from state delegations.[38] Membership in the network is rather informal; participants can be present at one negotiation meeting joining the group's activities there, and then miss the next one.[39] Simultaneously, they can be part of another TAN active at the international climate conferences, including Climate Action Now, the REDD+ Safeguards Working Group or the Indigenous Caucus.[40]

The Human Rights and Climate Change Working Group, among other TANs, was strongly involved in efforts to promote and institutionalize human rights in the climate regime at COP 16 in Cancun (Mexico) and at COP 21 in Paris (France). Members of the HRCCWG consider these negotiation sessions as reflecting incremental progress in the institutionalization of rights in climate governance.[41] The most important successes have been the following. First, in the preambulatory clauses of the Cancun Agreements, the UNFCCC member states recognized a Human Rights Council Resolution (HRC 10/4) on human rights and climate change, stating,

> the adverse effects of climate change have a range of direct and indirect implications for the effective enjoyment of human rights and that the effects of climate change will be felt most acutely by those segments of the population that are already vulnerable owing to geography, gender, age, indigenous or minority status, or disability.[42]

This means that there is a consensus that rights are adversely affected by climate change impacts. Second, in the operative part of the Cancun Agreements, member states announced, "parties should, in all climate-change related actions, fully respect human rights".[43] These actions, of course, refer to all climate policy initiatives introduced under the legal framework of the UNFCCC and the Kyoto Protocol. Third, the first procedural rights, also known as safeguards, were institutionalized as a requirement for the implementation of REDD+ programs in Annex 1 of the Cancun Agreements. This means that the rights to participation, information, transparency and free prior and informed consent need to be respected for local communities affected by the realization of REDD+ programs. The rights, knowledge and (land) ownership of Indigenous Peoples are particularly emphasized here.[44] Fourth, a review of the modalities and procedures of the CDM is underway. During COP 19 in 2013 in Warsaw and during the Intersessionals in 2014 and 2015 in Bonn, states (and CSOs) discussed stronger stakeholder consultation requirements and several references stating that activities under the CDM have to be carried out in accordance with human rights.[45] It can be expected that procedural rights will enter CDM policies in the future.[46] Finally, human rights have again been anchored in the preambulatory clauses of the Paris climate treaty in 2015. It stipulates that state parties should,

> when taking action to address climate change, respect, promote and consider their respective obligations on human rights, the right to health, the rights of indigenous peoples, local communities, migrants, children, persons with disabilities and people in vulnerable situations and the right to development, as well as gender equality, empowerment of women and intergenerational equity.[47]

Network strategies

The most successful strategy the HRCCWG employed for achieving its objectives has been to build friendly relations with state representatives at the climate negotiations. Making use of these receptive relations, the network receives access to negotiating texts and attempts to include human rights language in respective drafts asking state parties to introduce these drafts in closed negotiation sessions:

> Most of the actual negotiating meetings have been closed to me because I'm civil society but a lot of times I'll go and network with parties . . . and we're staying outside the door and if we have position papers on different things, we'll ask parties to introduce an item or . . . we write an email. . . [and ask to] consider it or anything. One of our . . . points is actually looking at the text, examining it, seeing places that we could think could be improved to better support the issues that we want them to support and then actually suggesting language. And parties actually take it up and are really excited about it. . . . I know several parties that I can just ask for: "What happens in that meeting?" or "could you possibly pass on to me the text?" You know or, "could you send it around?" You know, so that part is really through the relationship that you have with the parties and if the parties [change] you build new relationships with the parties. It's really great.[48]

Relations with state parties are particularly well established if the negotiators are open to human rights arguments due to their own liberal democratic state identity or increased pressure by domestic CSOs, or if they rely on the network's expertise and capacities like a number of developing countries do.[49] If the latter is the case, network members provide their expert knowledge and attend sessions on behalf of small delegations that cannot afford to travel to the negotiations with a large number of staff. This means, civil society actors become officially registered on respective state delegations and draft text suggestions can be introduced in meetings officially closed for non-state observers on their behalf:

> Many parties have so little capacities to follow half of the issues that it's actually very helpful to them if there is something drafted and they can either redraft it, interpret it or be able to kind of introduce it as it is. Their country position goes well with the language, they just didn't know, they didn't have time to actually formulate the language and so, you know, we are advocates but at the same time . . . technical experts on gender issues, you know, this has helped to influencing the agenda.[50]

To justify the use of human rights language in climate agreements, the HRCCWG frequently works with case studies emphasizing the adverse effects of climate change and climate policies on local people from Asia, Latin America and also Africa.[51] At the climate negotiations, case studies are usually presented during so-called side events[52] that run parallel to the meetings of state parties and are accessible to all, to governmental delegations but also to non-state observer groups, including civil society, the business community, representatives from religious organizations and academics. Cases on problematic rights situations are sometimes presented by locally affected people themselves, e.g. indigenous or agro-pastoralist groups. Some international partners of the HRCCWG have in the past sponsored representatives of these local groups to join international meetings, share their experiences and bring forward corresponding demands. TANs help to process these cases demonstrating that there are severe drawbacks in the body of rules and regulations of climate policies, such as REDD+ and the CDM, and to voice their demands for change in the modalities and procedures of such instruments.[53] In this way, local claims are fed into the international negotiation process. A representative from an international CSO summarizes their objective of transporting local claims to the international negotiation table with the following words:

> The way we do our research is really go into the field and speak with the people on the ground, do many many interviews, speak with everyone, not only the communities that are affected but of course also, you know, government and, . . . especially in a context where not everyone is able or, financially able or just generally can participate in these kind of high-level discussions and so, of course, in a COP like here there are many groups from different countries but still if you think about representation, these won't be the most disadvantaged people that will make it to these international negotiations. So I think this is also what we are trying to do with our work generally but also in this context is kind of bringing the voices of those that are not usually being heard to the international negotiations.[54]

Whereas the initial focus of the HRCCWG was on framing emission reductions as a human rights obligation, there is now a clear emphasis on response measures and their impact on local populations in developing countries.[55] Thus, procedural human rights, and especially the right to participation, lie at the heart of the network's activities.[56] Altogether, we can find a two-way process here: local advocates inform the policy-making process of states within the forum of an international organization and their decisions may, in return, possibly affect climate policy implementation at the local level.[57]

Enhanced outreach on the road to Paris

On the road to Paris, two key representatives of the HRCCWG initiated an inter-constituency alliance in order to combine the strengths of several civil society networks at the negotiations of a new climate treaty. The constituencies at the UNFCCC negotiations are clustered groups of officially registered CSOs sharing certain interests and acting as observers in the process, among them environmental CSOs (ENGO), Indigenous Peoples organizations (IPOs), youth CSOs (YOUNGO), or women and gender (WOMEN AND GENDER). Participation in a constituency comes with several advantages: it allows observers to make interventions at certain points in the state negotiation process, it facilitates the use of focal points for better coordination with the UNFCCC Secretariat and it enhances flexible information sharing.[58] Prior to the Paris negotiations, an inter-constituency alliance was established because most observer organizations shared some common concerns. Among them were the protection and fulfillment of human rights, in particular the right to health and food security, Indigenous Peoples' rights, a just transition of the workforce and the creation of decent jobs, gender equality and equal participation of women, as well as inter-generational equity.[59]

Besides initiating and coordinating the inter-constituency alliance, there were also other changes to the HRCCWG leading to a strengthening of its advocacy efforts. After they had commenced working mainly with environmental law organizations, representatives from Indigenous Peoples' and women's rights CSOs, the major non-governmental human rights players, Human Rights Watch (HRW) and Amnesty International (AI), came on board at the COP in Lima in 2014. A common Greenpeace-Amnesty statement, as well as a widely spread common press release by HRW and AI during the Paris negotiations, marked the arrival of the large human rights CSOs in the network.[60]

In addition to these new non-governmental partners from the human rights regime, state actors joined the working group as well. Among them were mainly IOs, like the OHCHR and UNICEF. Both took an active part in the network, further developed its strategy at the coordination meetings and engaged in awareness-raising activities targeting state delegates.[61]

Besides IOs, there were also some states taking an active part in the promotion of human rights in the climate agreement. Already prior to the negotiations, eighteen governments took action and initiated the *Geneva Pledge for Climate Action* calling for enhanced institutional interaction between the UNFCCC and the OHCHR and emphasizing that human rights obligations need to be observed in all climate-relevant actions. Among the committed states were mostly Latin American countries (e.g. Mexico, Peru and Costa Rica), many small island states (e.g. Maldives, Kiribati and Samoa) as well as a few European nations (e.g. France, Sweden and Ireland).[62]

Another great push for human rights in the climate regime came from increased media attention and a successful Twitter campaign. Under #Stand4Rights, the HRCCWG and several of their partners disseminated information regarding new versions of the negotiating text including human rights, spread the word on further awareness-raising actions and put pressure on governments who argued against rights in the climate agreement during the negotiations. An interesting example for the latter is the use of the #Stand4Rights to tweet a joint press release by Amnesty International and Human Rights Watch that blamed and shamed those countries that were severely blocking human rights language in the climate treaty at that point, namely the USA, Saudi Arabia and Norway.[63] Everyone on the mailing list of the HRCCWG was asked to re-tweet this press release, and the wide public attention around this led to immediate reactions of the opponents. Norway, for instance, released an official statement emphasizing that it supports human rights in the operative part of the treaty but not in Article 2 (which sets out the purpose of the agreement).[64]

Moreover, several high-quality media outlets, among them *The Guardian*, took up the issue because human rights as bracketed text were still part of the operative clauses of the agreement in the pre-final draft and thus remained one of the outstanding questions to be decided upon until the very end of COP 21.[65] Several representatives from the HRCCWG evaluated this as a significant success in public awareness raising on human rights in the context of climate change.[66]

Thus, prior to and in Paris, the network was benefitting from an enhanced outreach including combined strengths in the inter-constituency alliance, active participation of international governmental organizations and state actors as well as increased press and social media coverage. All of these aspects led to at least one success – the institutionalization of human rights in the preamble of the new climate agreement.

A boomerang pattern in climate negotiations?

What representatives of the HRCCWG have described as initial successes in institutionalizing human rights in the climate regime can best be explained with insights from scholarship on institutional interaction or regime interplay.[67] Institutional interaction means that the institutional development or effectiveness of one institution becomes affected by another institution.[68] Interaction can also occur across policy fields leading to either conflict or synergy. So far, the literature has predominantly focused on investigating cases of inter-institutional conflict.[69] Instances of interaction with synergetic effects have received less attention yet. One focus in the literature on institutional interaction is to identify causal mechanisms of influence exerted from one source institution to a specific target institution.[70] These comprise, first of all, cognitive interaction, or learning. Here, the source institution disposes of insights that it feeds into the decision-making process of the target institution.[71] Second, interaction through commitment means that the member states of a source institution have agreed upon commitments that might be relevant for the members of the target institution as well. If there is an overlap of membership, the commitments made in the source institution can lead to differing decision-making in the target institution.[72] Third, behavioral interaction comes into play if the source institution has obtained an output initiating behavioral changes that is meaningful for the target institution. In cases like this, the initiated changes in behavior can foster further behavioral changes.[73] And fourth, impact-level interaction is based on a situation of interdependence, in which a "functional linkage"[74] between the governance objectives of the institutions can be observed. If the source institution obtains an output that has an effect on the objectives of the source institution, this impact can also influence the objectives (and effectiveness) of the target institution.[75]

For developing a better understanding of institutional interaction between the human rights and the climate regime, the micro-macro link,[76] i.e. the mechanisms at play between the micro level of actors and the macro level of institutions, need to be further established. Here, constructivist international relations models highlighting the mutually constitutive character of actors and structures might be able to enrich rational choice-oriented institutionalist theories, i.e. scholarship on institutional interplay. Research on TANs[77] provides useful insights that help us understand how CSOs use information and established frames in one policy field to motivate (more powerful) IOs and their member states in a different policy field to change their policies.

The boomerang pattern describes a situation in a repressive state, in which channels between domestic CSOs and the norm-violating state are blocked and these CSOs decide to bypass the state government and provide information on rights violations to a TAN. Thereupon, the network mobilizes the human rights regime, including democratic states and IOs, using persuasion mechanisms. The regime, eventually, exerts pressure on the respective state to initiate a human rights change. Hence, what has departed from within a state, i.e. by oppositional groups, comes back like a boomerang from outside, i.e. the TAN and the human rights regime, and motivates the state government to institutionalize rights.[78]

Keck and Sikkink have developed a typology of tactics TANs use when employing persuasion, socialization and pressure mechanisms. In this context, they highlight (1) *information politics* understood as strategically using information, (2) *symbolic politics* as drawing on symbols and stories to highlight a situation to a target audience that might be geographically distant, (3) *leverage politics* as network actors being able to gain moral or material leverage over state actors and IOs, as well as (4) *accountability politics* referring to formerly adopted norms and policies of governmental actors and obligations to comply with them.[79]

At the climate conferences, a similar pattern can be observed: local CSOs provide information on rights infringements in climate policy implementation in certain states to advocacy networks. TANs, like the Human Rights and Climate Change Working Group, use this information to mobilize other actors of the human rights regime (*information politics*). At side events, for example, TANs encourage local actors to share their cases and stories from home countries to raise awareness about adverse human rights effects of both climate impacts and climate policies (*symbolic politics*). These cases are presented as instances of climate injustice in which local population groups who have contributed little to greenhouse gas emissions and have few resources to adapt cannot fully enjoy their human rights due to climate impacts or experience severe rights infringements due to climate policies. This creates moral leverage over states that have historically contributed to emissions and that are implementing climate policies in developing countries (*leverage politics*). Moreover, TANs persuade states to vote for an incorporation of human rights into climate agreements, and more particularly procedural rights into climate policies. Mechanisms of persuasion (and discourse) function according to a logic of appropriateness (or a logic of arguing) and are particularly successful with (often liberal democratic) state governments[80] that have already legally committed to human rights, understanding them as part of their state identity, e.g. European states like France, Sweden and Ireland (*accountability politics*). Actively engaged and in favor of rights institutionalization are also those states that are pressured from above through TANs and from below through domestic civil society organizations. These often are Latin American countries with strong CSO movements representing local communities' and Indigenous Peoples' concerns. Among them are Mexico, Peru, Costa Rica, Guatemala and Uruguay. Especially in those countries, a boomerang pattern can be observed as domestic CSOs pressure the government to change the modalities and procedures of climate policy implementation from inside, while TANs and the human rights regime exert pressure on the government from outside the country. Also in favor of rights institutionalization are small island states, such as the Maldives, Kiribati, Samoa and the Philippines, that fear severe climate change consequences for the citizens living in their territory. Some states (together with CSOs, IOs and other actors of the human rights regime) also try to pressure less democratic states to vote in favor of rights institutionalization, claiming that they will not fund climate policies with adverse right affects anymore, such as REDD+ and CDM programs. Thus, they use negative incentives or sanction mechanisms that function according to a logic of consequences.[81]

At least three aspects in this scenario are completely new to international relations research relating to the boomerang pattern. First, the COP is a transnational arena, in which states closely interact with CSOs. This might accelerate the boomerang pattern since information can be quickly and informally exchanged and strategically used. Second, due to very close interaction between TANs and states, civil society actors possibly deliver important text passages of final agreements that are being negotiated. This means boundaries between states and civil society at the international negotiations become increasingly blurred, contributing to further transformation of state sovereignty in this context. And third, in contrast to what empirical analyses on the boomerang pattern suggest so far, an institution will adopt these rights (and not the state itself). This might slow down the process because a consensus between different state actors has to be found. However, it might also lead to a transfer of, for instance, procedural rights to states that would not adopt them otherwise. Thus, integrating procedural rights into climate policies might also become a booster for improving the human rights situation of a particular state or it can lead to "democratic empowerment"[82] and foster further democratization processes.[83]

Cost-benefit calculations of state actors are an important mechanism explaining why states refrain from supporting human rights in the climate regime. Governments like the United States, for instance, have remained opposed to rights institutionalization in the operative part of the Paris Agreement[84] (and in previous negotiations) because they fear costly obligations and demands for compensation:

> And the US, for example, it's like they refuse to talk about human rights. Refuse. It's a non-starter. It's like a totally toxic kind of issue to bring to them. . . . If you follow the Human Rights Council discussions and dialogues, the US refuses to talk about climate change in the human rights regime as well. For exactly the same reason. It's like the loss and damage negotiations are playing out here for a reason, they won't talk about climate change in a human rights context because they don't want to be held liable for historic contributions to climate change.[85]

The strong emphasis on economic, social and cultural rights in the context of climate change impacts also leads to the US remaining highly skeptical regarding rights institutionalization in the climate regime.[86] African countries, in contrast, rather fear that conditionalities are being imposed on them when implementing climate policies funded by Annex I parties (i.e.

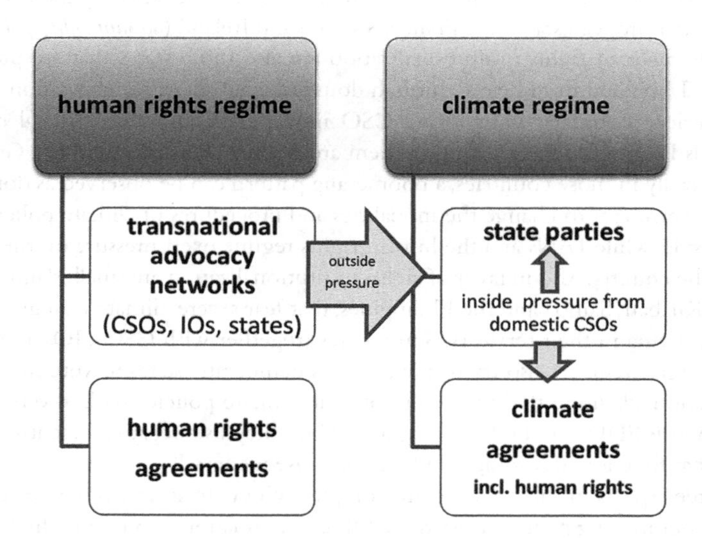

Figure 4.1 The boomerang pattern fostering institutional interaction

Source: Own compilation

developed states). They emphasize their state sovereignty and are concerned that deficiencies in their domestic rights situation could be exposed – and that the international community would interfere in their domestic affairs with the help of procedural rights in climate policies.[87]

In sum, actors like TANS and domestic CSOs, as well as mechanisms, such as information delivery, the boomerang pattern and persuasion, help us to understand institutional interaction between the human rights and the climate regime. They explain how cognitive interaction and interaction through commitment can be initiated. Behavioral interaction and impact-level interaction can probably rather be observed at later stages – when the Paris Agreement with its commitment to human rights is being implemented. And here, compliant state action will be most necessary.

Figure 4.1 displays the boomerang pattern fostering institutional interaction between the human rights and the climate regimes.

Conclusion

The objective of this chapter was to demonstrate how close state-society interaction can foster the institutionalization of human rights in the international climate regime. Those civil society actors and transnational advocacy networks collaborating intensively with state actors are much more likely to achieve moderate gains for the inclusion of human rights in climate politics. With the help of information dissemination, persuasion as well as inside and outside pressure mechanisms, they have convinced state representatives to recognize human rights in the climate regime. The empirical evidence gathered mainly from expert interviews at the international climate negotiations in Warsaw 2013 and in Paris 2015, primary documents and participatory observations of the HRCCWG strategic meetings suggests that in particular constellations, a boomerang pattern explains how local claims enter international negotiations. This is the case when information about local rights infringements in climate policy implementation is released by domestic civil society groups and comes back like a boomerang to the norm-violating government at international conferences. This has so far led to an acknowledgement of the human rights implications of climate change in the preambulatory clauses of the Cancun Agreements in 2010, a binding statement that all climate-related actions have to be carried out in accordance with human rights norms and an inclusion of procedural rights in REDD+ policies.[88] In Paris 2015, states agreed to include a reference to respecting, promoting and considering human rights in climate-relevant action in the preamble (UNFCCC 2015).[89] The focus on procedural rights upheld by many CSOs and TANs – compared to earlier attempts that have framed emission reductions as a human rights obligation – demonstrates that civil society operating *within the UNFCCC process* has become pragmatic, aiming at moderate changes and reforms in climate policies (as opposed to the more radical part of the climate justice movement operating exclusively *outside the UNFCCC process* protesting for an encompassing system change). The fact that civil society representatives even become registered on state delegations shows that demarcations between state and non-state activities at the climate conferences become blurry and state-society relations are in flux. The cooperative and often pragmatic approach of TANs has allowed for the strategic use of information and concerns from the local level to transform part of the international debate and to include human rights in climate agreements. Following upon that, the next problem to be addressed is whether rights institutionalization in climate agreements actually leads to rights implementation in climate action. This is a significant question that needs to be further investigated in future research on human rights in climate governance.

Notes

1 H. Bulkeley and P. Newell, *Governing Climate Change* (2010) 18.
2 On human rights advocacy, see also L. Wallbott, 'Rights, Representation, and Recognition: Practicing Advocacy for Women and Indigenous Peoples in UN Climate Negotiations' (this volume).
3 M. Keck and K. Sikkink, *Activists Beyond Borders* (1998) 9.
4 R. P. Hiskes, *The Human Right to a Green Future: Environmental Rights and Intergenerational Justice* (2009).
5 A. Schapper and M. Lederer, 'Climate Change and Human Rights: Mapping Institutional Inter-Linkages', 27 *Cambridge Review of International Affairs* (2014) 4, at 666–679.
6 W. Linda and A. Schapper, 'Negotiating By Own Standards? The Use and Validity of Human Rights Norms in UN Climate Negotiations', 17 *International Environmental Agreements: Politics, Law and Economics* (2017), at 209–228.
7 In this chapter, I understand institutionalization as the adoption of human rights norms. States can formally adopt norms due to strategic reasons without necessarily being convinced that this is an appropriate norm. The literature I refer to makes a conceptual differentiation between prescriptive status (institutionalization) and norm-consistent behavior (internalization). Thus, institutionalization has to be differentiated from behaving according to the norm. T. Risse, S. C. Ropp and K. Sikkink (eds.), *The Power of Human Rights: International Norms and Domestic Change* (1999).
8 See also D. D. Pugley, 'Rights, Justice, and REDD+: Lessons From Climate Advocacy and Early Implementation in the Amazon Basin' (this volume).
9 A. Gupta, 'Transparency Under Scrutiny: Information Disclosure in Global Environmental Governance', 8 *Global Environmental Politics* (2008) 2, at 1–7.
10 GA Res. 2200 A (XXI), 16 December 1966.
11 Gupta, *supra* note 9, at 3–4.
12 Schapper and Lederer, *supra* note 5.
13 See also A. Jegede, 'Protecting Indigenous Peoples' Land Rights in Global Climate Governance' (this volume) and B. Powless, 'The Indigenous Rights Framework and Climate Change' (this volume).
14 M. Keck and K. Sikkink, *supra* note 3; T. Risse, S. C. Ropp and K. Sikkink (eds.), *supra* note 7; T. Risse, S. C. Ropp and K. Sikkink (eds.), *The Persistent Power of Human Rights: From Commitment to Compliance* (2013).
15 C. Dany, 'Ambivalenzen der Partizipation. Grenzen des NGO Einflusses auf dem Weltgipfel zur Informationsgesellschaft', 19 *Zeitschrift für Internationale Beziehungen* (2012) 2, at 73.
16 T. Brühl, *Nichtregierungsorganisationen als Akteure internationaler Umweltverhandlungen: Ein Erklärungsmodell auf der Basis der situationsspezifischen Ressourcennachfrage* (2003) 186.
17 J. Steffek and P. Nanz, 'Emergent Patterns of Civil Society Participation in Global and European Governance', in J. Steffek, C. Kissling and P. Nanz (eds.), *Civil Society Participation in European and Global Governance* (2008) 1, at 29.
18 T. Squatrito, 'Opening the Doors to the WTO Dispute Settlement: State Preferences on NGO Access as Amici', 18 *Swiss Political Science Review* (2012) 2, at 1–24; J. A. Scholte, 'A More Inclusive Global Governance? The IMF and Civil Society in Africa', 18 *Global Governance* (2012) 2, at 185–206; I. Take, 'Legitimacy in Global Governance: International, Transnational and Private Institutions Compared', 18 *Swiss Political Science Review* (2012) 2, at 220–248; M. Bexell, J. Tallberg and A. Uhlin, 'Democracy in Global Governance: The Promises and Pitfalls of Transnational Actors', 16 *Global Governance* (2010) 1, at 81–101.
19 S. Bernstein, 'Legitimacy Problems and Responses in Global Environmental Governance', in P. Dauvergne (ed.), *Handbook of Global Environmental Politics* (2012) 147, at 162; K. Bäckstrand, 'Democracy and Global Environmental Politics', in P. Dauvergne (ed.), *Handbook of Global Environmental Politics* (2012) 507, at 519.
20 D. della Porta and L. Parks, 'Framing-Prozesse in der Klimabewegung: Von Klimawandel zu Klimagerechtigkeit', in M. Dietz and H. Garrelts (eds.), *Die internationale Klimabewegung: Ein Handbuch* (2013) 39, at 56; C. Görg and P. Bedall, 'Antagonistische Positionen: Die Climate-Justice-Koalition vor dem Hintergrund der Theorie gesellschaftlicher Naturverhältnisse', in M. Dietz and H. Garrelts (eds.), *Die internationale Klimabewegung: Ein Handbuch* (2013) 75, at 105; M. Dietz, 'Ergebnisse des Handbuchs: Verfassung, Einfluss und Zukunft der Klimabewegung', in M. Dietz and H. Garrelts (eds.), *Die internationale Klimabewegung: Ein Handbuch* (2013) 469, at 484.
21 V. De Lucia, 'Die Klimagerechtigkeitsbewegung und der hegemoniale Diskurs über Technologie', in M. Dietz and H. Garrelts (eds.), *Die internationale Klimabewegung: Ein Handbuch* (2013) 107, at 133.

22 Görg and Bedall, *supra* note 20, at 88–89.
23 There is also an established literature on climate justice stemming from normative political science including R. P. Hiskes, 'The Right to a Green Future: Human Rights, Environmentalism and Intergenerational Justice', 27 *Human Rights Quarterly* (2005) 4, at 1346–1364; S. Caney, 'Cosmopolitan Justice, Rights and Global Climate Change', 19 *Canadian Journal of Law and Jurisprudence* (2006) 2, at 255–278; R. P. Hiskes, *The Human Right to a Green Future: Environmental Rights and Intergenerational Justice* (2009); H. Shue, 'Human Rights, Climate Change and the Trillionth Ton', in D. G. Arnold (ed.), *The Ethics of Global Climate Change* (2011) 292, at 314; D. Moellendorf, 'Climate Change and Global Justice', 2 *Wiley Interdisciplinary Reviews* (2012) 131–143; D. Moellendorf, *The Moral Challenge of Dangerous Climate Change: Values, Poverty and Policy* (2014). It differentiates justice concerns between developing and developed states, societal groups, and today's and future generations.
24 della Porta and Parks, *supra* note 20, at 47.
25 A. Brunnengräber, 'Zwischen Pragmatismus und Radikalisierung: NGOs und soziale Bewegungen in der internationalen Klimapolitik', in M. Dietz and H. Garrelts (eds.), *Die internationale Klimabewegung: Ein Handbuch* (2013) 357, at 372.
26 T. Bernauer and C. Betzold, 'Civil Society in Global Environmental Governance', 21 *The Journal of Environment and Development* (2012) 1, at 62–66.
27 Görg and Bedall, *supra* note 20, at 94–95.
28 Bernauer and Betzold, *supra* note 26, at 63.
29 T. Böhmelt, V. Koubi and T. Bernauer, 'Civil Society Participation in Global Governance: Insights From Climate Politics', 53 *European Journal of Political Research* (2014) 1, at 19.
30 M. M. Betsill and E. Corell (eds.), *NGO Diplomacy: The Influence of Nongovernmental Organizations in International Environmental Negotiations* (2008).
31 Bernauer and Betzold, *supra* note 26, at 63.
32 T. Bernauer, T. Böhmelt and V. Koubi, 'Is There a Democracy-Civil Society Paradox in Global Environmental Governance?', 13 *Global Environmental Politics* (2013) 1, at 88–107.
33 R. Reitan, 'Coordinated Power in Contemporary Leftist Activism', in T. Olesen (ed.), *Power and Transnational Activism* (2011) 51, at 72.
34 Brunnengräber, *supra* note 25.
35 Görg and Bedall, *supra* note 20, 94–95.
36 Expert Interview, Representative of the Human Rights and Climate Change Working Group, COP 19 in Warsaw, 16 November 2013.
37 Expert Interview, Representative of an Indigenous Rights Organization, COP 19 in Warsaw, 16 November 2013.
38 Expert Interview, Representative from International CSO, COP 21 in Paris, 8 December 2016.
39 Expert Interview, Academic and Activist, COP 19 in Warsaw, 17 November 2013.
40 *Ibid.*
41 Expert Interview, Academic and Activist, COP 19 in Warsaw, 17 November 2013; Expert Interview, Representative of the Human Rights and Climate Change Working Group, COP 19 in Warsaw, 16 November 2013; Expert Interview, Representative of an Environmental Think Tank, COP 19 in Warsaw, 16 November 2013; Expert Interview, Representative from AIDA (Interamerican Association for Environmental Defense), via Skype after COP 21 in Paris, 4 February 2016; Expert Interview, Representative from International Organization, via Skype after COP 21 in Paris, 4 February 2016; Expert Interview, Representative of the OHCHR, at COP 21 in Paris, 10 December 2015.
42 UNFCCC/CP, *United Nations Framework Convention on Climate Change/ Decision 1/CP.16, the Cancun Agreements: Outcome of the Work of the Ad Hoc Working Group on Long-Term Cooperative Action Under the Convention*, FCCC/CP/2010/7/Add.1 (15 March 2011).
43 *Ibid.*
44 UNFCCC/AWGLCA, *United Nations Framework Convention on Climate Change/Ad Hoc Working Group on Long-Term Cooperative Action Under the Convention: Negotiating Text*, FCCC/AWGLCA/2010/14 (13 August 2010).
45 E. Filzmoser, 'Clean Development Mechanism', in Human Rights and Climate Change Working Group (ed.), *Summary of Rights-Related Developments at COP 19* (2013) 1, at 2.
46 J. Schade and W. Obergassel, 'Human Rights and the Clean Development Mechanism', 27 *Cambridge Review of International Affairs* (2014) 4, at 717–735.
47 UNFCCC, *Paris Agreement*, FCCC/CP/2015/L.9/Rev.1 (12 December 2015).

48 Expert Interview, Representative of a Women's Rights Organization, COP 19 in Warsaw, 15 November 2013.
49 *Ibid.*, Expert Interview, Academic and Activist, COP 19 in Warsaw, 17 November 2013; Wallbott and Schapper, *supra* note 6.
50 Expert Interview, Representative of a Women's Rights Organization, COP 19 in Warsaw, 15 November 2013.
51 Expert Interview, Representative of an Environmental Think Tank, COP 19 in Warsaw, 16 November 2013.
52 Expert Interview, Representative of an Indigenous Rights Organization, COP 19 in Warsaw, 16 November 2013.
53 Expert Interview, Representative of an Environmental Think Tank, COP 19 in Warsaw, 16 November 2013.
54 Expert Interview, Representative from an International CSO, COP 21 in Paris, 8 December 2015.
55 Expert Interview, Academic and Activist, COP 19 in Warsaw, 17 November 2013.
56 Expert Interview, Representative of the Human Rights and Climate Change Working Group, COP 19 in Warsaw, 16 November 2013; Expert Interview, Representative of an Environmental Think Tank, COP 19 in Warsaw, 16 November 2013.
57 Expert Interview, Representative of an Indigenous Rights Organization, COP 19 in Warsaw, 16 November 2013.
58 UNFCCC, *Nongovernmental Organization Constituencies* (2014), available online at https://unfccc.int/files/parties_and_observers/ngo/application/pdf/constituency_2011_english.pdf (last visited 24 June 2016).
59 Inter-constituency Proposal for Paragraph 15, circulated among the HRCCWG mailing group on 11 June 2015.
60 Expert Interview, Representative from International Organization, via Skype after COP 21 in Paris, 4 February 2016.
61 Expert Interview, Representative from OHCHR, at COP 21 in Paris, 10 December 2015.
62 *Geneva Pledge for Climate Action* (2015), available online at www.forestpeoples.org/sites/fpp/files/news/2015/02/Annex_Geneva%20Pledge.pdf.
63 Human Rights Watch, *Human Rights in Climate Pact Under Fire* (2015), available online at www.hrw.org/news/2015/12/07/human-rights-climate-pact-under-fire.
64 Government of Norway, *COP 21: Indigenous Peoples, Human Rights and Climate Change*, available online at www.regjeringen.no/no/aktuelt/cop21-indigenous-peoples-human-rights-and-climat-changes/id2466047/.
65 *The Guardian*, 'Climate Talk: Anger Over Removal of Human Rights Reference From Final Draft' (2015), available online at www.theguardian.com/global-development/2015/dec/11/paris-climate-talks-anger-removal-reference-human-rights-from-final-draft.
66 Expert Interview, Representative from International Organization, via Skype after COP 21 in Paris, 4 February 2016.
67 O. R. Young, *The Institutional Dimensions of Environmental Change: Fit, Interplay, and Scale* (2002); S. Oberthür and O. S. Stokke (eds.), *Managing Institutional Complexity: Regime Interplay and Global Environmental Change* (2011); See also A. Savaresi, 'Climate Change and Human Rights: Fragmentation, Interplay, and Institutional Linkages' (this volume).
68 T. Gehring and S. Oberthür, 'Introduction', in S. Oberthür and T. Gehring (eds.), *Institutional Interaction in Global Environmental Governance: Synergy and Conflict Among International and EU Policies* (2006) 1, at 6.
69 R. Andersen, 'The Time Dimension in International Regime Interplay', 2 *Global Environmental Politics* (2002) 3, at 98–117; F. Zelli, 'The Fragmentation of the Global Governance Architecture', 2 *Wiley Interdisciplinary Reviews: Climate Change* (2011) 2, at 255–270.
70 Gehring and Oberthür, *supra* note 68, at 6–7.
71 T. Gehring and S. Oberthür, 'The Causal Mechanisms of Interaction Between International Institutions', 15 *European Journal of International Relations* (2009) 1, at 133.
72 Gehring and Oberthür, *supra* note 71, at 136.
73 *Ibid.*, at 141–142.
74 Young, *supra* note 67.
75 Gehring and Oberthür, *supra* note 71, at 143–144.

76 B. Buzan, C. Jones and R. Little, *The Logic of Anarchy: From Neorealism to Structural Realism* (1993); Gehring and Oberthür, *supra* note 68.
77 Keck and Sikkink, *supra* note 3.
78 *Ibid.*, at 13.
79 *Ibid.*, at 16–25.
80 T. Risse and S. C. Ropp, 'Introduction and Overview', in T. Risse, S. C. Ropp and K. Sikkink (eds.), *The Persistent Power of Human Rights: From Commitment to Compliance* (2013) 3, at 16.
81 *Ibid.*, at 14.
82 A. Kaswan, 'Environmental Justice and Environmental Law', 24 *Fordham Environmental Law Review* (2013) 161.
83 J. R. May and E. Daly, *Global Environmental Constitutionalism* (2014).
84 At the Paris negotiations, the US actually spoke in favor of human rights but did not agree to them being included in article two outlining the purpose of the agreement; Human Rights Watch, *supra* note 63.
85 Expert Interview, Coordinator of the Human Rights and Climate Change Working Group, COP 19 in Warsaw, 16 November 2013.
86 See generally P. Alston, 'Putting Economic, Social and Cultural Rights Back on the Agenda of the United States', in W. F. Schulz (ed.), *The Future of Human Rights: U.S. Policy for a New Era* (2008) 120.
87 See also Wallbott and Schapper, *supra* note 6.
88 UNFCCC/CP, *supra* note 42.
89 UNFCCC, *supra* note 47.

5

Rights, representation and recognition

Practicing advocacy for women and Indigenous Peoples in UN climate negotiations

Linda Wallbott

Introduction

The overall question that this chapter aims to tackle is how and why effective human rights advocacy occurs. Thus, it sheds light on the lobbying practices, contestatory effects and experiences of different stakeholder groups in the negotiations of the United Framework Convention on Climate Change (UNFCCC). To this end, it links International Relations scholarship with relevant insights from Political Theory and Political Sociology. Empirically, it draws on interviews with Indigenous Peoples, women's rights activists, negotiators and observers of the UN process between 2010 and 2014.

In the first part, I will develop the conceptual and analytical framework in which theoretical accounts of representation/representativity (that may be either constituency or audience centered) will be linked with the concept of recognition. Recognition can be accredited to actors but also to specific (normative) claims, arguments and institutions. On this basis, I will then conceptualize human rights advocacy as a socio-political practice of seeking recognition, e.g. in climate negotiations. This will be followed by reviewing the actual relevance of "talking and doing human rights" in the context of global climate change and describing the areas of common concern in term of the functional, political-institutional and legal linkages between both issue areas.

In the empirical part, I will illustrate the conceptual approach by reconstructing the forms and modes through which human rights claimants have staged their concerns and have been recognized or rejected as valid representatives in the climate negotiations of the United Nations. Therein I focus on the practices of and responses toward representatives of two transboundary stakeholder groups, namely Indigenous Peoples and women. The analysis is supplemented by secondary literature and review of primary policy documents. It finds that members of the two constituencies mentioned have experienced different scales and scopes of (in)formal recognition and that inter-institutional developments matter in this regard. Overall, the chapter emphasizes the dynamics and plural forms of representation/representativity and the inconclusive character of global politics, i.e. climate governance. Furthermore, it outlines lessons learned and possible ways forward.

Representation, representativity and recognition: conceptual remarks

Representation in international negotiations

International negotiations face a particular challenge when it comes to their representational quality. Technical experts and heads of states get together to formulate coordinated and collective policy measures. Non-state actors exert their influence in the hallways of the conference venues and sometimes even as members of official delegations. Hence, chains of democratic delegation are oftentimes indirect, which could be considered a problem for standard theoretical accounts of political representation, as these build on normative principles of authorization (of a representative by those who would be represented) and accountability (of the representative to those represented).[1] Relatedly, particularly transnational non-governmental organizations are often criticized for their presumable deficits in legitimacy and democracy, as they were neither officially delegated by nor accountable to those they claim to speak for.[2] However, if we focus solely on this type of democratic justification of political representation we run the risk of missing other cases of influence that claim representativity and that are, on the one hand, effective, but that, on the other hand, do not fulfill the normative criteria mentioned above. To counter this potential blind spot in identifying political representation, it has been argued that the concept of representation has "a robust nonnormative descriptive sense, that is, it describes facts about the political world without necessarily appealing to normative standards of legitimacy or justice".[3] Instead, so the argument goes, political representation in a broad sense can be captured by asking whether a "relevant audience" *recognizes* a person as representative. In this alternative approach, the relevance of an audience is defined through the function of representation. It is assumed that representativity is performed with view to a specific purpose or function, e.g. voting on laws, proposing regulation on trade issues or advocating for environmental protection. Depending on the purpose and the level of action, the relevant audience can be identified, e.g. national legislative or decision-makers at the UN level. In other words, the audience are "the relevant parties before whom the Representative claims to stand in for the Represented and act as defined by the Function".[4]

While it is not the aim of this chapter to defend either one of these contrasting approaches, it seems worthwhile to assess the implications of the audience-centered perspective, not least to be able to draw conclusions on the strategies and effectiveness of different non-state groups, e.g. Indigenous Peoples and women, in international climate negotiations. To this end, it will be useful to continue with a closer look at the concept of recognition.

Recognition

Basically, recognition involves the acknowledgement of a class of claims.[5] Recognition can relate to the claims themselves and to the actors who voice them.

If we further specify the interpersonal understanding of recognition as a mutual process of social interaction, it becomes clear that an actor is recognized when a positive evaluation (beyond mere identification) concerning a particular feature of this actor is expressed. Also, it "implies that you bear obligations to treat [this actor] in a certain way, that is, you recognize specific normative status of the other person, e.g. as a free and equal person".[6]

It has been debated whether "mutuality" is indeed a defining element of recognition or whether also a less ambitious instance of "adequate regard" would suffice.[7] Proponents of the former position assume that recognition can only take place between subjects that can engage in such a process interactively. It thus includes persons and groups. The latter understanding is broader in the sense that it conceives of the possibility that also non-human entities can be objects of recognition. This brings such things as animals or inanimate nature into the picture even though these are not able to practice recognition themselves, or to act alike.[8] Furthermore, in this perspective, recognition can also be granted to an argumentative linkage between separate norm sets or issue areas. In that case, the applicability of one norm (set) to a previously unrelated issue area is newly acknowledged.[9] Basically, I like to claim that institutions, being social constructions, can also be objects of (mis)recognition, because the scale and frequency of such instances impact on the logic and stability (or even survival) of an institution. Thus, as a result of repeated misrecognition, e.g. norm violation without social or legal sanctioning or failure to support organizational functioning, an institution may become irrelevant and decay over time. In contrast, repetitive public recognition of an institution's validity, e.g. through written or oral statements and organizational support, can strengthen normative coherence and institutional robustness. Relatedly, I would argue that also (mis)recognition through an institution is possible. For institutions not only regulate behavior, but they "always express – as well as reinforce – underlying attitudes of those who designed or keep on reproducing them".[10] And, after all, every endeavor of political resistance has the goal of institutional (normative just like organizational) change and targets the general bindingness of institutionalized values, which stretch beyond the inclination of individual agents acting within these institutions.

Recognition has been conceptualized as being intrinsically linked to identity, following Axel Honneth,[11] or to social status, as developed by Nancy Fraser.[12] The latter model emphasizes an actor's position or status in society and the relevance of social institutions and interaction for determining modes of recognition. In Fraser's view, politics of recognition and redistribution may interact and need to be aligned to achieve social justice, but they can be analyzed separately. Thus, one could capture the situation of women of color by assessing their social status along with the general norms that guide the working of the economy, e.g. acceptance of un- or underpaid work. For Honneth, in contrast, it would be more appropriate to understand instances of perceived injustice as a failure to experience recognition. Thus, different reasons motivate the call for recognition. Next to emotional or identity-related factors that may speak more to the question of whether a social setting is appropriate, consequentialist instrumental causes can also be important drivers of claims for recognition.[13] Thus, the outlook to acquire material gains – for example through the power over land and natural resources (a seminal topic in all climate-related action) – can be a motivational factor to voice such claims, just as is the aim to gather symbolic resources like status and reputation.

Overall, denying someone their claim to be recognized as a social or political actor with their own rights concerns – and thus not perceiving of them as relevant – can thus be an expression of unjust social conditions. Misrecognition – be it through deliberate action or ignorance – can invoke different reactions on the part of the actors whose claims for recognition have not been met.[14] They may feel "overlooked, unvalued or even denigrated".[15] Similarly, it has been argued that politics of (mis)recognition lead to struggles and conflicts when a group demands justice in the sense that "other groups give it public acknowledgement for some feature it possesses, for which it thinks that it deserves recognition".[16] In response unrecognized actors may give up their claims, they may adapt their behavior to fit within the frame of established social action, or they may stick to their claims and fight for their resonance, possibly even with violent means.[17]

Human rights advocacy as socio-political practice

Given the previous outlines, in this chapter I consider human rights advocacy as a socio-political practice of seeking recognition. The concept of practices has been increasingly scrutinized by scholars from International Relations and Political Sociology[18] who consider the performative character of everyday politics. They agree that global politics are part of interrelated human practices[19] and that we can derive at a better understanding of political and social spaces through an analysis of these practices.

Following the review of studies on social movements,[20] we could expect different phases of such practices of collective action within the UNFCCC. In that stream of literature, it is assumed that, initially, advocates from different associations and grassroots groups engage in the process of identity formation. In this first phase, they "make diffuse, value-laden, nonnegotiable demands" that are probably accompanied by mass protest actions.[21] They articulate their interests and target the public sphere with the purpose of gaining recognition as a (new) collective actor. Thus, practices include above all "expressive action"[22] and direct participation. Once the identity of the movement has been formed and political recognition has been achieved, the practices that are employed are likely to change. A formalization of previously given, rather loose network structures sets in "and representation replaces direct forms of participation".[23] In this second phase, the rationality could be described as instrumental (instead of expressive) and accompanied by routinization and institutionalization. Strategic practices are employed over identity-based ones and learning processes involve "goal-rational adaptation to political structures".[24] In this phase, claims are likely to be more pragmatic in the sense that they "become susceptible to negotiation and political exchange".[25]

However, like Cohen and Arato, I assume that such a two-phase model sketches a rather ideal type of development from the social to the political. Real empirics will probably not see such a linear evolvement or a linear deviation from identity-based motivations to instrumental motivations. Instead, it is more interesting in analytical terms to reveal possible tensions between identity and strategy and to try to understand the internal processes and conflicts within a group. As such, there is not *one* movement. For example, the climate justice movement comprises a set of distinct networks with different rights claims that might nevertheless collaborate (see below). Yet, the model is useful as a starting point to bring together the different conceptual shades of recognition, representation and activism and to make them susceptible for empirical analysis.

But which kind of substance should be acknowledged from the perspective of human rights advocates in the context of international climate negotiations? What is the purpose of action beyond being recognized as a political actor? To answer this question, I will outline, in the next section, the general linkages between climate change and human rights before turning to the more specific claims of Indigenous Peoples and women rights groups.

Linkages between climate change and human rights

Traditionally, the UNFCCC has been characterized by a rather technocratic understanding of climate change. Thus, its original mandate of 1992 calls on states "to stabilize greenhouse gas concentration at a level that would prevent dangerous anthropogenic interference with the climate system".[26] As this is rather straightforward, one could ask why international negotiators and national policy-makers should bother with assuming any kind of responsibility for human rights issues. In the following, and as developed elsewhere,[27] I will spell out different types of linkages that reveal the relevance of human rights – in empirical but also normative terms – for climate politics. These are functional and political-institutional/legal in character.

First, in functional terms, climate change materializes on the ground as a threat to life, to the right to food, the right to water, the right to health, and the right to adequate housing. In collective terms, it also challenges the right to self-determination.[28] These trade-offs on individual and community rights are the result of sea level rise, droughts, extreme rainfall events, floods and tropical storms, inundation, erosion and saltwater intrusion. They impact people's everyday livelihoods – particularly of those societal groups that are oftentimes at the risk of marginalization anyway, such as children, ethnic minorities, the elderly or persons with disabilities.[29] But the consequences of climate change also threaten the infrastructure and political stability of societies, above all in the countries of the so-called Global South.[30] Another form of functional linkage between human rights and climate change can occur as a side effect of implementing concrete policy measures, either to mitigate, such as the Clean Development Mechanism (CDM), or to deal with climate change impacts, such as Community-Based Adaptation. Here, we might find infringements of procedural rights as well as of property rights, of rights to food, water and housing and of means of subsistence, to name but a few.[31]

Second, relations between the protection of human rights and the preservation of a high environmental quality have been raised and increasingly recognized as an area of concern in international politics over the past four and a half decades. The linkage was included in the Stockholm Declaration on the Human Environment in 1972 – a milestone in the development of international environmental law – when states agreed that both natural and man-made aspects of the environment were essential to human well-being "and to enjoyment of basic human rights", of both current and future generations.[32] However, because states would in the years to come avoid binding language, this level of argumentation was not upheld in the negotiations of the UNFCCC. The debate was taken up again only at the beginning of the new millennium, when Inuit brought before the Inter-American Court of Human Rights a petition, in which they accused the United States (US) of violating their international human rights obligations because they did not reduce their national greenhouse gas emissions.

Strong institutional linkages were also developed through the actions of the Human Rights Council which, in 2008, adopted its first resolution on "Human rights and climate change".[33] Herein, it requested the Office of the United Nations High Commissioner for Human Rights (OHCHR) – in consultation with the Intergovernmental Panel on Climate Change (IPCC), the Secretariat of the UNFCCC and other stakeholders – to conduct a detailed empirical analysis on the relationship between climate change and human rights. The study was carried out against the scientific portfolio of the Fourth Assessment Report of the IPCC[34] and published in January 2009. While any argument on clear causality was omitted in the document, it acknowledged the probability of rights infringements. Also, it clarified that state duties (for the realization of human rights) are not limited through territorial borders[35] but that states should instead enhance their cooperation to put human rights into practice. The OHCHR report was subsequently taken up in the Human Rights Council resolution 10/4,[36] and the OHCHR, for its part, further specified and deepened its approach to the issue by advocating for a "human rights-based approach to climate change negotiations, policies and measures".[37] It demanded that the design of all climate policies and programs should be in line with fundamental human rights standards and that their fulfillment shall be the main "objective" of these actions.[38] It can be argued that by this claim the OHCHR aimed at redefining the traditional technocratic mandate of the UNFCCC. Also, the Human Rights Council continued its engagement by requiring in its resolution 19/10[39] the appointment of an independent expert "on the issue of human rights obligations relating to the enjoyment of a safe, clean, healthy and sustainable development". In his second annual report to the Human Rights Council in 2014, the appointed expert, US scholar John Knox, stressed that international conventions – such as the International Covenant on Civil and Political Rights (ICCPR), the International Covenant on Economic, Social and Cultural Rights (ICESCR), the Convention on the Rights of Child (CRC), the European and American Conventions on Human Rights, and the African Charter on Human Rights and People's Rights – implied procedural and substantive obligations related to the enjoyment of a safe and healthy environment that needed to be respected. In other words, he created an inter-institutional argumentative space[40] that defied in-silo politics of supposedly unrelated processes at an international scale.

As this outline suggests, participants in climate negotiations can build on different rationales and arguments when claiming recognition for human rights. In moral terms, human rights agreements set standards for appropriate behavior of state action and emphasize the need for corresponding day-to-day practices. In the end, they generate reasons for safeguarding individual and collective well-being in the form of social, economic and political freedom. Consequential climate politics would thus need to end certain behavior, e.g. emitting greenhouse gases, but they would also need to imply proactive measures such as technological and financial support for mitigation and adaption (and the respect for human rights therein). Furthermore, a human rights perspective is fruitful in disclosing chains of responsibility between different sets of actors (and not only between industrialized and non-industrialized countries, as the original setup of the UNFCCC could suggest). For human rights norms are the most widely recognized value canon globally and imply the formal and moral equality of everyone. And, even though human rights might be regarded "as primarily ethical demands",[41] human rights language builds a bridge to justiciability of claims, e.g. in terms of compensation mechanisms,[42] and calls upon addressees to go beyond particularist interest-based narratives.

In other words, human rights operationalize different aspects of climate justice as one particular instance of social justice, be it between states, between generations or within societies. They bring to the fore the subject-character of their holders as political and legal agents. Bringing human rights into the game could thus be regarded as just the other and timely side of the coin that climate change is an issue of the Anthropocene.[43] We live in a geological epoch that is characterized by the unprecedented impact of human activities on the earth's ecosystem – a fact that has been accepted as contemporary common knowledge.[44] If we acknowledge this, then it is obvious that climate change and related impacts are social and political concerns as well. Thus, we obviously also must take the described connections between climate governance and human rights seriously. To shed further light on this linkage, I will detail two specific cases of human rights advocacy (as seeking recognition) in UN climate negotiations. These are the practices of Indigenous Peoples and women's rights organizations.

Indigenous and women's rights advocacy as socio-political practice: changing the climate?

Generally, non-state actors can participate in UNFCCC negotiations as accredited observers representing the interests of particular societal groups. To be able to make oral interventions at the negotiations in their own right they need to be recognized as an official constituency by the Secretariat of the Convention. Whereas the early days of the UNFCCC have been characterized by strong engagement of business and environmental organizations, other stakeholder groups have entered the debate over the past decade. As of July 2015, the UNFCCC has accredited more than 1500 observer organizations[45] and has recognized their setup through nine constituencies: business and industry (BINGO), environment (ENGO), youth (YOUNGO), farmers (on a provisional basis), local government and municipal authorities (LGMA), research and independent (RINGO), trade union (TUNGO), Indigenous Peoples (IPO) and women and gender. At the 21st Conference of the Parties (COP) in Paris in 2015, civil society networks built an inter-constituency alliance to advocate for intergenerational equity, gender equality and human rights. The alliance – that was renamed into *Coalition for Rights* a couple of years later to be more inclusive – includes human rights activists, youth and gender groups and Indigenous Peoples' representatives.

Indigenous Peoples

Indigenous Peoples became increasingly visible in the UNFCCC process in the context of negotiating the mitigation instrument "Reducing Emissions from Deforestation and Forest Degradation and the role of conservation, sustainable management of forests and enhancement of forest carbon stocks in developing countries" (REDD+) that had been introduced to the agenda – in a slimmer version – in 2005, through a joint submission of Costa Rica and Papua New Guinea and that was adopted in 2010 in the Cancun Agreements. After some back and forth negotiations, the Agreements also recognized social safeguards in the decision on REDD+ and called for "[t]he full and effective participation of relevant stakeholders, in particular, indigenous peoples and local communities". It furthermore included the mandate to demonstrate:

> Respect for the knowledge and rights of indigenous peoples and members of local communities, by taking into account relevant international obligations, national circumstances and laws, and noting that the United Nations General Assembly has adopted the United Nations Declaration on the Rights of Indigenous Peoples.[46]

This result was considered a big success by Indigenous Peoples' rights advocates, even though the safeguards were neither operational nor legally binding and the United Nations Declaration on the Rights of Indigenous Peoples (UNDRIP), a comprehensive document on collective human rights which had been adopted by the UN General Assembly in 2007,[47] was only "noted". Also, observers acknowledged that continuous lobbying activities of Indigenous Peoples were instrumental for achieving this outcome.

Representation and recognition

Whereas Indigenous Peoples were recognized as an official constituency to the UNFCCC already in 2001, they raised concerns on the impact of climate-related policies particularly with view to the REDD+ instrument. During the process of expanding the policy from its initial form of 2005 to the more expanded version of 2010,[48] the instrument's potential impacts on biodiversity and forest-dwelling communities, including Indigenous Peoples, were increasingly debated. Major concerns were that Indigenous Peoples' rights could be violated, and that their traditional relationships to territories and their dependence on forests as means for subsistence could be negatively affected through REDD+. The perceived misrecognition related, thus, to issues of identity and social status alike.

Claiming representativity of indigenous interests on the side of those participating in the negotiations was simultaneously framed in terms of rootedness in local contexts, as well as of the idea of a transnational movement[49] that would bring together those who have come to share stigmatizing historical experiences and structural positioning within their respective nation-states. Notably, there is no universal formal definition of an indigenous person (individual) or Indigenous Peoples (collective). Consequently, the scope of beneficiaries of indigenous rights is determined in line with common understanding that has evolved through a study by former Special Rapporteur of the Sub-Commission of Discrimination and Protection of Minorities, José R. Martínez Cobo.[50] Whereas no approximation to the collective term has been established to date, two conditions apply with view to the individual dimension: first, an indigenous person can self-define as indigenous. But there are limits to individual self-identification. Thus, the second provision of the individual definition holds that the person needs to be *recognized* as indigenous by the community with which he or she identifies. Hence, we can see that this is a strong example of how identity, mutuality, recognition and representation are interlinked. At the same time, with view to recognition on the international scale, different dynamics of indigenous representativity have been at stake. In the UNFCCC, Indigenous Peoples have from the beginning of their engagement invoked strong – maybe even stereotypical – images of a particular relationship of their cultures with nature. This has always been a crucial factor for the group's identity but also for its standing as a political actor. Hence, with view to REDD+ it turned out that the advocacy strategies of Indigenous Peoples in the negotiations started to resonate particularly well with delegates when states began to attribute a specific function to the representatives' claims. Indigenous Peoples' rights were thus increasingly framed instrumentally, to be a matter of securing the success and sustainability of implementing and monitoring REDD+ on the ground.[51] It shows that not only the relevant audience was identified with view to its function – with UNFCCC state parties having been targeted by indigenous rights' activists as they had to adopt the rights officially. Also, the representatives and their claims themselves were recognized when their purpose, or: the added value, began to show. As one interviewee put it:

> They [national governments] also recognized that the topics they are talking about are really territories of indigenous peoples. . . . They know very well the situation related to the forests in their countries, they know that indigenous peoples have customary claims to these territories. . . . It [the collaboration of Indigenous Peoples] is also one thing to ensure that it [REDD+] succeeds.[52]

Advocacy goals and practices

What Indigenous Peoples had called for was to counter what they perceived to be substantial and procedural forms of misrecognition. They thus advocated for a strong rights language to be included in the REDD+ agreement (but subsequently also in other climate governance areas) and for rights-based approaches to be respected in preparing and implementing concrete action on the ground. In terms of routinized action at the UNFCCC, Indigenous Peoples worked toward the recognition of their rights through three main practices that related to fostering the recognition of specific knowledge, actors and norms. First, Indigenous Peoples invoked their intangible resources like knowledge and experience in stewarding natural resources, and succeeded in disseminating know-how within their community and in reaching out to negotiators. Inter-institutional processes were crucial in this regard, such as the comprehensions that could be drawn from the Indigenous Peoples Forum on Climate Change (IIPFCC), the Accra Caucus of 2008, and from a special session on carbon trading and REDD+ that was held at the UN Permanent Forum on Indigenous Issues (UNPFII) in 2008. This process contributed to enhancing their own understanding of the international 'game', and at the same time led to a (perceived) 'professionalization' of the agents themselves. Relatedly, and second, the positioning of supportive individuals was instrumental, e.g. Vicky Tauli-Corpuz, a person who was equally recognized within the global indigenous community and as a reputable state delegate in the UNFCCC. She served as chair of the UNPFII between 2005 and 2010, had a history as an indigenous and environmental activist, was part of the Philippine delegation and co-chair of the Subsidiary Body that negotiated REDD+ in the UNFCCC. Being recognized by different audiences as legitimate representative she could navigate between institutional settings and perform different roles in the interest of rights promotion.

Third, indigenous rights' activists strengthened their arguments by reminding and educating parties repeatedly of their obligations stemming from UNDRIP, and by emphasizing lessons learned from other UN processes, e.g. the Convention on Biological Diversity, which has traditionally been much more open to indigenous concerns than has the UNFCCC.[53] After all, the international community had already recognized Indigenous Peoples' substantive rights claims, including collective rights to lands, territories and resources, and cultural rights as well as their procedural rights, including through the provision of free, prior and informed consent (FPIC). In this sense, Indigenous Peoples stressed the applicability of those norms to the climate realm and attempted to import their normative force into the climate context.

Ultimately, the combination of these strategies with a diversification of framings that was attributed to the role of Indigenous Peoples in the governance of tropical forests facilitated the recognition of their rights in the Cancun Agreements.

Women

Another strong case in lobbying for the recognition of human rights in the UNFCCC has emerged around the issue of gender. Two groups in particular have been working to raise attention for gender climate justice which relates to the (under)representation of women in climate politics at national and international levels. It has two main areas of concern: the strong exposure of women to the impacts of climate change and their participation in political processes.[54] These groups are "GenderCC – Women for Climate Justice" and the "Global Gender and Climate Alliance (GGCA)", with the former having been instrumental in preparing the establishment of the women's constituency in 2011.

Women's rights groups, along with other civil society actors, have thus over the past years strongly welcomed the overall advance of women and gender causes in UNFCCC negotiations. Recent examples include the establishment of a Gender Day at COP18 (in contrast, a Forest Day had at that time already been around for many years), the Lima Work Programme on Gender of 2014[55] and provisions included in the outcome of the negotiations on a new climate agreement in 2015. Thus, a preambular clause of the Paris Agreement that encompasses also a response to rights claims by other constituencies, e.g. Indigenous Peoples, reads:

> Acknowledging that climate change is a common concern of humankind, Parties should, when taking action to address climate change, respect, promote and consider their respective obligations on human rights, the right to health, the rights of indigenous peoples, local communities, migrants, children, persons with disabilities and people in vulnerable situations and the right to development, as well as gender equality, empowerment of women and intergenerational equity.[56]

Even though this provision is, at this point in time, not enforceable on its own or independently from other provisions of the agreement, activists claim that it provides "climate justice activists with the determination and drive needed to push for the loose COP21 framework to reflect the needs of countries most impacted by climate disruption".[57]

Representation and recognition

GenderCC was launched in 2003 at COP9 when some organizations including LIFE and WECF (Women in Europe for a Common Future) informally discussed "whether the issue of 'gender' should be given more attention at climate change negotiations. There was a strong sense that further networking and collaboration was needed".[58] Today, membership to GenderCC is open to individual experts and activists, groups and organizations alike, and the network claims representativity as "one of the largest membership-based organisations working in this field".[59] At COP13, GenderCC was officially mandated by the informal women's caucus to submit the application for a women and gender constituency to the UNFCCC Secretariat.

GGCA came into existence in 2007 through the coming together of four global development, environmental and women's institutions (UNDP, IUCN, WEDO and UNEP). The initial goal was "to ensure that climate change policies, decision-making, and initiatives at the global, regional and national levels are gender responsive".[60] As of January 2015, GGCA included more than 90 UN agencies and intergovernmental and non-governmental organizations, and can thus build on some previous recognition of its members in front of the targeted audience, the UNFCCC and its member states.[61]

The parallel existence of these two gender justice groups is an example of the diversity of claimants and understandings of representativity that are possibly even within one subsection of a broader movement (here, the focus on women's rights within the inter-constituency alliance). Thus, GenderCC and GGCA differ with regards to identifying the major instances of misrecognition of women's interests in global climate politics.

GenderCC holds the view that the overall socio-political and economic global system is unjust. An interviewee characterized the governance approach of GenderCC as truly "feminist" in the sense that climate politics should not only aim at responding to existing gender roles but should instead challenge these roles and historically grown "patriarchal" structures. These pertain to the social and inert construction of gender images and (in)justice perceptions at a global scale, including the historically grown acceptance toward and impacts of traditional division of labor between men and women worldwide. On the other hand, GGCA rather challenges the policy-specific instances of gender injustice as they materialize through the implementation of given instruments and approaches.[62] Thus, one major concern of GGCA is to increase the share of women in party delegations.[63]

Advocacy goals and practices

The two groups display commonalities but also differences concerning their understandings of gender justice and its relation to climate change. On the one hand, both groups share the conviction that gender climate justice should be realized not merely between states as international justice, but also as a matter of intrasocietal and intergenerational justice. However, GenderCC and GGCA also differ with a view to the response measures they promote and to the practices they employ to realize this interest. To a certain extent, these differences reflect the diverging assumptions that both groups make regarding what they perceive as the dominant form of misrecognition which women experience in climate governance.

In line with its radical critique of complementary instances of exclusion from socio-economic and political institutions on the one hand and vulnerability to the impacts of climate change on the other hand, GenderCC favors "a shift from dominant, market-based mechanisms to people-centred ones".[64] Thus, existing mechanisms should be reformed to include e.g. grievance and safeguards systems (e.g. for the CDM). At a more fundamental level, GenderCC strives for a new course of the global economy and the rejection of offsetting and market-based instruments and risk technologies,[65] for "we can see that the same causes exist more or less for gender disparities and gender discrimination and the climate problem".[66] In other words, GenderCC ultimately seeks recognition for a new normative foundation of global politics.

In contrast, GGCA has the goal to improve the contemporary climate process and its outcomes within the existing meta-governance structure and institutional setup to achieve "gender-sensitive" practices.[67] Thus, it favors gender mainstreaming, that is "engendering" existing and future climate policies through incorporation of gender language in UNFCCC decisions, corresponding standards, indicators and gender-sensitive distribution of benefits.[68] Correspondingly, GGCA declares as its working areas "policy, advocacy, capacity-building, women's leadership, climate finance, and knowledge-generation".[69] Also, it emphasizes the relevance of including more women in party delegations to the international negotiations – thereby enhancing the representativity of delegations in terms of gender composition – through financial support via the Women Delegates Fund, which has been set up with the support of the Finnish and Icelandic governments.

In comparison, the goals of GenderCC can be characterized as being more radical and revolutionary, for ultimately they aim at the transformation of the dominant economic and cultural practices that determine different levels of vulnerability, inequality and injustice. GGCA, instead, opts for more pragmatic adjustments to the UNFCCC and a stronger institutionalist perspective of recognition.

However, these different goals still come with relatively similar advocacy practices. Both networks have, like those of Indigenous Peoples, established official organizational platforms to develop and promote their position. Also, they host web presences, side events and network meetings and engage in trainings, e.g. knowledge dissemination, at local and national levels. In this sense, the groups have acknowledged that support for their claims can also come from outside the negotiations and that the broader public might also be considered a relevant audience in this regard.

Differences, commonalities and lessons learned

Looking at the cases in a more comparative perspective, which generalizations can be drawn? Both advocacy networks, Indigenous Peoples and women, claim to speak on behalf of a transnational constituency for which, ultimately, no clear and formal chain of democratic representation (in terms of authorization and accountability) exist. However, it is possible to capture the political representation that is at stake by drawing on alternative conceptual accounts that focus more on the interaction between the claimant and the relevant audience, in this case the UNFCCC and negotiating parties.

For both cases, recognition is highly relevant with a view to social status (Fraser), even though the issue of identity seems to be more salient for Indigenous Peoples as compared to the women's rights groups. Similarly, it has been crucial for both cases to seek formal recognition as a UNFCCC constituency, a status that is granted by the Secretariat, to increase visibility and acceptance. Thereby, their interests were acknowledged as being relevant to the negotiation process, they were attributed with some political rigor and the claimants themselves were recognized as political actors. On the other hand, by engaging in the UNFCCC process, both the women's and indigenous rights groups recognize the authority and relevance of this forum, its processes and other participants of the negotiations. Hence, we can observe that mutuality of recognition is a central feature of (inclusive) international negotiations. In the further policy development, this reciprocity was formally substantiated as parties and COPs recognized the agency and the validity of claims of these two groups, e.g. through referencing the UNDRIP or including language on gender equality.

When it comes to the practices employed, both groups engage strongly in capacity building within their own peer group and in interacting with national delegates. But outreach and the creation of bonds across their own constituencies also occur. Thus, the above-mentioned outcome of COP21 was attributed to the strategic coalition building of the inter-constituency alliance. In this sense, it can be concluded that the establishment of policy linkages is triggered by strategic linkages between advocacy groups and their mutual recognition as potential allies.

On the other hand, both groups have also displayed instances of a rationalization of their own claims over time, as conceptualized in the two-phase model of social movements' activities above. Thus, Indigenous Peoples deviated from a very strong normative language, which dominated their advocacy in the early REDD+ negotiations, to develop more technical and policy-specific lines of argumentation. Ultimately, what was recognized in the negotiations was their instrumental contribution to REDD+ by functioning as managers of forests - and not an alternative indigenous cosmology as compared to the Western, natural science perspective on the environment.

Similarly, it had become clear over time for GenderCC that different types of intervention might be required for different addressees:

> When I am here, in the process, it does not make much sense to stand up every five minutes and to spread radical statements, because people here cannot really handle it. But you can write a paper so that those who are interested can read about it. If I want to engage in the process, I must give in to the rules of the game to a certain extent. And I can only trigger marginal changes.[70]

In this sense and in line with the audience-centered approach to recognition, it could be argued that human rights advocates are aware that they need to be recognized as authoritative speakers who hold issue-specific knowledge that the negotiators consider relevant. In other words, both groups have been considered suitable objects of recognition for features that are deemed relevant for the successful implementation of climate policies. Otherwise, claims of linking human rights and climate politics do not (instantly) resonate. This observation has also been further confirmed at the micro level of negotiations through interviews with female delegates of different ages and from different regions and party delegations. Asked whether they have experienced stereotyping in a male-dominated setting, the majority responded that they were taken seriously as soon as they could display their issue-specific expertise. However, at this point it remains an open conceptual and empirical question, whether this understanding of political representation tends to reinforce existing power structures as it might foreclose radical opposition from within. In the same line of argument, it is clear that this chapter exclusively dealt with non-state actors inside the negotiations. It would be interesting to compare how the findings relate to those groups of activists, who, either deliberately or because they have not been recognized, stay outside the formal institutions and do not or cannot, in this sense, cooperate with the official process.

Finally, politics are always inconclusive, and with the recognition of one group comes new conflicts and possibly ruptures of the movement. As Cohen and Arato[71] put it:

> The rights achieved by movements stabilize the boundaries between lifeworld, state, and economy; but they are also the reflection of newly achieved collective identities, and they constitute the condition of possibility of the emergence of new institutional arrangements, associations, assemblies, and movements.

Conclusion

In this chapter I have introduced an analytical framework for the study of human rights practices within international negotiations which aligned Political Theory, Political Sociology and

International Relations scholarship on representation and recognition. This approach was used to understand the emergence of societal claims for human rights within UN climate negotiations. Specifically, I analyzed the advocacy practices of Indigenous Peoples and women's rights groups. Future research may wish to analyze in more detail the specific instances of accepting or refusing claims of representativity of non-state actors through state delegates and international bureaucracies. Also, it would be worthwhile to further scrutinize the impact of concepts, e.g. in answering the question why the issue of women's rights has been included through the concept of "gender equality" and not "gender justice".

In terms of practical relevance of the contribution, the chapter outlined different types of linkages between the human rights and climate change domains and highlighted the necessity of including those affected by climate change impacts. Granting substantial and procedural rights to these groups is one avenue for (more) effective and legitimate climate governance and for acknowledging the political quality of this anthropogenic challenge. However, it should also be noted that the relevance of recognition does not end at the exit sign of the UN conference venue. Rather, the contrary is the case. As a magnitude of interviews for this research has shown and as we know from other work, stakeholder engagement at the international level mostly has the ultimate goal to change a persistent situation of misrecognition in nation-states. But recognizing rights claimants as political actors and possibly integrating their demands in policy setup could also be a deliberate strategy to co-opt critics, to silent protest and continuous struggles for recognition. Thus, once the implementation (or, in contrast, the omission) game starts, further analyses are required to understand the challenges and dynamics of norm change on the ground.

Notes

1 H. F. Pitkin, *The Concept of Representation* (1967).
2 V. Collingwood and L. Logister, 'State of the Art: Addressing the INGO "Legitimacy Deficit"', 2 *Political Studies Review* (2005) 175.
3 A. Rehfeld, 'Towards a General Theory of Political Representation', 1 *The Journal of Politics* (2006) 21, at 2.
4 *Ibid.*, at 6.
5 A. Sen, 'Elements of a Theory of Human Rights', 4 *Philosophy and Public Affairs* (2004) 315, at 343.
6 Stanford Encyclopedia of Philosophy, *Recognition* (2013), available online at https://plato.stanford.edu/entries/recognition/.
7 A. Laitinen, 'On the Scope of "Recognition": The Role of Adequate Regard and Mutuality', in H.-C. S. Busch and C. Zurn (eds.), *The Philosophy of Recognition: Historical and Contemporary Perspectives* (2010) 319.
8 Stanford Encyclopedia of Philosophy, *supra* note 5.
9 L. Wallbott, 'Keeping Discourses Separate: Explaining the Non-Alignment of Climate Politics and Human Rights Norms by Small Island States in UN Climate Negotiations', 4 *Cambridge Review of International Affairs* (2014) 736.
10 *Ibid.*
11 A. Honneth, *The Struggle for Recognition: The Moral Grammar of Social Conflicts* (1996).
12 N. Fraser, 'Recognition without Ethics?', 2(3) *Theory, Culture & Society* (2001) 21.
13 S. Herr, *Regulierte Rebellen: Zum Einfluss von Anerkennung auf die Normakzeptanz nichstaatlicher Gewaltakteure* (2015).
14 E. Ringmar, 'The International Politics of Recognition', in T. Lindemann and E. Ringmar (eds.), *The International Politics of Recognition* (2016) 3.
15 S. Thompson, *The Political Theory of Recognition: A Critical Introduction* (2006) 8.
16 *Ibid.*, at 3; see also Sen, *supra* note 4, at 343.
17 See Honneth, *supra* note 9; For an International Relation's perspective on this see T. Lindemann, *Causes of War: The Struggle for Recognition* (2010).
18 I. B. Neumann, 'Returning Practice to the Linguistic Turn: The Case of Diplomacy', 3 *Millennium – Journal of International Studies* (2002) 627; E. Adler and V. Pouliot, 'International Practices', 3 *International Theory* (2011) 1; D. Bigo, 'Pierre Bourdieu and International Relations: Power of Practices, Practices of Power', 3 *International Political Sociology* (2011) 225.

19 T. R. Schatzki, 'Introduction: Practice Theory', in T. R. Schatzki, K. Knorr-Cetina and E. Savigny (eds.), *The Practice Turn in Contemporary Theory* (2001) 10, at 11.
20 J. L. Cohen and A. Arato, *Civil Society and Political Theory* (1992) 556–561.
21 *Ibid.*, at 556.
22 *Ibid.*
23 *Ibid.*
24 Cohen and Arato, *supra* note 18, at 557.
25 *Ibid.*
26 Art. 2, *United Nations Framework Convention on Climate Change*, FCCC/INFORMAL/84, (1992).
27 L. Wallbott and A. Schapper, 'Negotiating By Own Standards? The Use and Validity of Human Rights Norms in UN Climate Negotiations', *International Environmental Agreements* (2017) 209.
28 Office of the United Nations High Commissioner for Human Rights (OHCHR), *Report of the Office of the United Nations High Commissioner for Human Rights on the Relationship Between Climate Change and Human Rights, Human Rights Council*, A/HRC/10/61 (15 January 2009), at 8–15.
29 *Ibid.*, at 15–18.
30 D. Smith and J. Vivekananda, *A Climate of Conflict: The Links Between Climate Change, Peace and War* (2007).
31 A. Schapper and M. Lederer, 'Introduction: Human Rights and Climate Change: Mapping Institutional Inter-Linkages', 4 *Cambridge Review of International Affairs* (2014) 666; J. Schade and W. Obergassel, 'Human Rights and the Clean Development Mechanism', 4 *Cambridge Review of International Affairs* (2014) 717.
32 UN, *Report of the United Nations Conference on the Human Environment*, A/CONF.48/14 (1972), at para 1.
33 Human Rights Council (HRC), *Resolution 7/23: Human Rights and Climate Change, Adopted at the 41st Meeting*, A/HRC/RES/7/23 (18 March 2008).
34 Intergovernmental Panel on Climate Change (IPCC), *Fourth Assessment Report: Climate Change. Synthesis Report* (2007).
35 J. H. Knox, 'Linking Human Rights and Climate Change at the United Nations', 33 *Harvard Environmental Law Review* (2009) 477, at 478.
36 HRC, *Resolution 10/4: Human Rights and Climate Change, Adopted at the 41st Meeting*, A/HRC/RES/10/4 (25 March 2009).
37 OHCHR, *Applying a Human Rights-Based Approach to Climate Change Negotiations, Policies and Measures* (2010).
38 *Ibid.*
39 HRC, *Resolution 19/10: Human Rights and the Environment, Adopted at the 19th Session*, A/HRC/RES/19/10 (19 April 2012).
40 L. Wallbott, 'Indigenous Peoples in UN REDD+ Negotiations: "Importing Power" and Lobbying for Rights Through Discursive Interplay Management', 1 *Ecology and Society* (2014) 21.
41 Sen, *supra* note 4, at 319.
42 A. Schapper, 'Der globale Klimawandel aus menschenrechtlicher Perspektive', in P. Dannecker and B. Rodenberg (eds.), *Klimaveränderung, Umwelt und Geschlechterverhältnisse im Wandel – neue interdisziplinäre Ansätze und Perspektiven* (2014) 48.
43 See also L. Wallbott, 'The Practices of Lobbying for Rights in the Anthropocene Era: Local Communities, Indigenous Peoples and International Climate Negotiations', in P. Pattberg and F. Zelli (eds.), *Environmental Politics and Governance in the Anthropocene: Institutions and Legitimacy in a Complex World* (2016) 213.
44 C. N. Waters *et al.*, 'The Anthropocene Is functionally and Stratigraphically Distinct From The Holocene', 351(6269) *Science* (2016) 137; B. Crew, *Scientists Just Declared the Dawn of a New, Human-Influenced Epoch: Welcome to the Anthropocene* (2016), available online at www.sciencealert.com/scientists-just-declared-the-dawn-of-a-new-human-influenced-epoch.
45 United Nations Framework Convention on Climate Change (UNFCCC), *Parties and Observers*, available online at http://unfccc.int/parties_and_observers/items/2704.php.
46 UNFCCC, *Cancún Agreements*, FCCC/CP/2010/7/Add.1 (2010), at 26.
47 United Nations General Assembly, *United Nations Declaration on the Rights of Indigenous Peoples*, A/RES/61/295 (13 September 2007).
48 For more detail, see Wallbott, *supra* note 38.
49 S. J. Anaya, *Indigenous Peoples in International Law*, 2nd ed. (2004).

50 United Nations Permanent Forum on Indigenous Issues (UNPFII), *The Concept of Indigenous Peoples.* Background Paper. PFII/2004/WS.1/3 (19–21 January 2004).
51 For more detail, see Wallbott, *supra* note 38.
52 Interview with indigenous peoples' representative to the UNFCCC, 25 May 2012, Bonn.
53 Interview with indigenous peoples' representative to the UNFCCC at 25 May 2012, Bonn.
54 GenderCC, Gender@UNFCCC, available online at http://gendercc.net/genderunfccc.html.
55 UNFCCC, *Draft decision/-CP.20* (2014).
56 UNFCCC, *Adoption of the Paris Agreement*, U.N. Doc. FCCC/CP/2015/L.9/Rev.1 (12 December 2015).
57 J. Olson, *Human Rights Are Women's Rights: Gender in the COP21 Climate Negotiations* (2015), available online at www.sierraclub.org/compass/2015/12/human-rights-are-women-s-rights-gender-cop21-climate-negotiations.
58 Gender CC, History, available online at http://gendercc.net/who-are-we/history.html.
59 Ibid.
60 Global Gender and Climate Alliance (GGCA), *Our Primary Goal*, available online at http://gender-climate.org/.
61 GGCA, *Who Are We*, available online at http://gender-climate.org/who-we-are/.
62 Expert interview, GenderCC representative, June 2014, intersessional in Bonn (own translation).
63 GenderCC, *Gender@UNFCCC* (2016), available online at http://gendercc.net/genderunfccc.html.
64 H. Röhr, 'Engendering the Climate Change Negotiations: Experiences, Challenges, and Steps Forward', 1 *Gender and Development* (2009) 19, at 26.
65 GenderCC, *Market Based Mechanisms*, available online at http://gendercc.net/genderunfccc/topics/market-based-mechanisms.html.
66 Expert interview, GenderCC representative, June 2014, intersessional in Bonn (own translation). See also the network's vision statement, available online at www.gendercc.net/about-gendercc.html.
67 Women's Environment & Development Organization, available online at www.wedo.org/category/themes/sustainable-development-themes/climatechange/ggca.
68 Aguilar, *Engendering REDD Workshop (GGCA, IUCN, WOCAN)* (2009), available online at http://wocan.org/sites/drupal.wocan.org/files/Engendering%20REDD%20Workshop.pdf; L. Schalatek, 'Zwischen Geschlechterblindheit und Gender Justice', in A. Brunnengräber (ed.), *Zivilisierung des Klimaregimes: NGOs und soziale Bewegungen in der nationalen, europäischen und internationalen Klimapolitik* (Wiesbaden: VS Verlag für Sozialwissenschaften, 2011) 135 at 154; see also GGCA, *Launch of the Global Gender and Climate Alliance*, available online at www.wedo.org/wp-content/uploads/global-gender-and-climate-alliance.pdf.
69 GGCA, *What We Do*, available online at http://gender-climate.org/what-we-do/.
70 Expert interview, GenderCC representative, June 2014, intersessional in Bonn (own translation).
71 Cohen and Arato, *supra* note 18, at 562.

Part II
International framework

6

State responsibility for human rights violations associated with climate change

Margaretha Wewerinke-Singh

Introduction

International human rights law is *prima facie* relevant to climate change because climate change and its associated impacts have an adverse effect on the enjoyment of internationally recognised human rights. Indeed, the link between climate change and human rights has been articulated in multilateral forums, by various human rights treaty bodies, and by the Conference of the Parties (COP) to the United Nations Framework Convention on Climate Change (UNFCCC).[1] Notably, however, in the various statements of international bodies linking climate change with human rights, no reference is made to 'violations' of human rights. This raises questions about the premise that the purpose of human rights law is, to quote the European Court of Human Rights (ECtHR), '[to guarantee] not rights that are theoretical or illusory but rights that are practical and effective'.[2]

This chapter demonstrates how existing norms of international law can be employed to establish State responsibility for acts and omissions that lead to dangerous climate change and associated violations of human rights, or for human rights violations resulting from measures to respond to climate change.[3] This is done through an analysis of the law of State responsibility; the nature of States' obligations to prevent human rights violations and to take measures to ensure the realisation of human rights at home and abroad; and questions related to causation and proof of damages. The conclusion elaborates on the potential role of the law of State responsibility in strengthening the legal protection offered by international law to peoples and individuals affected by climate change.

State responsibility for human rights violations associated with climate change

The law of State responsibility is important for answering legal questions related to climate change and human rights because it contains 'the general conditions under international law for the State to be considered responsible for wrongful actions or omissions, and the legal consequences which flow therefrom'.[4] This general law builds on the doctrine expressed by the Permanent Court in the *Factory at Chorzów* case that 'it is a principle of international law, and even a general conception of law, that any breach of an engagement involves an obligation to make reparation'.[5] Today, the law of State responsibility is stated authoritatively in the 'Articles on the Responsibility of States for Internationally Wrongful Acts' ('ARS') produced by the International Law Commission ('ILC').[6]

The relevance of the general rules of State responsibility to human rights obligations has been expressly and widely recognised by international human rights bodies,[7] and examples of cases where human rights bodies relied on these rules for the interpretation of human rights treaties are increasingly numerous.[8] Academic literature similarly suggests that the law of State responsibility and international human rights law are mutually reinforcing.[9] From an international human rights law perspective, the right to a remedy is a substantive right. This right is protected under customary international law[10] and expressed in human rights treaties in various forms.[11] The right to a remedy exists not only *ex post facto* but also when there is a threat of a violation.[12] Accordingly, the general law of State responsibility can be understood as providing a structure through which redress for human rights violations can be obtained by States on behalf of the victims of the violation, or directly by victims themselves.

Establishing state responsibility

The law of State responsibility is based on the principle of independent responsibility of States. This principle basically means that each State is responsible for its own conduct. The principle follows from the constituent elements of an internationally wrongful act of a State listed in Article 2 of the ARS, which states that a State has committed an internationally wrongful act when an action or omission:

a is attributable to the State under international law; and
b constitutes a breach of an international obligation of the State.[13]

Whether or not certain acts or omissions are attributable to a State is determined by reference to the rules on attribution. These rules exist because States can rarely, if ever, guarantee the conduct of all private persons or entities on its territory.[14] It is important to get to grips with these rules for the purpose of establishing State responsibility for climate change–related conduct that affects the enjoyment of human rights: after all, a large part of the greenhouse gases that cause climate change are emitted by entities other than States: corporations that exploit fossil fuels, utility companies that produce electricity, enterprises that manufacture products, airlines and car companies that allow travel, and producers and consumers who supply and demand these products and services.

The rules on attribution are expressed in Articles 4–11 of the ARS. The general rule of attribution is contained in Article 4 (entitled 'Conduct of organs of a State'), which provides that

> [t]he conduct of any organ shall be considered an act of that State under international law, whether the organ exercises legislative, executive, judicial or any other functions, whatever position it holds in the organisation of the State, and whatever its character as an organ of the central Government or of a territorial unit of the State.[15]

The Commentaries clarify that the reference to a State organ in Article 4 extends to organs of government 'of whatever kind or classification, exercising whatever functions, and at whatever level in the hierarchy'.[16] This rule operates similarly, if not identically, in international human rights law: the UN Human Rights Committee (HRC), for example, has found violations of the

Covenant that were attributable to central government and its legislature, federal governments, municipal authorities, judicial authorities, police and security forces and various types of State agents.[17] The type of conduct that is generally attributable to a State as a consequence of these rules includes national legislation, decisions of the judiciary or administrative measures.[18]

It is worth emphasising that the general rule of attribution reflected in Article 4 of the ARS allows omissions to be attributed to States (that is, a failure on the part of the State's organs or agents to carry out an international obligation).[19] The Commentaries to the ARS stress that '[c]ases in which the international responsibility of a State has been invoked on the basis of an omission are at least as numerous as those based on positive acts, and no difference in principle exists between the two'.[20] Further, the Commentaries clarify that whether an act of a State involves an act or an omission, '[w]hat is crucial is that a given event is sufficiently connected to conduct ... which is attributable to the State.'[21]

The scope for attribution is extended even further through the rule that an internationally wrongful act may consist of several acts and omissions that cumulatively amount to a breach of obligations.[22] In the ARS, this is expressed in Article 15 which states that State responsibility can arise from a 'breach consisting of a composite act'.[23] The breach has to extend over the entire period 'starting with the first of the actions or omissions of the series and lasts for as long as these actions or omissions are repeated and remain not in conformity with the international obligation'.[24]

Together, these rules on attribution suggest that a contextual analysis of a State's conduct and the obligations by which it is bound is the most appropriate method for determining whether a human rights violation has occurred. Such an analysis could take account of a range of conduct as attributable to the State – from information reported to the Conference of the Parties to the UNFCCC to its national legislation and regulatory framework, energy subsidies, trade policies and the extent of assistance provided and received in accordance with technology transfer and financial obligations – to determine whether this conduct is in accordance with its international human rights obligations. The sections below set out the standards against which such conduct should be analysed.

The scope and nature of states' human rights obligations related to climate change

Before discussing the scope of States' obligations under international human rights law, it is important to consider the sources from which these obligations emerge. The key point to highlight in this regard is that all States are bound by a wide range of human rights obligations that demand the protection of civil and political as well as economic, social and cultural rights.

First of all, the UN Charter contains more than a dozen references to human rights, proclaims the realisation of human rights as one of the main purposes of the organisation and provides that Member States shall cooperate to take joint and separate action with the UN to promote respect for and observance of human rights.[25] The Universal Declaration of Human Rights (UDHR) can be understood as an authoritative interpretation of the substantive rights referred to in the UN Charter.[26] Widely ratified human rights treaties provide additional human rights obligations for States. For example, the International Covenant on Civil and Political Rights (ICCPR)[27] has 168 State parties, which include all States listed in Annex I to the UNFCCC, and dozens of States located in areas where climate change is forecast to have serious negative impacts on human life and livelihoods.[28] The vast majority of States have also ratified the International Covenant on Economic, Social and Cultural Rights (ICESCR),[29] with 164 State parties.[30] The number of ratifications of international human rights treaties has risen rapidly in recent years, with all UN Member States having ratified at least one core human rights treaty and 80 per cent having ratified four or more.[31] The effect of the consolidation of human rights norms through various sources of international law is that the norms contained in the UDHR are applicable across different fields of international law as customary norms binding on all States.[32]

As regards the interpretation of human rights treaties, we must note that the emphasis is, in Nowak's words, '[e]ssentially . . . on interpreting treaties . . . in the light of their object and purpose'.[33] And for human rights treaties, the main object and purpose is guaranteeing the enjoyment of the rights protected in those treaties.[34] As discussed below, this presses in favour of a broad interpretation of the substantive human rights that are affected by the adverse effects of climate change.

Obligations to prevent human rights violations

Perhaps the most important human rights obligations in the context of climate change are obligations to take measures to prevent future harm. Such obligations are important not only to prevent or mitigate a range of adverse effects of climate change that would affect the enjoyment of human rights, but also to allow for the establishment of State responsibility for climate change–related human rights violations that might already be occurring as a result of past emissions. This section peruses an analysis of States' obligations related to the right to life as protected under Article 6 of the ICCPR and numerous other human rights instruments to illustrate the scope and nature of these obligations.

In its General Comment No. 6 on the Right to Life, the HRC states explicitly that the right to life 'is a right which should not be interpreted narrowly'.[35] This reflects the position of all regional and international human rights bodies with respect to the scope of the right to life. For example, the Inter-American Court of Human Rights (IACtHR) has stated that the

> fundamental right to life includes, not only the right of every human being not to be deprived of his life arbitrarily, but also the right that he will not be prevented from having access to the conditions that guarantee a dignified existence.[36]

And in *SERAC v Nigeria*, the African Commission on Human and Peoples' Rights (ACHPR) found a violation of the right to life based on 'unacceptable' levels of 'pollution and environmental degradation'.[37] Commentators understand the right as protecting the ability of each individual to 'have access to the means of survival; realize full life expectancy; avoid serious environmental risks to life; and to enjoy protection by the State against unwarranted deprivation of life'.[38]

Article 6(1) of the ICCPR generates two categories of obligations: a prohibition of the arbitrary deprivation of life, and an obligation to take positive measures to ensure that right, including measures to ensure its protection in law.[39] As regards the obligation to take legislative measures, the HRC has found that the law's protection is required against a wide variety of threats, including infanticide committed to protect a woman's honour,[40] killings resulting from the availability of firearms to the general public,[41] and the 'production, testing, possession, deployment and use of nuclear weapons'.[42] At the European level, the ECtHR similarly holds that the States' legislative and administrative framework must protect against a wide variety of threats to human life,[43] including environmental damage.[44] It seems safe to assume that in a similar vein, climate change–related threats must be mitigated through effective legislation in order to protect human life. According to Nowak, a violation of the obligation to protect the right to life by law can be assumed 'when State legislation . . . is manifestly insufficient as measured against the actual threat'.[45]

However, the positive obligations of States under the right to life go beyond an obligation to take legislative measures.[46] For example, the HRC has taken the view that the right requires that States take 'measures to reduce infant mortality and to increase life expectancy, especially in adopting measures to eliminate malnutrition and epidemics'.[47] Moreover, it has stressed that these positive obligations will only be fully met if States protect individuals against violations by its agents as well as violations committed by private persons or entities likely to prejudice the enjoyment of Covenant rights.[48] In a similar vein, the IACtHR found in the landmark case *Velásquez Radríguez v Honduras*[49] that State responsibility for the violation had arisen 'not because of the act [of abduction and killing] itself, but because of the lack of due diligence to prevent the violation or to respond to it as required by the Convention'.[50] This line of jurisprudence suggests that States are obliged to take measures to prevent human rights violations resulting from the actions of private persons that cause climate change, including fossil fuel companies and other polluting industries.[51]

ECtHR jurisprudence, starting with *Osman v UK*, suggests that the standard of care required in relation to a risk, of which the State had actual or presumed knowledge, is one of reasonableness:

> The Court does not accept . . . that the failure to perceive the risk to life in the circumstances known at the time or to take preventive measures to avoid that risk must be tantamount to gross negligence or wilful disregard of the duty to protect life. . . . Such a rigid standard must be considered to be incompatible with the requirements of [the right to life]. . . . [H]aving regard to the [fundamental] nature of [the right], it is sufficient for an applicant to show that the authorities did not do all that could be reasonably expected of them to avoid a real and immediate risk to life of which they have or ought to have knowledge.[52]

Actual or presumed knowledge of the climate change–related risks may arise from the UNFCCC, the reports of the IPCC and other scientific studies, as well as from affected communities' efforts to draw attention to these risks.

The case of *Tatar C. v Roumanie*[53] illustrates the overlap between States' obligations to prevent human rights violations and 'due diligence' obligations arising from the precautionary principle as embodied in international environmental law, including the UNFCCC.[54] In its ruling, the Court stressed that even in the absence of scientific probability regarding a causal link, the existence of a 'serious and substantial' risk to health and well-being of the applicants imposed on the State 'a positive obligation to adopt adequate measures capable of protecting the rights of the applicants to respect for their private and family life and, more generally, to the enjoyment of a healthy and protected environment'.[55]

States' prevention obligations under international human rights law may however go further than parallel 'due diligence' obligations under international environmental law. It is clear from the interpretative practice of human rights bodies that States are not only obliged to assess potential risks to human life, but must also respond to any 'serious and substantial' risk with measures 'designed to secure respect' for human rights, and 'capable of protecting [those rights]'.[56] In other words, States do not have the discretion to prioritise policy objectives such as the protection of particular industries over mitigation and other response measures that would avert the serious and substantial risks posed by climate change to human life. Moreover, these response measures must themselves be compatible with States' obligations to respect and ensure human rights. This means, amongst other things, that all States must reconcile obligations to protect peoples and individuals against the adverse effects of climate change with co-existing obligations to realise the rights of those who have obtained negligible benefits from emission-producing activities.

Through their focus on equity and common but differentiated responsibilities and respective capabilities (CBDRRC), the UNFCCC and its associated instruments encourage States to fulfil their human rights obligations by taking science-based mitigation measures without perpetuating existing inequalities. Although the principle of CBDRRC applies exclusively to relations between States, it shares with international human rights law the objective of achieving substantive equality.[57] The lack of legally binding emission reduction commitments in the recently adopted Paris Agreement underscores the importance of substantive human rights obligations to mitigate climate change in a manner that is fair and equitable. At the same time, the Paris Agreement provides a procedural framework that could shed light on States' compliance with these human rights obligations. The reference to human rights in the Preamble of the Agreement could catalyse further information on, and review of, the overall human rights implications of States' mitigation actions.

Obligations to ensure the realisation of human rights at home and abroad

In addition to obligations to prevent future harm, international human rights law imposes obligations on States to ensure the progressive realisation of human rights within the State's own territory as well as internationally. This section discusses such obligations and their relevance in the context of climate change, taking the right to health as an example.[58]

The right to health, as protected under Article 12 of the ICESCR and numerous other human rights instruments, is a right that States are obliged to progressively realise. The Committee on Economic, Social and Cultural Rights (CESCR) has emphasised that although 'the right to health is not to be understood as a right to be *healthy*',[59] it nonetheless creates States' obligations.[60] These obligations are understood as including 'immediate obligations. . . [to] . . . guarantee that the right will be exercised without discrimination of any kind' and to take steps 'towards the full realization' of the right that 'must be deliberate, concrete and targeted towards the full realization of the right to health'.[61]

To clarify the content of States' obligations, the CESCR has used a respect-protect-fulfil typology of obligations that arise from the right to health.[62] It understands the obligation to 'respect' the right as 'an obligation of States to respect the freedom of individuals and groups to preserve and to make use of their existing entitlements'.[63] The CESCR has interpreted the right to health as requiring respect for the right to health of a people within a State's territory and in other States,[64] entailing an obligation 'to refrain from unlawfully polluting air, water and soil, e.g. through industrial waste from State-owned facilities'.[65] Accordingly, the right could be violated by actively engaging in 'activities that harm the composition of the global atmosphere or arbitrarily interfere with healthy environmental conditions'.[66] Moreover, the CESCR has explicitly stated that the right to health obliges States to ensure that international instruments, presumably including climate change–related agreements, 'do not adversely impact upon the right to health'.[67]

The obligation to protect the right to health involves 'the preservation of existing entitlements or resource bases', including through regulation,[68] in accordance with the UN Charter and applicable international law.[69] States must accordingly adopt measures against environmental and occupational health hazards[70] and national policies to reduce and eliminate air, water and soil pollution.[71] Moreover, States must prevent 'encroachment on the land of indigenous peoples or vulnerable groups',[72] 'ensure food availability, regulation of food prices and subsidies, and rationing of essentials while ensuring producers a fair price',[73] and prevent private enterprises from engaging in environmental pollution 'especially that which contaminates the food chain'.[74]

In the context of climate change, the right to health also appears to entail an obligation to regulate private actors in order to achieve and uphold emission limitation and reduction standards,[75] and to adopt and implement 'laws, plans, policies, programmes and projects that tackle the adverse effects of climate change'.[76]

The CESCR further directs that States must give 'sufficient recognition to the right to health in the national political and legal systems, preferably by way of legislative implementation',[77] and must allocate 'a sufficient percentage of a State's available budget ... to the right to health'.[78] This illustrates what the CESCR describes as the obligations to 'fulfil' the right to health; a positive obligation that is triggered 'whenever an individual or group is unable, for reasons beyond their control' to enjoy the right 'by the means at their disposal'.[79] It basically requires that the State 'be the provider', which 'can range anywhere from a minimum safety net, providing that it keeps everyone above the poverty line appropriate to the level of development of that country, to a full comprehensive welfare model'.[80] Again, this obligation has an extraterritorial dimension: States are required to 'facilitate access to essential health facilities, goods and services in other countries, wherever possible and [to] provide the necessary aid when required'.[81] In the context of climate change, this is interpreted as an obligation on high-income States to facilitate access to essential health services as well as assistance to adapt to climate change in low-income States.[82]

The parallel obligations of developed States contained in Article 4 of the UNFCCC and reaffirmed in the Paris Agreement[83] could serve as a bottom line in the interpretation of these obligations. And again, the procedural framework established under the Paris Agreement could serve to shed light on compliance with these obligations. The Agreement specifically requires developed States to communicate information related to the fulfilment of their finance obligations, while other States providing resources are encouraged to communicate such information.[84] This information is to inform the global stocktaking process aimed at reviewing States' 'collective progress towards achieving the purpose of [the Agreement] and its long-term goals',[85] including the goals of '[m]aking finance flows consistent with a pathway towards low greenhouse gas emissions and climate resilient development' and '[i]ncreasing the ability to adapt to the adverse impacts of climate change and foster climate resilience and low greenhouse gas emissions development, in a manner that does not threaten food production'.[86] To meet these goals and fulfil parallel obligations under international human rights law, developed States would need to scale up funding to assist developing States in taking the resilience-building and adaptation actions required to ensure the realisation of human rights.[87]

Legal consequences of state responsibility

When a State actually violates its human rights obligations, State responsibility is established '*as immediately as* between the two [or more] States'.[88] This rule indicates that the legal consequences of State responsibility arise automatically once a State violates a human rights obligation, irrespective of whether any victim of the violations actively seeks a remedy for the damage or harm suffered. This section spells out the legal consequences of an internationally wrongful act, once it occurs.

Cessation of wrongful conduct

The basic principle governing the legal consequences of wrongful conduct (or what the ARS call the 'content' of State responsibility) is that a State that commits an internationally wrongful act 'must, so far as possible, wipe-out all the consequences of the illegal act and re-establish the situation which would, in all probability, have existed had that act not been committed'.[89]

The emphasis on restoring the situation to what it was before the wrongful act was committed reflects the broader objective of compliance with obligations, which is emphasised in the ARS through the codification of the continued duty of performance,[90] and of the duty to cease the wrongful act (if it is ongoing)[91] in two separate articles. Together, these provisions make it clear that the law of State responsibility is not a liability system with the primary or exclusive goal of providing injured persons with compensation. The Commentaries further emphasise that compliance with existing obligations is a prerequisite to the restoration and repair of the legal relationships affected by the breach.[92] The duty of cessation further comprises an obligation to offer appropriate assurances and guarantees of non-repetition where the circumstances require.[93]

International human rights law similarly recognises that adequate and effective remedies for violations 'serve to deter violations and uphold the legal order that the treaties create'.[94] The duty of cessation has been characterised by the HRC as 'an essential element of the human right to a remedy' that entails an obligation 'to take measures to prevent the recurrence of a violation', including through changes in the State Party's laws or practice if necessary.[95] The ACHPR's findings in *SERAC v Nigeria* illustrate that in the human rights context, the duty to offer appropriate assurances and guarantees of non-repetition may reinforce existing procedural rights; when violating a range of human rights, Nigeria had incurred 'secondary' obligations to provide 'information on health and environmental risks and meaningful access to regulatory and decision-making bodies to communities likely to be affected by oil operations'.[96]

The consequences for States that incur these types of obligations based on climate change–related wrongful conduct could be drastic, particularly where the violation involves not a single act, but a series of wrongful acts and omissions. To meet its obligation of cessation, a State may need to make changes to significant parts of its laws, regulatory system and levels of assistance requested from, or provided to, other States in order to restore compliance with the substantive obligation that was violated. For example, a State may need to withdraw fossil fuel subsidies, adopt new regulations and policies to phase out fossil fuels, and bring all existing regulations and policies in line with emission reduction goals that reflect its highest possible ambition as well as CBDRRC.[97] In a similar vein, a developed State might be under an immediate obligation to scale up funding for mitigation, adaptation and capacity-building actions in developing States to restore compliance with its human rights obligations.

Realising victims' right to a remedy

The second set of obligations arising from an internationally wrongful act centre around an obligation to make full reparations for the injury caused by the wrongful act.[98] Injury is understood as including any material or moral damage caused by the act[99] and includes 'the injury resulting from and ascribable to the wrongful act' rather than 'any and all consequences' flowing from it.[100] This makes it clear that there must be a link between the wrongful act and some injury in order for there to be an obligation of reparation. However, the causal requirement inherent in the link is not the same in relation to every breach[101] and can be established even when the wrongful conduct was only one of several factors that contributed to the injury.[102]

Where the obligation breached relates to the prevention of harm, the link between injury and the breach is likely to involve consideration of the extent to which the harm was a reasonably foreseeable consequence of the action taken.[103] Based on the reports of the IPCC, input from affected communities and the definition of 'climate change' in Article 1 of the UNFCCC, a broad range of climate change–related risks and harm could be considered as reasonably foreseeable consequences of climate change and the human activities that are known to cause it. As far as evidential requirements are concerned, the principle of effectiveness may require shifting at least part of the risk of uncertainty to the State where it can be established with a reasonable degree of certainty that specific injury has occurred as a result of global warming.[104] Moreover, as Werksman suggests, the correlation between greenhouse gas emissions, atmospheric chemistry and global warming has probably

> been demonstrated with sufficient confidence that it seems unlikely that an adjudicator would require a complainant, in order to obtain relief, to demonstrate what would not be possible – that a specific emission of greenhouse gases by State S directly caused the specific impact in State I.[105]

All this means that existing evidence may well be sufficient to substantiate claims for reparation for climate change–related State conduct that constitutes a violation of international human rights law. As the science of attribution evolves, the chances that the victims of such wrongful conduct will be able to ascertain their entitlement to reparations should further increase. In addition, where State responsibility is invoked through individual complaint procedures under human rights treaties, victims have usually been identified in a claim's admissibility stage. In such cases, a link between State conduct and the individual's situation will already have been established once the case reaches the reparations stage.[106]

Once the duty to make full reparations has been triggered, the scope of the injury has to be established. This will be a fact-sensitive exercise which will require significant interpretation of complex evidence related to risks and probabilities. However, the law of State responsibility does provide some clear road signs for determining the nature and amount of reparations due. The first is the principle that no reduction or attenuation of reparation will be made for any concurrent causes.[107] The duty to make reparations is similarly unaffected by a responsible State's ability to pay,[108] or by a claimant's inability to determine the quantity and value of the losses suffered.[109] In other words, the duty of the responsible State to make full reparations for the injury is unqualified in general international law.[110]

The understanding of the right to a remedy as a substantive human right implies that the focus of the duty to make reparations for a breach of international human rights law lies squarely on restoring the rights of victims, insofar as victims of the violation can be identified. Where it is not certain whether an individual qualifies as a victim of the breach, uncertainty could be addressed in accordance with the human rights principle *in dubio pro libertate et dignitate*. Furthermore, irrespective of whether victims can be identified, the content of the obligation must reflect the aim of re-establishing the *status quo ante*.[111] The wide range of remedies awarded for human rights violations (including restitution, compensation, rehabilitation, and measures of satisfaction such as public apologies, public memorials, guarantees of non-repetition and, more importantly, changes in relevant laws or practices) reflect the potential for constructing remedies for climate change damage that are consistent with human rights objectives.[112] These remedies should materialise through bottom-up processes: individuals and communities affected by climate change themselves are in the best position to identify and develop suitable remedies for violations of their human rights. Thus, in cases where a State plans to invoke the law of State responsibility on behalf of affected communities, consultative processes will be needed to ensure that reparation claims accurately reflect the demands of those communities.

Concluding remarks

This chapter has demonstrated that States' obligations under international human rights law could provide a basis for State action in the context of climate change, as well as for State responsibility claims related to climate change and associated human rights violations. In a nutshell, international human rights law requires climate action that not only reflects States' maximum efforts to combat climate change, but also leads to a fair distribution of mitigation and adaptation burdens at local, domestic and global levels. Moreover, all States must take measures to prevent human rights from being violated in the context of response measures. The law of State responsibility is automatically triggered once a State breaches any of these obligations.

Whether a State has breached its human rights obligations through climate change–related conduct needs to be established on a case-by-case basis, taking into account the effects of its conduct on the enjoyment of human rights at home and abroad and the foreseeability of those effects. Hereby the States' obligations under the UNFCCC and the Paris Agreement could be taken as bottom lines for human rights obligations related to international cooperation and assistance.[113]

Once a breach of obligations has occurred, the responsible State must first and foremost restore compliance with the obligations that were violated. In other words, a State whose legislative framework or conduct is not in accordance with a human rights obligation incurs an obligation to bring its laws and practices in line with the relevant obligation. Moreover, the State must take measures to prevent future breaches of the obligation. And where the unlawful conduct – such as a State's failure to take adequate measures to prevent loss of life associated with the adverse effects of climate change – has actually caused harm, the State also incurs an obligation to make full reparations for the injury. These reparations must be directed to the beneficiaries of the obligation, which usually means the victims of the human rights violation.

The scope and nature of appropriate remedies might be relatively easy to establish where the violation concerns localised damage to individuals or communities, such as harm resulting from land grabbing or the exclusion of vulnerable communities from adaptation programmes. However, where the unlawful conduct relates to the impact of climate change *per se* on the enjoyment of human rights, the severity and scale of damage and the virtually limitless number of potential victims will trigger difficult questions related to causation, proof and victimhood. These questions will be complicated by the fact that multiple States might be responsible for the same damage. In such cases, the effectiveness of the law of State responsibility is hinged on the extent to which States cooperate to give effect to the victims' right to a remedy and to restore the rule of law. With liability and compensation excluded, at least for now, from the scope of the Warsaw Mechanism on Loss and Damage Associated with Climate Change,[114] it would seem opportune to explore the role of human rights bodies in facilitating such cooperation. Meanwhile, the human rights community could work with affected communities to develop guidance on the types of remedies that might be appropriate for various climate change–related human rights violations.

Notes

1 *United Nations Framework Convention on Climate Change 1992*, 1771 UNTS 107 (UNFCCC). See Decision 1/CP.16, UN Doc. FCCC/CP/2010/7/Add.1 (10 December 2010), at preambular para. 7 and ch. I, para. 8. See also *Paris Agreement on Climate Change*, UN Doc. FCCC/CP/2015/L.9/Rev.1 (12 December 2015) ('Paris Agreement'), at preambular para. 11.

2 ECtHR, *Airey v. Republic of Ireland*, Appl. no 6289/7, Judgment of 9 October 1979. All ECtHR decisions are available online at http://hudoc.echr.coe.int.

3 As the focus of this chapter is on human rights, it does not specifically discuss State responsibility for non-compliance with the UNFCCC. On concurrent responsibility under the two international regimes, see Wewerinke, 'The Role of the UN Human Rights Council in Addressing Climate Change', 7 *Human Rights and International Legal Discourse (HRILD)* (2014) 21–23.

4 J. Crawford, *The International Law Commission's Articles on State Responsibility: Introduction, Text, Commentaries* (2002) 31.
5 *Case Concerning the Factory at Chorzów (Germany v Poland)*, 1927 PCIJ Series A, No. 17, at 29.
6 ILC, *Articles on the Responsibility of States for Internationally Wrongful Acts, Report of the ILC on the Work of Its 53rd Session, Official Records of the General Assembly, 56th Session*, Supplement No. 10, UN Doc A/56/10 (2001), at chapter IV.E.2 ('ILC ARS'). On the legal authority of the Articles, see J. Crawford and S. Olleson, 'The Continuing Debate on a UN Convention on State Responsibility', 54 *International and Comparative Law Quarterly (ICLQ)* (2005) 959, at 968 and 971 (pointing out that 'there is an ongoing process of consolidation of the international rules of State responsibility as reflected in the Articles' with the Articles 'performing a constructive role in articulating the secondary rules of responsibility'). *Cf.* D.D. Caron, 'The ILC Articles on State Responsibility: the Paradoxical Relationship Between Form and Authority', 96 *American Journal of International Law (AJIL)* (2002) 857, at 867 (arguing that the Articles are 'similar in authority to the writings of highly qualified publicists').
7 *Patrick Coleman v Australia*, Communication No. 1157/2003, UN Doc. CCPR/C/87/D/1157/2003, para. 6.2; *S Jegatheeswara Sarma v Sri Lanka*, Communication No. 950/2000, UN Doc. CCPR/C/78/D/950/2000, para. 9.2; *Humberto Menanteau Aceituno and Mr. José Carrasco Vasquez (represented by counsel Mr. Nelson Caucoto Pereira of the Fundación de Ayuda Social de las Iglesias Cristianas) v Chile*, Communication No. 746/1997, UN Doc. CCPR/C/66/C/746/1997, para. 5.4. See also, the Individual concurring opinion of Kurt Herndl and Waleed Sadi in *Cox v Canada*, Communication No. 539/1993, UN Doc. CCPR/C/52/D/539/19930.
8 For a clear example, see IACtHR, *The Mayagna (Sumo) Awas Tingni Community v Nicaragua*, Judgment (31 August 2001), at para. 154; IACtHR, *Marino López et al. v Colombia*, Judgment (Merits) (31 March 2011), at para. 42. See also D. McGoldrick, *The Human Rights Committee* (1991), at 169. All IACtHR decisions are available online at www.corteidh.or.cr/index.php/en/jurisprudencia.
9 D. McGoldrick, 'State Responsibility and the International Covenant on Civil and Political Rights', in M. Fitzmaurice and D. Sarooshi (eds.), *Issues of State Responsibility Before International Judicial Institutions* (2004), at 199; E.W. Vierdag, 'Some Remarks About Special Features of Human Rights Treaties', 25 *Netherlands Yearbook of International Law (NYIL)* (1994) 119, at 135.
10 See *Basic Principles and Guidelines on the Right to a Remedy and Reparation for Victims of Gross Violations of International Human Rights Law and Serious Violations of International Humanitarian Law*, GA Res. 60/147 (16 December 2005), at Annex, Principles 1(b), 2 and 3 (pertaining to gross violations of international human rights law and international crimes), and 11. See also D. Shelton, *Remedies in International Human Rights Law*, 2nd ed. (2010), at 103.
11 For an overview of global and regional human rights treaties that incorporate the right to a remedy, see Shelton, *supra* note 10, at 113–20. See also, Crawford, *The ILC Articles*, *supra* note 4, at 95, paras 3–4.
12 Shelton, *supra* note 10, at 104ff.
13 Art. 2 ILC ARS.
14 J. Crawford, 'The ILC's Articles on Responsibility of States for Internationally Wrongful Acts: a Retrospect', 96 *American Journal of International Law (AJIL)* (2002) 874, at 879.
15 Art. 4(2) ILC ARS clarifies that '[a]n organ includes any person or entity which has that status in accordance with the internal law of the State'. See also R.B. Lillich *et al.*, 'Attribution Issues in State Responsibility', 84 *Proceedings of the Annual Meeting (American Society of International Law)* (1990) 51, at 52 (pointing out that the principle that 'a State may act through its own independent failure of duty or inaction when an international obligation requires state action in relation to non-State conduct' is reflected in all codifications and restatements of the law on State responsibility).
16 ILC ARS, Commentary to Article 4, para. 5.
17 McGoldrick, 'State Responsibility', *supra* note 9, at fn 74–83 and accompanying text, containing citations.
18 *Ibid.*
19 This formulation was used in Article 1 of the Outcome Document of the Third Committee of the 1930 Hague Conference, reproduced in Yearbook of the United Nations 1956, Vol. II, p. 225, Document A/CN.4/96, Annex 3.
20 Crawford, *The ILC Articles*, *supra* note 4, at 35 and at Commentary to Article 2, para. 4.
21 *Ibid.*, at Commentary to Article 2, paras 5–6.
22 See, for example, ECtHR, *Paul and Audrey Edwards v. United Kingdom*, Appl. No. 46477/99, Judgement of 14 March 2002, at para. 64.
23 Art. 15(1) ILC ARS.

24 Art 15(2) ILC ARS. See also ECtHR, *Ireland v. United Kingdom*, Appl. No. 5310/71, Judgment of 18 January 1978 at para. 159, in which the ECtHR discussed the concept of a 'practice incompatible with the Convention'.
25 1 UNTS XVI. See especially Arts 1, 55 and 56.
26 Universal Declaration of Human Rights 1948, GA Res. 217 A (III), A/810 (UDHR), 10 December 1948, at preambular recitals 6 and 7. See also, Proclamation of Teheran, *Final Act of the International Conference on Human Rights, Teheran, 22 April to 13 May 1968*, UN Doc. A/CONF. 32/41 3 (1968) (stating that the UDHR 'states a common understanding of the peoples of the world concerning the inalienable and inviolable rights of all members of the human family and constitutes an obligation for all members of the international community'). See further, De Schutter *et al.*, 'Commentary to the Maastricht Principles on Extraterritorial Obligations of States in the Area of Economic, Social and Cultural Rights', 34 *Human Rights Quarterly (HRQ)* (2012) 1084, at 1092 and N. Jayawickrama, *The Judicial Application of Human Rights Law: National, Regional and International Jurisprudence* (2002) 30.
27 International Covenant on Civil and Political Rights 1966, 999 UNTS 171 (ICCPR).
28 For ratification status, see http://treaties.un.org/Pages/Treaties.aspx?id=4&subid=A&lang=en (last visited 1 September 2016).
29 International Covenant on Economic, Social and Cultural Rights 1966, 993 UNTS 3 (ICESCR).
30 For ratification status, see http://treaties.un.org/Pages/Treaties.aspx?id=4&subid=A&lang=en (last visited 1 September 2016).
31 See website of the UN OHCHR, www.ohchr.org/EN/HRBodies/Pages/HumanRightsBodies.aspx (last visited 17 August 2016).
32 *Legality of the Threat or Use of Nuclear Weapons*, Advisory Opinion, 8 July 1996, ICJ Reports (1996) 241, para. 79. See also M. Salomon, 'Deprivation, Causation, and the Law of International Cooperation', in M. Langford *et al.* (eds.), *Global Justice, State Duties: The Extraterritorial Scope of Economic, Social and Cultural Rights in International Law* (2012) 259, at 304 (arguing that '[t]oday, the existence of a customary international law principle to respect and observe human rights in the main, which can be said to apply to basic socio-economic rights, is increasingly difficult to refute). *Cf.* J. Crawford, *Brownlie's Principles of Public International Law*, 8th ed. (2012) 21.
33 M. Nowak, *U.N. Covenant on Civil and Political Rights: CCPR Commentary*, 2nd ed. (2004) XXIV. See also A. Orakhelashvili, 'Restrictive Interpretation of Human Rights Treaties in the Recent Jurisprudence of the European Court of Human Rights', 14 *European Journal of International Law (EJIL)* (2003) 529.
34 M. Nowak, *Introduction to the International Human Rights Regime* (2003) 65. See also IACtHR, *Other Treaties Subject to the Advisory Jurisdiction of the Court (Article 64 ACHR)*, Advisory Opinion, 24 September 1982, at para. 24 and IACtHR, *The Effect of Reservations on the Entry into Force of the ACHR (Articles 74 and 75)*, Advisory Opinion, 24 September 1982, at para. 27.
35 UN Human Rights Committee (UNHRC), *General Comment No. 6: The Right to Life (Article 6)*, available online at www.refworld.org/docid/45388400a.html, at paras 1, 5 (quote from para. 1).
36 IACtHR, *Villagrán-Morales, et al., v Guatemala*, Judgment, 19 November 1999, at para. 144.
37 ACHPR, *SERAC and Another v Nigeria*, Decision (Merits), 27 October 2001, at para. 67. See also, ECtHR, *Oneryildiz v Turkey*, Appl. no. 48939/99, Judgment of 30 November 2004, at 115.
38 Ramcharan, 'The Concept and Dimension of the Right to Life', in B. G. Ramcharan (ed.), *The Right to Life in International Law* (1985) 1, at 7.
39 See UNHRC, *General Comment*, *supra* note 35, at paras 3–5; Nowak, *U.N. Covenant*, *supra* note 33, at 105.
40 UNHRC, *Concluding Observations on Paraguay* (1995), UN Doc. CCPR/79/Add 48, 5 April 995, at para. 16.
41 UNHRC, *Concluding Observations on the United States* (1995), UN Doc. CCPR/C79/Add 50, 7 April 1995, at para. 17.
42 UNHRC, *General Comment No. 14: Nuclear Weapons and the Right to Life (Article 6)*, available online at www.refworld.org/docid/453883f911.html (last visited 15 March 2017), at paras 6–7. See also *Nowak, U.N. Covenant, supra* note 33, at 126; S. Joseph, J. Schultz and M. Castan, *The International Covenant on Civil and Political Rights: Cases, Materials, and Commentary*, 2nd ed. (2005) 185.
43 See, for example, ECtHR, *Osman v United Kingdom*, Appl. no. 48939/99, Judgment of 30 November 2004, para. 115; ECtHR, *Ilhan v Turkey*, Appl. no. 22277/93, Judgment of 20 December 2011, para. 91; ECtHR, *Kilic v Turkey*, Appl. no. 22492/93, Judgment of 28 March 2000, para. 62; and ECtHR, *Mahmut Kaya v Turkey*, Appl. no. 22535/93, Judgment of 19 February 1998, para. 85.

44 See ECtHR, *Oneryildiz v Turkey*, *supra* note 37, para. 79 (concerning the State's failure to prevent a possible explosion of methane gas from a garbage dump under the authority of the City Council).
45 Nowak, *U.N. Covenant*, *supra* note 33, at 123. See also Ramcharan, *supra* note 38, at 20.
46 See *General Comment No. 6: The Right to Life (Article 6)*, *supra* note 35, para. 5.
47 *Ibid.*, para. 5. In Nowak's opinion, the HRC's interpretation of Art. 6(1) shows 'not only a willingness to innovate, but also a resolute application of the premises derived from Art. 6, whereby the right to life is not to be interpreted narrowly and States parties are obligated to take positive measures to ensure it.' Nowak, *U.N. Covenant*, *supra* note 33, at 127.
48 UNHRC, *General Comment No 31: The Nature of the General Legal Obligation Imposed on States Parties to the Covenant*, UN Doc. CCPR/C/21/Rev1/Add 13 (26 May 2004), para. 8.
49 IACtHR, *Velásquez Radríguez v Honduras*, Judgment, 29 July 1988.
50 *Ibid.*, para. 172.
51 See also *Trail Smelter (United States v Canada)*, 11 March 1941, 3 UN Reports of International Arbitration Awards 1905, at 1965. Although this case concerned injury caused by fumes, the principle that all States are obliged to take measures to prevent injury 'established by clear and convincing evidence' has a more general application to all sorts of environmental damage, including damage to the world's common spaces. See J. Charney, 'Third State Remedies for Environmental Damage to the World's Common Spaces', in F. Francioni and T. Scovazzi (eds.), *International Responsibility for Environmental Harm* (1991) 149.
52 *Osman v United Kingdom*, *supra* note 43, paras 115–16. See also ECtHR, *Budayeva et al v Russia*, Appl. No. 48939/99, Judgment of 30 Nov 2004, para. 132.
53 ECtHR, *Tatar c. Roumanie*, Appl. no. 67021/01, Judgment of 5 July 2007.
54 UNFCCC, Art. 3(1).
55 *Tatar c. Roumanie*, *supra* note 53, paras 107–24.
56 ECtHR, *X. and Y. v Netherlands*, Appl. no. 8978/80, Judgment of 26 March 1985, para. 23.
57 See C. Segger *et al.*, 'Prospects for Principles of International Sustainable Development Law After the WSSD: Common But Differentiated Responsibilities, Precaution and Participation', 12 *Review of European, Comparative and International Environmental Law (RECIEL)* (2003) 54, at 57 (explaining that differentiated responsibility 'aims to promote substantive equality between developing and developed States within a regime, rather than mere formal equality').
58 For an examination of climate change and economic, social and cultural rights, see generally S. Jodoin and K. Lofts (eds.), *Economic, Social, and Cultural Rights and Climate Change: A Legal Reference Guide* (2013), available online at https://environment.yale.edu/content/documents/00004236/ESC-Rights-and-Climate-Change-Legal-Reference-Guide.pdf?1386877062.
59 CESCR, *General Comment No. 14: The Right to the Highest Attainable Standard of Health (Article 12)*, UN Doc. E/C 12/2000/4 (11 August 2000), para. 8.
60 *Ibid.*, paras 30–45.
61 *Ibid.*, para. 30.
62 *Ibid.*, para. 15. This typology is based on the work of the former Special Rapporteur on the Right to Food, Asbjørn Eide. See Eide, 'Economic, Social and Cultural Rights as Human Rights', in A. Eide, C. Krause and A. Rosas (eds.), *Economic, Social and Cultural Rights*, 2nd edn (2001) 9 at 23–25.
63 See A. Eide, 'The Right to an Adequate Standard of Living Including the Right to Food', in A. Eide, C. Krause and A. Rosas (eds.), *Economic, Social and Cultural Rights*, 2nd edn (2001) 133 at 142.
64 *General Comment No. 14*, *supra* note 59, at para. 39.
65 *Ibid.*, para. 34.
66 P. Hunt and R. Khosla, 'Climate Change and the Right to the Highest Attainable Standard of Health', in S. Humphreys (ed.), *Climate Change and Human Rights* (2009) 238, at 252.
67 *General Comment No. 14*, *supra* note 59, at para. 39.
68 See Eide, 'The Right to an Adequate Standard of Living', *supra* note 63, at 143. See also *General Comment No. 14*, *supra* note 59 at para. 35.
69 *General Comment No. 14*, *supra* note 59 at para. 39.
70 *Ibid.*, paras 36–37.
71 *Ibid.*
72 Eide, 'The Right to an Adequate Standard of Living', *supra* note 63, at 143.
73 *Ibid.*, at 144.
74 CESCR, *Concluding Observations on the Russian Federation*, UN Doc E/C.12/1/Add.13 (20 May 1997), at para. 38.

75 Hunt and Khosla, *supra* note 66, at 252. Inspiration may also be drawn from the *Minors Oposa* decision of the Philippine Supreme Court, which decided that on the basis of the rights to health and ecology contained in the Philippines Constitution, the Philippine government had to protect the population against the impacts of rainforest logging activities. *Juan Antonio Oposa et al. v the Honorable Fulgencio Factoran Jr., Secretary of the Department of the Environment and Natural Resources et al.*, Supreme Court of the Philippines, G.R. No. 101083 (Phil).
76 Hunt and Khosla, *supra* note 66, at 255.
77 *General Comment No. 14*, *supra* note 59, at para. 36.
78 See, for example, CESCR, *Concluding Observations on Chile,* UN Doc E/C.12/1988/SR.13, para. 12 and CESCR, *Concluding Observations on North Korea*, UN Doc E/C.12/1987/SR.22, paras 5 and 17. See also, C. Doebbler and F. Bustreo, 'Making Health an Imperative of Foreign Policy: The Value of a Human Rights Approach', 12 *Health and Human Rights Journal (HHR)* (2010) 47, at 53.
79 CESCR, *General Comment No. 12: Substantive Issues Arising in the Implementation of the International Covenant on Economic, Social and Cultural Rights: The Right to Adequate Food (Art. 11)*, UN Doc. E/C 12/1999/5 (12 May 1999), at para. 15.
80 See Eide, 'The Right to an Adequate Standard of Living', *supra* note 63, at 145.
81 *General Comment No. 14*, *supra* note 59, at para. 39. See also Doebbler and Bustreo, *supra* note 78, at 53 (stating that the right to health 'encourages a world order in which donor states can point out human rights obligations to recipient countries, while recipient countries can point out the duties to cooperate to ensure human rights, including the obligations for providing adequate resources that are incumbent upon donor countries').
82 Hunt and Khosla, *supra* note 66, at 252.
83 Paris Agreement, Article 9(1).
84 Paris Agreement, Article 9(5).
85 Paris Agreement, Article 14(1).
86 Paris Agreement, Article 2(1)(c) and (b).
87 See, for example, Callaghan, 'Climate Finance After COP21: Pathways to Effective Financing of Commitments and Needs,' a report for Investor Watch (December 2015) at 10 (pointing out that '[o]n the evidence of the INDCs, financial requirements of developing nations to meet proposed actions far exceeds (probably by a factor of five times) the USD 100 billion' pledged by developed countries. See further M. Wewerinke-Singh and C. Doebbler, 'The Paris Agreement: Some Critical Reflections on Process and Substance', 39(4) *University of New South Wales Law Journal* (2016) 114.
88 *Phosphates in Morocco (Italy v France)*, 1938 PCIJ Series A/B, No. 74 at 10, para. 48.
89 *Case Concerning the Factory at Chorzów (Germany v Poland)*, *supra* note 55, at 47at 47
90 ILC ARS, Art. 29.
91 ILC ARS, Art. 30(a). The treatment of cessation as a distinct legal consequence of an internationally wrongful acts is a relatively novel development; previously, cessation was considered as part of the remedy of satisfaction. See also, Shelton, *Remedies in International Human Rights Law*, *supra* note 10 and Crawford, *The ILC Articles*, *supra* note 4, at 68, para. 114 and Crawford, *Brownlie's Principles of Public International Law*, *supra* note 3232567567.
92 ILC ARS, Commentary to Art. 30, para. 1.
93 ILC ARS, Art. 30(b) and Commentary to Art. 30, para. 1.
94 Shelton, *Remedies in International Human Rights Law*, *supra* note 10, at 99.
95 UNHRC, *General Comment No 31*, *supra* note 48, paras 16–17.
96 *SERAC and Another v Nigeria*, *supra* note 37, para. 69.
97 This requirement matches the procedural obligation under the Paris Agreement of submitting 'nationally determined contributions' that reflect these objectives. See Paris Agreement, Article 4(3).
98 ILC ARS, Art. 31(1).
99 ILC ARS, Commentary to Art. 31, para. 5.
100 *Ibid.*, para. 10.
101 *Ibid.*
102 See, for example, *United States Diplomatic and Consular Staff in Tehran*, Judgment, 24 May 1980, ICJ Reports (1980) 3.
103 Shelton, *Remedies in International Human Rights Law*, *supra* note 10, at 89, 99.
104 *Ibid.*, at 50, 317 (stating that the burden of uncertainty or lack of proof may sometimes shift to the State to uphold the deterrent function of remedies for human rights violations).

105 J. Werksman, 'Could a Small Island Successfully Sue a Big Emitter? Pursuing a Legal Theory and a Venue for Climate Justice', in M. B. Gerrard and G. E. Wannier (eds.), *Threatened Island Nations: Legal Implications of Rising Seas and a Changing Climate* (2012) 409, at 412.

106 The 'victim requirement' is one of the admissibility criteria that need to be met before a particular judicial or quasi-judicial human rights body can consider the merits of an international complaint for an alleged human rights violation. For example, Art. 34 of the ECHR provides that only complainants who are directly affected by the alleged breach of the ECHR have a right to complain about the violation before the ECtHR. Art. 1 of the Optional Protocol of the ICCPR and Art. 2 of the Optional Protocol to the ICESCR contain a similar requirement.

107 See, for example, *United States Diplomatic and Consular Staff in Tehran*, *supra* note 102.

108 See J. Crawford, *State Responsibility: The General Part* (2013) 481.

109 ECtHR, *Mentes et al. v Turkey,* Appl. no. 58/1996/677/867, Judgment of 24 July 1998, at para. 106 (stating that there should be some pecuniary remedy, but 'since the applicants have not substantiated their claims as to the quantity and value of their lost property with any documentary or other evidence, the Government have not provided any detailed documents, and the Commission has made no findings of fact in this respect, the Court's assessment of amounts to be awarded must, by necessity, be speculative and based on principles of equity').

110 ILC ARS, Commentary to Art. 31, para. 12 (stating that 'international practice and the decisions of international tribunals do not support the reduction or attenuation of reparation for concurrent causes'). See also Crawford, *State Responsibility: The General Part*, *supra* note 108, at 496.

111 *Case Concerning Application of the Convention on the Prevention and Punishment of the Crime of Genocide (Bosnia and Herzegovina v Serbia and Montenegro)*, Judgment, 26 February 2007, ICJ Reports (2007) 43; *Ahmadou Sadio Diallo (Guinea v DRC)*, Judgment, 30 November 2010, ICJ Reports (2010) 639; *Velásquez Radríguez v Honduras*, *supra* note 49.

112 UNHRC, *General Comment No 31*, *supra* note 48, at para. 16.

113 See further Wewerinke, *supra* note 3 at 12ff.

114 UNFCCC, *COP Decision 1/CP.21*, UN Doc. CP/2015/10/Add.1 (29 January 2016), para. 52. See further M. J. Mace and R. Verheyen, 'Loss, Damage and Responsibility After COP21: All Options Open for the Paris Agreement', 25 *Review of European, Comparative and International Environmental Law (RECIEL)* (2016) 197.

7

Climate change impacts

Human rights in climate adaptation and loss and damage

Sven Harmeling

Climate change impacts on human rights: state of science

The fact that anthropogenic climate change impacts are taking place today, are felt by millions of people around the world, and happen in areas which are key to realising human rights is undisputed. Various IPCC reports, including the most recent 5th Assessment Report, have built their conclusions from thousands of studies from across the globe. The impacts of climate change are also recognised by major human rights stakeholders.[1]

The 5th Assessment Report of the IPCC provides evidence for a range of adverse climate change impacts. A specific chapter of the Working Group II (Impacts, Adaptation and Vulnerability) compiles scientific findings in relation to human security.[2] It concludes, *inter alia*, that

- Human security will be progressively threatened as the climate changes (robust evidence, high agreement);
- Climate change will compromise the cultural values that are important for community and individual well-being (medium evidence, high agreement);
- Indigenous, local and traditional forms of knowledge are a major resource for adapting to climate change (robust evidence, high agreement);
- Climate change will have significant impacts on forms of migration that compromise human security (medium evidence, high agreement).

With regard to human rights, the IPCC authors also discuss the relationship between human security and human rights, preferring the terminology of human security, *inter alia* because "framing the issue of rights specifies minimum standards that apply universally, and such rights are often not realized in national and international law and practice or neglect the harm or rights of nonhuman species", whereas "Human security by contrast is inclusive of political, sociocultural, and economic rights, rather than legal rights (CHS, 2003), which are instrumental to its achievement (Bell, 2013)." It also concludes that research on climate change risks to human rights examines legal issues in policy, litigation and compensation.[3]

With the strengthened temperature target of the Paris Agreement "to pursue efforts to limit global warming to 1.5°C",[4] which was facilitated through an intense science-policy process under the so-called First Periodic Review and Structured Expert Dialogue (SED) in the UNFCCC that underpinned the arguments that 2°C warming levels already entail risks way beyond a "safe" level, more attention is being paid now to understanding the risks from climate change impacts at a 1.5°C increase level compared to higher levels. Schleussner et al. provide an "assessment of key impacts of climate change at warming levels of 1.5°C and 2°C, including extreme weather events, water availability, agricultural yields, sea-level rise and risk of coral reef loss". Although the focus of the report is on the natural science impacts and not the socio-economic impacts which eventually are what matters from a human rights perspective, their "results reveal substantial differences in impacts between a 1.5°C and 2°C warming that are highly relevant for the assessment of dangerous anthropogenic interference with the climate system".[5] Key areas of concern with significantly higher risks include heat extremes, the risk of tropical yield reductions, coral reef bleaching and subtropical water scarcity, all areas where it is imaginable how they might adversely impact on human rights.[6] Tropical and dry subtropical regions emerge as hotspots, with increased risks for humans due to the combination with limited adaptive capacity and high exposure in many countries in the region.[7] The SED also concluded, "[R]egional food security risks are significantly different between 1.5°C and 2°C of warming".[8] The differential impacts between 1.5°C and 2°C with regard to human rights have also been addressed by OHCHR in a specific submission to the Human Rights Council.[9]

Thus, from the perspective of climate change impacts, the intention of the 1.5°C limit is also a positive message for the question to what extent climate change impacts might undermine human rights. However, it is clear that in many regions impacts already hit many people hard and will still do so at a 1.5°C warming level. But a future at much higher warming levels is very likely to make the fulfilment of human rights, amongst other large-scale earth system changes, much more difficult.

Major future IPCC reports, which will address issues of concern for key human rights, are being compiled in the next years. "Global Warming of 1.5°C", an IPCC special report to be released in October 2018, will discuss the interrelated threats of climate change, sustainable development and poverty eradication, with consideration of ethics and equity, and look into limits of adaptive capacity and irreversible impacts, but also reduced risks at a 1.5°C warming level compared to 2°C or higher temperature scenarios.[10] Also, two special reports on climate change and land-related issues,[11] both to be released in 2019, looking at a broad set of relevant issues such as food security, sustainable land management and greenhouse gases, as well as the one on climate change and oceans and the cryosphere[12] – particularly relevant e.g. regarding how sea level rise might affect the loss of territory and have other associated impacts – address key areas of concern from a human rights perspective.

General human rights obligations in relation to tackling climate impacts

Advancing the integration of human rights in climate action is an important task. It is a question of justice, as human rights are universal legal guarantees of protection and those who have contributed the least to the problem are regarded as especially vulnerable.[13]

However, it is also argued that "integrating human rights in climate actions will necessitate higher levels of ambition and improve mitigation and adaptation strategies by making them more effective and inclusive".[14]

As the UNFCCC is the major international policy forum related to the adaptation to and loss and damage from climate change impacts, it is important to provide a more in-depth analysis of the extent to which human rights have featured in UNFCCC work and to identify opportunities for stronger attention in the future, which will be the focus of this article.

The principal human rights obligations in relation to climate change impacts are well established, and also have to be seen in conjunction with obligations under the UNFCCC, including the Paris Agreement which is now into force. As the Office of the United Nations High Commissioner for Human Rights (OHCHR) expresses,

> States (duty-bearers) have an affirmative obligation to take effective measures to prevent and redress these climate impacts, and therefore, to mitigate climate change, and to ensure that all human beings (rights-holders) have the necessary capacity to adapt to the climate crisis.[15]

As underlined by the OHCHR[16] and also pointed out by CARE/CIEL,[17] States have obligations to protect those whose rights are affected by climate change, with priority given to groups that are particularly vulnerable. These obligations exist independent from the question of the cause. That means that e.g. States who have hardly contributed anything to the emissions that cause climate change are still bound by this obligation, and they need to take action in light of the risks of climate change to their citizens. That is one of the reasons why the debate about human rights as a framework for adaptation action has also been contested by some developing countries, fearing they might commit to obligations they might not be able to keep with the harsh reality of climate change. However, this brings us to the obligations of countries to assist those who are particularly vulnerable, both from a human rights perspective – the obligation to support others in the fulfilment of human rights – and from a climate policy perspective – the obligation for developed countries to assist developing countries in meeting the costs of adaptation, as enshrined in Article 4.4. of the Convention.

In recent years, various statements and publications have strengthened the underpinnings of the human rights and climate change impacts link, including the International Bar Association[18] (2014) and Mary Robinson Foundation[19] (2016).

Human rights, adaptation, and loss and damage under the UNFCCC

Human rights in the Paris Agreement

Under the UNFCCC, human rights remained undervalued after Copenhagen, despite the recognition that "Parties should, in all climate-related actions, fully respect human rights", contained in the Cancun Agreements.[20] In 2015, in the run-up to Paris, human rights received increasing attention. The Geneva Pledge for Human Rights in Climate Action,[21] released during the Geneva UNFCCC session in February 2015, sent a strong signal of voluntary action in a situation where the integration of human rights obligations in the Paris Agreement faced significant scepticism and resistance by certain Parties in the UNFCCC.

Eventually, all governments agreed on referencing human rights in the preamble of the Paris Agreement in a way that it reaffirms Parties' respective human rights obligations and that they should "respect, promote and consider" these when taking action to address climate change.[22] It further mentions specifically the right to health and the rights of Indigenous Peoples. By referring to "people in vulnerable situations", the preamble also implicitly refers to climate change impacts, where obviously adaptation is a key approach to take.

Furthermore, gender equality and the empowerment of women receive special mention. Mayer argued that the "main added value of this provision is its insertion in a treaty rather than a COP decision", constituting a unique step as human rights references were missing in the UNFCCC and the Kyoto Protocol, despite not creating any new human rights obligations.[23]

The operational relevance of this preambular provision remains subject to scholarly discussion, and obviously further operational steps taken by the UNFCCC process building on this preamble would increase the likelihood of its explicit application. In the course of the debates before and after Paris, for example, observers demanded to set up specific work programmes as next steps, e.g. to "monitor and assess the progress in integrating human rights into all aspects of climate action".[24] A group of human rights organisations suggested various steps to integrate human rights in the post-Paris negotiations under the APA.[25]

In addition to giving practical relevance to the preambular provisions through follow-up actions under the UNFCCC, scholars see the connection with the Sustainable Development Goals (SDGs) as particularly relevant. They argue that, based on a contextual analysis especially of the SDGs, "states have properly integrated a human rights dimension into key operative provisions of the Paris Agreement, albeit indirectly."[26]

Of particular relevance to this context is Article 7.5 of the Paris Agreement, which sets out some key principles that should be applied when undertaking adaptation action, in particular in relation to gender, participation and transparency, and special consideration of vulnerabilities. Furthermore, it demands that adaptation action "should be based on and guided by the best available science and, as appropriate, traditional knowledge, knowledge of indigenous peoples and local knowledge systems".[27]

This Article was also the place where the integration of a human rights reference was discussed for some time during the negotiations in the lead-up to COP21. The draft negotiation text that came out of the last preparatory session before COP21 (ADP) included in what was then Article 4.3 human rights in brackets, indicating the lack of agreement at that stage.[28] That human rights was subject to differences in opinion is also on public record.[29]

One of the concerns raised by a delegate from a developing country was lack of certainty of the implications of including human rights here, fearing the country would be expected to deliver on all human rights even if climate change would make it much more difficult to fulfil the government's human rights obligations. Another point in debate was the fact that a human rights reference was discussed in various sections of the draft Paris Agreement in October 2015.[30]

However, also Articles 7.1 and 7.2 add context to human rights and adaptation, through establishing a global goal on adaptation and its linkages to sustainable development and through highlighting the important role of adaptation for protecting people, livelihoods and ecosystems. This is also reflected in proposals to operationalise the adaptation goal by ensuring "human security in a well below 2°C world by the end of the century".[31]

Human rights in specific UNFCCC adaptation work streams

It is important to look beyond the Paris Agreement provisions to understand if and how human rights obligations and frameworks have played a role in the various adaptation work streams under the UNFCCC.

Adaptation Committee

The UNFCCC Adaptation Committee (AC) is the key institution mandated to promote coherence on adaptation under the Convention. It was set up through the Cancun Agreements.

Its key functions are to

1 Provide technical support to the Conference of the Parties;
2 Enhance sharing of information on adaptation on all levels;
3 Promote synergies and an entry point for engagement with national, regional and international organizations and networks to advance adaptation action;
4 Provide information and recommendations to the COP on the support of adaptation actions;
5 Stream information by Parties on monitoring and review of adaption actions for possible needs and gaps to recommend further actions.

The AC started is work in 2012 and has implemented various activities through its work plans. Until Paris, human rights and the role that adaptation can play in this context did not feature at all.[32] The current work plan, revised after COP21, covers 2016 to 2018. In light of the relevance of the Paris Agreement, it is noticeable that the AC framed its implementation "with the overall aim of contributing to the global goal on adaptation as contained in Article 7.1 of the Paris Agreement, and the principles guiding adaptation action as contained in Art. 7.5 of the Paris Agreement".[33] Against this background, the further implementation of the current work plan provides a number of opportunities to bring the adaptation and human rights context into sharper focus:

- The AC's workstream on providing guidance to the Parties on adaptation planning could take up supportive measures with regard to addressing human rights.
- The AC also envisages promoting the "exchange of information, particularly as it relates to the most vulnerable people, sectors and regions" through focused events.
- The Adaptation Technical Examination Process (A-TEP) provides an opportunity to consider more in-depth the linkages between adaptation and human rights.[34]
- In light of the stronger attention of the SDGs towards human rights, activities are planned in 2018 concerning national adaptation goals/indicators and how they relate to indicators/goals for sustainable development and for disaster risk reduction in the context of the Sendai Framework for Disaster Risk Reduction.

Nairobi Work Programme

The Nairobi Work Programme on Impacts, Adaptation and Vulnerability (NWP) was established in 2005 at COP11 in Nairobi as a five-year work programme.[35] Under the Subsidiary Body for Scientific and Technological Advice (SBSTA), it focuses on the generation and exchange of knowledge and experience related to climate change impacts, adaptation and vulnerability. The programme was later extended with various activities and also shifting approaches in terms of its focus areas.[36] This section centres on technical activities in focus areas which are particularly relevant to specific human rights.

Indigenous and traditional knowledge

In 2014, the Nairobi Work Programme undertook dedicated activities in relation to "available tools for the use of indigenous and traditional knowledge and practices for adaptation, needs of local and indigenous communities and the application of gender-sensitive approaches and tools for adaptation". The main discussions and findings from that expert meeting have been captured in a report. Although this does not address human rights specifically, it contains several mentions of the rights of local/indigenous communities/holders of indigenous and traditional knowledge.[37] In the context of a discussion on good practice examples, it also states that local, indigenous and traditional knowledge should be "subject to the application of a rights-based approach and protected under an appropriate intellectual property rights regime" (p. 9). However, it does not spell out further what this would actually mean.

Water

The Human Right to Water and Sanitation (HRWS) was recognised by the United Nations General Assembly on 28 July 2010. Over the years, the NWP addressed the climate change and water context through a number of activities.

In 2011, the UNFCCC Secretariat produced a technical paper on water and climate change impacts and adaptation strategies which provides information on links between climate change and freshwater resources and on adaptation to climate change in the water sector.[38] Though it refers to examples of actions, such as "advocacy support for rights of vulnerable people" (p. 28), "advocacy on rights to livelihood resources" (p. 28) and "women's rights" (p. 26), it does not refer to the right to water and lacks any approach to a rights-based framework.

In 2012, the NWP organised a specific workshop on water and climate change impacts and adaptation strategies. The official UNFCCC report of the workshop highlights the findings in four different areas, (a) observational data and their interpretation for understanding climate change impacts on water resources; (b) assessment of climate change impacts on water resources and on related sectors and ecosystems; (c) adaptation planning and practices related to water resources at different levels; and (d) communication, stakeholder engagement, knowledge sharing and management.[39] There is a notable absence of any mention of human rights in the meeting report as well as the main presentations given (including those from CSOs [Civil Society Organisations]). The NWP report, however, reflects awareness of some key aspects of the human rights debate, regarding the need for broad stakeholder participation and the attention to indigenous knowledge, as well as discussing the impacts of climate change on other areas of human concern (such as health and gender equality).

In 2014, Parties and NWP observer organisations were invited to submit their good practice experience related to adaptation planning on various issues, including water. The official UNFCCC synthesis report was published in late 2015. Case studies referred to provided information concerning the impacts of climate change on water and other rights-relevant areas linked to water, including health and human settlements. However, the synthesis does not include any reference to the right to water framework.[40]

Recently, the Secretariat published a synopsis which provides a useful overview of the main findings from the activities undertaken in previous years in relation to water.[41] It lacks any reference to the right to water.

Health

The right to health is the economic, social and cultural right to a universal minimum standard of health, to which all individuals are entitled. "Countries that have ratified international legal instruments relating to climate change and/or the right to health are obliged to implement them and to translate their obligations into national law", as OHCHR argues in a submission to NWP on the linkages between climate change and the right to health.[42] It distinguishes climate change risks on health by various categories, including the increase in premature deaths and disease, diminished well-being, occupational health risks and displacement.

The 2014 submission process on good practice experience in relation to adaptation planning also included health aspects. Case studies referred to in the synthesis report provided information concerning the impacts of climate change on health and highlights the "serious health risk to societies, especially their vulnerable groups (e.g. the elderly, infants and young children, chronically ill people and people who work outdoors)" (p. 11).[43] However, here also any mention of the key rights to health is missing.

Recently, the Secretariat published a synopsis which provides a useful overview of the main findings from the activities undertaken in previous years related to health.[44] It lacks any reference to the right to health.

Ecosystems

A healthy environment of which ecosystems are a fundamental pillar is also seen as a "prerequisite for the enjoyment of human rights (implying that human rights obligations of States should include the duty to ensure the level of environmental protection necessary to allow the full exercise of protected rights)".[45] A recent report by OHCHR to the UN Human Rights Council gives additional weight to this argument by discussing the dependence of human rights on biodiversity and ecosystem services, as well as human rights obligations relating to the conservation and sustainable use of biodiversity.[46] Thus, it is also worth considering to what extent activities under the NWP in relation to ecosystems have addressed the role of human rights.

In 2011, the UNFCCC Secretariat published an information note on ecosystem-based approaches to adaptation.[47] Drawing upon that and inputs from governments and observer organisations, in 2013, a technical workshop on ecosystem-based approaches was held in Tanzania. Amongst others, it highlights the benefits of ecosystem-based approaches as contributing to livelihood sustenance and food security, sustainable water management, protection of Indigenous Peoples and local communities etc. However, the concept of human rights is not referred to.[48]

SBSTA agriculture work programme

In light of the significance of climate change impacts on agriculture and thereby on the right to adequate food, this section will look at the UNFCCC's work in relation to agriculture. The fact that climate change has an impact on the right to adequate food, and proposals for how to address this, including inside the UNFCCC, have been out there for many years.[49] A report by the UN Special Rapporteur on the Right to Food to the UN General Assembly from 2015 provides substantial analysis on the linkages, and concludes that

> in order to eradicate hunger and ensure the full realization of the right to food, more must be done to develop relevant, effective mitigation and adaptation policies and a human rights approach must be adopted as a means of achieving climate justice.[50]

Under the Subsidiary Body for Scientific and Technological Advice (SBSTA), governments in the UNFCCC began to more substantially discuss agriculture in 2011, based on the outcome of the Ad Hoc Working Group on Long-Term Cooperative Action (AWGLCA). Bickersteth also noted that "the significance of this move from LCA to SBSTA was that agriculture could be explored in a more politically neutral environment, by focusing on the scientific and technical aspects of the sector in relation to climate change."[51] However, agriculture negotiations continued to be difficult.[52] At least, in June 2014, governments agreed on a work programme focused on adaptation in agriculture with a series of submissions and in-session workshops during UNFCCC sessions in 2015/2016. These addressed experience in relation to, *inter alia*, early warning systems and contingency plans, risk and vulnerability assessments to agricultural systems, adaptation measures, and measures to enhance productivity.[53] At the end of these activities, governments were unable to reach further conclusions on the way forward at COP22, but agreed on joint work at COP23.

However, the reports from the work programme activities, in particular from the in-session workshops, provide useful insights. The reports from the four workshops imply that the right to adequate food is largely absent from the debates. Only civil society organisations referred to it in a way that received coverage in the workshop reports.[54] Even organisations like the FAO or WFP did not explicitly refer to it.[55] In 2015, observers also noted a lack of attention to the specific needs of small-scale farmers and vulnerable populations, which are a core group to consider from a right-to-adequate-food perspective.[56]

National adaptation plans and nationally determined contributions

As a follow-up to the National Adaptation Programmes of Action (NAPA), Parties to the UNFCCC, agreed at COP16 to establish a process for the design and support of National Adaptation Plans (NAPs), with initial guidelines agreed at the subsequent COP in Durban, South Africa.[57] Apart from some references to participatory consultations, these do not reflect aspects of specific importance in the human rights and adaptation context. Similarly, the more detailed guidelines, prepared by the Least Developed Countries Expert Group (LEF) in 2012, do not provide any reference to human rights and only go a little bit further in terms of describing expectations towards how to conduct participatory processes.[58] This is in stark contrast to other (international financial) institutions with more elaborated implementation guidelines (in general, beyond NAPs), as Hirsch and Lottje point out.[59]

Various supplementary guidelines have been prepared by the UN, bilateral development cooperation and other organisations, although not specifically from a human rights perspective.[60]

Although since then various support programmes have been set up to assist developing countries, in particular LDCs, in preparing NAPs, most of the countries still seem to be in the design stage, as only a few NAPs have been formally submitted to the UNFCCC.[61] Thus, there is an obvious gap on the global policy level to link human rights with national adaptation planning.

A recent analysis proposed three criteria for a human rights–based approach to National Adaptation Plans:

> 1 The NAP process must include a human rights audit to ensure that human rights standards (the rights to food, water, etc.) are guaranteed and human rights principles (participation, empowerment, non-discrimination, equality, transparency, accountability) are respected.
>
> 2 A human rights-based understanding of vulnerability requires that not only vulnerable sectors or regions, but also vulnerable population groups are identified in the context of vulnerability and risk analysis.
>
> 3 The inclusion and involvement of the people or its representatives is the third key element that can be derived. True civil society participation must apply to the entire planning process and moreover to ensure that potentially vulnerable groups – i.e. indigenous people, minorities and women – are not excluded.[62]

This analysis, which looked at the German government's support in relation to NAPs, also implies that partially it comes down to the existing support programmes to omit a targeted approach on human rights, even though respective donor governments are officially committed to them. Eventually, the responsibility to address human rights in their NAPs and other national plans lies with the respective developing country governments, but the lack of targeted support also misses opportunities to build their capacity in doing so. Although the NAP process is still in an early stage, the conclusion by the authors that many NAP processes and the applied methodological approaches do not reflect a consistent human rights–based approach[63] seems plausible, given the lack of international policy guidance.

With regard to the Nationally Determined Contributions (NDCs) which countries submitted to the UNFCCC under the Paris Agreement, the process to consider more detailed guidance for these regular contributions is ongoing after Paris. An initial analysis of certain Intended NDCs (INDCs) undertaken prior to COP21 found that 17 countries referred to human rights as a principle guiding the implementation of NDCs, whereas 7 countries referred to human rights when describing the national context.[64] In 2017 and 2018, governments are expected to finalise the negotiations on future guidance for the design of NDCs, including on adaptation communications pursuant to Article 7.10 and 7.11. These could advance the issue of human rights in the context of the scope of the adaptation communication, the preparatory process in-country, technical guidelines, and synthesis reporting by the UNFCCC Secretariat.[65]

Human rights in the loss and damage debate

The basic linkage between human rights and climate change is the same for loss and damage as for adaptation: Climate change impacts already adversely effect, and will do so even more (with the extent depending *inter alia* on the level of global temperature increase and associated climatic changes) key areas which are fundamental to fulfilling human rights. For example, CARE/CIEL argued in 2015 that "such loss and damage to lives, livelihoods, property and culture threaten the human rights of the individuals and communities most vulnerable to, but least responsible for, climate change".[66] Van der Geest and Warner (2015) argued that "Limits to adaptive capacity will affect human rights and well-being across sectors: water and food security, culture and identity, sovereignty, the economy, infrastructure, etc." in relation to empirical research.[67]

For both adaptation and mitigation there is also the need to pay particular attention to those segments of the population which are particularly vulnerable and, due to marginalisation, poverty and other reasons, often face substantial difficulties to claim the fulfilment of their human rights. However, whereas adaptation basically refers to limiting adverse effects through preventive measures, addressing loss and damage entails the fundamental recognition that by far not all impacts will be avoided, even if adaptation was to be stepped up immediately in many risk-prone areas, and therefore responses to the unavoided damage need to be considered. This has also been attested by case studies in various developing countries.[68] Mitigation, adaptation, and loss and damage are therefore connected in one conceptual framing, where the level of mitigation determines the necessary level of adaptation, and the more mitigation and adaptation, the less residual damage.[69] Durand and Huq (2015)[70] and also Vulturius and Davis (2016)[71] provided a more detailed overview of the discussions on defining loss and damage.

However, one aspect which receives particular attention from a human rights perspective is that of remedy to harm caused. OHCHR summarises that "the Universal Declaration of Human Rights, the International Covenant on Civil and Political Rights, and other human rights instruments require States to guarantee effective remedies for human rights violations", and therefore demands "accountability and effective remedy for human rights harms caused by climate change".[72] As a specific form of remedy, compensation for losses and damages has often been at the forefront of the political debate around loss and damage, partially because it was an important element of one of the earliest proposals for an international mechanism on loss and damage put forward in 2008 by AOSIS,[73] but also because of the resistance of basically all developed countries (and probably also certain emerging economies) to engage in a compensation discussion, despite their undisputable high responsibility for the causes of climate change. Compensation continued to feature through 2015 in the run-up to the finalisation of the Paris Agreement, including through demands by Least Developed Countries (LDCs) to negotiate a compensation regime,[74] which could have been a cooperative approach to tackle compensation, and by observer organisations (e.g. CARE/CIEL 2015). The fact that the Paris Agreement's COP decision excludes the use of its Article 8 as a "basis for any liability or compensation" is unlikely to be the end of the compensation debate, but certainly sets limits under the UNFCCC of which strategies could deliver compensation to those who suffer from loss and damage (as part of a comprehensive package of measures). However, the Warsaw International Mechanism (WIM) might still play an important role e.g. for helping to understand and assess the potential scale of loss and damage.[75]

The focus of this section, however, will not be on a definitional debate or the role of compensation, but on considering how human rights have been reflected within the UNFCCC architecture on loss and damage.

In terms of the UNFCCC architecture, the loss and damage debate is still a relatively new one. From the inclusion of the term in the Bali Action Plan in 2007, it took intense subsequent political negotiations to agree a loss and damage work programme (COP17, 2011)[76] which built the ground towards establishing the WIM to address loss and damage associated from climate change impacts at COP 19 (2013). An initial work plan was approved in 2014 and has been under implementation in 2016 and 2016. COP22 concluded in a first review of the WIM and adopted a framework for a more comprehensive five-year work plan of the WIM.[77]

None of the loss and damage COP decisions include an explicit reference to human rights, which is not surprising from a political perspective, given how limited the engagement with human rights concepts and terminologies have been within the UNFCCC overall. In the years preceding the WIM establishment, substantial technical work was undertaken:

- A literature review on a range of approaches to address loss and damage, with the concept of human rights being absent;[78]
- Summarising key outcomes and discussions from four regional workshops held in 2012 as part of the UNFCCC work programme on loss and damage, the official report lacks any mentioning of human rights;[79]
- Technical papers on slow-onset events, with the concept of human rights being absent;[80]
- Technical paper on non-economic losses,[81] which considers the fundamental right to life in the context of climate change–related loss of life (pages 23 and 26); argues that "rights-based ethical frameworks are also distinct from welfare and well-being frameworks" (pages 13 and 16); addresses the loss of sovereignty as a potential violation of the right to self-determination (page 29); and discusses the pros and cons of economic valuations of non-economic losses including with regard to concerns from a human rights perspective (page 41); and
- Technical paper on gaps in existing institutional arrangements,[82] which considered that "at the global level, a range of non-economic assessment areas, such as displacement and human mobility, climate change and human rights and loss of cultural heritage, is starting to emerge" (page 5); referred to the emerging work of various institutions (including OHCHR) in developing "guidance and advocacy in relation to assessing the impacts of climate change on human rights" (page 31).

Overall, this work provided little proof for a stronger, systematic attention to the human rights and loss and damage context.

The WIM has been set up to broadly pursue three core functions: (a) to enhance knowledge and understanding, (b) to strengthen dialogue, coordination and synergies, and (c) to enhance action and support.[83] At least until now, it has been set up to work in a catalytic manner, with an Executive Committee (ExCom) being the main governance body, with some recently established task forces and technical expert groups (on non-economic losses; forced displacement; comprehensive risk management), but without its own dedicated means to undertake operational work or even fund actions to address loss and damage, a situation which also has caused criticism in particular from developing countries, but also civil society.

When the ExCom of the Warsaw International Mechanism started to implement the initial work plan[84] in 2015, this plan did not reflect an explicit focus on human rights based on some of the previous technical work. However, it included a specific Action Area 1 to

> Enhance the understanding of how loss and damage associated with the adverse effects of climate change affect particularly vulnerable developing countries, segments of the population that are already vulnerable owing to geography, socioeconomic status, livelihoods, gender, age, indigenous or minority status or disability, and the ecosystems that they depend on, and of how the implementation of approaches to address loss and damage can benefit them. (page 7)

This lens on the particularly vulnerable has been agreed as a cross-cutting issue for the new five-year work plan through decision 3/CP.22 from COP22. It provides an important focus which can also initiate more human rights–related work across the WIM's work plan. In particular, the task force on forced displacement and the expert group on non-economic losses should pay more attention to human rights. The displacement task force only commenced its work in 2017, based on the Terms of Reference agreed by the ExCom in 2016. In particular, its provision to "Identify legal, policy and institutional challenges, good practices, lessons learned" provides an important starting point from a rights-based perspective[85] (page 2). The draft work plan of the non-economic losses expert group does not reflect a specific approach to human rights.[86]

When the ExCom substantively engages in the implementation of the five-year work plan approved at COP23, it can take this important opportunity to strengthen its approach towards human rights, across the work plan.

Human rights in relevant adaptation funds

Adaptation Fund

The Adaptation Fund (AF) was formally established through Decision 1/CP.13 at the UN climate conference in Bali in 2007. From its outset, it focused on the funding of concrete adaptation projects and programmes, following the demands from developing countries to be provided with more funds to invest into concrete adaptation measures. Initially, its main source was carbon credits obtained from the 2% share of proceeds from certificates provided to Clean Development Mechanism (CDM) projects. However, with constantly low certificate prices in the last years, very few resources could be generated from this source, and it was left to a number of developed countries to provide grant donations to the Adaptation Fund to at least match some of the demands from developing countries for adaptation resources. A particular innovative feature has been its direct access modalities.

As of December 2017, the Adaptation Fund Board was able to allocate a total USD 462 million, mostly on the funding of projects and programmes (about 10% has been spent on operational expenses such as Secretariat costs, AFB meetings etc.).[87] According to the AF website, there are now 73 countries with approved Adaptation Fund initiatives with an expected 5.48 million beneficiaries in developing countries.[88] These projects in particular target sectors/areas like agriculture and food security, disaster risk reduction and coastal zone management.

The Adaptation Fund developed its approach towards environmental and social issues over time, adopting an Environmental and Social Policy in 2013[89] and a gender policy in 2016[90]; which came late as a number of projects were already under implementation. However, its Environmental and Social Policy has been regarded as a "good model" by influential human rights experts.[91]

The Environmental and Social Policy of the Adaptation Fund, in its amended version from March 2016, includes human rights as a specific category. Para 15 states that "Projects/programmes supported by the Fund shall respect and where applicable promote international human rights".[92] It furthermore addresses a number of other key issues more specifically, including gender equality and women's empowerment, prioritisation of marginalised and vulnerable groups etc. These general provisions are spelled out further in a guidance document to be applied by accredited entities to the AF.[93] In addition to providing short background on the Universal Declaration on Human Rights and key HR bodies, this guidance document identifies three possible elements that may be considered when assessing the projects and programmes regarding human rights:

- The IE may provide an overview of the relevant human rights issues, then the host country of the project/programme is cited in any Human Rights Council Special Procedures;
- Human rights issues should be an explicit part of consultations with stakeholders during the identification and/or formulation of the project/programme; and
- Even if the country or countries where the project/programme will be implemented is not a Party to any of the nine core international human rights treaties,[94] compliance with UDHR, at a minimum, will be monitored.[95]

Under the AF, it is mandatory that the implementing entities prepare for each project that has the potential to cause environmental and social risks, an "environmental and social assessment that identifies any environmental or social risks, including any potential risks associated with the Fund's environmental and social principles set forth above".[96]

More recently, the AFB also started to approve specific technical grants for complying with the Environmental and Social Policy and Gender, as well as South-South cooperation grants.[97] The call for proposals under this readiness programme outlined the objectives, which is to "build their capacity in environmental and social risk as well as gender-related risk management by tapping into external expertise through short-term consultancies".[98] In addition, at its 28th meeting in October 2016, the AF Board decided to establish a new Ad Hoc Complaint Handling Mechanism (ACHM), complementing the existing risk management framework.[99] Such a step has also been asked for by experts.[100]

What will be most interesting from a practical perspective is whether the projects funded through the AF will eventually support the fulfilment of human rights, a matter which deserves a closer look once a larger number of AF projects is fully completed.

Least Developed Countries Fund (LDCF)

The Least Developed Countries Fund was established through the Marrakesh Accords in 2001 and since then has been operated by the Global Environment Facility (GEF) upon guidance and regular reviews by the COP. Its focus from the beginning was on assisting Least Developed Countries in adaptation matters, first in relation to the National Adaptation Programmes of Action (NAPA) and related capacity building, and more recently also in the design of longer-term National Adaptation Plans which emerged from the Cancun Agreements in 2010.

According to the GEF as of October 2017, 50 LDCs had accessed a total of $1,176.3 million for 199 projects" in support of NAPA implementation.[101] Of those projects which provided information of the estimated number of beneficiaries, LDCF funding of USD 794 million seeks to reduce the vulnerability of over 20 million people.[102] Though the LDCF has continuously received new pledges, the available resources do not match the requests for funding from recipient countries, with about USD 7 million available in October 2016 compared to additional requests of about USD 300 million.[103] In terms of the sectorial approaches, the largest single share of resources addresses agriculture (about 30%), followed by adaptation in the area of natural resource management, coastal zone management and climate information services.[104]

In the beginning of this decade, the GEF took several steps to strengthen its approach to environmental and social risk assessments.[105] In 2011, the GEF approved the "GEF Policy on Agency Minimum Standards on Environmental and Social Safeguards".[106] In 2013, the GEF approved "GEF Agency Action Plans on Environmental and Social Safeguards & Gender Mainstreaming", and in 2015 the "Agency Minimum Standards on Environmental and Social Safeguards" were updated. A specific Gender Mainstreaming Policy was adopted in 2012.[107] Although these were not specifically designed for the LDCF, they are supposed to be applied to all GEF-funded activities. However, it is noteworthy that there seem to be few explicit analyses of the LDCF in relation to human rights available.

The minimum standards only make explicit reference to human rights in the context of Indigenous Peoples. Undoubtedly, this is an important area in light of the GEF's work and investments in the relation to environmental protection in areas inhabited by Indigenous Peoples. But an agency like the GEF with decades of experience could be expected to have a more encompassing approach to human rights. At least, the minimum requirements stipulate that "Agency conducts Environmental and Social Impact Assessments of proposed projects to help ensure their environmental and social soundness and sustainability", and that there is the duty to involve relevant stakeholders, including affected groups, in project preparation as early as possible, make their views known and take them into account, and continue consultations throughout project implementation.[108]

Therefore, one needs to conclude that the LDCF does not comply with the need to "Include a general commitment to respect all human rights", as it is the case in the AF and as requested by the UN Special Rapporteur on human rights.[109] Of course this does not mean that human rights overall are necessarily entirely absent of what countries and agencies propose to the LDCF for receiving funds, as shows e.g. a project example from Vanuatu. This aims, amongst others, for a result that "Regional, national, local and traditional governance systems are strengthened, respecting and upholding human rights, especially women's rights in line with international standards".[110] However, on a global policy level, the LDCF provisions lag behind e.g. the Adaptation Fund in this regard.

Green Climate Fund

The Green Climate Fund was set up following the Copenhagen Accord and the Cancun Agreements.[111] It is on its way to becoming the largest single fund for financing adaptation action in vulnerable developing countries. After years of preparation and now endowed with about

USD 10 billion of committed resources, it has now become fully operational and has started to approve first funding tranches for a number of project proposals.

It intends to allocate about 50% of its resources to adaptation, and of that at least half to particularly vulnerable developing countries, including LDCs, small island developing states (SIDS) and African countries. Of the USD 2.7 billion which was committed by end of 2017, 31% would go to adaptation, with another 29% labelled as cross-cutting (adaptation and mitigation) by the GCF Secretariat.[112] It is estimated that through these projects over 159 million people would benefit from an increase in climate resilience (and, from total GCF funding, 1 billion tonnes of CO_2 be avoided). The GCF's key intended impacts in relation to adaptation include (a) enhanced livelihoods of the most vulnerable people, communities, and regions, (b) increased health and well-being, and food and water security, (c) resilient infrastructure and built environment to climate change threats, and (d) resilient ecosystems, which also address key human rights areas which are at risk of being harmed by climate change impacts.[113]

In terms of environmental and social safeguards, the GCF Board decided, after intense discussions, to apply the International Finance Corporation (IFC) eight Performance Standards on an interim basis.[114] These address a number of areas which are relevant to key human rights, including environmental and social risks assessments; community health, safety and security; and Indigenous Peoples. In terms of explicit reference to human rights, this is limited to the aspects on Indigenous Peoples and community health. The standards on environmental and social risks also address "Engagement with affected communities or other stakeholders throughout funding proposal cycle. This includes communications and grievance mechanisms."[115]

A civil society submission from 2014 outlined a number of key areas of work for the GCF, including in relation to the GCF's policies on accreditation as well as on environmental, social and gender safeguards.[116] At that point, the experts argued that "the social and environmental safeguards are not sufficiently detailed to ensure a 'do no harm' result, and gender considerations are currently missing or vague".[117]

Step by step, it seems that the GCF is also becoming more active in terms of engaging with the wider expert community when drafting policies relevant in this regard. Most recently, the GCF has been engaging in a process to develop its Environmental and Social Management System (ESMS), "an overarching framework for achieving improvements in environmental and social outcomes while addressing any unintended adverse impacts in all the GCF-financed activities".[118] The consultation draft from December 2016 contained as one guiding principle:

> Human rights. All activities supported by the GCF will be designed and implemented in a manner that will promote universal respect for, and observance of, human rights for all. Actions to support human rights will be consistent with the principles contained in the Universal Declaration of Human Rights. The GCF will require entities to ensure that the supported activities do not cause, promote, contribute to, perpetuate or exacerbate human rights violations.[119]

The consultation draft (October 2017) of the Green Climate Fund's Gender Equality and Social Inclusion Policy and Action Plan 2018-2020 explicitly defines a rights-based approach and highlights a human rights approach as a guiding principle.[120]

In addition, the guiding principles contain more specific aspects related to *inter alia* stakeholder engagement and disclosure, gender-sensitive approach, and environmental and social sustainability. For the first time, this would give human rights a more comprehensive standing in the GCF procedures. Although the final outcome of the process towards ESMS could not be foreseen at the time of finalising this article, an in-depth reaction by a number of civil society organisations reflected concerns that the ESMS might be concluded without the GCF detailing the GCF's own Environmental and Social Standards, as the IPC Performance Standards are regarded "not adequate for long-term use by the GCF". This is also relevant to human rights with regard to them being included in the ESMS as guiding principles, but without fully embedding them at the core of the environmental and social safeguards.[121]

Conclusions

Climate change is happening and has its effects all over the world. Its impacts affect areas which are key to human rights, and adaptation to climate change can reduce and minimise some of these impacts. But there is also a growing recognition that loss and damage from climate change impacts has to be faced increasingly.

Better integrating human rights in climate actions is essential, both from the perspective of climate justice, but also in terms of achieving better results of climate action. In the run-up to Paris and the conclusion of the Paris Agreement, the human rights and climate change context received increasing attention, not the least because of the growing impacts, eventually resulting in the inclusion in the preambular section of the Agreement. Article 7 related to adaptation includes additional elements which are important from a human rights perspective.

The analysis of previous work under the UNFCCC, including in key human rights areas, such as water, health, agriculture etc., overall reveals an absence of the human rights concept. This is true for much of the adaptation-related work under various work streams, as well as in relation to the emerging loss and damage discussion under the Warsaw International Mechanism. Many of these activities were missed opportunities from the perspective of integrating human rights, as on substance the climate change impacts on areas of key concern to the fulfilment of human rights, as well as measures to adapt to those impacts, were considered. Apart from the Adaptation Fund which has to be regarded as the human rights pioneer under the UNFCCC-related climate funds, human rights have received limited attention. In its next steps towards an environmental and social management system, the Green Climate Fund may incorporate some of the experience of the Adaptation Fund.

In terms of going forward – apart from the need to drive emission reduction ambition upwards in order to keep global warming within 1.5°C and to avoid climate disruption that could wipe out all hopes of human rights enjoyment – it will be important to address the human rights concept in the context of future adaptation and loss and damage work more coherently. An important role here lies also with the UNFCCC Secretariat, as for example there is no absence of inputs from OHCHR into the UNFCCC. The case for human rights has been made, but it requires a forward-looking uptake for an effective implementation which serves the needs of vulnerable populations and countries.

Notes

1 See e.g. OHCHR, *Understanding Human Rights and Climate Change* (2015), Submissions of the Office of the United Nations High Commissioner for Human Rights to the 21st Conference of Parties to the United Nations Framework Convention on Climate Change, available online at www.ohchr.org/Documents/Issues/ClimateChange/COP21.pdf.

2 Adger *et al.*, 'Human Security', in Field *et al.* (eds.), *Climate Change 2014: Impacts, Adaptation, and Vulnerability, Part A: Global and Sectoral Aspects, Contribution of Working Group II to the Fifth Assessment Report of the Intergovernmental Panel on Climate Change* (2014) 755, available online at www.ipcc.ch/pdf/assessment-report/ar5/wg2/WGIIAR5-Chap12_FINAL.pdf.

3 *Ibid.*, at 759.

4 UNFCCC, *Paris Agreement* (2015).
5 Schleussner *et al.*, 'Differential Climate Impacts for Policy-Relevant Limits to Global Warming: The Case of 1.5°C and 2°C', 7 *Earth System Dynamics (ESD)* (2016) 327.
6 *Ibid.*
7 *Ibid.*
8 UNFCCC, *Report on the Structured Expert Dialogue on the 2013–2015 Review: Note by the Co-Facilitators of the Structured Expert Dialogue*, UN Doc. FCCC/SB/2015/INF.1 (4 May 2015) 108.
9 OHCHR, *The Effects of Climate Change on the Full Enjoyment of Human Rights* (30 April 2015), available online at http://www4.unfccc.int/Submissions/Lists/OSPSubmissionUpload/202_109_130758775867568762-CVF%20submission%20Annex%201_Human%20Rights.pdf.
10 IPCC, *Sixth Assessment Report (AR6) Products, Outline of the Special Report on 1.5°C Decision*, Decision IPCC/XLIV-4 (2016).
11 IPCC, *Special Report on Climate Change, Desertification, Land Degradation, Sustainable Land Management, Food Security, and Greenhouse Gas Fluxes in Terrestrial Ecosystems*, report on progress, available online at www.ipcc.ch/report/sr2/.
12 IPCC, *Special Report on Climate Change and the Oceans and the Cryosphere*, report on progress, available online at www.ipcc.ch/report/srocc/.
13 OHCHR, *Understanding Human Rights and Climate Change*, *supra* note 1.
14 See *ibid.* Similarly, see also CARE and CIEL, *Climate Change: Tackling the Greatest Human Rights Challenge of Our Time* (9 February 2016), available online at http://careclimatechange.org/publications/human-rights-climate-change/.
15 OHCHR, *Key Messages on Human Rights and Climate Change*, available online at www.ohchr.org/Documents/Issues/ClimateChange/KeyMessages_on_HR_CC.pdf.
16 OHCHR, *Report of the Office of the United Nations High Commissioner for Human Rights on the Relationship Between Climate Change and Human Rights*, UN Doc. A/HRC/10/61 (15 January 2009).
17 CARE/CIEL, *supra* note 14.
18 International Bar Association, *Achieving Justice and Human Rights in an Era of Climate Disruption* (2014), available online at www.ibanet.org/Document/Default.aspx?DocumentUid=0F8CEE12-EE56-4452-BF43-CFCAB196CC04.
19 Mary Robinson Foundation, *Incorporating Human Rights Into Climate Action, Version 2*, May 2016, available online at www.mrfcj.org/wp-content/uploads/2016/05/Incorporating-Human-Rights-into-Climate-Action-Version-2-May-2016.pdf.
20 UNFCCC, *Cancun Agreements*, Decision 1/CP.16, UN Doc. FCCC/CP/2010/7/Add.1 (5 March 2011).
21 *Geneva Pledge for Human Rights in Climate Action* (February 2015), available online at www.forestpeoples.org/sites/fpp/files/news/2015/02/Annex_Geneva%20Pledge.pdf.
22 UNFCCC, *Paris Agreement*, *supra* note 4.
23 Mayer, 'Human Rights in the Paris Agreement', 6 *Climate Law* (2016) 109.
24 CARE/CIEL, *supra* note 14.
25 CARE *et al.*, *Submission to the Ad Hoc Working Group on the Paris Agreement (APA)* (2016), available online at http://climaterights.org/wp-content/uploads/2015/11/CrossCuttingPrinciples-jointAPASubmission-preCOP22-all-1.pdf.
26 P. G. Ferreira, *Did the Paris Agreement Fail to Incorporate Human Rights in Operative Provisions? Not If You Consider the 2016 SDGs*, CIGI Papers No.113 (October 2016), available online at www.cigionline.org/sites/default/files/documents/Paper%20no.113.pdf, at 9.
27 UNFCCC, *Paris Agreement*, *supra* note 4.
28 UNFCCC, ADP Group, *Draft Agreement and Draft Decision on Workstreams 1 and 2 of the Ad Hoc Working Group on the Durban Platform for Enhanced Action* (version of 23 October 2015), available online at http://unfccc.int/files/bodies/application/pdf/ws1and2@2330.pdf.
29 IISD, 'Bonn Highlights', 12(649) *Earth Negotiations Bulletin (ENB)* (22 October 2015) 1.
30 Author's personal observations.
31 A. Magnan and T. Ribera, 'Global Adaptation After Paris: Climate Adaptation and Mitigation Cannot Be Uncoupled Science', 352(6291) *Science* (2016) 1280. See also ActionAid, CARE and WWF, *Global Goal on Adaptation: From Concept to Practice* (10 November 2016), available online at http://careclimatechange.org/publications/global-goal-adaptation-concept-practice/.
32 See e.g. UNFCCC, *Three-Year Workplan of the Adaptation Committee* (2015), available online at http://unfccc.int/files/adaptation/cancun_adaptation_framework/adaptation_committee/application/pdf/work_plan_final.pdf (last visited 31 March 2017); UNFCCC, *Elaboration of Activities Included in the*

Adaptation Committee's Workplan Under Three Workstreams in Order to Enhance Overall Coherence, UN Doc. FCCC/SB/2013/2 (2013).

33 UNFCCC, *Revised Workplan of the Adaptation Committee for 2016–2018* (2016), available online at http://unfccc.int/files/adaptation/cancun_adaptation_framework/adaptation_committee/application/pdf/20160308_wp_revised.pdf.

34 UNFCCC, *Concept Note on the Technical Examination Process on Adaptation*, UN Doc. AC/2016/3 (27 February 2016).

35 UNFCCC, *Five-Year Programme of Work of the Subsidiary Body for Scientific and Technological Advice on Impacts, Vulnerability and Adaptation to Climate Change*, Decision 2/CP.11, UN Doc. FCCC/CP/2005/5/Add.1 (30 March 2006).

36 UNFCCC, *Chronology – Nairobi Work Programme*, available online at http://unfccc.int/adaptation/workstreams/nairobi_work_programme/items/3916.php.

37 UNFCCC, *Report on the Meeting on Available Tools for the Use of Indigenous and Traditional Knowledge and Practices for Adaptation, Needs of Local and Indigenous Communities and the Application of Gender-Sensitive Approaches and Tools for Adaptation*, Note by the Secretariat, UN Doc. FCCC/SBSTA/2014/INF.11* (8 May 2014).

38 UNFCCC, *Water and Climate Change Impacts and Adaptation Strategies*, Technical paper, Un Doc. FCCC/TP/2011/5 (22 November 2011).

39 UNFCCC, *Report on the Technical Workshop on Water and Climate Change Impacts and Adaptation Strategies*, Note by the Secretariat, UN Doc. FCCC/SBSTA/2012/4 (13 September 2012) 2.

40 UNFCCC, *Water Resources, Climate Change Impacts and Adaptation Planning Processes: Overview, Good Practices and Lessons Learned* (2016), available online at http://unfccc.int/files/adaptation/application/pdf/3_synopsis_water.pdf.

41 *Ibid.*

42 OHCHR, Response to UNFCCC Secretariat request for submissions on: Nairobi Work Programme on impacts, vulnerability and adaptation to climate change – health impacts, including occupational health, safety and social protection (FCCC/SBSTA/2016/2, para 15(a)(i)). The response is available at www.ohchr.org/EN/Issues/HRAndClimateChange/Pages/UNFCCC.aspx.

43 UNFCCC, *Human Health and Adaptation: Understanding Climate Impacts on Health and Opportunities for Action, Synthesis Paper by the Secretariat*, UN Doc. FCCC/SBSTA/2017/2 (3 March 2017).

44 UNFCCC, *Human Health and Adaptation Planning Processes: Overview, Good Practices and Lessons Learned* (2016), available online at http://unfccc.int/files/adaptation/application/pdf/7_synopsis_health.pdf.

45 UNEP, *Human Rights and the Environment*, available online at http://web.unep.org/divisions/delc/human-rights-and-environment.

46 Human Rights Council (HRC), *Report of the Special Rapporteur on the Issue of Human Rights Obligations Relating to the Enjoyment of a Safe, Clean, Healthy and Sustainable Environment, Note by the Secretariat*, UN Doc. A/HRC/31/52 (1 February 2016).

47 UNFCCC, *Ecosystem-Based Approaches to Adaptation: Compilation of Information, Note by the Secretariat*, UN Doc. FCCC/SBSTA/2011/INF.8 (16 December 2011).

48 UNFCCC, *Report on the Technical Workshop on Ecosystem-Based Approaches for Adaptation to Climate Change, Note by the Secretariat*, UN Doc. FCCC/SBSTA/2013/2 (10 May 2013).

49 See e.g. Columbia Law School – Human Rights Institute, 'Climate Change and the Right to Food: A Comprehensive Study', in Heinrich Böll Foundation (eds.), 8 *Publication Series on Ecology* (2009); C. Bals, M. Windfuhr and S. Harmeling, *Climate Change, Food Security and the Right to Adequate Food* (2008), available online at https://germanwatch.org/de/download/2798.pdf.

50 UN General Assembly, *Interim Report of the Special Rapporteur on the Right to Food*, UN Doc. A/70/287 (5 August 2015).

51 CDKN Global, S. Bickersteth, *OPINION: The Current Climate of Agriculture in the UNFCCC* (19 April 2013), available online at https://cdkn.org/2013/04/the-current-climate-of-agriculture-in-the-unfccc/?loclang=en_gb.

52 See e.g. M. Kaplan, German Development Institute (DIE), *Agriculture in the International Climate Negotiations: Supporting Sustainable Development or Just Dubious Emission Reductions?* Briefing Paper 16/2012 (2012), available online at www.die-gdi.de/uploads/media/BP_16.2012.pdf.

53 UNFCCC, *Report of the Subsidiary Body for Scientific and Technological Advice on Its Fortieth Session, Held in Bonn From 4 to 15 June 2014*, UN Doc. FCCC/SBST A/2014/2 (18 July 2014).

54 See UNFCCC, *Workshop on the Identification and Assessment of Agricultural Practices and Technologies to Enhance Productivity in a Sustainable Manner, Food Security and Resilience, Considering the Differences in Agroecological Zones and Farming Systems, Such as Different Grassland and Cropland Practices and Systems*, Report by the Secretariat, UN Doc. FCCC/SBSTA/2016/INF.6 (30 August 2016); UNFCCC, *Workshop on the Identification of Adaptation Measures, Taking Into Account the Diversity of the Agricultural Systems, Indigenous Knowledge Systems and the Differences in Scale as Well as Possible Co-Benefits and Sharing Experiences in Research and Development and on-the-Ground Activities, Including Socioeconomic, Environmental and Gender Aspects*, Report by the secretariat, UN Doc. FCCC/SBSTA/2016/INF.5 (30 August 2016).

55 See e.g. A. Maybeck, FAO, *Assessment of Risk and Vulnerability of Agricultural Systems*, presentation at SBSTA42 in-session workshop on agriculture (3 June 2015), available online at http://unfccc.int/files/land_use_and_climate_change/agriculture/application/pdf/fao_vulnerabilityrisksbonn2015_v4.pdf (last visited 31 March 2017). See also World Food Programme, *Experiences in Supporting Early Warning and Contingency Planning* (3 June 2015), available online at http://unfccc.int/files/land_use_and_climate_change/agriculture/application/pdf/1_wfp_unfccc_bonn_ew_cp_workshop_presentation.pdf.

56 A. Ceinos and T. Rawe, Care International, *Small-Scale Food Producers Are Hungry for Action on Climate Change* (9 June 2015), available online at http://news.trust.org/item/20150609111618-si8t4.

57 UNFCCC, *Initial Guidelines for the Formulation of National Adaptation Plans by Least Developed Country Parties*, Annex, UN Doc. FCCC/CP/2011/9/Add.1 (11 December 2011).

58 LDC Expert Group, UNFCCC, *National Adaptation Plan: Technical Guidelines for the National Adaptation Plan Process* (December 2012), available online at http://unfccc.int/files/adaptation/cancun_adaptation_framework/application/pdf/naptechguidelines_eng_high__res.pdf.

59 T. Hirsch und C. Lottje, *Klimapolitik im Faktencheck: Armutsorientierung und Bürgerbeteiligung in der Nationalen Anpassungsplanung (NAP)* (2016), available online at www.deutscheklimafinanzierung.de/wp-content/uploads/2016/11/Analyse-62-Klimapolitik-im-Faktencheck.pdf.

60 NAP Central, UNFCCC, *Supplementary Materials to the NAP Technical Guidelines*, available online at http://www4.unfccc.int/nap/Guidelines/Pages/Supplements.aspx.

61 NAP Central, UNFCCC, *National Adaptation Plans*, available online at http://www4.unfccc.int/nap/Pages/national-adaptation-plans.aspx.

62 C. Lottje, T. Hirsch and S. Minninger, *Poverty Orientation and Civil Society Participation in National Adaptation Plans (NAP)* (8 February 2017), available online at www.germanclimatefinance.de/2017/02/08/poverty-orientation-civil-society-participation-national-adaptation-plans-nap/.

63 Hirsch and Lottje, *Klimapolitik im Faktencheck*, *supra* 59, at 24.

64 Human Rights and Climate Change Working Group, *Integrating Human Rights Into the Paris Commitments – (I)NDCs*, available online at http://climaterights.org/our-work/unfccc/human-rights-in-indcs/.

65 CARE *et al.*, *Submission to the Ad Hoc Working Group*, *supra* note 25.

66 CARE/CIEL, *supra* note 14.

67 K. Van der Geest and K. Warner, 'Editorial: Loss and Damage From Climate Change: Emerging Perspectives', 8(2) *International Journal of Global Warming (IJGW)* (2015) 133.

68 K. Warner, K. Van der Geest and S. Kreft, 'Pushed to the limit: Evidence of Climate Change-Related Loss and Damage When People Face Constraints and Limits to Adaptation', 11 *United Nations University – Institute for Environment and Human Security (UNU-EHS)* (2013).

69 Loss and Damage in Vulnerable Countries Initiative, *Framing the Loss and Damage Debate: A Conversation Starter by the Loss and Damage in Vulnerable Countries Initiative*, advance version (August 2012), available online at http://loss-and-damage.net/download/6530.pdf, at 4.

70 Durand and Huq, *Defining Loss and Damage: Key Challenges and Considerations for Developing an Operational Definition* (2015), available online at www.icccad.net/wp-content/uploads/2015/08/Defininglossanddamage-Final.pdf.

71 Vulturius and Davis, *Defining Loss and Damage: The Science and Politics Around One of the Most Contested Issues Within the UNFCCC* (2016), Discussion Brief, available online at www.sei-international.org/mediamanager/documents/Publications/Climate/SEI-DB-2016-Loss-and-damage-4-traits.pdf.

72 OHCHR, *Key Messages on Human Rights and Climate Change*, *supra* note 15, at 2.

73 Alliance of Small Island States (AOSIS), *Multi-Window Mechanism to Address Loss and Damage From Climate Change Impacts* (2008), available online at http://unfccc.int/files/kyoto_protocol/application/pdf/aosisinsurance061208.pdf.

74 LDC, *Submission by Nepal on Behalf of the Least Developed Countries Group on the ADP Co-Chairs' Non Paper of 7 July 2014 on Parties Views and Proposal on the Elements for a Draft Negotiating Text*

(21 October 2014), available online at www4.unfccc.int/submissions/Lists/OSPSubmissionUpload/39_99_130584499817551043-Submission by Nepal ADP_21 Oct 2014.pdf.

75 M. J. Mace and R. Verheyen, 'Loss, Damage and Responsibility After COP21: All Options Open for the Paris Agreement', 25(2) *Review of European, Comparative & International Environmental Law (RECIEL)* (2016) 197.

76 UNFCCC, *Report of the Conference of the Parties on Its Seventeenth Session, Held in Durban From 28 November to11 December 2011, Addendum, Part Two: Action Taken by the Conference of the Parties at Its Seventeenth Session*, UN Doc. FCCC/CP/2011/9/Add.2 (15 March 2012).

77 UNFCCC, *Decisions 3/CP.22 and 4/CP.22, Marrakech Climate Change Conference* (November 2016), available online at http://unfccc.int/meetings/marrakech_nov_2016/meeting/9567/php/view/decisions.php#c.

78 UNFCCC, *A Literature Review on the Topics in the Context of Thematic Area 2 of the Work Programme on Loss and Damage: A Range of Approaches to Address Loss and Damage Associated With the Adverse Effects of Climate Change*, Note by the secretariat, UN Doc. FCCC/SBI/2012/INF.14 (15 November 2012).

79 UNFCCC, *Report on the Regional Expert Meetings on a Range of Approaches to Address Loss and Damage Associated With the Adverse Effects of Climate Change, Including Impacts Related to Extreme Weather Events and Slow Onset Events*, Note by the secretariat, UN Doc. FCCC/SBI/2012/29 (19 November 2012).

80 UNFCCC, *Slow Onset Events*, Technical paper, UN Doc. FCCC/TP/2012/7 (26 November 2012).

81 UNFCCC, *Non-Economic Losses in the Context of the Work Programme on Loss and Damage*, Technical paper, UN Doc. FCCC/TP/2013/2 (9 October 2013).

82 UNFCCC, *Gaps in Existing Institutional Arrangements Within and Outside of the Convention to Address Loss and Damage, Including Those Related to Slow Onset Events*, Technical paper, UN Doc. FCCC/TP/2013/12 (4 November 2013).

83 UNFCCC, *Warsaw International Mechanism for Loss and Damage Associated With Climate Change Impacts*, available online at http://unfccc.int/adaptation/workstreams/loss_and_damage/items/8134.php (last visited 31 March 2017).

84 UNFCCC, *Report of the Executive Committee of the Warsaw International Mechanism for Loss and Damage Associated With Climate Change Impacts*, UN Doc. FCCC/SB/2014/4 (24 October 2014).

85 UNFCCC, *Terms of References of the Task Force on Displacement* (2016), available online at http://unfccc.int/files/adaptation/groups_committees/loss_and_damage_executive_committee/application/pdf/tor_task_force.pdf.

86 UNFCCC, *Summary of Proceedings of the First Meeting of the Expert Group on Non-Economic Losses* (2016), available online at http://unfccc.int/files/adaptation/groups_committees/loss_and_damage_executive_committee/application/pdf/summary_of_proceedings_nels_eg_2016_09_16_2000.pdf.

87 Ethics and Finance Committee of Adaptation Fund, *Adaptation Fund Trust Fund: Financial Report Prepared by the Trustee (as at 30 June 2016)*, Doc. No. AFB/EFC.19/11 (4–7 October 2016).

88 Cumulative project information contained at the following website: www.adaptation-fund.org.

89 Adaptation Fund Board, *Environmental and Social Policy* (November 2013), available online at www.adaptation-fund.org/wp-content/uploads/2015/09/Environmental-Social-Policy-approved-Nov2013.pdf.

90 Adaptation Fund Board, *Gender Policies and Action Plan* (March 2016), available online at www.adaptation-fund.org/wp-content/uploads/2016/04/OPG-ANNEX4_Gender-Policies-and-Action-Plan_approved-in-March-2016.pdf.

91 J. Knox, *Human Rights and Safeguards in the New Climate Mechanism Established in Article 6, Paragraph 4 of the Paris Agreement* (3 May 2016), available online at www.ohchr.org/Documents/Issues/Environment/Letter_to_SBSTA_UNFCCC_May2016.pdf.

92 Adaptation Fund Board, *Environmental and Social Policy*, *supra* note 89, at 6.

93 Adaptation Fund, *Guidance Document for Implementing Entities on Compliance With the Adaptation Fund Environmental and Social Policy* (June 2016), available online at www.adaptation-fund.org/wp-content/uploads/2016/07/ESP-Guidance_Revised-in-June-2016_Guidance-document-for-Implementing-Entities-on-compliance-with-the-Adaptation-Fund-Environmental-and-Social-Policy.pdf.

94 OHCHR, *The Core International Human Rights Instruments and Their Monitoring Bodies*, available online at www.ohchr.org/EN/ProfessionalInterest/Pages/CoreInstruments.aspx (last visited 31 March 2017).

95 Adaptation Fund, *Guidance Document*, *supra* note 93, at 8–9.

96 Adaptation Fund Board, *Environmental and Social Policy*, *supra* note 89, at para. 30.

97 Adaptation Fund, *Projects Table View*, available online at www.adaptation-fund.org/projects-programmes/project-information/projects-table-view/.

98 Adaptation Fund, *Call for Technical Assistance Grand – ESP and Gender* (2016), available online at www.adaptation-fund.org/apply-funding/grants/call-technical-assistance-grant-proposals-esp-gender/.
99 Adaptation Fund Board, *Report of the 28th meeting of the Adaptation Fund Board*, Doc. No. AFB/B.28/9 (21 December 2016), available online at www.adaptation-fund.org/wp-content/uploads/2016/12/AFB-B-28-report_final_approved-1.pdf.
100 Carbon Market Watch, *Social and Environmental Accountability of Climate Finance Instruments*, policy brief (September 2015), available online at http://carbonmarketwatch.org/wp-content/uploads/2015/09/SOCIAL-AND-ENVIRONMENTAL-08-web-.pdf#page=12.
101 GEF, Progress Report on LDCF and SCCF , Doc. No. GEF/LDCF.SCCF.23/03 (13 November 2017), available online at https://www.thegef.org/sites/default/files/council-meeting-documents/EN_GEF.LDCF_.SCCF_.23.03.Rev_.01_Progress_Report_LDCF_SCCF.pdf.
102 *Ibid.*
103 *Ibid.*
104 *Ibid.*, at 13.
105 For an overview, see also A. Johl and Y. Lador, *A Human Rights-Based Approach to Climate Finance* (February 2012), available online at http://library.fes.de/pdf-files/iez/global/08933.pdf.
106 GEF, *Agency Minimum Standards on Environmental and Social Safeguards*, Doc. No. SD/PL/03 (draft of April 3, 2011), last updated on 19 February 2015, available online at www.thegef.org/sites/default/files/documents/Policy_Environmental_and_Social_Safeguards_2015.pdf.
107 GEF, *Policy on Gender Mainstreaming*, Doc. No. SD/PL/02 (1 May 2012), available online at www.thegef.org/sites/default/files/documents/Gender_Mainstreaming_Policy-2012_0.pdf (last visited 31 March 2017).
108 GEF, Agency Minimum, *supra* note 106.
109 Knox, *Human Rights*, *supra* note 91.
110 UNDP and GEF, *Adaptation to Climate Change in the Coastal Zone in Vanuatu*, Project Document (2014), available online at https://info.undp.org/docs/pdc/Documents/VUT/Vanuatu%20LDCF%20Prodoc%20Final.docx.
111 See Climate Focus, *Green Climate Fund and the Paris Agreement* (February 2016), www.climatefocus.com/sites/default/files/GCF%20and%20Paris%20Brief%202016.new_.pdf, at 1.
112 Green Climate Fund (GCF), *Projects*, available online at www.greenclimate.fund/projects/portfolio.
113 GCF, *Funding*, available online at www.greenclimate.fund/funding/proposal-approval.
114 GCF, *Annex III: Interim Environmental and Social Safeguards of the Fund*, Doc. No. GCF/B.07/11 (2014), available online at www.greenclimate.fund/documents/20182/114264/1.7_-_Environmental_and_Social_Safeguards.pdf/e4419923-4c2d-450c-a714-0d4ad3cc77e6.
115 *Ibid.*, at 36 (page 1 of document).
116 CIEL, *Civil Society Submission to the Green Climate Fund on Accreditation, Safeguards and Fiduciary Standards* (17 March 2014), available online at http://ciel.org/Publications/Safeguards_CSOs_17Mar2014.pdf.
117 *Ibid.*, at 2.
118 GCF, *Call for Public Inputs: Green Climate Fund Environmental and Social Management System (ESMS)* (15 December 2016), available online at www.greenclimate.fund/documents/20182/24913/DCP_15-12-2016_-_GCF_Environmental_and_Social_Management_System.pdf/3b05eb86-4636-4179-aa1d-ebfcec34037e.
119 *Ibid.*, at p. 12.
120 GCF, Call for Public Inputs: Draft of Green Climate Fund's Updated Gender Equality and Social Inclusion Policy and Action Plan. https://www.greenclimate.fund/documents/20182/24913/DCP_27-10-2017_-_Draft_Gender_Equality_and_Social_Inclusion_Policy.pdf/4a6c90dc-e263-4b94-adf7-aa3a88c03eb2
121 CSO submission from 24 February 2017 to the ESMS Call for Public Inputs [on file with the author].

8

Human rights and climate displacement and migration

Alice Thomas

Introduction and overview

As early as 1990, the Intergovernmental Panel on Climate Change (IPCC) concluded that one of the most significant impacts of climate change would be on human mobility.[1] Climate change is anticipated to both trigger forced displacement and contribute to voluntary migration both within countries and across international borders, in some cases permanently. While at present there is insufficient data regarding the full extent to which climate change is contributing to human mobility, there is ample evidence that events directly linked to climate change are already acting to influence population movements. For example, climate change–related effects on the Arctic – which is warming at twice the rate of the rest of the planet – are already forcing Alaskan native communities to retreat inland.[2] While the future magnitude of displacement and migration owing to climate change will depend on both the extent to which governments successfully act to curb greenhouse gas emissions and the ability of vulnerable communities to successfully adapt, it is likely that in the absence of ambitious action, hundreds of millions of people could be uprooted due to climate change in the coming decades. The threat that this presents to the enjoyment of human rights triggers an obligation of states to take action to protect at-risk populations:

> States have obligations at the national level to take adaptation actions to protect their vulnerable populations from the effects of climate change, and at the international level to cooperate in order to facilitate the protection of vulnerable communities wherever they are located.[3]

The threat that climate change presents to human rights is amplified for displaced persons, as well as for migrants who resort to irregular migration as strategy to cope with climate-related disasters or stress. Displaced persons and undocumented migrants are highly vulnerable to exploitation, abuse and other human rights violations, including lack of access to adequate shelter, food, safe drinking water, healthcare and education.

Yet, at present, existing national and international laws and policies for protecting the human rights of refugees, internally displaced persons (IDPs) and migrants are insufficient and ill-suited to protect those uprooted by climate change–related effects. Moreover, to the extent that these frameworks are designed largely to protect the human rights of people who are *already* on the move, they do not sufficiently emphasize the obligations of governments to proactively prevent or minimize climate displacement risk and associated threats to human rights.

Given the shortcomings and gaps in the existing national and international framework for protecting the human rights of those facing climate change–related displacement and migration, this chapter discusses several recent initiatives that provide important opportunities to design more effective, practical and politically-viable strategies. As discussed below, rather than aspiring to a new international agreement or convention, a more effective approach will entail the adoption of voluntary regional and bilateral arrangements, as well as national laws and policies to more effectively prevent, minimize and address the threat to human rights arising from climate change–related human mobility.

Evidence that climate change is already dislocating vulnerable communities, and the fact that the future effects of climate change will undoubtedly have a far greater impact on displacement and migration, clearly implicate the human rights obligations of states to proactively address this issue. The scale of future displacement will depend on the extent to which governments act now to substantially reduce carbon emissions, build the resilience of vulnerable communities, mitigate climate risk and adapt to climate change effects.

Climate change impacts on displacement and migration: human rights implications

The links between climate change and human mobility

The most direct pathway between climate change and human mobility is forced displacement resulting from the increasing incidence and changing intensity of extreme weather events associated with climate change.[4] As such, disasters resulting from extreme weather are often equated with displacement from climate change, despite the fact that it is not always possible to link an individual weather event to climate change (see Challenges to Addressing Climate-Change Related, Cross-Border Movements). Climate-related disasters such as floods and tropical cyclones not only threaten the rights to life and health, but also trigger displacement by destroying homes, assets and livelihoods. Between 2008 and 2015, 203.4 million people – or an average of 25.4 million people per year – were displaced by rapid-onset disasters brought on by natural hazards, primary weather-related hazards such as floods, hurricanes and tropical cyclones.[5] (Notably, the number of people displaced by disasters in 2015 was twice the number of people displaced by violence.)[6] Not included in this figure were those uprooted by slow-onset, weather-related hazards such as droughts, as there are no global estimates available for drought-related displacement and migration.[7]

Other more gradual processes of environmental degradation associated with climate change, such as increased temperatures, more erratic rainfall patterns, decreased water availability and more frequent or intense drought, threaten to undermine agricultural livelihoods and food and water security, thereby prompting the movement of people. Poor and underdeveloped regions of the globe are particularly at risk, such as West Africa's Sahel region, where 80 percent of the population is reliant upon natural resource-dependent livelihoods to survive.[8]

Melting of glaciers brought on by climate change will also impact human mobility. For example, the melting of Himalayan glaciers threatens the drinking water supplies of hundreds of millions of people in Asia. The melting of the Greenland ice sheet and the Antarctic ice shelf are particularly concerning due to the magnitude of their potential impact on sea level rise.[9]

Accelerated coastal erosion, saltwater inundation of freshwater sources and sea level rise contribute to permanent changes to land and ecosystems upon which human settlements depend, thereby rendering some areas uninhabitable. Most at risk are populations living in river deltas, the inhabitants of low-lying coastal areas and island atolls, and Arctic communities. In such areas, climate change adverse effects are more likely to lead to permanent displacement.

The potential of climate change to permanently displace the inhabitants of certain Small Island Developing States (SIDS) has been long recognized.[10] Island nations like the Maldives, Tuvalu and Kiribati, which lie between one and three meters (three to nine feet) above sea level, have repeatedly called upon the United Nations (UN) member states to act immediately to address the threat that climate change presents to their human rights (and their very existence) by mitigating greenhouse gas emissions.[11] Leaders of the Alliance of Small Island States (AOSIS), a coalition of over 40 small-island and low-lying coastal countries, have issued numerous declarations expressing their grave concern regarding scientific evidence that the effects of human-induced climate change are happening faster than anticipated and are already being felt and will further intensify. The declarations urge UN member states "to consider and address the security implications of climate change including violation of territorial integrity, more frequent and severe climate-related disasters, threats to water and food security, increased natural resource scarcity, and forced displacement."[12]

In the Arctic, which is warming at twice the rate of the rest of the planet,[13] the dramatic changes to the climate system are already forcing vulnerable communities to permanently abandon their villages. In the United States, numerous Native Alaskan communities already are being forced to retreat inland due to the rapid rise in temperatures along with permafrost melt, accelerated coastal erosion, increased storm surge and loss of sea ice.[14]

Finally, it is important to note that one of the challenges to building political will among countries to develop an international protection framework for those forced from their countries due to climate change–related effects is a lack of understanding of the scale of the problem. Unlike estimates of internal, disaster-related displacement, at present there is no comprehensive and systematic data collection and analysis on cross-border movements associated with climate change. Going forward, there is an urgent need to develop methods for collecting such data.[15] Compounding the problem is the complexity around accurately predicting climate change–related future effects, such as the rate of sea level rise, resulting in a frustrating lack of urgency despite more recent scientific evidence that climate change is happening faster than anticipated.

Climate change, armed conflict and displacement

Experts have also pointed to the propensity of climate change–related effects, such as decreased water availability, arable land or grazing grounds, to contribute to social unrest, violence or armed conflict, thereby triggering displacement. While there is limited evidence that changes in climate directly trigger armed conflict, studies show that climate change acts as a "threat multiplier" by compounding underlying socio-economic, ethnic or political tensions.[16] For example, in pastoral societies in Africa and other resource-dependent economies, changes in rainfall can enhance the risk of localized violent conflict, especially in the absence of institutions for managing conflict.[17] In fact, many of the factors that are linked with a higher risk of violent conflict (e.g., slow rates of economic growth, low per capita incomes and economic shocks) are themselves sensitive to climate change and variability.[18]

Conversely, where conflict is resource-based, sound resource management has the potential to contribute to peacebuilding by channeling competing interests over resources into non-violent resolutions. There is evidence, for example, that trans-boundary water cooperation can provide a basis for inter-state cooperation on a variety of contentious matters and that basin-wide institutional development focused on sharing resources can lower conflict potential.[19]

Moreover, ongoing or recent armed conflict enhances vulnerability and undermines the adaptive capacity of governments and populations by weakening government institutions, exacerbating poverty and constraining livelihoods.[20] The most recent IPCC report concludes that

> [V]iolent conflict undermines human security and the capacity of individuals, communities and states to cope with changes. These observations suggest, with high confidence, that where violent conflict emerges and persists the capacity to adapt to climate change is reduced for affected populations.[21]

Given the links between state fragility, climate change vulnerability and conflict, the IPCC report identifies the effect of climate change on conflict and insecurity as one of several "emergent risk areas."[22]

Finally, it has been noted that climate change mitigation and adaption activities can also increase the risk of armed conflict and exacerbate the vulnerabilities of certain groups. This is especially so where climate change mitigation or adaption measures alter distribution of, or access to, natural resources.[23] For example, rapid expansion of biofuels production has been linked to land grabbing, land disposition and social conflict.[24] This suggests the need for more focus on building the adaptive capacity of fragile and conflict-ridden states, as risks to human rights abuses are perhaps highest in this context.

Climate displacement, vulnerability and human rights

Many of the factors that render certain geographic regions and groups more vulnerable to climate change likewise exacerbate their vulnerability to human rights threats associated with displacement and migration. For example, climate change impacts on permanent displacement in low-lying island nations and in the Arctic (discussed above) are demonstrative of the acute climate vulnerability of indigenous groups, whose way of life is inextricably tied to their land and/or natural resources.

Not only are certain groups more vulnerable to being uprooted by climate change–related effects, but they are also more vulnerable to protracted displacement or to risks associated with irregular migration once on the move. This is evident in recent case studies indicating that causes of social vulnerability such as poverty, gender, age or discrimination can contribute to prolonged, recurrent or permanent displacement, despite the common assumption that most people displaced by disasters return home as soon as possible once the hazard has passed. During Hurricane Katrina in the United States, for example, many poor minorities whose homes were destroyed by the storm never returned.[25] In the wake of Typhoon Haiyan in the Philippines, poor, landless households living in informal settlements along the coast were at greater risk of eviction and protracted displacement.[26] Moreover, in the aftermath of disasters, the specific protection and assistance needs of women and children, the elderly, persons with disabilities, the poor and those belonging to socially marginalized groups (e.g., migrants, Indigenous Peoples, minorities) are often not sufficiently addressed. In such crises, pre-existing vulnerabilities and patterns of discrimination are often enhanced, leaving already marginalized and vulnerable groups at heighted risk of human rights abuses.[27]

Regardless of the cause, displaced people face an increased risk of human rights violations stemming from their lack of access to adequate shelter, food, health services, jobs and social safety nets, as well as from other adverse material, social and psychological effects commonly associated with displacement.[28] Although moving to a safer location (e.g., via planned or spontaneous evacuation) provides temporary protection from immediate risk of harm, prolonged displacement exacerbates existing vulnerabilities and generally leads to a number of serious protection, humanitarian and human rights challenges.[29] As the Special Rapporteur on the Human Rights of IDPs has recognized,

> As levels of displacement rise in the context of climate change, the urgency of finding long-lasting solutions for affected populations and avoiding the precariousness, marginalization and instability associated with situations of protracted displacement, will become a national, and potentially regional, security imperative.[30]

At the same time, it is important to recognize that when well planned and voluntary, migration can serve as an effective adaptation strategy for populations facing climate change–related disasters or stress. Migration to urban areas, or to pursue economic opportunities abroad, by one or more family members not only relieves the pressure on the household by reducing the burden on limited resources, but also provides an important source of additional income to those left behind via remittances. However, given the current lack of available safe and legal pathways for international labor migration, migrants from climate-vulnerable countries may be forced to resort to irregular migration, leaving them at a heightened risk of trafficking, abuse and other human rights violations.

Gaps in the current international and national frameworks for addressing climate change impacts on human mobility

Despite the well-recognized threat to the full enjoyment of human rights that climate change–related displacement and migration poses to vulnerable people around the globe, at present there is no comprehensive international legal framework for protecting people who are forcibly displaced or who migrate to another country in the context of climate change. As discussed below, those uprooted due to disasters, environmental factors or climate change effects do not fit neatly within existing national and international laws and policies for protecting traditional refugees, migrants and IDPs. While there have been some efforts to extend refugee-like protections to climate-displaced persons on a regional level, distinctions regarding both the cause and voluntariness of the movement, which are critical under refugee law and trigger a duty to protect, are less relevant, if not impossible to ascertain, in the case of people who move in the context of climate change.

Challenges to addressing climate change–related, cross-border movements

Limitation on the applicability of refugee law

Despite the fact that people who flee their countries due to disasters and other climate change–related effects may be in "refugee-like" situations, refugee law, which prohibits the forced return of a person to a country of persecution, provides limited protection in such situations.[31] The 1951 UN Convention Relating to the Status of Refugees and its protocol (Refugee Convention) protects people who are forced to flee their countries owing to a "well-founded fear of being persecuted for reasons of race, religion, nationality, membership of a particular social group or political opinion."[32] As such, it does not extend to people fleeing their countries due to climate change–related effects. A decision by the High Court of New Zealand rejecting a Kiribati man's request for asylum as a "climate change refugee" is illustrative of the limitations of refugee law in protecting those forced to flee abroad due to climate change.[33] In that case, Ioane Teitiota argued he should be entitled to protection as a refugee because rising sea levels and environmental hazards caused by climate change were endangering his life on Kiribati, a low-lying island nation in the South Pacific.[34] The Court concluded that Teitiota's claim fell short of the Refugee Convention legal criteria because he was unable to show that by returning to Kiribati, he would suffer "a sustained and systematic violation of his basic human rights such as right to life . . . or the right to adequate food, clothing and housing."[35]

Several regional instruments in Africa and Latin America extend the Refugee Convention to those fleeing events "seriously disturbing public order" or other forms of "generalized violence."[36] While theoretically these regional conventions could be applied to protect those forced to flee large-scale climate change impacts that arguably disturb public order, their application in such contexts is largely untested and likely to serve only as a partial response to the challenges of climate change–induced displacement and migration.[37] Given that many of the signatory countries to the Organization for African Unity Convention already host large numbers of refugees, there may be limited appetite to expand this protection more broadly to displacement caused solely by climate change, as opposed to mixed movements where both conflict and disasters act as a dynamic push factor (such as was the case in the 2011 famine in Somalia).

A further challenge to applying traditional laws and policies related to refugees and asylum seekers to those who move across a border in the context of climate change is determining whether the movement is voluntary or forced. The question of voluntariness is important given the legal distinction that has developed around voluntary economic migrants on the one hand (who are understood to have willingly migrated to another country for reasons of economic opportunity and whose migration is governed by rules relating to territorial sovereignty), and asylum seekers and refugees on the other, who are forced to cross an international border due to persecution and therefore recognized as in need of international protection.[38] In the case of those who migrate due to more gradual climatic changes, the anticipatory nature of the movement and the perceived element of choice in their decision to move means that from a legal and political standpoint, they are likely to be equated with economic migrants, even if migration is the option of last resort (which is likely to be the case for inhabitants of SIDS).

There are a significant number of instances where temporary humanitarian protections have been extended on a bilateral basis to receive persons from, or to avoid their deportation to, countries experiencing natural disasters and other humanitarian emergencies. However, an analysis of the law, relevant institutions and operational responses pertinent to the protection and assistance of cross-border disaster-displaced persons reveals a general lack of preparedness leading to ad hoc responses in most cases.[39]

Defining a protection framework for those uprooted by climate change–related effects will also require identifying who is moving due to climate change as opposed to other factors. Yet given myriad external and internal factors that ultimately cause a person to move, establishing climate change as a sole or primary driver of displacement or migration is problematic. The first challenge is attributing the event – e.g., a storm, flood or drought – to climate change as opposed to a naturally occurring or human-made event. Scientific understanding of whether the intensity or likelihood of extreme weather events can be attributed to anthropogenic climate change is still evolving. Moreover, despite recent improvements to "event attribution science,"[40] in most cases available evidence linking specific events, hazards or changes that trigger human mobility to anthropogenic climate change will be scant. And in the case of weather events (as opposed to temperature rise), there are many other natural and human-made factors that also play a role.

Finally, human-made factors often play a role and contribute to the extent of displacement resulting from climate change–related hazards. For example, the historic 2010 floods in Pakistan, which displaced nine million people, were reportedly made worse by rapid deforestation in high mountain areas and the uncontrolled expansion of informal settlements into river basins.[41] (However, it is important to note that the contribution of human factors that act to increase the risk of displacement from climate-related hazards is relevant to the obligation of states to take proactive measures to mitigate such risks and thereby safeguard human rights.)[42]

Gaps in laws and protections relating to migration

With respect to migratory movements, human rights law provides robust protection for migrants who are moving for a range of reasons and under an array of circumstances. The International Covenant on Civil and Political Rights and the International Covenant on Economic, Social, and Cultural Rights guarantee civil, political, social, economic and cultural rights to all individuals without discrimination, thereby ensuring such fundamental rights extend to migrants. Several other conventions applicable to migrant workers also contain important provisions reaffirming the human rights of migrants.[43] In addition, the conference of the Parties to the UN Framework Convention on Climate Change (UNFCCC) has also recognized that "Parties should, when taking action to address climate change, respect, promote and consider their respective obligations on human rights, [including the rights of] migrants."[44]

However, given that climate change–induced migrants are among many categories of vulnerable migrants in need of human rights protection, the Special Rapporteur on the human rights of migrants has suggested that rather than create a specific category of climate migrants, coherent policies regarding the rights of all migrants should be developed "which take[] into account the myriad circumstances which lead people to migrate, including the need for human rights protections, in particular for those who are 'induced' or 'forced' to migrate."[45] Thus, despite the fact that increasing numbers of people from poor and vulnerable countries (such as those in sub-Saharan Africa) who have little to no responsibility for climate change could increasingly be forced to resort to international migration as the only option to cope with climate change–related stress (a.k.a. "survival migration"), under current definitions they are not entitled to any protections beyond what are available to economic migrants.

Given the unclear status of those who cross borders in the context of climate change and the fact that they do not fit neatly into existing definitions, it is not surprising that, at present, no international agency or institution has been given the mandate to protect climate change–induced migrants. Although both the Office of the UN High Commissioner for Refugees (UNHCR) and the International Organization for Migration (IOM) have led much of the research and advocacy in this area, the lack of institutional responsibility or a UN mandate creates challenges in efforts to develop strategies and marshal political will and resources to better protect climate change–induced migrants.

Challenges to addressing climate change–related internal displacement

Despite the focus on the gap in the international legal frameworks for addressing so-called climate refugees who will be forced to abandon their countries due to climate change–related effects, experts agree that the vast majority of people uprooted by climate change will be internally displaced.[46] The term "internal displacement" refers to situations in which individuals and groups are (1) forced or obliged to leave and remain away from their homes, but (2) remain within the borders of their own countries.[47] It is important to note that national governments (as opposed to UNHCR or the international community) bear primary responsibility for protecting those who are forcibly displaced or who migrate within their own countries.

In recognition of the specific vulnerabilities and human rights challenges that confront persons who are forcibly displaced within their countries, the Representative of the Secretary General on the Human Rights on Internally Displaced Persons, Francis Deng, developed the UN Guiding Principles on Internal Displacement, which were presented to the Human Rights Commission in 1998.[48] Drawn from humanitarian, human rights, and refugee law, the Guiding Principles outline the assistance and protection obligations of national governments with respect to persons

> forced or obliged to flee or to leave their homes or places of habitual residence, in particular as a result of or in order to avoid the effects of armed conflict, situations of generalized violence, violations of human rights or natural or human-made disasters, and who have not crossed an internationally recognized state border.[49]

Two regional instruments have been adopted by African nations that reinforce the protection obligations outlined in the Guiding Principles: the 2009 Kampala Convention for the Protection and Assistance of Internally Displaced Persons in Africa (Kampala Convention) and the 2006 Great Lakes Protocol on the Protection and Assistance to Internally Displaced Persons.[50] The Inter-Agency Standing Committee (IASC) Operational Guidelines on Human Rights and Natural Disasters also outline the human rights protections that humanitarian workers should seek to uphold in natural disaster situations.[51]

The Guiding Principles obligate governments to protect and assist those forced to flee not only conflict and persecution, but also "natural or man-made disasters."[52] The Kampala Convention goes further to obligate states to protect and assist "persons who have been internally displaced due to natural or human made disasters, *including climate change*."[53] Yet neither of these instruments proposes a satisfactory method for national governments to determine who or when someone is displaced by a climate change–related hazard or effect as opposed to a "disaster."[54]

In the case of climate change, the hazards resulting therefrom will vary widely in terms of intensity, scale, degree and timeframe. While the protection obligations are clear in the case of those displaced by large-scale, extreme events like acute flooding, superstorms and protracted drought, which are likely to overwhelm local coping capacities and give rise to a "disaster," they are far murkier in the case of those affected by other climate change–related hazards, like decreased seasonal rains, gradual coastal erosion or sea level rise. The latter occurrences may not be recognized as "disasters." Whether and at what point the hazard becomes significant enough to trigger movement is likely to be highly subjective.

In addition, many of the more complex effects of climate change on displacement do not align with the bulk of experience and practice regarding IDPs, which has largely focused on responding to sudden, often *en masse*, forced movements in situations of armed conflict and large-scale disasters. While most governments have civil protection authorities responsible for protecting IDPs in such situations, in the case of smaller-scale or more slowly unfolding climate-related hazards, it is unclear which government authorities are responsible for preventing and managing displacement. In short, existing laws, policies and operational practices for protecting the human rights of IDPs will need to be adapted and expanded to adequately protect those who are internally displaced due to climate change–related effects.

Opportunities to develop a new framework to more effectively protect the human rights of persons uprooted in the context of climate change

In the past several years, numerous initiatives have emerged that present an opportunity to develop new and more suitable approaches to prevent, minimize and address climate change–related displacement and migration, as well as to support, where appropriate, migration and planned relocation as an adaptation strategy. Rather than pursuing the lofty – and in many ways, impractical – goal of securing international agreement on a new international convention to address so-called climate refugees, these initiatives seek to build upon existing regional and bilateral arrangements related to displacement, migration and complimentary protection. Recognizing the climate impacts on internal displacement and the likelihood that increasing numbers of people living in at-risk areas will need to be supported to move out of harm's way, other initiatives and guidance have emerged that center on the development of national laws, policies and strategies for planned relocation.

Developing a protection framework using this piecemeal approach may appear less than ideal to the extent that it lacks the consistency and coherence of an international framework. On the other hand, as acknowledged by many governments during the Nansen Initiative consultations (discussed below), when it comes to climate change impacts on displacement and migration, a "one size fits all approach" may not be appropriate. As discussed above, climate change will affect different regions of the globe in myriad ways and contribute to myriad forms of human movement (e.g., anticipatory, temporary, circular and permanent movements; forced displacement, multi-causal migration or something in between). The need for regional and/or bilateral approaches, combined with national-level strategies, is supported by the complex, and often multi-causal, link between climate change and human movements.

Finally, from a practical standpoint, climate-related initiatives will need to adapt to the political realities within and between states. Solutions for West African states may not be well suited for Central American nations or Pacific Island states. Rather, solutions will need to not only build on existing agreements, policies, tools and measures for addressing displacement and migration but also be integrated across numerous other sectors, including climate change adaptation, disaster risk reduction, agricultural policy and land use planning.

Initiatives to address cross-border movements

The Nansen Initiative and its successor, the Platform on Disaster Displacement

The most comprehensive initiative to date aimed at addressing gaps in the international legal framework has been the Nansen Initiative on Cross-Border Displacement in the Context of Disasters and Climate Change (Nansen Initiative) launched in 2012 by Norway and Switzerland. Recognizing the resistance among governments of extending the Refugee Convention to a new category of "climate refugees," as well as the implausibility of attaining a new international convention or agreement on disaster- and climate-displaced persons, the Nansen Initiative sought to build consensus among governments using a bottom-up, state-led consultative approach.[55] Notably, the Initiative is aimed at cross-border, "forced" movements and, as such, extends more broadly to all disasters (whether they be climate-related, geological or human-made) given the involuntary nature of the movement and the recognized need for complementary protection in such contexts.

Over three years, the Nansen Initiative convened a series of regional consultations across the globe. The goal of these consultations was to convene government representatives, experts and members of civil society to examine the current and anticipated impacts of disasters and climate change on human movement in their region, explore the existing landscape of laws, policies and "tools" for preventing, minimizing and addressing migration and displacement and identify how they might be adapted to the context of disasters and climate change.[56]

The culmination of the process was the adoption in October 2015 by more than 100 governments of the Agenda for the Protection of Cross-Border Displaced Persons in the Context of Disasters and Climate Change (Nansen Protection Agenda).[57] The Nansen Protection Agenda consolidates legal standards and protection mechanisms to provide a framework for measures at the local, regional, sub-regional, national and international levels applicable to "disaster-isplaced" persons. The Protection Agenda recommends three areas that can serve to enhance coordination and action between these multiple levels of authority. First, it recommends multilateral efforts to collect data and share knowledge on cross-border disaster displacement. For example, international organizations and agencies are encouraged to share technical advice and offer operational support to governmental authorities. Second, the Protection Agenda encourages the application of humanitarian protection measures for cross-border, disaster-displaced persons through the harmonizing of approaches at the sub-regional level. Lastly, the Agenda recommends enhancing coordination with respect to disaster displacement risk management in the country of origin; more specifically, it suggests that human mobility considerations be integrated into national laws, policies and strategies related to disaster risk reduction and climate change adaptation in recognition that internal, disaster-related displacement, when not managed properly, can lead people to seek assistance abroad.

The Nansen Initiative has since been replaced by the Platform on Disaster Displacement (PDD), which, with the support of Germany, Bangladesh and a handful of other countries, will seek to move forward with concrete actions to implement the Nansen Protection Agenda.[58] In late 2016, the PDD adopted a Strategic Plan and Work Plan aimed at undertaking activities outlined in the Nansen Protection.[59]

The implementation of the PDD Strategic Plan and Work Plan, which will be supported by governments, UN agencies including UNHCR and IOM, academic institutions and civil society organizations, presents an important opportunity not only to collect and share relevant data, but also to adopt novel measures and strategies for enhanced human rights protections for people who are displaced or migrate due to climate change–related effects.

The UN Framework Convention on Climate Change and its climate displacement task force

The issue of how to address displacement and migration arising from the negative effects of climate change has also been taken up by state parties to the 1992 UN Framework Convention on Climate Change.[60] Pursuant to a decision adopted in Cancun, Mexico, in 2010, parties to the Convention were invited to "enhance action on [climate change] adaptation . . . by undertaking. . . [m]easures to enhance understanding, coordination and cooperation with regard to climate change induced displacement and planned relocation, where appropriate, at the national, regional and international levels."[61] This decision – known as the Cancun Adaptation Platform – clearly identifies measures to address climate displacement as a form of adaptation, thereby linking it to adaptation funds established under the UNFCCC, such as the Green Climate Fund.[62]

Climate change–related displacement and migration have also been identified under the UNFCCC as a form of "loss and damage." The concept of loss and damage arose in acknowledgement of the fact that despite commitments of the developed country parties to the UNFCCC to stabilize greenhouse gas (GHG) concentrations in the atmosphere and to help developing countries adapt to climate change, some climate change impacts will be "unavoidable." These unavoidable or residual impacts are referred to under the UNFCCC as "loss and damage." In 2012, the conference of the parties to the UNFCCC (COP) adopted a decision during the Doha Climate Change Conference that further defined the concept of loss and damage and explicitly identified displacement as a potential form of loss and damage by acknowledging the need to "advance the understanding of and expertise on loss and damage, which includes, inter alia. . . [h]ow impacts of climate change are affecting patterns of migration, displacement and human mobility."[63]

In 2013, the COP established the Warsaw International Mechanism for Loss and Damage (WIM) "to address loss and damage associated with impacts of climate change, including extreme events and slow-onset events, in developing countries that are particularly vulnerable to the adverse effects of climate change."[64] The implementation of the functions of the WIM is led by an Executive Committee (WIM ExComm), under the guidance of the COP. Given the recognition by the COP that climate change impacts on displacement and migration may constitute a form of loss and damage, the WIM ExComm's initial two-year workplan included among its activities "[e]nhanc[ing] understanding of and expertise on how the impacts of climate change are affecting patterns of migration, displacement and human mobility; and the application of such understanding and expertise."[65] In addition, the decision reached by the COP in Paris in December 2015 requested the WIM ExComm to establish a task force to work with existing bodies and expert groups "to develop recommendations for integrated approaches to avert, minimize and address displacement related to the adverse impacts of climate change."[66] The 22nd COP in Morocco approved an indicative five-year rolling workplan of the WIM ExComm, which includes among one of its five workstreams "[m]igration, displacement and human mobility, including the task force on displacement," and activities under this workstream are now being developed.[67]

The ongoing work of the WIM ExComm – and the establishment of the climate displacement task force – present further opportunities to develop strategies for preventing, minimizing and addressing climate-related displacement and migration, both internally and on an inter-state level. However, given many of the political issues tied up in loss and damage, along with the small size (approximately 14 members)[68] and hence limited capacity of the task force, it is likely that any action under the UNFCCC will proceed slowly and potentially be mired in other issues. However, what is important about including climate change–related displacement and migration under the UNFCCC is the opportunity to share knowledge and consolidate best practices to share with governments. Of particular importance will be increasing understanding and potentially developing more reliable estimates for the potential magnitude of displacement that can be linked to temperature rises above 1.5°C. This could present a key advocacy message for states regarding their obligation under human rights law to limit temperature rise to 1.5°C consistent with their obligation to avoid loss and damage.

The inclusion of climate change–related displacement and migration under the UNFCCC also provides opportunities to address the issue at the national level. In order to help Least Developed Countries (LDCs) identify activities necessary to adapt to climate change, the UNFCCC calls on LDCs to adopt National Adaptation Programs of Action (NAPAs). NAPAs are intended to "provide a process for [LDCs] to identify priority activities that respond to their urgent and immediate needs to adapt to climate change – those for which further delay would increase vulnerability and/or costs at a later stage."[69] These NAPAs present an opportunity for states to adopt measures and strategies to support migration as an adaptation strategy and to reduce climate displacement risk, including by implementing laws, policies and procedures for planned relocation of at-risk communities.

The UN New York Declaration for Refugees and Migrants

As efforts continue on actions to better protect those displaced across borders due to disasters and other climate change–related effects, the numbers of traditional refugees and asylum seekers has continued to rise rapidly.[70] By the end of 2015, UNHCR reported that the number of forcibly displaced persons worldwide was at historically high levels totaling 65.3 million, including 21 million refugees, three million asylum seekers or people in refugee-like situations, and more than 40 million internally displaced people (IDPs).[71] The number of migrants is also at an all-time high, having surpassed 244 million by the end of 2015.[72]

In order to address this global crisis, in September 2016 the UN General Assembly hosted a high-level summit to address large movements of migrants and refugees. The outcome of the summit was the adoption by all 193 member states of the New York Declaration for Refugees and Migrants.[73] Importantly, the Declaration acknowledges the complex drivers of displacement and migration in the 21st century, among them, climate change.

> Some people move in search of new economic opportunities and horizons. Others move to escape armed conflict, poverty, food insecurity, persecution, terrorism, or human rights violations and abuses. Still others do so in response to the adverse effects of climate change, natural disasters (some of which may be linked to climate change), or other environmental factors. Many move, indeed, for a combination of these reasons.[74]

In order to address the crisis, the New York Declaration calls for the development of two global compacts: a global compact for safe, orderly and regular migration (GCM) and a global compact on refugees (GCR).[75] In recognition of the lack of a specific legal regime concerning migrants (as opposed to refugees), the goal of the GCM is to set out a range of principles, commitments and understandings regarding international migration in all its dimensions.[76] In contrast to the GCR, it will articulate, for the first time, a comprehensive framework for migration and be developed through a member state–driven process.[77] The GCM will be elaborated through a process of intergovernmental negotiations starting in early 2017 and will culminate in an intergovernmental conference on international migration in 2018, at which the GCM will be presented for adoption.[78]

The development of the GCM over the next few years presents an opportunity for states to develop principles, policies and processes to address those uprooted in the context of climate change using a human rights–based approach that will be incorporated into a global agreement on migration. Not only does the Declaration specifically identify climate change as a driver of human movement, but it also includes a commitment by states to address such drivers including by "combating environmental degradation and ensuring effective responses to natural disasters and the adverse impacts of climate change."[79] It further includes a commitment by nations to assist migrants in countries that are experiencing natural disasters and notes the Nansen Initiative Protection Agenda.[80]

Moreover, the New York Declaration reiterates and reaffirms in numerous places the human rights of refugees and migrants, and it includes a commitment by states to fully protect the human rights of all refugees and migrants, regardless of status, and to fully respect international human rights law in their response.[81] With respect to commitments for migrants, the Declaration includes a commitment by states to "protect[] the safety, dignity and human rights and fundamental freedoms of all migrants, regardless of their migratory status, at all times," and to "cooperate closely to facilitate and ensure safe, orderly and regular migration."[82] Among the inexhaustive list of 24 issues that could be included in a GCM is "the effective protection of the human rights and fundamental freedoms of migrants, including women and children, regardless of their migratory status, and the specific needs of migrants in vulnerable situations."[83]

Finally, the Declaration encourages deeper interaction among governments and civil society to find responses to the challenges and opportunities presented by international migration and invites civil society, as well as the private sector, the diaspora and migrant organizations, to contribute to the process of developing the GCM.[84]

Nonetheless, there are some potential limitations on what can be achieved through the GCM. Though the process will be government-led, the extent to which non-state actors, such as civil society or migrants themselves, will be actively engaged in the development of the GCM remains to be seen. In order to fulfill their obligations to allow public participation of migrants and other climate change–affected communities, the process will need to include safeguards and mechanisms to ensure public participation. In addition, the final GCM will need to include more specific obligations to which states can be held accountable, such as granting work permits to migrants and opening up more migration opportunities for people from the most climate-vulnerable countries. Moreover, given the fact that the New York Declaration and GCM are aimed at addressing "large movements of refugees and migrants," it is not entirely clear what types of events or situations will be considered large enough to trigger their commitments. As discussed above, many of the climate change–related drivers of human mobility are likely to unfold slowly, or sporadically, and it is difficult to predict the types of movement (with respect to both numbers of people and over what timeframe) they will precipitate. One hopes, however, that many of the overall commitments in the New York Declaration to respect the human rights of migrants and displaced people will be better respected going forward than at present.

Opportunities to address climate change–related internal displacement

The Special Rapporteur on the issue of human rights obligations relating to the enjoyment of a safe, clean, healthy and sustainable environment has pointed out, "each State has an obligation to protect those within its jurisdiction from the harmful effects of climate change."[85] This includes the obligation to adopt a legal and institutional framework that assists those within their jurisdiction to adapt to the unavoidable effects of climate change. It also includes a duty to protect against "foreseeable environmental impairment of human rights whether or not the environmental harm itself violates human rights law, and even whether or not the States directly cause the harm."[86] While the UN Guiding Principles and other national and regional laws and instruments related to internal displacement provide a principled basis for extending human rights protections to those facing internal displacement due to climate change, unfortunately, only a limited number of governments have adopted binding laws on the treatment of IDPs.[87] Even where they have, implementation is often weak. Thus, far more concerted effort is needed on behalf of national governments to develop and implement laws and policies related to IDPs that specifically incorporate people displaced by climate change effects (including large- and small-scale, recurrent, cumulative, sudden- and slow-onset effects).

Building upon traditional IDP laws and policies, national governments will also need to go further to consider strategies for preventing and minimizing climate-related displacement and migration, or where appropriate, supporting migration and planned relocation as adaptation strategies. A notable shortcoming of using traditional laws, policies and operational practices on internal displacement as a starting point is that they are, for the most part, focused on responding to people already uprooted or on the move and thus fail to take advantage of the enormous opportunities to avoid climate change–related displacement.

Although there is evidence that climate change is already dislocating vulnerable communities, the future effects of climate change threaten to have a far greater impact on displacement. The scale of future displacement will depend on the extent to which governments act now to substantially reduce carbon emissions, build the resilience of vulnerable communities, mitigate climate risk and adapt to climate change effects. Thus, any strategy aimed at addressing climate change–related displacement must place a premium on avoiding or minimizing displacement, i.e., *on prevention*.

Also needed are laws, policies, processes and institutional arrangements for relocating populations at risk of displacement from climate change–related affects, which are entirely absent. Recognizing the need for a normative framework to address climate change–related displacement within states, a group of climate change experts and international lawyers came together in 2013 to develop the Peninsula Principles on Climate Displacement Within States (Peninsula Principles).[88] Drawn heavily from human rights principles, the Peninsula Principles outline a set of guidelines, principles and processes for the planned, internal relocation of communities away from at-risk, climate-vulnerable areas, such as low-lying coastal areas. The Peninsula Principles' particular value is in articulating the rights of "climate-displaced persons," especially their right to remain in place, as well as to initiate and undertake planned relocation. The Peninsula Principles also offer institutional planning guidance for many aspects of internal displacement, such as participation and consent of affected individuals, land identification and post-displacement return.[89]

Given the need for a more comprehensive approach to managing cross-border displacement that considers root causes, the Nansen Protection Agenda also includes strengthening the management of disaster displacement risk in the country of origin among its three priority areas for action. According to the Nansen Protection Agenda, specific measures include the need to integrate human mobility within disaster risk reduction (DRR), climate change adaptation strategies and other relevant development processes. In addition, the Agenda emphasizes the need to improve the use of planned relocation for populations living in high-risk areas as a means to anticipate disaster displacement. It further prioritizes the importance of ensuring that relevant laws and policies on disaster risk management specifically address the needs of IDPs.[90]

In short, addressing disaster or climate displacement risk in the country of origin will require adopting new strategies that are both far more contextual and better integrated across other sectors and that focus on climate displacement risk as its starting point. It is only by focusing on climate change–related risk that national and local governments will be able to take advantage of opportunities to implement more effective displacement prevention measures. Potential avenues for doing so include (a) improving understanding of climate displacement risk as a factor for hazard exposure, vulnerability and resilience; (b) emphasizing displacement avoidance/mitigation strategies; (c) integrating displacement risk considerations into relevant laws, regulations and policies at the national, sub-national and local levels, including disaster risk reduction, land use planning, natural resource management and climate change adaptation; (d) developing laws and institutional arrangements for managing climate displacement risk; and (e) empowering local governments and communities with the resources and tools to manage and mitigate climate displacement risk.[91]

Conclusion

In the past several years, numerous initiatives and venues have emerged that present an opportunity for governments to both cooperate to address cross-border movements stemming from climate change and to develop new laws, policies and processes at the national level for protecting the human rights of vulnerable communities who face displacement and dislocation due to climate change. At present, more comprehensive data on the extent of human mobility associated with climate change is sorely lacking. Nonetheless, governments must seize these opportunities. Unlike displacement from war and persecution, there are enormous opportunities to avert and minimize the foreseeable harms associated with climate change adverse effects – including forced displacement. But with climate change adverse effects unfolding more quickly than anticipated, governments – with the support and input of academics, the private sector, civil society and affected communities themselves – must do so now. The risk of displacement and dislocation facing hazard-prone coastal communities, low-lying island nations, communities in the Arctic, those living in fragile and conflict ridden states and those in developing and least-developed countries who rely on natural resources to survive are evident. Governments must act urgently to meet their obligation to protect them and others from the threat to the full enjoyment of human rights that climate change presents.

Notes

1 IPCC, *Climate Change: The IPCC Scientific Assessment* (1990), available online at www.ipcc.ch/ipccreports/far/wg_I/ipcc_far_wg_I_full_report.pdf, at 103.
2 R. Bronen, *Climate-Induced Displacement of Alaska Native Communities* (30 January 2013), available online at www.brookings.edu/research/climate-induced-displacement-of-alaska-native-communities; C. Mele and D. Victor, *Reeling From the Effects of Climate Change, Alaskan Village Votes to Relocate* (19 August 2016), available online at www.nytimes.com/2016/08/20/us/shishmaref-alaska-elocate-vote-climate-change.html?_r=0.
3 UN Human Rights Council, *Report of the Special Rapporteur on the Issue of Human Rights Obligations Relating to the Enjoyment of a Safe, Clean, Healthy and Sustainable Environment*, UN Doc. A/HRC/31/52 (1 February 2016), at para 83.
4 IPCC, *Climate Change 2014: Impacts, Adaptation, and Vulnerability, Working Group II Contributions to the Fifth Assessment Report of the Intergovernmental Panel on Climate Change* (2014), available online at www.ipcc.ch/pdf/assessment-report/ar5/wg2/WGIIAR5-Chap12_FINAL.pdf, at 767.
5 Internal Displacement Monitoring Centre (IDMC), *Global Report on Internal Displacement* (May 2016), available online at www.internal-displacement.org/globalreport2016/pdf/2016-global-report-internal-displacement-IDMC.pdf, at 8.
6 *Ibid.*, at 14.
7 *Ibid.*, at 47.
8 UN Environment Programme, *Livelihood Security: Climate Change, Migration and Conflict, in the Sahel* (2011), available online at http://postconflict.unep.ch/publications/UNEP_Sahel_EN.pdf.
9 Royal Geographic Society, *Glacial Environments – How Will Melting Glaciers Affect People Living in Other Countries?* Available online at www.rgs.org/OurWork/Schools/Teaching+resources/Key+Stage+3+resources/Glacial+environments/How+will+melting+glaciers+affect+people+living+in+other+countries.htm.
10 IPCC, *Climate Change 2014: Impacts, Adaptation, and Vulnerability, Working Group II Contribution to the Fifth Assessment Report of the Intergovernmental Panel on Climate Change* (2014) 346.
11 CIEL, *Malé Declaration on the Human Dimension of Global Climate Change* (14 November 2007), available online at www.ciel.org/Publications/Male_Declaration_Nov07.pdf (noting that the fundamental right to an environment capable of supporting human society and the full enjoyment of human rights is recognized in the constitutions of over 100 states and in several international instruments; expressing concern that "climate change has clear and immediate implications for the full enjoyment of human rights including *inter alia* the right to life, the right to take part in cultural life, the right to use and enjoy property, the right to an adequate standard of living, the right to food, and the right to the highest attainable standard of physical and mental health").
12 AOSIS, *Alliance of Small Island States Leaders' Declaration* (27 September 2012), available online at http://aosis.org/wp-content/uploads/2012/10/2012-AOSIS-Leaders-Declaration.pdf.

13 In November 2016, temperatures in Alaska were reported to be 36°F above normal. C. Mooney and J. Samenow, *The North Pole Is an Insane 36 Degrees Above Normal as Winter Descends* (17 November 2016), available online at www.washingtonpost.com/news/energy-environment/wp/2016/11/17/the-north-pole-is-an-insane-36-degrees-warmer-than-normal-as-winter-descends/?utm_term=.db53a98aa3d2.
14 R. Bronen, *Climate-Induced Displacement of Alaska Native Communities* (30 January 2013), available online at www.brookings.edu/research/climate-induced-displacement-of-alaska-native-communities; Mele and Victor, *supra* note 2.
15 There are, in fact, numerous instances in which people affected by disasters (both weather-related and other forms of natural disasters, such as earthquakes or volcanic eruptions) have crossed a border in search of safety and/or protection and assistance in another country. Based on available data, Africa along with Central and South America, in particular, have seen the largest number of incidences of cross-border disaster displacement. In recent decades, at least 50 countries have received or refrained from returning people in the aftermath of disasters. Nansen Initiative, *Agenda for the Protection of Cross-Border, Displaced Persons in the Context of Disasters and Climate Change* (October 2015), available online at https://nanseninitiative.org/wp-content/uploads/2015/02/PROTECTION-AGENDA-VOLUME-1.pdf, at 6.
16 IPCC, *Climate Change 2014: Impacts, Adaptation, and Vulnerability, Working Group II Contributions to the Fifth Assessment Report of the Intergovernmental Panel on Climate Change* (2014) 771.
17 *Ibid.*, at 772–773.
18 *Ibid.*
19 *Ibid.*, at 775.
20 *Ibid.*, at 774–775.
21 *Ibid.*, at 774 (references omitted).
22 *Ibid.*, at 1060.
23 *Ibid.*, at 773.
24 S. Vigil, 'Displacement as a Consequence of Climate Change Mitigation Policies', 49 *Forced Migration Review*, available online at www.fmreview.org/climatechange-disasters/vigil.html, at 43.
25 Bronen, *supra* note 14, at 767.
26 Oxfam, *Beyond Safe Land: Why Security of Land Tenure Is Crucial for the Philippines' Post-Haiyan Recovery* (August 2014), available online at www.oxfam.org/sites/www.oxfam.org/files/file_attachments/bp-beyond-safe-land-security-tenure-philippines-110814-en.pdf.
27 UN Human Rights Council, *Report of the Special Representative to the Secretary-General on the Human Rights of Internally Displaced Persons, Walter Kälin, Addendum: Protection of Internally Displaced Persons in Situations of Natural Disasters*, UN Doc. A/HRC/10/13/Add.1 (5 March 2009), at para 3.
28 *UN Special Rapporteur on the Human Rights of Internally Displaced Persons, Protection of and Assistance to Internally Displaced Persons*, UN Doc. A/66/285 (9 August 2011), at para 36.
29 UN Special Rapporteur on the human rights of internally displaced persons, *supra* note 28, at 74–75.
30 *Ibid.*
31 W. Kälin and N. Schrepfer, *Protecting People Crossing Borders in the Context of Climate Change Normative Gaps and Possible Approaches* (February 2012), available online at www.unhcr.org/4f33f1729.pdf.
32 UNHCR, Convention Relating to the Status of Refugees, 1951, 189 UNTS 150, at Art. 1(2); UNHCR, Protocol Relating to the Status of Refugees, 1967, 606 UNTS 267.
33 New Zealand Supreme Court, *Teitiota v. The Chief Executive of the Ministry of Business, Innovation and Employment* (26 November 2013), NZHC 3125.
34 *Ibid.*, at paras 13–15, and at para 21(41).
35 *Ibid.*, at para 54.
36 Organization of African Unity, Convention Governing the Specific Aspects of Refugee Problems in Africa, 10 September 1969, 1001 UNTS 45, at Art. 1, para 2; Inter-American Commission on Human Rights, Cartagena Declaration on Refugees, 22 November 1984, in Annual Report of the Inter-American Commission on Human Rights, OAS Doc OEA/Ser.L/V/II .66/doc.10, rev. 1, 190–93 (1984–85), available online at http://hrlibrary.umn.edu/instree/cartagena1984.html.
37 Kälin and Schrepfer, *supra* note 31, at 34.
38 UNOHCHR, *Report of the Special Rapporteur on the Human Rights of Migrants*, UN Doc. A/67/299 (13 August 2012), at para 59.
39 Nansen Initiative, *supra* note 15, at 6.

40 See e.g., National Academies of Science, Engineering, and Medicine, *Attribution of Extreme Weather Events in the Context of Climate Change* (2016). The report found clear links between anthropogenic climate change and heat waves, droughts and intense rain- and snowstorms. The report found less evidence linking climate change to tornadoes, hurricanes and wildfires.
41 A. Thomas and R. Rendon, *Confronting Climate Displacement: Learning From Pakistan's Floods* (November 2010), available online at www.alnap.org/resource/6077, at 3–4.
42 UN Human Rights Council, *supra* note 3, at paras 37–38.
43 UNOHCHR, *supra* note 38, at paras 54–56.
44 UNFCCC, *Adoption of the Paris Agreement*, U.N. Doc. FCCC/CP/2015/L.9/Rev. 1 (12 December 2015), at preamble.
45 UNOHCHR, *supra* note 38, at para 65.
46 Brookings Institution, *Climate Change and Internal Displacement* (2014), available online at www.brookings.edu/wp-content/uploads/2016/06/Climate-Change-and-Internal-Displacement-October-10-2014.pdf.
47 1998 UN Guiding Principles on Internal Displacement, *Introduction*, at para 2.
48 *Ibid.*
49 UNOCHA, *Guiding Principles on Internal Displacement*, UN Doc. E/CN.4/1998/53/Add.2 (1998), at para 2.
50 African Union Convention for the Protection and Assistance of Internally Displaced Persons in Africa (Kampala Convention), 22 October 2009; International Conference on the Great Lakes Region, Protocol on the Protection and Assistance to Internally Displaced Persons, 30 November 2006.
51 *Inter-Agency Standing Committee (IASC) Operational Guidelines on Human Rights and Natural Disasters* (January 2011), available online at www.ohchr.org/Documents/Issues/IDPersons/OperationalGuidelines_IDP.pdf.
52 Brookings Institution, *supra* note 46.
53 African Union Convention for the Protection and Assistance of Internally Displaced Persons in Africa, *supra* note 50, at Art. 5, para 4.
54 As defined by the UN International Strategy for Disaster Reduction (UNISDR), a "disaster" is "[a] serious disruption of the functioning of a community or a society causing widespread human, material, economic or environmental losses which exceed the ability of the affected community or society to cope using its own resources." UNISDR, *Terminology* (30 August 2007), available online at www.unisdr.org/we/inform/terminology (last visited 16 May 2016). But see 16th Congress of the Republic of the Philippines, *Act Protecting the Rights of Internally Displaced Persons and Penalizing the Acts of Arbitrary Internal Displacement* (19 August 2014), Senate Bill 2368 (defining "internal displacement" to include "the involuntary movement or forced evacuation or expulsion of any person who flee or leave their homes or places of habitual residence, within the national borders, as a result of or in order to avoid or minimize the effects of . . . natural, human-induced and man-made *hazards*"; emphasis added).
55 Nansen Initiative, *supra* note 15, at 6.
56 *Ibid.*
57 *Ibid.*
58 Platform on Disaster Displacement (PDD) (2017), available at www.disasterdisplacement.org.
59 PDD, *Strategic Framework 2016 to 2019* (24 April 2016), available online at http://disasterdisplacement.org/wp-content/uploads/2015/03/24042016-Strategic-Framework-Platform-on-Disaster-Displacement-DRAFT.pdf.
60 UN Conference on Environment and Development (UNCED), *United Nations Framework Convention on Climate Change (UNFCCC)*, 1771 UNTS 107 (9 May 1992).
61 UNFCCC, *Report of the Conference of the Parties on Its Sixteenth Session, Addendum, Part Two: Action Taken by the Conference of the Parties at Its Sixteenth Session*, Decision 1/CP.16, UN Doc. FCCC/CP/2010/7/Add.1 (15 March 2011), at para 14(f).
62 *Ibid.*, at paras 95–137.
63 UNFCCC, *Report of the Conference of the Parties on Its Eighteenth Session, Addendum, Part Two: Action Taken by the Conference of the Parties at Its Eighteenth Session*, Decision 3/CP.18, Approaches to address loss and damage associated with climate change impacts in developing countries that are particularly vulnerable to the adverse effects of climate change to enhance adaptive capacity, UN Doc. FCCC/CP/2012/8/Add.1 (28 February 2013), at para 7(a)(vi).
64 UNFCCC, *Report of the Conference of the Parties on Its Nineteenth Session, Addendum, Part Two: Action Taken by the Conference of the Parties at Its Nineteenth Session*, Decision 2/CP.19, Warsaw international mechanism for loss and damage associated with climate change impacts, UN Doc. FCCC/CP/2013/10/Add.1 (31 January 2014), at para 1.

65 *UNFCCC Report of the Executive Committee of the Warsaw International Mechanism for Loss and Damage Associated With Climate Change Impacts, Annex III – Specific Actions Undertaken in the Nine Action Areas of the Initial Two Year Workplan of the Executive Committee of the Warsaw International Mechanism*, U.N. Doc. FCCC/SB/2016/3 (14 October 2016), available online at http://unfccc.int/resource/docs/2016/sb/eng/03.pdf, at para 13.
66 UNFCCC, *supra*, note. 44, at para 50.
67 UNFCCC, *supra*, note 65, Annex I – Indicative framework for the five-year rolling workplan of the Executive Committee of the Warsaw International Mechanism.
68 *See* UNFCCC, *Task Force on Displacement*, available online at http://unfccc.int/adaptation/groups_committees/loss_and_damage_executive_committee/items/9978.php.
69 *See* UNFCCC, *National Adaptation Programmes of Action (NAPAs)*, available online at http://unfccc.int/national_reports/napa/items/2719.php.
70 This is due in part to the conflict in Syria but also to protracted conflict in places like Afghanistan, Somalia and Iraq. These are only the latest additions to the millions more refugees around the globe that continue to live in protracted displacement situations – some for lifetimes, having been born into displacement in refugee camps – with no prospects for a solution. V. Turk and M. Garlick, 'From Burdens and Responsibilities to Opportunities: The Comprehensive; Refugee Response Framework and a Global Compact on Refugees', 28 *International Journal of Refugee Law* (2016) 4, at 656.
71 UNHCR, *Global Trends: Forced Displacement in 2015* (20 June 2016), available online at www.unhcr.org/statistics/country/576408cd7/unhcr-global-trends-2015.html.
72 GA Res. 71/1, 19 September 2016, at para 3.
73 *Ibid.*
74 *Ibid.*, at para 1.
75 The New York Declaration defines "large movements of refugees and migrants" as follows: "Large movements" may be understood to reflect a number of considerations, including the number of people arriving, the economic, social and geographical context, the capacity of a receiving state to respond and the impact of a movement that is sudden or prolonged. The term does not, for example, cover regular flows of migrants from one country to another. "Large movements" may involve mixed flows of people, whether refugees or migrants, who move for different reasons but who may use similar routes. GA Draft Res. A/71/L.1 (13 September 2016), at para 6.
76 *Ibid.*, at Annex II, para 2.
77 UNHCR, *The New York Declaration for Refugees and Migrants: Quick Guide*, available online at www.unhcr.org/57e4f6504.pdf.
78 GA Res, *supra* note 72, at Annex II, para 9.
79 *Ibid.*, at para 43.
80 *Ibid.*, at para 50.
81 *Ibid.*, at paras 5, 6.
82 *Ibid.*, at para 41.
83 *Ibid.*, at Annex II, para 8(i).
84 GA Res, *supra* note 72, at para 61; Annex II, para 15.
85 UN Human Rights Council, *supra* note 3, at para 68.
86 *Ibid.*, at para 37.
87 *See* IDMC, *IDP Laws and Policies: A Mapping Tool*, available online at www.internal-displacement.org/law-and-policy.
88 Displacement Solutions, *The Peninsula Principles on Climate Displacement Within States* (19 August 2013), available online at http://displacementsolutions.org/peninsula-principles.
89 *Ibid.*, Principles 9–11, at 19–24.
90 IDMC collects annual global estimates of people displaced by disasters.
91 See M. Burn, M. Gerrard, M. Leighton and A. Thomas, *Report of the International Bar Association on Human Rights and Climate Change* (2017) (publication pending).

9

Climate change under regional human rights systems

Sumudu Anopama Atapattu[*]

Introduction

Climate change has been described as the greatest threat to human rights in the 21st century.[1] While scientists, policy makers, politicians and lawyers have all debated, discussed and analyzed climate change from various perspectives, litigation relating to climate change is rare, especially at the international level.[2] There are several reasons for this. First, international environmental litigation *per se* is quite rare. States prefer to settle their disputes through diplomatic channels or by other methods than to resort to litigation. Second, in relation to many environmental issues, it is hard to establish causation.[3] Third, the main objective of environmental law is to prevent damage, because it is virtually impossible to repair environmental damage once it has already taken place. Finally, the remedies available under international law are often ill-suited and inadequate for environmental damage, especially climate change. The first environmental case before the International Court of Justice was decided only in the 1990s,[4] although a few prior cases had implicated environmental issues.[5] Since then, a few environmental cases have come before the ICJ[6] – an environmental case that implicated human rights issues was withdrawn by the parties although their memorials offer insights into the link between human rights and the environment.[7]

Regional human rights systems, on the other hand, have been the foci of action relating to environmental issues. They have developed a rich jurisprudence relating to environmental rights, interpreting existing rights expansively and creatively. This development, coupled with constitutional developments and judgments from national courts, have led some scholars to contend that a human right to a healthy environment may be emerging under international law.[8]

The link between climate change and human rights has received considerable attention in recent years.[9] That climate change can undermine the enjoyment of protected rights is no longer disputed, although framing climate change as a human rights issue is a recent development. As the Office of the High Commissioner for Human Rights noted in its report on the relationship between climate change and human rights:

> Climate change–related impacts, as set out in the assessment reports of the Intergovernmental Panel on Climate Change, have a range of implications for the effective enjoyment

of human rights. The effects on human rights can be of a direct nature, such as the threat extreme weather events may pose to the right to life, but will often have an indirect and gradual effect on human rights, such as increasing stress on health systems and vulnerabilities related to climate change–induced migration.[10]

The Special Rapporteur on Human Rights and Environment, John Knox, also has endorsed the link between human rights and climate change.[11] The latest endorsement of this link came from the Paris Agreement on Climate Change that called upon parties, *inter alia*, to respect, promote and *consider* their respective human rights obligations when taking action to address climate change.[12] While this formulation seems to be a watered-down version of the typology of obligations under human rights law,[13] the Paris Agreement is the first global environmental treaty to explicitly include a provision on human rights.[14] Human rights obligations with respect to climate change extend to both mitigation measures and adaptation measures.[15]

Because climate change is the result of day-to-day action by 7.5 billion human beings, Peel and Osofsky argue that virtually all litigation could be taken as "climate litigation."[16] However, climate litigation in countries like the United States and Australia has a direct link to climate change "by addressing either the greenhouse gas emissions that cause the problem (mitigation-related litigation) or the predicted impacts of climate change on ecosystems, communities, and infrastructure (adaptation-related litigation)."[17] They identify four categories of cases in this regard: (1) litigation with climate change as the central issue; (2) litigation with climate change as a peripheral issue; (3) litigation with climate change as one motivation but not raised as an issue; and (4) litigation with no specific climate change framing but with implications for mitigation or adaptation.[18] While this classification is helpful, often the lines are blurred and there is an overlap between (1) and (4) above.

While litigation often puts the focus on the victim and the perpetrator, it can serve a wider purpose. As Hunter points out, litigation strategies could influence international negotiations and draw attention to the problem.[19] The Inuit petition, discussed below, succeeded in doing just that by highlighting the plight of victims of climate change. Whether it has had an impact on climate negotiations is, of course, rather questionable.

This chapter seeks to explore how climate change issues have been addressed under the main regional human rights systems in Europe, Africa and the Americas. The chapter proceeds in five sections. The second section begins with a brief overview of the three regional human rights systems in the world and the jurisprudence developed by them in relation to environmental rights. The third section is devoted to a discussion of the seminal case filed before the Inter-American Commission of Human Rights in 2005 by the Inuit of the US and Canada against the United States[20] – the then highest emitter of greenhouse gases. It will also discuss the Athabaskan petition filed against Canada – currently pending before the Inter-American Commission. The fourth section discusses other action that regional human rights bodies have taken in relation to climate change. The fifth section will offer some concluding thoughts about environmental litigation in general and climate litigation in particular.

Regional human rights systems and development of environmental rights

Currently three regions of the world have a human rights system in place: Europe, the Americas and Africa.[21] While there have been other regional efforts, nothing akin to the three regional systems exists in other parts of the world. For example, while Asian countries adopted the Asian Human Rights Charter in 1998 and recommended the adoption of a regional Convention and the establishment of a Commission or Court that can receive complaints, this has yet to materialize.[22] The Asian Human Rights Commission is an independent non-governmental body "which seeks to promote greater awareness and realisation of human rights in the Asian region."[23] The ASEAN Human Rights Declaration[24] was adopted in 2012 and includes economic, social and cultural rights, including the right to a safe, clean and sustainable environment.[25] The ASEAN Intergovernmental Commission on Human Rights (AICHR), as the overarching institution, is entrusted with the task of promoting and protecting human rights in the ASEAN region. It, however, lacks the mechanisms available in the three regions described earlier to receive individual complaints. The Arab Charter on Human Rights was adopted in 2004 by the League of Arab Nations.[26] Similar to the ASEAN Declaration, the Arab Charter also recognizes a right to a healthy environment as a component of the right to an adequate standard of living: "Every person has the right to an adequate standard of living for himself and his family, which ensures their well-being and a decent life, including food, clothing, housing, services and the right to a healthy environment."[27] Under Article 45, an Arab Human Rights Committee is established consisting of seven members elected by the parties to the Charter. While state parties are required to submit reports to the Committee,[28] it does not recognize an individual right of petition.

While there have been some initiatives to establish other regional mechanisms,[29] a large portion of the globe currently remains outside regional human rights mechanisms. This section will briefly describe the three human rights systems and their jurisprudence related to environmental rights, which may be useful in relation to future climate litigation.[30] As with all international tribunals where individual right of petition is recognized, local remedies have to be exhausted before claimants can resort to these regional bodies.

European Court of Human Rights

The oldest of the three regional systems, the European Commission and Court of Human Rights, were established by the European Convention on Human Rights and Fundamental Freedoms adopted in 1950 within the framework of the Council of Europe.[31] It seeks to supplement the UN Human Rights system.[32] Rather than wait for the UN to adopt a treaty to formalize the commitments embodied in the Universal Declaration of Human Rights, the Council of Europe decided to "take the first steps for the collective enforcement of certain of the rights stated in the Universal Declaration."[33] A two-tier system was established with a Commission and a Court of Human rights, which were later fused into one. Judge Burguenthal summarizes the significant role played by the Court:

> Over time, the European Court of Human Rights for all practical purposes has become Europe's constitutional court in matters of civil and political rights. Its judgments are routinely followed by the national courts of the states parties to the Convention, their legislatures, and their national governments. The Convention itself has acquired the status of domestic law in most of the states parties and can be invoked as such in their courts.[34]

The European Convention is the only regional human rights instrument not to contain a specific right to a healthy environment, although it has developed a rich body of jurisprudence on environmental rights by interpreting existing rights, especially Article 8 on the Right to Private and Family Life.[35] The Council of Europe has debated the issue whether the European Convention should be amended to include a distinct right to a healthy environment but decided against it on the ground that existing rights are sufficient to articulate environmental rights.[36] Shelton calls this a "backlash" against recognizing environmental rights:

> Also at the regional level, the preambles to European Community directives often state their aim "to protect human health and the environment." Yet, the Council of Europe's Committee of Ministers has repeatedly rejected proposals from the Parliamentary Assembly to add a protocol to the European Convention on Human Rights that would provide a right to a safe and healthy environment to the regional guarantees. The European Social Charter mentions only the right to a safe working environment.[37]

The first case that reflected the convergence between human rights and the environment to reach the Court was *Powell v. United Kingdom*[38] in 1990. It dealt with the impact of excessive noise levels from the Heathrow airport on citizens living in the vicinity. While the court held that the applicants' rights under Article 8 had been infringed, the Court was of the view that it was justified based on the role that the Heathrow airport plays in the UK economy and the measures taken by the United Kingdom to abate the noise were within the margin of appreciation. A similar decision was reached by the Court in *Hatton v. United Kingdom*.[39] The Court was more open to the possibility of a violation of Article 8 in *Lopez Ostra v. Spain*.[40] Mrs. Ostra and family lived in a town in Spain that had a heavy concentration of leather industries. Due to fumes and smells emanating from tanneries, many people, including Mrs. Ostra's family, developed health problems. Expert evidence submitted in court showed that the symptoms displayed by Mrs. Ostra's daughter were consistent with living in a highly polluted area. The daughter's pediatrician recommended that she be moved from the area. The Court found a violation of Article 8 on the right to private and family life. Since then, a long line of cases has affirmed that environmental pollution and degradation can lead to a violation of protected rights.[41]

Since 1990, the European Court has developed a significant body of jurisprudence on environmental rights.[42] While most cases dealt with damage that has already taken place, in *Taskin v. Turkey*,[43] the Court "found a violation with respect to the operation of a gold mine, despite the absence of any accidents or incidents with the mine. The mine was deemed to present an unacceptable risk."[44] As noted earlier, the Court has developed several rules in relation to environmental rights: a fair balance should be struck between competing interests of the individual and the community as a whole; states enjoy a certain margin of appreciation; and courts will generally give deference to the decisions taken by public authorities. The Court has also applied principles of international environmental law, including procedural rights, and referred to judgements of other regional courts and international bodies, including the ICJ.[45]

A case that is imminent to be filed in the European Court of Human Rights involves a coal mine in Kosovo, the villagers of Hade and the World Bank. Community leaders say that they will take their dispute to the European Court "after a World Bank decision failed to end 12 years of displacement caused by a coal mine expansion."[46] Although the case is not directly related to climate change, because it involves a coal mine, it can be taken as indirectly related to climate change and could have implications for how relocations involving coal mines are treated by human rights bodies. The Word Bank Inspection Panel decided that these villagers have been left "in limbo" for more than a decade after being relocated many times but failing to be provided with new housing and services.[47] Ironically, the Bank is considering underwriting another coal power plant in the same area.[48] This will lead to more displacement of people, not to mention the emission of more greenhouse gases contributing to climate change. If the Bank were serious about addressing climate change, it should not fund any more coal power plants around the world. Given the number of recent climate cases at the national level in Europe,[49] it will probably be a matter of time before a climate case reaches the European Court of Human Rights.

The Inter-American system of human rights

The Inter-American Commission and Court of Human Rights were established under the American Convention on Human Rights, concluded in San Jose in 1969 and adopted within the Organization of American States (OAS). The American Declaration of Human Rights, adopted in 1949, preceded the Convention. While the Convention included only civil and political rights, other rights have been added through protocols. The San Salvador Protocol includes a right to a healthy environment: "1. Everyone shall have the right to live in a healthy environment and to have access to basic public services. 2. The States Parties shall promote the protection, preservation, and improvement of the environment."[50] By ratifying the Convention, states also accept the jurisdiction of the Commission to hear cases brought against them by individuals.[51] Only the Commission can refer petitions to the Court, and individuals have no standing before the Court. The Inter-American Court has developed considerable jurisprudence on environmental rights and was the first regional body to recognize the property rights of indigenous groups. It also developed the free, prior and informed consent principle in relation to Indigenous Peoples as discussed below.

Mayagna (Sumo) Awas Tingni Community v. Nicaragua[52] was the first time that the collective property rights of Indigenous Peoples were recognized. In the case of *Mary and Carrie Dann v. US*,[53] the Inter-American Commission held that any determination with regard to indigenous lands must "be based on fully informed consent of the whole community, meaning that all members be fully informed and have the chance to participate."[54] In *Maya Communities of the Toledo District v. Belize*,[55] the Commission held that Belize violated property rights of Indigenous Peoples by granting concessions with the lands "without effective consultations with and the informed consent of the Maya people."[56] The Commission stressed that the duty to consult is a fundamental component of the state's obligations with regard to communal property rights and that consultations should be held with the goal of obtaining consent.

The case of *Saramaka People v. Suriname*[57] involved resource concessions that the Suriname government had granted to private companies within the territories of Saramaka people without obtaining their consent or even consulting them. The Court held that Suriname had violated the rights of Saramaka people to judicial protection and property rights and failed to have effective mechanisms to protect them from acts that violate their rights to property. However, the Court noted that these property rights are not absolute and the State has the right to restrict property rights in the interests of society. This is similar to the jurisprudence developed by the European Court. These restrictions must be: previously established by law, necessary, proportionate and with the aim of achieving a legitimate objective in a democratic society. In addition, such restrictions cannot violate the right of Indigenous Peoples to survival. In order to do so, the Court prescribed a series of safeguards: (a) states must ensure effective participation of the affected parties; (b) states must guarantee that the affected people will receive a reasonable benefit from the project; and (c) prior to granting the concession, environmental and social impacts must be carried out to mitigate any negative effects.

Furthermore, participation must be in line with their customs and traditions; states have a duty to disseminate and receive information; and consultations must be in good faith, culturally appropriate and have the intent to reach an agreement. In the case of large-scale development projects that could impact the survival of indigenous people, states must obtain their free, prior and informed consent.[58] This case endorses very important principles and sheds light on what "consultation" entails. The Court limited the application of FPIC to large-scale development projects that may threaten the survival of indigenous people.[59] The need for consultations and the principle of free, prior and informed consent was endorsed by the Court again in *Kichwa People of Sarayaku v. Ecuador*.[60]

African system of human rights

The baby of the three regional systems – the African system – consisted of only the African Commission of Human Rights until recently. It was established by the Banjul Charter on Human and Peoples' Rights adopted by the African Union in 1981. The Protocol establishing the African Court of Human Rights[61] came into force in January 2005.[62] The African Charter explicitly recognizes a right to a healthy environment: "All peoples shall have the right to a general satisfactory environment favorable to their development."[63]

In the seminal case of *Ogoniland*,[64] the Commission elaborated on the scope of Article 24 and what the right to environment means:

> The right to a general satisfactory environment, as guaranteed under Article 24 of the African Charter or the right to a healthy environment, as it is widely known, therefore imposes clear obligations upon a government. It requires the State to take reasonable and other measures to prevent pollution and ecological degradation, to promote conservation, and to secure an ecologically sustainable development and use of natural resources. Article 12 of the International Covenant on Economic, Social and Cultural Rights (ICESCR), to which Nigeria is a party, requires governments to take necessary steps for the improvement of all aspects of environmental and industrial hygiene. The right to enjoy the best attainable state of physical and mental health enunciated in Article 16(1) of the African Charter and the right to a general satisfactory environment favourable to development (Article 16(3)) already noted obligate governments to desist from directly threatening the health and environment of their citizens.[65]

The Commission further elaborated on the right to environment:

> Government compliance with the spirit of Articles 16 and 24 of the African Charter must also include ordering or at least permitting independent scientific monitoring of threatened environments, requiring and publicising environmental and social impact studies prior to any major industrial development, undertaking appropriate monitoring and providing information to those communities exposed to hazardous materials and activities and providing meaningful opportunities for individuals to be heard and to participate in the development decisions affecting their communities.[66]

Thus, a distinct right to a healthy (or satisfactory) environment, far from being vague and indeterminate, imposes clear obligations on states to prevent pollution, promote conservation and secure sustainable development. To achieve those objectives, states have to undertake monitoring, require the preparation of environmental and social impact studies,[67] and provide information and meaningful opportunities for individuals to participate in development decisions.

Having discussed the existing regional human rights systems and their environmental rights jurisprudence briefly, let us now turn to a discussion of actual cases on climate change that have been brought before these regional bodies. As can be imagined, these cases are quite rare.

Cases on climate change

The Inter-American Commission of Human Rights was the first regional institution to specifically recognize the property rights of Indigenous Peoples.[68] Although the Inuit petition, discussed below, was dismissed, the Inter-American system has since then explicitly recognized the link between climate change and human rights and held a hearing on the subject in 2007.

The Inuit petition[69]

The petition brought by the Inuit Circumpolar Conference on behalf of Inuit of US and Canada against the US in the Inter-American Commission of Human Rights is the first case that utilized the human rights framework to litigate climate change. Even though the case itself was dismissed, it was successful in other respects. It succeeded in giving a human face to the problem,[70] as well as highlighting that consequences of climate change were already taking place and were affecting real communities and their traditional way of life. The petition helped dispel the belief held by many that climate change would affect some unknown community in a distant future.

In their petition to the Inter-American Commission, the Inuit people of Canada and the United States together with the Inuit Circumpolar Conference alleged that the US, as the highest contributor to greenhouse gases, is responsible for the damage that is being caused to the Arctic. The changes in the weather in the Arctic region are adversely affecting their traditional way of life, which, in turn, is affecting their human rights. The full petition, 175 pages in length, was complete with photographs, scientific information and data from the 2004 Arctic Climate Impact Assessment and signed by over 60 Inuit people.[71] The petition alleged that the US is by far the largest emitter of greenhouse gases and "thus bears the greatest responsibility among nations for causing global warming."[72]

The petition summarized the climate change impacts on the Inuit people as follows: because of the changes in the climate, it has become harder for the Inuit to predict weather, which is important for their hunting practices, travel and food preservation methods. The Inuit culture, which is intrinsically linked to snow and ice, is being increasingly threatened by the changing climate:

> Like many indigenous peoples, the Inuit are the product of the physical environment in which they live. The Inuit have fine-tuned tools, techniques and knowledge over thousands of years to adapt to the arctic environment. They have developed an intimate relationship with their surroundings, using their understanding of the arctic environment to develop a complex culture that has enabled them to thrive on scarce resources. The culture, economy and identity of the Inuit as an indigenous people depend upon the ice and snow.[73]

The changing climate has given rise to many changes in the Arctic environment. Sea ice, which is a critical resource for the Inuit for travel, harvesting and communication, has become thinner and is freezing later and thawing earlier, thus affecting the everyday life of the Inuit. Moreover, the quantity, quality and timing of snowfall have also changed. Lack of igloo-quality snow is causing problems to travelers and is also resulting in loss of igloo building knowledge, which is an important component of Inuit culture.

Humans are not the only species being affected by a changing climate. Marine species that depend on sea ice are likely to decline, with some even facing extinction. Other species that are important for harvesting are moving to new locations, exacerbating travel problems for the

Inuit people. In addition, higher temperatures and sun intensity have increased the risk of health problems such as sunburn, skin cancer and cataracts. Moreover, it is predicted that new diseases like West Nile Virus will pose health risks. The petition noted that while current impacts in the Arctic are severe, the projected impacts are expected to be much worse.[74]

The petitioners noted that several principles of international law are relevant here. United States is a member of the Organization of American States and has signed the American Declaration of the Rights and Duties of Man. Moreover, it has ratified the International Covenant on Civil and Political Rights and has signed the International Covenant on Economic, Social and Cultural Rights. In addition, the US is bound by international environmental law obligations, including the obligation not to cause transboundary harm, and is a party to the United Nations Framework Convention on Climate Change (UNFCCC) under which it has committed to implementing policies to return to its 1990 levels of greenhouse gas (GHG) emissions.

The petitioners argued that the impacts of climate change, caused by acts and omissions by the US, violate the Inuit's human rights protected by the American Declaration. These include the rights to culture, property, health, life, physical integrity, security, means of subsistence, and residence, the freedom of movement and the inviolability of the home.[75] Despite the ratification of the UNFCCC, the US had repeatedly declined to take steps to regulate its GHG emissions. The petitioners alleged that the US was the largest contributor to GHG emissions in the world, and with full knowledge that its action is radically transforming the Arctic environment, "the United States has persisted in permitting the unregulated emission of greenhouse gases from within its jurisdiction into the atmosphere."[76]

The petitioners requested the Commission to: (a) make an onsite visit to investigate the harm suffered by the Inuit; (b) hold a hearing; (c) prepare a report declaring that the US is internationally responsible for violating the rights in the American Declaration; (d) recommend that the US adopt mandatory measures to limit its emissions; (e) establish and implement a plan, in coordination with the affected Inuit, to protect Inuit culture and resources; (f) establish and implement a plan to provide assistance necessary to adapt to the impacts of climate change that cannot be avoided; and (g) provide any other relief that the Commission considers appropriate and just.[77]

The Commission dismissed the petition without prejudice, stating that the information provided was insufficient to make a determination.[78] However, at the petitioners' request, the Commission held a hearing to address matters relating to global warming and human rights.[79] Sheila Watt-Cloutier of the Inuit Circumpolar Conference, Martin Wagner of Earthjustice and Daniel Magraw of the Center for International Environmental Law attended the hearing.[80] Although the Commission did not give a decision on the merits of the case, the petition broke new ground in international law. As Osofsky notes:

> The Inuit petition serves as an important example of creative lawyering in both substance and form. It reframes a problem typically treated as an environmental one through a human rights lens, and moves beyond the confines of U.S. law to a supranational forum. In so doing, the petition lies at the intersection of two streams of cases occurring at multiple levels of governance: (1) environmental rights litigation and petitions and (2) climate litigation and petitions.[81]

The petition also exemplified the challenges inherent in this type of litigation. Given that climate change is a global problem, it involves multiple actors in every state, so establishing causation is challenging. Moreover, given the disproportionate contribution of states to causing the problem, the current state responsibility principles are ill-suited to apply to climate change. Another issue is the time lag between emissions and the consequences. In addition, damage to future generations cannot be captured within this framework. Finally, what would be an appropriate remedy in this situation? For example, how can one compensate for loss of culture?[82] It is thus no surprise that the Inuit petition was dismissed by the Inter-American Commission.

At the hearing held by the Commission on climate change and human rights, questions raised by the Commissioners offer "perhaps the best indication of the challenges that future litigation over human rights violations as consequence of climate change will face:"[83] (1) how to attribute or divide responsibility among states in the region and even those who are not in the region; (2) how the rights violations suffered by the Inuit could be linked to acts or omissions of specific states; (3) whether petitioners had exhausted domestic remedies; and (4) are there any examples of good practice undertaken by states that could guide the Commission?[84]

Of course, there is no logical reason why we cannot base state responsibility on the common but differentiated responsibility principle (CBDR).[85] After all, the international community was willing to base their obligations in relation to climate change on this principle – a huge breakthrough in international law.[86] If obligations are based on the CBDR principle, the legal regime that comes into play when a violation of that obligation has taken place[87] should also be based on that principle. While this makes perfect sense in a perfect world, given how controversial the CBDR principle has become, it is highly unlikely that states will be willing to accept it in relation to liability as well.

Athabaskan petition[88]

Resembling the Inuit petition very closely, the Arctic Athabaskan Peoples petition to the Inter-American Commission on Human Rights argues that black carbon pollution from Canada is harming the Arctic environment and the ecosystems of the Arctic region. The Arctic is warming twice as fast as the rest of the world is and black carbon emissions from Canada is a major contributor to this warming. They argued that because black carbon is a potent climate warming agent and is emitted near the Arctic, it has a significantly higher climate warming impact. Arctic warming is dramatically changing the Arctic environment and damaging its natural resources. This, in turn, is having a huge impact on Athabaskan peoples and their way of life. As one Special Rapporteur concluded in 2005: "the effects of global warming and environmental pollution are particularly pertinent to the life chances of Aboriginal people in Canada's North, a human rights issue that requires urgent attention at the national and international levels."[89]

The petitioners argued that "the rapid warming and melting in Athabaskan Lands, caused in significant part by Canada's failure to reduce black carbon emissions, violates a number of the Arctic Athabaskan peoples' fundamental rights."[90] These included the right to culture, property, means of subsistence and health. The petition argued that while black carbon is considered a "short lived" pollutant, it is also identified as a particularly potent pollutant in regions of ice and snow. There are many steps that Canada can take to reduce its black carbon emissions that could have a significant impact on reducing the warming in the region. Arctic warming and melting is adversely affecting this indigenous group's ability to pass their cultural knowledge to future generations. This knowledge has been developed over millennia, but due to the warming and melting, the weather is no longer predictable and the behavior of wildlife has become so erratic that elders are no longer confident about teaching the younger generation their traditional ways.

The petitioners further noted that the American Declaration guarantees their right to culture, which has been affirmed by the Inter-American Court of Human Rights in several cases – the failure to prevent environmental damage to indigenous lands can cause "catastrophic damage" to Indigenous Peoples. For Indigenous Peoples, land is closely linked to their cultural practices, customs and language. It has repeatedly recognized that the relationship of Indigenous Peoples with their territory is "crucial for their cultural structures and requires special measures under international human rights law in order to guarantee their physical and cultural survival."[91] They further alleged that their right to subsistence is being affected by changing weather patterns, forest fires and unpredictable storms and other severe weather events. These events are also affecting their right to property guaranteed under the American Declaration. Moreover, the warming has led to a loss of traditional food. This loss has adversely affected the health of Athabaskan peoples, as they can no longer obtain food through traditional hunting, fishing and gathering; they have to supplement their diets with store-bought food, which is less healthy and leads to an increased prevalence of chronic diseases. Their water quality has also been affected, and the likelihood of disease and injury due to dangerous conditions has caused psychological stress.

The petitioners requested the Commission to (a) investigate and confirm the harms suffered by the Athabaskan people as a result of Arctic warming and melting; (b) prepare a report declaring that Canada's failure to take steps to substantially reduce its black carbon emissions violates the rights recognized in the American Declaration of Rights and Duties of Man; and (c) recommend that Canada take steps to limit black carbon emissions and protect Athabaskan culture and resources from the effects of Arctic warming and melting. This petition is currently pending before the Commission.

This case is very similar in scope to the Inuit petition. It remains to be seen how the Inter-American Commission will handle the petition. Unlike the Inuit petition, because the Athabaskan petition requests Canada to take action domestically to reduce black carbon, which will have a noticeable impact in the Arctic region, there is a possibility that the Commission will consider the petition favorably. Moreover, unlike in 2007 when the Inuit petition was considered, there is much more attention paid to climate change by the international community as well as by the Inter-American Commission. Scientists are warning of even more dire consequences than predicted before, so it would be hard for the Commission to turn a blind eye to this petition.

Other action relating to climate change by regional bodies

Ironically, barely 10 years after dismissing the Inuit petition on the ground of insufficient information, the Inter-American Commission issued a press release expressing its concern on the impact of climate change on human rights prior to the climate conference in Paris in 2015.[92] It noted that:

> The Inter-American Commission on Human Rights (IACHR) expresses its concern regarding the grave harm climate change poses to the universal enjoyment of human rights. Ahead of the upcoming Conference of Parties to the United Nations Framework Convention on Climate Change (COP21), to be held in Paris December 7–8, the IACHR urges the Member States of the Organization of American States (OAS) to work to ensure that any climate agreement reached there incorporates human rights in a holistic manner.[93]

The Commission noted that climate change poses a threat to all three pillars of sustainable development and in particular undermines efforts to eradicate poverty in the region. It reminded states that at COP 16 in Cancun, the parties agreed to respect human rights in all actions relating to climate change. It also recalled the latest report of the Intergovernmental Panel on Climate Change (IPCC) and the dire consequences on people if we fail to limit the temperature increase to 2° Celsius. It highlighted the consequences of climate change on human rights:

> Climate change affects human rights in different ways. The consequences of climate change lead to deaths, injuries, and displacement of individuals and communities because

> of disasters and events such as tropical cyclones, tornadoes, heat waves, and droughts. The Inter-American Commission has received hundreds of cases related to conflicts over land and water and threats to food sovereignty which evidence that climate change is a reality that is affecting the enjoyment of human rights in the region.[94]

Moreover, it noted that people who are already in vulnerable situations would be disproportionately affected and their vulnerabilities will multiply. Furthermore, the Commission recognized that Indigenous Peoples are especially at risk:

> Climate change will most severely affect the lives of those who are already more vulnerable and whose human rights have been affected, including women, children, rural communities, the elderly, and people living in poverty. In addition, climate change has a special impact on indigenous peoples, whose lands and natural resources come under direct threat. Some individuals and communities will be forced to migrate, and those who do not have the opportunity to do so may end up trapped in situations of environmental risk.[95]

The Organization for American States, the parent organization of the Inter-American human rights system, adopted a resolution on climate change and human rights in 2008.[96] Recalling that the adverse consequences of climate change may have a negative impact on the enjoyment of human rights, OAS instructed the Commission to contribute "to the efforts to determine the possible existence of a link between adverse effects of climate change and the full enjoyment of human rights,"[97] coordinating its efforts with the United Nations Human Rights Council and the Office of the United Nations High Commissioner for Human Rights.[98]

Another way that the Inter-American Commission on Human Rights has been used in relation to climate change is in respect of projects that are certified as falling within the clean development mechanism (CDM) under the Kyoto Protocol.[99] Some of these CDM projects, while supposedly contributing to sustainable development and reducing greenhouse gases, have resulted in massive human rights violations.[100] One such project is the Barro Blanco hydropower project in Panama. This project would have resulted in displacing indigenous communities, inundating homes and agricultural land, and destroying cultural sites. No consultations were held with these communities and the project was registered as a CDM project in 2011. However, NGOs complained to the CDM Executive Board, and the Inter-American Commission held a hearing on the issue.[101] Due to the impact on human rights of the people in the area, the Panamanian government withdrew registration of this project as a CDM project, the first time a government has done so.[102]

The Bajo Aguan biogas recovery project in Honduras has a similar narrative. This project is directly associated with the murder, disappearance and torture of dozens of campesinos. Over 70 organizations sent an open letter to the UK government asking it to withdraw CDM funding for this project, which was associated with massive human rights violations. Complaints were sent to the CDM Executive Board as well. In this case, the Inter-American Commission on Human Rights conducted an onsite visit and held a hearing on the alleged murder of human rights defenders. The Commission observed that the situation there "continues to be highly worrisome."[103] This is an effective way of using existing human rights institutions without actually resorting to litigation.

Similar to the Inter-American Commission on Human Rights, the African Commission on Human and Peoples' Rights has adopted several resolutions on climate change and human rights. In its resolution adopted in 2009, the Commission referred to its own mandate to promote and protect the rights under the African Charter, especially the right of peoples to economic, social and cultural development and the right of peoples to a satisfactory environment favorable to their development.

It also referred to the UN Declaration on the Rights of Indigenous Peoples, African Convention on the Conservation of Nature and Natural Resources, and the Convention on Biological Diversity. The Resolution urged the Assembly of Heads of State and Government of the African Union

> to ensure that human rights standards, safeguards, such as the principle of free, prior and informed consent, be included in any adopted legal text on climate change as preventive measures against forced relocation, unfair dispossession of properties, loss of livelihoods and similar human rights violations.[104]

It also called upon the Assembly to ensure that special measures of protection of vulnerable groups are included in any international agreement on climate change. These groups include children, women, indigenous groups, the elderly, victims of natural disasters and conflicts. Furthermore, the Commission decided to carry out a study on the impact of climate change on human rights in Africa.

In its resolution on the same topic adopted in 2016, the Commission noted that the implementation of the UNFCCC and the Paris Agreement should adequately reflect the African perspective on human rights, especially the right to a general satisfactory environment favorable to their development, the right to development, and the right to health. It expressed concern about the failure of developed countries to comply with their obligations to take the lead in mitigation "while creating enabling conditions for African countries to realise their right to sustainable development and to adapt to climate change."[105] It further expressed its concern that lack of cooperative action, including technology transfer and financial assistance for mitigation and adaptation, seriously undermined the capacity of African governments to safeguard human rights in Africa. It again urged member states to adopt special measures to protect vulnerable groups and requested the Working Group on Economic and Social Rights in collaboration with the Working Group on Extractive Industries, Environment and Human Rights Violations to undertake a "study on the impact of climate change on human rights in Africa" and to present it within two years.[106] In addition, the African Union adopted the African Strategy on Climate Change in 2014.[107]

Conclusion

Although only a couple of cases specifically on climate change have been filed before the regional human rights systems, the jurisprudence developed by them in relation to environmental rights is relevant for future cases. This jurisprudence also offers guidance to states when addressing issues relating to climate change, especially in relation to mitigation measures and adaptation measures. With regard to damage caused by historic emissions, the situation is a little murky because of the archaic legal principles that govern state responsibility. Because it is virtually impossible to establish the causal link between damage to people and the emissions of a particular state, contemporary human rights law does not offer much relief to victims of climate change. On the other hand, if courts were willing to adopt novel theories of liability similar to those adopted at the national level,[108] human rights law will be better equipped to offer relief to victims of climate change.

With regard to mitigation and adaptation measures that states adopt in response to climate change, the situation is very different. States are required to ensure that they do not violate the rights of their citizens when taking action on climate change. Procedural rights also play an important role here. Regional human rights bodies can play an important role in these situations, scrutinizing mitigation and adaptation measures that states take in response to climate change, whether they are dealing with emergencies such as severe weather events or slow-onset events such as sea level rise. The Special Rapporteur on Human Rights and the Environment identifies three distinct advantages of bringing human rights into a discussion on climate change:

> Bringing human rights to bear on climate change has three principal benefits. First, advocacy grounded in human rights can spur stronger action . . . Second, human rights norms clarify how States should respond to climate change . . . Third, human rights bodies can inform and improve climate policy by providing forums for issues concerning climate change and human rights that might otherwise be overlooked.[109]

It is encouraging that these regional human rights institutions have recognized the adverse impact of climate change on the enjoyment of rights protected under human rights law. As noted, adverse consequences of climate change have serious ramifications for the enjoyment of rights. Thus, regional human rights institutions can be useful to victims of adverse effects of climate change if the current archaic principles in relation to state responsibility can be discarded in favor of a more advanced system of liability that can accommodate a problem as complex as climate change. Human rights law operates vertically between duty bearers (states) and rights holders (citizens), and because climate change is a global problem, greenhouse gas emissions that give rise to climate change and its adverse consequences do not fall within the same polity.[110] Therefore, it is hard to apply a human rights framework to damage caused by climate change. In general, human rights law does not operate extraterritorially, and because countries in the global North are more responsible for the emission of greenhouse gases while countries and vulnerable communities in the global South are disproportionately affected by climate change,[111] a human rights framework alone cannot provide relief for victims of climate change.[112] Developing novel theories of liability similar to those developed at the national level together with a justice framework[113] might help these hapless victims.

It is hoped that regional human rights bodies, which have been at the forefront of developing environmental rights, will take up the challenge of climate change and provide relief to victims of climate change. With regard to mitigation measures and adaptation measures, it would be easier to apply a human rights framework because we would still be operating within the vertical framework of human rights. Thus, if a citizen's rights are violated as a result of measures taken in response to climate change, be it mitigation or adaptation, the state could be held accountable under human rights law. However, the principles developed by these regional bodies – balancing rights with societal benefits, margin of appreciation and giving deference to decisions by government authorities – would apply to these potential cases.

Notes

* The author draws from chapter 11 of her book on adjudicating climate change. See S. Atapattu, *Human Rights Approaches to Climate Change: Challenges and Opportunities* (2016).

1 See *Climate Change – the Greatest Threat to Human Rights in the 21st Century – John Knox, UN Special Rapporteur on Human Rights and the Environment*, available online at www.ucl.ac.uk/global-governance/ggi-events/john-knox-climatechange and Mary Robinson Foundation Climate Justice, *Position Paper: Human Rights and Climate Justice* (2014), available online at www.mrfcj.org/media/pdf/PositionPaper-HumanRightsandClimateChange.pdf.

See also *Human Development Report 2007/08: Fighting Climate Change: Human Solidarity in a Divided World*, available online at http://hdr.undp.org/sites/default/files/reports/268/hdr_20072008_en_complete.pdf which describes climate change as the defining human development issue of our times.

2 S. Atapattu, *Human Rights Approaches to Climate Change: Challenges and Opportunities* (2016) 267; M. G. Faure, Andre Nolkaemper and Amsterdam International Law Clinic, *Analyses of Issues to Be Addressed: Climate Change Litigation Cases* (2007).
3 See R. H. J. Cox, 'The Liability of European States for Climate Change', Case Note, 30 *Utrecht Journal of International and European Law* (2014) 78.
4 *Case Concerning Gabcokovo Nagymaros Project (Hungary v. Slovakia)*, 25 September 1997, ICJ Reports (1997).
5 *Nuclear Tests Cases (Australia and New Zealand v. France)*, ICJ Reports (1974); Request for an Examination of the Situation in Accordance with Paragraph 63 of the Court's Judgment of 20 December1974 in the Nuclear Tests Case (N.Z. v. Fr.), 22 September 1995, I.C.J. 288, 319–63.
6 *Pulp Mills on the River Uruguay (Argentina v. Uruguay)*, ICJ Reports (2010); *Legality of the Threat of Nuclear Weapons*, Advisory Opinion, ICJ Reports (1996).
7 *Case Concerning Aerial Herbicide Spraying (Ecuador v. Colombia)*, ICJ Reports (2013).
8 See D. Hunter, J. Salzman and D. Zaelke, *International Environmental Law and Policy* (2015) at chapter 18.
9 There are numerous scholarly writings and reports on the topic. See, generally, S. Humphreys, *Human Rights and Climate Change* (2010); J. Knox, 'Climate Change and Human Rights', 50 *Virginia Journal of International Law* (2009) 795; P. Stephens, 'Applying Human Rights Norms to Climate Change: The Elusive Remedy', 21 *Colorado Journal of International Environmental Law and Policy* (2010) 49; A. Sinden, 'Climate Change and Human Rights', 27 *Journal of Land, Resources & Environmental Law* (2007) 255; D. Bodansky, 'Climate Change and Human Rights: Unpacking the Issues', 38 *Georgia Journal of International and Comparative Law* (2010) 511; and citations in Atapattu, *supra* note 2, at 75, fn 27.
10 *Report of the Office of the United Nations High Commissioner for Human Rights on the Relationship Between Climate Change and Human Rights*, A/HRC/10/61 (15 January 2009), at para 92.
11 J. Knox, *Report of the Special Rapporteur on the Issue of Human Rights Obligations Relating to the Enjoyment of a Safe, Clean, Healthy and Sustainable Environment*, A/HRC/31/52 (1 February 2016); Atapattu, *supra* note 2, chapter 3.
12 UNFCCC, *Paris Agreement*, available online at https://unfccc.int/resource/docs/2015/cop21/eng/l09.pdf, at Preamble (emphasis added).
13 See S. Atapattu, 'Climate Change, Human Rights and COP 21: One Step Forward and Two Steps Back or Vice Versa?' XVII *Georgetown Journal of International Affairs* (2016), 47.
14 See J. Knox, 'The Paris Agreement as a Human Rights Treaty', in D. Roser (ed.), *Human Rights in the 21st Century* (forthcoming). The Aarhus Convention was the first environmental treaty to include a right to a healthy environment, but this is a regional treaty.
15 Atapattu, *supra* note 2, chapter 5.
16 J. Peel and H. M. Osofsky, *Climate Change Litigation: Regulatory Pathways to Cleaner Energy* (2015) 4–5.
17 *Ibid.*, at 5. They point out that cases at the edge of these categories – such as fracking – are harder to classify. One could add cases that deal with damage caused by climate change, such as the Inuit case, discussed below, as another category of cases that deals with damage caused by historic emissions.
18 *Ibid.*, at 8.
19 D. Hunter, 'The Implications of Climate Change Litigation: Litigation for International Environmental Law-making', in W. Burns and H. Osofsky (eds.), *Adjudicating Climate Change: State, National and International Approaches* (2009) 357.
20 Inuit petition, *infra* note 71, and accompanying text.
21 See O. de Schutter, *International Human Rights Law* (2010), 898.
22 *Asian Human Rights Charter*, available online at www.refworld.org/pdfid/452678304.pdf. The Charter contains a provision on environmental protection and sustainable development but is framed as an objective and not as a right.
23 ESCR-Net, *Asian Human Rights Commission*, available online at www.escr-net.org/member/asian-human-rights-commission-ahrc.
24 Association of Southeast Asian Nations, *Human Rights Declaration*, available online at www.asean.org/storage/images/ASEAN_RTK_2014/6_AHRD_Booklet.pdf.
25 *Ibid.*, Art. 28(f).
26 *Arab Charter on Human Rights*, available online at http://hrlibrary.umn.edu/instree/loas2005.html.
27 *Ibid.*, Art. 38.
28 *Ibid.*, Art. 48.
29 See K. Hay, *A Pacific Human Rights Mechanism: Specific Challenges and Requirements*, available online at www.victoria.ac.nz/law/research/publications/about-nzacl/publications/special-issues/hors-serie-volume-viii,-2008/Hay.pdf.

30 Veronica de law R. Jaimes, 'Climate Change and Human Rights Litigation in Europe and the Americas', 5 *Seattle Journal of Environmental Law* (2015) 165.
31 See Schutter, *supra* note 21, at 899. He refers to the procedure for complaints established under the Convention as "revolutionary" – it recognized inter-state applications resembling *actio popularis* that allows for collective enforcement. The other revolutionary feature was the ability of individuals to seek a remedy before an international tribunal although initially they were not allowed to petition the court directly, which remained the prerogative of the Commission. The Commission and the Court were fused into one body by Protocol No 11 adopted in November 1998, and individuals were given direct standing before the court, *ibid.*, at 901.
32 D. Anton and D. Shelton, *Environmental Protection and Human Rights* (2011) 336.
33 *Ibid.*, at 336, quoting T. Burguenthal 'The Evolving International Human Rights System', 100 *American Journal of International Law* (2006) 783, at 792–794.
34 *Ibid.*, 337.
35 See O. Pedersen, 'European Environmental Human Rights and Environmental Rights: A Long Time Coming?', 21 *Georgetown International Environmental Law Review* (2008) 73.
36 Council of Europe, *Manual on Human Rights and the Environment* (2006), available online at www.echr.coe.int/LibraryDocs/DH_DEV_Manual_Environment_Eng.pdf.
37 D. Shelton, 'Whiplash and Backlash – Reflections on a Human Rights Approach to Environmental Protection, 13 *Santa Clara Journal of International Law* (2015) 11, at 18–19, footnotes omitted. [*Whiplash*] See also Pedersen, *supra* note 35.
38 *Powell v. United Kingdom*, 172 Eur. Ct. H.R. (ser. A) (1990) 18–20, and Pedersen, *supra* note 35, at 85.
39 *Hatton v. United Kingdom*, 2003-VIII Eur. Ct. H. R. 189, 228 (2003) and Pedersen, *supra* note 35.
40 Application No 16798/90, 20 European Human Rights Reports 277 (1994) (Eur. Ct. H.R.).
41 See Hunter *et al.*, *supra* note 8, at 1350; Anton and Shelton, *supra* note 32, chapter 7.
42 See D. Shelton, 'Human Rights and the Environment: What Specific Environmental Rights Have Been Recognized?', 35 *Denver Journal of International Law* & Policy (2006) 129.
43 *Taşkin v. Turkey*, 2004-X Eur. Ct. H.R. 185 (2004).
44 Shelton, *Whiplash*, *supra* note 37, at 25.
45 *Ibid.*
46 See K. Mathiesen, *Kosovan Villagers to Take Coal Mine Woes to Human Rights Court* (19 December 2016), available online at www.climatechangenews.com/2016/12/19/kosovan-villagers-to-take-coal-mine-woes-to-echr/.
47 *Ibid.*
48 *Ibid.*
49 These cases include Urgenda (the Netherlands), available online at www.urgenda.nl/en/climate-case/legal-documents.php; Klimaatzaak (Belgium), see J. M. Klein, *Lawsuit Seeks to Force Belgian Government to Take Action Against Climate Change* (2015), available online at http://blogs.law.columbia.edu/climatechange/2015/06/08/lawsuit-seeks-to-force-belgian-government-to-take-action-against-climate-change/; Klima Seniorinnen (Switzerland), available online at http://klimaseniorinnen.ch/wp-content/uploads/2016/10/161024_summary_swiss-climate-case_def.pdf; and Magnolia case (Sweden), see *Youth Sue Swedish State on Climate Grounds*, available online at www.generosity.com/community-fundraising/youth-sue-swedish-state-on-climate-grounds—2. An interesting case currently on appeal in German courts relates to a farmer from the Peruvian Andes who is suing the German energy company RWE, "demanding the firm take responsibility for its CO_2 emissions and help reduce the risk of flooding" in the Andes. See *Andean Farmer Demands Climate Justice in Germany*, available online at http://glacierhub.org/2017/02/02/andean-farmer-demands-climate-justice-in-germany/.
50 *Additional Protocol to the American Convention on Human Rights in the Area of Economic, Social and Cultural Rights* (Protocol of San Salvador), OAS Treaty Series No 69, available online at www.oas.org/juridico/english/treaties/a-52.html, Art. 11.
51 Anton and Shelton, *supra* note 32, at 340, referring to Burguenthal, *supra* note 33.
52 See IACtHR, *Mayagna (Sumo) Awas Tingni Community v. Nicaragua* (2011). All IACtHR decisions available online at www.corteidh.or.cr/index.php/en/jurisprudencia.
53 See IACtHR, *Mary and Carrie Dann v. US* (2002).
54 *Ibid.*
55 See IACtHR, *Maya Indigenous Community of the Toledo District v. Belize* (2004).

56 See T. Ward, 'The Right to Free, Prior and Informed Consent: Indigenous Peoples' Participation Rights Within International Law', 10 *Northwestern Journal of International Human Rights* (2011) 54, at 63.
57 See IACtHR, *Saramaka People v. Suriname* (2007).
58 See Ward, *supra* note 56, at 64, who believes that this case clearly set a precedent within the Inter-American system.
59 The language of the *UN Declaration on Rights of Indigenous Peoples (UNDRIP)* in relation to FPIC, however, is not so restrictive, available online at www.un.org/esa/socdev/unpfii/documents/DRIPS_en.pdf.
60 IACtHR, *Kichwa Indigenous People of Sarayaku* (27 June 2012).
61 *Protocol to the African Charter on Human and People's Rights on the Establishment of the African Court on Human and Peoples' Rights* (9 June 1998), available online at www.achpr.org/instruments/court-establishment/.
62 See Anton and Shelton, *supra* note 32, at 353.
63 *African Charter on Human and Peoples' Rights*, available online at www.achpr.org/instruments/achpr/, at Art. 24.
64 African Commission on Human and Peoples' Rights, *The Social and Economic Rights Action Center and the Center for Economic and Social Rights v. Nigeria* (2001), available online at http://www1.umn.edu/humanrts/africa/comcases/155-96.html.
65 *Ibid.*, at para 52.
66 African Commission on Human and Peoples' Rights, *supra* note 64, at para. 53.
67 In the *Pulp Mills Case*, *supra* note 6, the ICJ stated that transboundary environmental impact assessment is now part of customary international law.
68 See discussion in section 2.2.
69 This section draws from author's book: S. Atapattu, *Human Rights Approaches to Climate Change: Challenges and Opportunities* (2016).
70 Because climate change originated as an environmental problem, people still tend to consider it as such. Recognizing that there are actual victims of climate change was important to highlight that adverse impacts of climate change were having a significant impact on peoples' lives.
71 *The Petition to the Inter American Commission on Human Rights Seeking Relief From Violation Resulting From Global Warming Caused by Acts and Omissions of the United States*, Summary of the Petition (2005), available online at http://earthjustice.org/sites/default/files/library/legal_docs/petition-to-the-inter-american-commission-on-human-rights-on-behalf-of-the-inuit-circumpolar-conference.pdf. *[Inuit petition]*
72 *Ibid.*, at 9.
73 *Ibid.* at 1.
74 *Ibid.*, at 4.
75 *Ibid.*, at 5.
76 *Ibid.*, at 7.
77 *Ibid.*, at 7–8.
78 See J. Gordon, 'Inter-American Commission to Hold Hearing After Rejecting Inuit Climate Change Petition', 7 *Sustainable Development Law & Policy* (2007), available online at http://digitalcommons.wcl.american.edu/cgi/viewcontent.cgi?article=1239&context=sdlp.
79 Earthjustice, *Inter-American Commission on Human Rights to Hold Hearing on Global Warming* (2007), available online at http://earthjustice.org/news/press/2007/inter-american-commission-on-human-rights-to-hold-hearing-on-global-warming.
80 See *Testimony of Martin Wagner Before the Inter-American Commission on Human Rights* (2007), available online at http://earthjustice.org/news/press/2007/global-warming-human-rights-gets-hearing-on-the-world-stage.
81 See H. M. Osofsky, 'The Inuit Petition as a Bridge? Beyond Dialectics of Climate Change and Indigenous Peoples' Rights', in W. C. G. Burns and H. M. Osofsky (eds.), *Adjudicating Climate Change: State, National, and International Approaches* (2009) 273.
82 See Atapattu, *supra* note 2, at 274–281.
83 M. Chapman, 'Climate Change and the Regional Human Rights Systems', 10 *Sustainable Development Law & Policy* (2010) 37.
84 *Ibid.*, at 38. See also A. Parsons, 'Human Rights and Climate Change: Shifting the Burden to the State?', 9 *Sustainable Development Law & Policy* (2009) 22, who believes that "a human rights approach to climate change may be hardest to implement in countries that need it most."

85 See L. Rajamani, *Differential Treatment in International Law* (2006); S. Atapattu, *Emerging Principles of International Environmental Law* (2006), chapter 5; Atapattu, *supra* note 2, at 281.
86 See United Nations Framework Convention on Climate Change (UNFCCC), Art. 3 and Kyoto Protocol to the UNFCCC.
87 See International Law Commission (ILC), *Draft Articles on Responsibility of States for Internationally Wrongful Acts* (2001), available online at http://legal.un.org/ilc/texts/instruments/english/draft_articles/9_6_2001.pdf.
88 *Petition to the Inter-American Commission on Human Rights seeking Relief From the Violations of Rights of Arctic Athabaskan Peoples Resulting From Rapid Arctic Warming and Melting Caused by Emissions of Black Carbon From Canada* (2013), available online at http://earthjustice.org/sites/default/files/AAC_PETITION_13-04-23a.pdf. [Athabaskan petition]. See also S. Khan, 'Connecting Human Rights and Short-Lived Climate Pollutants: The Arctic Angle' (this volume).
89 Athabaskan petition summary, *Ibid.*
90 *Ibid.*
91 *Ibid.*
92 *IACHR Expresses Concern Regarding Effects of Climate Change on Human Rights* (December 2015), available online at www.oas.org/en/iachr/media_center/PReleases/2015/140.asp.
93 *Ibid.*
94 *Ibid.*
95 *Ibid.*
96 Organization of American States, Human Rights and Climate Change in the Americas, General Assembly, AG/RES. 2429 (XXXVIII-O/08), adopted at the fourth plenary session (June 3, 2008).
97 *Ibid.*
98 *Ibid.*
99 Article 12 of the Kyoto Protocol established the clean development mechanism. See *Kyoto Protocol to the United Framework Convention on Climate Change* (1997), available at http://unfccc.int/resource/docs/convkp/kpeng.pdf.
100 Atapattu, *supra* note 2, at 134.
101 *Ibid.*, at 135.
102 A. Chatziantoniou and K. Alford-Jones, *Panama Withdraws Problematic Barro Blanco Dam Project From CDM Registry, Center for International Environmental Law* (12 January 2016), available online at www.ciel.org/panama-withdraws-problematic-barro-blanco-dam-project-cdm-registry/.
103 Atapattu, *supra* note 2, at 136.
104 African Commission on Human and Peoples' Rights, Resolution, (2009), at para 1.
105 African Commission on Human and Peoples' Rights, *supra* note 104, at preamble.
106 *Ibid.*
107 *African Strategy on Climate Change* (2014), available online at www.un.org/en/africa/osaa/pdf/au/cap_draft_auclimatestrategy_2015.pdf.
108 See S. Atapattu, 'Climate Change, Differentiated Responsibilities and State Responsibility: Devising Notel Legal Strategies for Damage Caused by Climate Change', in B. Richardson, Y. Le Bouthillier, H. McLeod-Kilmurray and S. Wood, (eds.), *Climate Law and Developing Countries: Legal and Policy Challenges for the World Economy* (2009) 37; See R. P. Murray. '*Sindell v. Abbott Laboratories:* Using the Market Share Approach to DES Causation', 69 *California Law Review* (1981) 1179; S. Lawson, 'The Conundrum of Climate Change Causation: Using Market Share Liability to Satisfy the Identification Requirement in *Native Village of Kivalina v. Exxonmobil Co*', 22 *Fordham Environmental Law Review* (2011) 433; J. Kilinski, 'International Climate Change Liability: A Myth or Reality?', 18 *Journal of Transnational Law and Policy* (2009) 401, where she notes that "a liability compensation scheme similar to that in *Sindell v. Abbott Laboratories* may be ideal in the realm of climate change" (footnotes omitted).
109 Human Rights Council, *Report of the Special Rapporteur on the Issue of Human Rights Obligations Relating to the Enjoyment of a Safe, Clean, Healthy and Sustainable Environment* (1 February 2016), A/HRC/31/52.
110 See J. Knox, 'Climate Change and Human Rights', 50 *Virginia Journal of International Law* (2009) 163.
111 See S. Atapattu, 'Justice for Small Island Nations: Intersections of Equity, Human Rights, and Environmental Justice', in R. Abate (ed.), *Climate Justice: Case Studies in Global and Regional Governance Challenges* (2016) 299.
112 *Ibid.*
113 *Ibid.*

10

From Copenhagen to Paris at the UN Human Rights Council

When climate change became a human rights issue

Felix Kirchmeier and Yves Lador

Introduction

Over the last decade, the evidence that climate change already has and will continue to have an impact on the enjoyment of human rights has become an increasingly important issue on the agenda of the Human Rights Council (HRC) and of its Special Procedures.

This discussion progressively established a rather broad consensus, acknowledging the impact of climate change on the enjoyment of human rights, as well as the contributions that human rights–based approaches can bring to the adaptation to and the mitigation of climate change.

Also interesting has been the outreach efforts made by the HRC and its Special Procedures toward the negotiations at the UN Framework Convention on Climate Change. It indicates an awareness that actions on human rights need to cut across institutional boundaries.

This process at the HRC has been carried out mainly by States that are most immediately affected by climate change. The Maldives first introduced the issue in 2007. In 2011, other States, including Bangladesh and the Philippines, carried the resolution further. Members of the Climate Vulnerable Forum (VCF), a group of 20 States directly concerned with the impact of climate change, have also been very active in making the bridge between the HRC and the negotiation of the Paris Agreement. They were supported by intergovernmental organizations, such as the South Center, which provided an analysis[1] of what could be at stake for Developing States in such a bridge.

Of course, these attempts to emphasize the need for urgent action by States to address the transboundary impact of greenhouse gas (GHG) emissions had to overcome quite some resistance.

This progress was also promoted and made possible by the contributions of a number of civil society organizations,[2] as well as academics. They provided expertise, strategic advice and opportunities for dialogue among States. In Geneva, three initiatives around the work of the HRC on climate change brought organizations together. The Geneva Interfaith Forum on Climate Change, Human Rights and Environment (GIF) was established in 2010 with organizations representing a large spectrum of faiths; the Geneva climate change consultative group came out of regular informal NGO meetings on climate change hosted by the Friedrich-Ebert-Stiftung (FES), which formalized itself as a group, the GeCCco, in 2014. INTLawyers, based in Geneva, echoed concerns about equitable development from networks of the global South, in particular from Africa. All of these contributions made individually or collectively are an integral part of the process and would deserve a specific study. They are not well documented, and in this paper, we will be able to mention only those initiatives that were made public and directly involved in the general process of the negotiation.

The aim of this paper is to provide basic factual information about some of the key events of this process, which saw the issue of human rights and climate change from doubts and even denial to an acknowledgement at both the UN Human Rights Council and the UNFCCC, in particular, at the adoption of the Paris Agreement.[3]

Opening the discussion

First call (2007)

It's interesting to note that the very first reference to human rights and climate change in the HRC documents did not come from one of the members of the Council, but from one of the UN HRC Special Procedures.[4]

The first to have referred to climate change in his report was Paul Hunt, Special Rapporteur on the right to health, who in his 2007 *Report of the Special Rapporteur on the right of everyone to the enjoyment of the highest attainable standard of physical and mental health* called on the "Human Rights Council to urgently study the impact of climate change on human rights generally and the right to the highest attainable standard of health in particular".[5]

Later, several other Special Procedures followed and also called on the Council to consider the linkages between climate change and other issues such as displacement, migration, extreme poverty and the right to food, adequate housing, drinking water and sanitation.

The first HRC resolution (2008)

On 28 March 2008, at the seventh session of the HRC, the Maldives, together with 78 co-sponsors from all regional groups, tabled a resolution on "Human Rights and Climate Change", which was adopted by consensus by the Human Rights Council (Resolution 7/23).[6]

Resolution 7/23, for the first time in an official UN resolution, stated explicitly that climate change "poses an immediate and far-reaching threat to people and communities around the world and has implications for the full enjoyment of human rights" (Preambular §1).

The members of the Council agreed to give further consideration by mandating the Office of the High Commissioner for Human Rights (OHCHR) to prepare a detailed analytical study on the relationship between climate change and human rights.

The origin of this action was the *Malé Declaration*, adopted by representatives of the Small Island Developing States, who met in Male' from 13 to 14 November 2007, where they had solemnly requested:

> The Office of the United Nations High Commissioner for Human Rights to conduct a detailed study into the effects of climate change on the full enjoyment of human rights, which includes relevant conclusions and recommendations thereon, to be submitted prior to the tenth session of the Human Rights Council.[7]

> The United Nations Human Rights Council to convene, in March 2009, a debate on human rights and climate change.[8]

This 2008 HRC 7/23 Resolution was the first of a long list.

Other relevant discussions and reports

Olivier de Schutter, the Special Rapporteur on the right to food, followed the first step made by Paul Hunt, the Special Rapporteur on the right to health, and identified the impacts of climate change that fell within the scope of his mandate.[9]

On 3 September, the HRC Social Forum 2008[10] also devoted one of its discussions about global challenges to climate change.

In October, the OHCHR, while preparing the study mandated by the Human Rights Council (Resolution 7/23), also held a one-day consultation meeting on the relationship between climate change and human rights.

Besides these UN bodies, the International Council on Human Rights Policy (ICHRP), an independent think-tank based in Geneva and focusing on trends and emerging human rights issues, published its own study titled "Climate Change and Human Rights: A Rough Guide".[11] It was a first mapping of a range of research agendas, assessing the adequacy of human rights to the larger justice concerns raised by anthropogenic climate change and by the strategies devised to address it. It pointed to areas where climate change will have direct and indirect human rights impacts, and where human rights principles might sharpen policy-making on climate change, including in the two core policy areas of adaptation and mitigation.

Preparing for COP15 in Copenhagen (2009)

A landmark report

The Report of the Office of High Commissioner on Human Rights on "the relationship between Climate Change and Human Rights", as requested by the HRC in its Resolution 7/23, came out as a landmark report. It was the first UN analytical study on the relationship between climate change and human rights. It was submitted to the Council's 10th session in March 2009.

The OHCHR report outlined the many ways climate change undermines a range of internationally recognized human rights, in particular the rights to life, adequate food, safe and drinkable water, health, adequate housing and self-determination. It also pointed to the concerns about the rights of specific vulnerable groups like women, children and Indigenous Peoples.

The report also addressed the human rights implications of climate change–induced displacement and conflicts, as well as the human rights implications of the measures taken to address climate change.

It referred to the factual elements identified by the IPCC reports and underlined the basis of the UNFCCC process in pursuing response measures (mitigation and adaptation), such as the recognition of an unequal burden of impact between states and regions, as well as the need for an equity principle, formulated as "common but differentiated responsibilities (CBDR)".

It analyzed the concepts of climate justice and of "common yet differentiated responsibilities" and their implication for the obligations of high-income countries to provide assistance to particularly affected regions. On such a basis, the study also discussed international cooperation as a legal obligation "of a state towards other states, as well as towards individuals".[12] Unfortunately, this characterization of the binding nature of obligations to cooperate was rejected in the Council resolution in order to secure consensus.[13]

Building on the OHCHR report

Another report received some attention from the press and confirmed the work of the OHCHR. The Global Humanitarian Forum[14] issued its own report, "The Anatomy of a Silent Crisis",[15] with the aim to give a human face to climate change, at a time when it was still mainly synonymous with a white bear. It documented the most critical impacts of climate change on human society worldwide, namely on food, health, poverty, water, human displacement and security. It highlighted the massive socio-economic implications of those impacts, in particular, that the worst affected are the world's poorest groups, who cannot be held responsible for the problem.

In the perspective of the discussion of the OHCHR report at the coming HRC 10th session, two organizations held an expert meeting[16] to set out approaches on how the linkages between climate change and human rights might be made operational or institutionalized, and how human rights aspects could be mainstreamed into the Copenhagen Climate Conference. Its conclusions were forwarded to countries for possible use in the Council debate in March 2009 to follow up the previous 2008 Resolution 7/23.

The second HRC resolution on climate change

Based on the key points of the OHCHR report, a second resolution on climate change and human rights was adopted by the Human Rights Council on 25 March 2009 (Resolution 10/4),[17] noting that climate change has both direct and indirect implications for the enjoyment of human rights and recognizing that vulnerable individuals and communities will be the most acutely affected.

In this Resolution, the Council also affirmed that "human rights obligations and commitments have the potential to inform and strengthen international and national policy making in the area of climate change, promoting policy coherence, legitimacy and sustainable outcomes" (Preambular §10).

The HRC decided to hold a panel discussion on the relationship between climate change and human rights at its following session, in order to contribute to the realization of the goals of the Bali Action Plan (the road map providing the framework for a two-year negotiation process under the UNFCCC) and to make the summary available to the Conference of the Parties to the UNFCCC for its consideration.

The HRC's first panel discussion on human rights and climate change

The panel discussion requested by Resolution 10/4 took place on 15 June 2009, during the HRC 11th session. After hearing experts from different backgrounds, Member States and Observers generally agreed that climate change has implications for a wide range of internationally protected human rights, that climate-vulnerable countries are most at risk and that the human rights impacts do not fall evenly across a given population, but affect first the more vulnerable groups. Several States even argued that such issues could be considered as a form of "climate injustice".[18]

The discussion is furthered outside the HRC

This discussion was continued outside the HRC, the following week, in a workshop during the 2009 Global Humanitarian Forum, chaired by Mary Robinson, Former President of Ireland and Former UN High Commissioner for Human Rights. The workshop acknowledged that the climate change and human rights linkage had moved from the periphery to the mainstream. It underlined the added value of human rights by bringing a focus on the most vulnerable, who are paying the price for the activities of the most privileged and by providing arenas where such issue can be raised, like it has been done at the Inter-American Commission on Human Rights.[19]

The HRC special procedures foreseeing COP15 in Copenhagen

In the run-up to COP15 to be held in Copenhagen, a group of 20 UN Special Procedures mandate holders issued a joint statement, entitled "An Ambitious Climate Change Agreement Must Protect the Human Rights of All",[20] in which they argued that "A weak outcome of the forthcoming climate change negotiations threatens to infringe upon human rights". Rising sea levels, increasing ocean and surface temperature and extreme weather events like storms droughts and cyclones have and will continue to have a range of direct and indirect implications for the enjoyment of human rights.

They called on mitigation and adaptation policies to be developed in accordance with human rights norms and reminded the Conference that the adverse effects of climate change are felt most acutely in the poorest countries of the world.

Finally, they urged participants at the Copenhagen Climate Change Conference

> to step up their efforts to achieve a new agreement that prevents further climate change, protects affected individuals from its adverse impact and leads to the formulation of global and national mitigation and adaptation responses based on internationally recognized human rights norms and standards.

On her side, the Special Rapporteur on the right to adequate housing, Raquel Rolnik, had announced her intention to consider the impact of climate change in her annual report to the UN General Assembly. This was welcomed by the Human Rights Council in its March Resolution 10/4, which also encouraged other relevant Special Procedure mandate holders to give consideration to the issue of climate change within their respective mandates, having in mind the upcoming COP15.

Her report was published in the summer 2009 for the session of the General Assembly.[21] It provided an overview of the scope and severity of climate change, its implications with extreme weather events and its impact on urban and rural areas, including unplanned and unserviced settlements, on human mobility and on small islands and low-lying coastal zones. It also outlined the relevant international human rights obligations in connection to the right to housing. The Special Rapporteur also urged States to uphold their human rights obligations when mitigating climate change and adapting to its inevitable impacts.

It gave a particular attention to urban areas, seen as key players both in the generation of greenhouse gases and in strategies to reduce emissions and also as places needing urgent action to reduce the vulnerability of urban dwellers to the impact of climate change.[22]

She reminded all States to comply with their commitments to the global atmosphere by reducing their harmful warming emissions, with industrialized countries having a leading role to play in this reduction of emissions levels and in supporting developing countries in pursuing low-carbon development paths.[23]

In parallel to this report, Special Rapporteur Rolnik also made the first country visit, clearly addressing climate change, with a mission in February 2009 to the Maldives, a country directly affected and threatened to disappear by sea level rise. A preliminary note was issued in March 2009[24] for the HRC session and the final report was published in January 2010.[25]

The main purposes of the mission was to examine the difficulties encountered in the post-tsunami reconstruction process and the impact of climate change on the right to adequate housing. According to what she witnessed, climate change has aggravated and will amplify some of the problems linked with Maldives characteristics, which include land scarcity and vulnerability of the islands to natural phenomena. The Special Rapporteur stated that there is an international responsibility to urgently support adaptation strategies for the impact of climate change on Maldives.

That same year, Walter Kälin, the Representative of the Secretary-General on the human rights of internally displaced persons, devoted the thematic part of his report to the General Assembly to climate change and displacement.[26] He reaffirmed the normative framework that protects persons displaced by the effects of climate change and described the human rights challenges when protecting internally displaced persons in the context of climate change (evacuations, prohibition of return, permanent relocations and durable solutions).

The Special Rapporteur on the right to safe drinking water and sanitation, Catarina de Albuquerque, produced for the COP15 a first version of a position paper, to be finalized in 2010 and updated for the following COP16 in Mexico.

All these documents from the Special Procedures mandate holders provided background material to their joint statement to COP15.

Bridging with the UNFCCC at COP15

In order to inform the UNFCCC about the work done by the human rights bodies, a side event[27] was organized in June at the UNFCCC Bonn Climate Talks, preparing for COP15 at the end of the year.

On 7 to 18 December 2009, the 15th Conference of the Parties of the United Nations Framework Convention on Climate Change (UNFCCC COP15) took place in Copenhagen, Denmark. Its outcome was quite below all expectations, raising severe concerns about the capacity of multilateralism to deliver in front of global challenges.

Reflection after Copenhagen (2010)

The failure of the Copenhagen Conference on Climate Change had of course some impacts on the work of the human right bodies. For example, no resolution on climate change was adopted at the HRC in 2010. But it did not prevent the reflection to continue, as climate change impacts remained persistent on the ground.

The HRC 2010 social forum

It is only in the HRC Resolution 13/17 on "The Social Forum", adopted on 25 March, that climate change was addressed in 2010.[28] The HRC Social Forum brings together States representatives and a broad range of other stakeholders, especially from developing countries.

The HRC decided that the Social Forum 2010 would consider the adverse effects of climate change on the full enjoyment of human rights, including the right to life and economic, social and cultural rights, and the international assistance and cooperation needed to address these impacts, including those on the most vulnerable groups, particularly women and children.

The Social Forum 2010, chaired by Ambassador Laura Dupuy, the Permanent Representative of Uruguay, met in Geneva on 4 to 6 October and heard several round tables on setting the scene about the adverse effects of climate change on the full enjoyment of human rights; on measures and actions addressing the impact of climate change on the full enjoyment of human rights (including on most vulnerable groups, particularly women and children); and on going forward with a rights-based approach to climate change.

In her concluding remarks, the Chairperson-Rapporteur of the 2010 HRC Social Forum noted that the discussions reaffirmed the findings of HRC Resolution 10/4 that "human rights obligations and commitments have the potential to inform and strengthen international and national policy-making in the area of climate change, promoting policy coherence, legitimacy and sustainable outcomes" (Preambular §10).

In its report,[29] the Social Forum recommended that the Human Rights Council establish a new mechanism, which could take the form of a Special Rapporteur or Independent Expert dedicated to human rights and climate change; that it continues holding an annual discussion with the view to tracking the rapidly evolving impacts of climate change on human rights; and that the 16th Conference of the Parties in Cancun be informed of the deliberations of the 2010 Social Forum.

It also recommended that REDD and REDD+ programs adopt a more rights-based approach, create legal awareness programs along with other support programs for Indigenous Peoples that may be affected by REDD programs, and improve participatory and access-to-justice provisions; that a human rights–based approach be applied to intellectual property and technology transfer in order to facilitate adaptation and even mitigation efforts; and finally that a mechanism be put in place for measuring performance on climate change, which should include human rights indicators to create awareness and promote sustainable development options.

The special procedures cover more and more areas

Special Procedures mandate holders also continued to give a strong attention to the impact of climate change on the rights and issues they monitor.

Walter Kälin reiterated his recommendations about the nexus between climate change and displaced persons, in his 2010 report.[30]

Olivier de Shutter, the Special Rapporteur on the right to food, reported on how agro ecology improves resilience to climate change and thus contributes to the protection of the right to food.[31]

The Special Rapporteur on adequate housing, Rachel Rolnik, as a component of the right to an adequate standard of living, continued to report on the impact of climate change in her 2010 report.[32]

The Independent Expert (later Special Rapporteur) on extreme poverty, Magdalena Sepúlveda Carmona, explained that environmental degradation disproportionally affects people living in extreme poverty in her 2010 report.[33]

On her side, the Special Rapporteur on the right to safe drinking water and sanitation produced the definitive version of her position paper[34] for the UNFCCC COP16 to be held in December 2010 in Mexico.

The Special Rapporteur called on the UNFCCC negotiators to not forget that improved water resource management should be a central component of climate change adaptation strategies. She provided a range of recommendations to guide the climate negotiations and climate policy more broadly. She underlined that the rights to water and sanitation must be properly and adequately reflected within the agreement to be reached by COP16, as well as in processes beyond COP16, to ensure any successful climate mitigation and adaptation.

Three country visits also generated concerns about the impact of climate change

The Independent Expert on the question of human rights and extreme poverty, Magdalena Sepúlveda Carmona, undertook a mission to Viet Nam from 23 to 31 August 2010.[35] She witnessed that even though the authorities have strong policies to combat poverty, there is a high risk of groups that had progressed falling back into poverty as a consequence of natural disasters, which may only become more frequent with climate change.[36] Therefore, the Independent Expert called on the international community to continue providing to Viet Nam the necessary funds for climate change mitigation and adaptation measures.[37]

The Special Rapporteur on the right to food, Olivier De Schutter, made the second country visit relating to climate change with his mission to Syria from 29 August to 77 September 2010. His report titled "Drought and Climate Change: The Need for a Human Rights-Based Response" was published in January 2011.[38]

He reported that the effects of climate change in the country are already evident from the cycles of drought, which have shortened from a cycle of 55 years in the past to the current cycle of seven or eight years. In this context, he noted the importance of human rights standards and principles, particularly individual empowerment, community participation in decision-making processes, equality and non-discrimination, and accountability mechanisms.[39]

The government has taken measures. However, two factors have limited its ability to react effectively. First, the Government was slow to recognize the scale of the problem and thus to take all the measures required and to call upon foreign assistance. Second, there was a lack of capacity.[40] He urges expanding international cooperation, including financial and technical support, with the Government of the Syrian Arab Republic when addressing the impact of drought and climate change.[41]

De Schutter also undertook the third 2010 country visit referring to climate change when he went to China from 15 to 23 December 2010. He published a preliminary note on the mission in February 2011[42] and the final report in January 2012,[43] for its presentation at the 19th session of the Human Rights Council in March 2012.

He found that while the agricultural system in China has achieved impressive results over the past 30 years, it must now focus on becoming more resilient to climate-related shocks and on making more efficient use of scarce resources.[44] He called on the authorities to use this opportunity in some of the concerned regions to engage with nomadic herders to improve their security of land tenure and to combat climate change through participatory agro ecological methods.[45]

The work of the Independent Expert on extreme poverty received support with the report, "Exposed – The Human Rights of the Poor in a Changing Global Climate",[46] describing the general links between global climate change and extreme poverty. Climate change, including weather extremes and the accelerated degradation of lands, will impact certain vulnerable groups, with particular severity for people living in poverty, resulting in the violation of a range of human rights. The report calls the Parties of the UNFCCC to use human rights as guiding principles for any international agreement, as well as for domestic measures taken in response to climate change.

Beside the Special Procedures, a new track was opened with the report titled "Climate Change in the Work of the Committee on Economic, Social and Cultural Rights".[47] The CESCR can play a particularly prominent role in finding remedies for the expected harms that the most vulnerable communities will suffer, as the Committee oversees most of the rights at the core of the debate on climate change and human rights. The CESCR is likely to be increasingly challenged by the issue of climate change, as some of the State reports it receives will likely point to climate change–related environmental degradations as causes for non-compliance with human rights obligations.

A political turning point (2011–2012)

No substantial HRC resolution on climate change was adopted during 2010, and in 2011, it is only at its September session that the Human Rights Council adopted its third resolution on "human rights and climate change", Resolution 18/22.[48]

The resolution changes hands

This time, Resolution 18/22 was tabled by the Philippines and Bangladesh and not by the Maldives. But the Maldives remained among the 43 co-sponsors of the resolution.

The Maldives had joined another group of countries[49] tabling another resolution addressing the larger issue of human rights and the environment. The objective was to integrate climate change in its wider environmental context and to cover all its dimensions. The hope was also to initiate stronger action by the HRC, which was indeed achieved in March 2012 with the creation of the mandate of the Independent Expert (and afterwards in 2015 of a Special Rapporteur), on the issue of human rights obligations relating to the enjoyment of a safe, clean, healthy and sustainable environment.[50]

The countries that continued to table the resolution on human rights and climate change wanted to keep the HRC's attention focused on it, including its economic and social aspects. They also remained interested by having a special procedure at some point that would be devoted to this issue only, as climate change–induced events are increasingly affecting territories and populations.

Resolution 18/22 asked the OHCHR to convene a seminar to

> (a) To convene, prior to the nineteenth session of the Human Rights Council, a seminar on addressing the adverse impacts of climate change on the full enjoyment of human rights, with a view to following up on the call for respecting human rights in all climate change–related actions and policies, and forging stronger interface and cooperation between the human rights and climate change communities.
>
> *(§2.a)*

Reflections for a HRC at a crossroad

Considering the new situation in the HRC, now with two tracks, one on climate change and one on environment, an expert meeting[51] was organized on 25 to 27 January 2012, with the aim to bring together State representatives, academics and civil society organizations to identifying concrete inroads for human rights to both debates and to propose future action within the Human Rights Council.

Despite climate change being also an environmental issue, the HRC debates have treated as distinct topics "human rights and climate change", and "human rights and environment", which led to separate Council resolutions with the possibility of separate mechanisms to advance the respective linkages.

The expert meeting considered various possibilities regarding the nature of any future mandate: the establishment of a "Human Rights and Environment" mandate and/or a "Human Rights and Climate Change" mandate, or a combination of both. In view of the political and financial constraints in the HRC, the discussion focused especially on the question of whether merging the initiatives into one mandate was a better or more feasible option.[52]

The HRC seminar on human rights and climate change

The Human Rights Council seminar on human rights and climate change requested by Resolution 18/22 was held 23–24 February 2012, in the Palais des Nations. In the conclusions of the seminar, it was highlighted that all the experts had acknowledged the adverse impacts of climate change on people and communities and on their human rights.

Some statements even claimed that "climate change can be considered as the single most important threat to food security in the future" and that "climate change is the biggest threat to the enjoyment of a safe and healthy environment".

The role of Special Procedures in bridging the gap between the human rights and climate change communities was welcomed. Special Procedures can clarify legal issues, carry fact-finding missions on the ground and develop legal instruments responding to some of the challenges. It was also proposed that an HRC Special Procedure be specifically devoted to the issue of climate change. It could address the problem as a whole and more coherently and also facilitate the work of other concerned Special Procedures.

A summary report[53] was addressed to the June session of the Council and made available to the 18th session of the Conference of Parties to the UNFCCC in November 2012 in Doha (COP18).

Special procedures continue to document the issue

Special Procedures continued to include in their annual reports to the UN HRC and the UN General Assembly information about the impact of climate change on human rights.

The Special Rapporteur on the human rights of internally displaced persons, Chaloka Beyani, in August 2011 continued the work of his predecessor and devoted the thematic part of his report to the UN General Assembly to the "Protection of and assistance to internally displaced persons".[54]

His report explored the linkages between climate change and internal displacement from a human rights perspective. He argued that climate change is already acting as "an impact multiplier and accelerator".

> In addition to its negative impact on social and economic rights, which will itself provoke some displacement, climate change, interacting with other pressures or social and political factors, will exacerbate the risk of conflicts, which could then act as a driver of further displacement.[55]

Therefore, he considered that "durable solutions for displaced populations should be part of national adaptation plans, and local and national capacity-building programs, and be supported by funds made available for adaptation measures".[56]

In January 2012, he included further considerations in his annual report[57] to the HRC.

> The erosion of livelihoods, in part provoked by climate change, is considered a key "push" factor for the increase in rural to urban displacement and migration, most of which is likely to be to urban slums and informal settlements offering precarious living conditions. The Special Rapporteur believes that the urban dimensions of climate-change-induced displacement should be a key consideration in medium- and long-term national development strategies, as well as adaptation.[58]

In August 2012, the Special Rapporteur on the human rights of migrants, François Crépeau, presented his report to the General Assembly on climate change and migration.[59]

He underlined that with climate change, the rate and scale of migration could be multiplied.[60] But, "although environmental transformations experienced as a result of climate change may contribute to migratory movements, environmental migration, like every kind of migration, is essentially a complex, multicausal phenomenon which may be driven by a multiplicity of push-and-pull factors".[61] "Therefore, only with precise knowledge of the scope and nature of environmental migration will States be able to develop and agree upon common policies in this regard".[62] What is required, however, is a more concerted and concrete application of the existing human rights law to the situation of climate change–induced migrants.[63] But one category of climate change–induced migrants that international law needs to consider urgently is the one of the inhabitant of low-lying island States.[64] To date, the international legal framework appears to be largely inadequate to address their situation.[65]

Country visits to address concerns on climate change were also conducted. The Special Rapporteur on the human right to safe drinking water and sanitation, Catarina de Albuquerque, made two visits in 2012 to threatened Pacific Small Island Developing States: Tuvalu, on 17–19 July 2012,[66] and Kiribati, on 23–26 July 2012.[67] The Special Rapporteur paid particular attention to the impacts of climate change on the enjoyment of the human rights to water and sanitation.

The atolls of Kiribati are only a few meters above sea level, which means that most of the people are directly exposed to extreme weather events intensified by climate change.[68] Therefore, the Special Rapporteur recommends that adaptation plans put the human rights to water and sanitation at the center to respond to people's actual needs without discrimination and that international assistance to adaptation also be based on such plans.[69]

Tuvalu has attracted international attention as one of the states most vulnerable to potential "disappearance" due to climate change, facing risks to their sovereignty or existence.[70] Climate change will further increase vulnerability to climatic events such as cyclones and drought and hence exacerbate water scarcity, saltwater intrusions, sea level rise and frequency of extreme weather events.[71] The Special Rapporteur saw little evidence that affected populations, including women and children, were informed of or given opportunities to participate in discussions on the impacts of climate change and policy-making related to them.[72] This disconnection should be addressed in order for Tuvaluans to adapt to climate change and tackle its adverse effects in ways that would enable them to make their own decisions.[73] At the same time, she called on those countries most responsible for the current climate change situation to comply with their legal obligations to prevent or remedy the impacts of climate change on the human rights of individuals and communities[74] and to take immediate action to assist Tuvalu and small island States with possible adaptation measures, as well as planning for potential scenarios in the very near future.[75]

The Special Rapporteur on the right to food, Olivier de Schutter, visited Canada on 6–16 May 2012. In the part of his report[76] examining specific problems faced by Indigenous Peoples,[77] he notes that accessing traditional foods includes now the impacts of climate change on migratory patterns of animals and on the mobility of those hunting them.

The Special Rapporteur also went in a mission to Cameroon on 16–23 July 2012.[78] He noted that the coastal regions, as well as the Sahelian regions in the north of Cameroon, are particularly hard hit by climate change.[79] He recommended to the government to develop a program to introduce structural improvements in the northern region, because of its vulnerability to climate change.[80]

The Special Rapporteur on internally displaced persons, Chaloka Beyani, conducted an official visit to the Maldives on 16–21 July 2011, during which he examined the situation of persons internally displaced as a result of the 2004 tsunami and studied issues related to risks of potential internal displacement in the future, including due to the effects of climate change.[81]

He found that climate change and other factors specific to the low-lying island environment of the Maldives were already affecting the livelihoods and rights of residents of many islands, including the rights to housing, safe water and health. Moreover, other factors, such as more frequent storms and flooding, coastal erosion, salination, overcrowding and the existential threat posed by rising sea levels, point to increased risks of potential internal displacement in the future.

Maximum tension at the HRC (2013)

The logical step after a seminar such as the one held early 2012 is to have a thematic resolution that takes stock of the discussion and of any progress made. But there was no follow-up in 2012. In March 2012, the HRC did adopt a resolution on human rights and the environment, creating a new mandate on this issue, addressing just minimal references to climate change but not making it a specific priority for the mandate.

It was only in June 2013 that Philippines and Bangladesh considered it possible to table a draft resolution on climate change. After a month of long and intense discussions and negotiations during the HRC session, just a few days before the Council was going to take action on the resolution, it had to be withdrawn, as the conditions for its adoption by consensus did not seem secured.

A split vote on such a resolution would be sending a very negative signal and could make a contentious issue of the human right and climate change nexus. The countries tabling the resolution did not want to take such a risk.

The difficulties encountered in the HRC were partly reflecting those of the UNFCCC process, as the final discussion in Paris for a possible new instrument was approaching, which had failed in Copenhagen. Resistance was coming from all sides, with speculations about how this or that formulation could perhaps have an influence on the discussions in the UNFCCC and with a rising level of mistrust on this issue among the regional groups in the HRC.

The process would resume only a year later.

For the Special Procedures in 2013, the annual report[82] of the Special Rapporteur on the human rights of internally displaced persons, Chaloka Beyani, includes a situation and a general issue relating to climate change.

The Special Rapporteur noted that Sudan continued to experience a variety of causes and contexts of internal displacements. These include displacement due to conflicts over resources because of climate change and natural disasters. He therefore urged the government to embrace a comprehensive framework for the protection of the human rights of IDPs by ratifying the Kampala Convention.[83]

Noting also the increased attention devoted to the adverse effects of climate change on potentially vulnerable groups, including women, the Special Rapporteur acknowledged that climate change impacts men and women differently at all stages, from preparedness to reconstruction.[84] He therefore recommended to all States to closely examine the gender dimensions of displacement linked to the effects of climate change, in order to identify specific vulnerabilities and good practices in gender-sensitive protection, assistance, adaptation, mitigation, relocation and reconstruction processes.[85]

A new momentum before Paris (2014–15)

The HRC resolution on climate change comes back

In 2014, the resolution was tabled again by the Philippines and Bangladesh at the June session of the HRC. The discussions and negotiations were rough during the session, but finally a wording was found that made it possible to adopt Resolution 26/27 without a vote.[86]

The resolution reiterated the concerns that "climate change has contributed to the increase of both sudden-onset natural disasters and slow-onset events, and that these events have adverse effects on the full enjoyment of all human rights" (§2).

The HRC also decided to include into its program of work for the 28th session a full-day discussion on two specific themes relating to human rights and climate change: on the realization of all human rights for all, in particular those in vulnerable situations, and on States' efforts to realize the right to food.

In addition, the Council addressed briefly the issue of the impact of climate change in the resolutions on human rights and the environment 25/21 (2014) and 28/11 (2015).

The HRC full-day discussion on climate change

The full-day discussion took place during the 28th HRC session in March 2015 on the basis of Resolution 26/27, requesting two panels on international cooperation and on the right to food. Depicting climate change as one the greatest human rights challenge of the 21st century, panelists and participants left no doubt as to the terrible reality of its impacts.

The right to food was identified as a right particularly affected by climate change and requiring concrete action, not only by the reduction of emissions but also by changing agricultural patterns. The Human Rights Council was called to assume responsibility for safeguarding those whose rights were undermined and destroyed by the impacts of climate change, as such impacts could exceed the capacity of many States to protect their people. A human rights–based approach demands the inclusion of climate justice, as well as international cooperation and solidarity, all of which are essential in supporting affected countries, including through finance and technology.[87]

The last resolution before Paris

In the June session of the HRC, the 2015 Resolution 29/15[88] on climate change and human rights reached consensus more easily and was co-sponsored by 110 countries. The resolution welcomed the establishment of the Climate Vulnerable Forum and welcomed the holding of the 21st Conference of the UNFCCC in Paris in December 2015 (COP21). It mandated the Office of the UN High Commissioner for Human Rights (OHCHR), together with other key stakeholders, to prepare a detailed analytical study on the relationship between climate change and the human right to health for the March 2016 HRC session. The resolution also called for a panel discussion on climate change, human rights and health to be held at the 31st session (March 2016) of the Human Rights Council.[89]

Special procedures make their voiced heard

The Special Rapporteur on human rights and the environment, John Knox, has been particularly active on the question of climate change, and in 2014 for the 25th session of the HRC in March, he issued a mapping report of statements made by human rights mechanisms regarding the numerous human rights that are threatened by climate change, as well as human rights obligations related to climate change.[90] In his report to the 28th session of the Human Rights Council in March 2015, Knox identified good practices, particularly on climate change.[91]

The Special Procedures also made several important joint advocacy efforts in view of the upcoming COP21 in Paris and the possible adoption of a new instrument.

On 17 October 2014, 28 Special Procedures wrote an "Open Letter"[92] to State parties to the UNFCCC that urged them "to adopt urgent and ambitious mitigation and adaptation measures to prevent further harm" and to include in the 2015 Paris climate agreement a commitment that "the Parties shall, in all climate change related actions, respect, protect, promote and fulfill human rights for all, and to launch a work program to ensure that human rights are integrated into all aspects of climate actions".

Five thematic Special Rapporteurs prepared a report for the Climate Vulnerable Forum, entitled "The Effects of Climate Change on the Full Enjoyment of Human Rights". On behalf of the Forum, the Philippines submitted the report on 1 May 2015 to the Conference of Parties to the UNFCCC. This report underlined that the increase of global warming also inevitably increases the negative impacts on the enjoyment of human rights. With this report in hand, the Forum urged the COP21 to adopt a more ambitious target in the climate agreement under negotiation to avoid the devastating effects of a rapidly warming planet on the basic rights of all persons.[93] Several Parties that were Members of the Climate Vulnerable Forum referred several times to this report during the negotiation in Paris for the inclusion of the target of 1.5°C in the Paris Agreement.

For the World Environment Day on 5 June 2015, the group of 28 Special Procedures issued a "Joint Statement"[94] drawing attention to the grave harm even a 2°C increase in average global temperature would cause to the enjoyment of human rights. The Special Rapporteurs urged climate negotiators in the UNFCCC to reach an agreement at COP21 in Paris concerning obligations human rights law places on States to protect and promote human rights in the context of climate change.

On 10 June, the Universal Rights Group,[95] together with the Climate Vulnerable Forum and the government of Costa Rica, organized an event to present and build on a Joint Statement (Friday 5 June) by UN Special Procedures drawing attention to the CVF-commissioned report on climate change and human rights submitted to the UNFCCC ahead of the UN Climate Change Conference at Bonn, Germany (1–12 June 2015).

The UNFCC COP20 in Lima and COP21 in Paris

At COP20 in Lima, the first UN Human Rights Day during a UNFCC COP was celebrated on the 10th of December. Such an initiative was notably supported by the Mary Robinson Foundation for Climate Justice (MRFCJ),[96] marking the day with meetings between ministers, ambassadors, experts and representatives of local and international civil society. Several organizations also held on that day a side event[97] to give an opportunity to UN Human Rights Special Rapporteurs to present their joint Open Letter to the UNFCCC made by 28 of them.

Between the COP20 in Lima and the COP21 in Paris, several inter-sessional meetings and negotiations were held. Usually such meetings take place in Bonn. But one of them, the eighth part of the second session of the Ad Hoc Working Group on the Durban Platform for Enhanced Action (ADP 2–8), met in Geneva on 8–13 February.

This meeting was clearly preparing the first drafting steps of what became the Paris Agreement at COP21. As most of the human rights UN bodies are in Geneva, as well an important number of diplomatic experts and NGOs, it provided a unique opportunity to discuss the linkages between climate change and human rights and foresee what should be included in the future agreement on this matter. Actually, the very first proposals were made by a number of States at the opening of the meeting in the Palais des Nations taking place in the room facing the one of the HRC.

Having a session in Geneva was not only beneficial for the governmental delegations. It was also an opportunity for NGOs from the two fields to work together.[98] The "NGO Working Group on Human Rights and Climate Change", created at the 2008 COP14 in Poznan to promote the inclusion of human rights in climate change actions, met with the GeCCco and the GIF. This joint work greatly helped consolidate the collaboration between the networks and would be pursued up to Paris.

It was also in Geneva, at the February 2015 Climate Justice Dialogue invited by the Mary Robinson Foundation (MRFCJ), that Costa Rica proposed to create the "Geneva Pledge for Human Rights in Climate Action",[99] a voluntary initiative for States to facilitate the sharing of best practices and knowledge between human rights and climate experts at the national level. The Geneva Pledge has since held regular discussions at HRC sessions.

For Paris and its COP21 in December 2015, the OHCHR made a submission:[100] "Understanding Human Rights and Climate Change", to inform the ongoing negotiations of the 21st Conference of the Parties to the UNFCCC. It contained the OHCHR's "Key Messages on Human Rights and Climate" highlighting the essential obligations and responsibilities of States and other duty-bearers (including businesses) and their implications for climate change–related agreements, policies and actions. It aims to foster policy coherence and help ensure that climate change mitigation and adaptation efforts are adequate, sufficiently ambitious, non-discriminatory and otherwise compliant with human rights obligations, considerations which should be reflected in all climate action.

Also in view of COP21, Human Rights Watch released its first report[101] on the impact of climate change on the human rights. It analyzed the situation of the people in the Turkana region of Kenya. It described the increased difficulty they face in getting water, as many water sources have dried out, making every day a struggle for survival. Women and girls often have to walk longer distances to dig for water in dry riverbeds. The report included key recommendations to the Kenyan authorities, as well as donor countries, on integrating climate change into development plans and including affected communities in the planning process.

At the COP21 in Paris, on 10 December, Human Rights Day, an important round table[102] was held on "Climate Change: One of the Greatest Human Rights Challenges of Our Time". It brought on the panel Mary Robinson, several Special Procedures, a civil society representative from the Pacific Islands and the OHCHR, who all called the State delegations to bring COP21 to a meaningful conclusion in terms of human rights.

The Paris Agreement

How much exactly have all these efforts impacted on the drafting of the Paris Agreement will remain an issue for discussion. But clearly some of the contributions of the HRC have been used as arguments or as references in the negotiations. Of course, the result of having human rights language mainly in the preamble of the Agreement is seen by some organizations and observers as too weak. Some parties were asking for an inclusion in the operative part, which would have ensured a more direct implementation.

Nevertheless, the Paris Agreement contains the strongest language on human rights in any environmental treaty to date. The agreement states that countries should "respect, promote and consider" human rights in their response to climate change and underlines the rights of Indigenous Peoples, women, migrants, children and those in vulnerable situations.[103]

Some of this wording sounds like an echo of the HRC resolutions. The variety of rights and issues referred to in the preamble is similar to the issues discussed at the HRC and covered by mandates of HRC Special Procedures.

Only the notion of intergenerational equity can be perhaps considered as not been addressed very clearly yet by the HRC.

The preamble also makes reference to crosscutting issues where the HRC cooperates with specialized institutions and agencies, such as the World Health Organization (WHO), the International Labour Organization (ILO), the United Nations Development Program (UNDP), the International Organization for Migrations (IOM) and the Food and Agriculture Organization (FAO).[104] These references to rights open a web of connections that can contribute usefully to the effective and correct implementation of the Agreement.

Another concern of the human rights bodies has also been taken into consideration. Mitigation and adaptation projects are at times the cause of human rights abuses.[105] Human rights experts have repeatedly reminded States that developing "green energy" alternatives for the benefit of climate stability should not be done at the expense, for example, of food security or be the cause of unlawful forced displacement. Funding for climate change, which will be an increasing issue in the coming years, must not fuel discrimination. It must be done with proper and effective consultation and participation of all affected populations.

Keeping in mind the influence that they have been able to have in this process from Copenhagen to Paris, human rights bodies, in particular the HRC and its Special Procedures, now will have to see how they can continue to struggle against climate change by supporting the implementation of the Paris Agreement, by easing the communication among bodies of both fields, by bringing facts about the impacts of climate actions on the ground and by evaluating progress made in preventing human rights violations relating to climate change.[106]

Notes

1 M. Wewerinke and V. P. Yu III, *Addressing Climate Change Through Sustainable Development and the Promotion of Human Rights*, Research Paper 34 (November 2010), available online at www.southcentre.int/wp-content/uploads/2013/05/RP34_Climate-Change-Sustainable-Development-and-Human-Rights_EN.pdf.

2 Among the very involved civil society organizations around the HRC on this issue, we can mention the Friedrich-Ebert-Stiftung (FES), the Center International for Environmental Law (CIEL), Earthjustice, The Universal Rights Group (URG), Franciscans International, the World Council of Churches, Brahma Kumaris, and INTLawyers.

3 United Nations Framework Convention on Climate Change (UNFCC), *Adoption of the Paris Agreement*, UN Doc. FCCC/CP/2015/L.9/Rev. 1 (12 December 2015).

4 The Special Procedures of the Human Rights Council are independent human rights experts (Special Rapporteur, Independent Expert, Working Group) with mandates to report and advise on human rights from a thematic or country-specific perspective. The system of Special Procedures is a central element of the United Nations human rights machinery and covers all human rights.

5 General Assembly (GA), *Report of the Special Rapporteur on the Right of Everyone to the Enjoyment of the Highest Attainable Standard of Physical and Mental Health*, UN Doc. A/62/214 (8 August 2007).

6 Human Rights Council (HRC) Res. 7/23, *Human Rights and Climate Change*, UN Doc. A/HRC/RES/7/23 (adopted without a vote on 28 March 2008).

7 Center for International Environmental Law (CIEL), *Male' Declaration on the Human Dimension of Global Climate Change* (14 November 2007), available at www.ciel.org/Publications/Male_Declaration_Nov07.pdf, at para 4.
8 *Ibid.*, at para 5.
9 O. D. Schutter, Report of the Special Rapporteur on the right to food, *Building Resilience: A Human Rights Framework for World Food and Nutrition Security*, UN Doc. A/HRC/9/23 (8 September 2008).
10 HRC, *Social Forum 2008*, Room XVII, Palais des Nations, Geneva, 1–3 September 2008.
11 International Council on Human Rights Policy, *Climate Change and Human Rights: A Rough Guide* (2008).
12 Un Office of the High Commissioner of Human Rights (OHCHR), *Report of the OHCHR on the Relationship Between Climate Change and Human Rights*, UN Doc. A/HRC/10/61 (15 January 2009). The report had benefited from written and oral submissions by over 30 States and 35 international agencies, national human rights institutions, NGOs and academic bodies.
13 Limon, 'Human Rights Obligations and Accountability in the Face of Climate Change', 38 *Georgia Journal of International and Comparative Law (GJICL)* (2010) 543.
14 The Global Humanitarian Forum was a non-profit foundation that was active from 2007 to 2010 and presided over by former United Nations Secretary-General Kofi Annan. The Forum intended to serve as an independent platform for debate and collaboration on global humanitarian issues.
15 Global Humanitarian Forum, *The Anatomy of a Silent Crisis* (2009), available online at www.ghf-ge.org/human-impact-report.pdf.
16 Expert Meeting on Human Rights and Climate Change, organized by the Geneva Office of the Friedrich-Ebert-Stiftung (FES) and the Center for International Environmental Law (CIEL), Château de Bossey, Geneva, on 22–24 January 2009.
17 HRC Res 10/4, *Human Rights and Climate Change*, UN Doc. A/HRC/RES/10/14 (adopted without a vote on 25 March 2009).
18 HRC, Human Rights Council Panel Discussion on the Relationship Between Climate Change and Human Rights -15 June 2009, *Summary of the Panel Discussion* (2009), available at www.ohchr.org/EN/Issues/HRAndClimateChange/Pages/Panel.aspx.
19 CIEL *et al.*, *Global Humanitarian Forum 2009, Parallel Workshop: Climate Change and Human Rights* (23 June 2009), available at www.fes-globalization.org/geneva/documents/HumanRights/23June09_Report_GHF_HR_CC.pdf.
20 OHCHR, *Joint Statement of the Special Procedure Mandate Holders of the Human Rights Council on the UN Climate Change Conference* (2009).
21 R. Rolnik, *Report of the Special Rapporteur on Adequate Housing as a Component of the Right to an Adequate Standard of Living, and on the Right to Non-Discrimination in This Context*, UN Doc. A/64/255 (6 August 2009).
22 *Ibid.*, at para 66.
23 *Ibid.*, at para 70.
24 R. Rolnik, *Report of the Special Rapporteur on Adequate Housing as a Component of the Right to an Adequate Standard of Living, and on the Right to Non-Discrimination in This Context*, Addendum, Preliminary note on the mission to Maldives, UN Doc. A/HRC/10/7/Add.4 (3 March 2009).
25 R. Rolnik, *Report of the Special Rapporteur on Adequate Housing as a Component of the Right to an Adequate Standard of Living, and on the Right to Non-Discrimination in This Context*, Addendum, Mission to Maldives (18 to 26 February 2009), UN Doc. A/HRC/13/20/Add.4 (11 January 2010).
26 W. Kälin, *Report of the Representative of the Secretary-General on the Human Rights of Internally Displaced Persons*, Addendum 1, 'Protection of Internally Displaced Persons in Situations of Natural Disasters', UN Doc. A/HRC/10/13/Add.1 (5 March 2009).
27 Side Event on 'Climate Change and Human Rights', organized by the Centre for International Environmental Law (CIEL) and the Friedrich-Ebert-Stiftung (FES), in Bonn, on 2 June 2009.
28 HRC Res. 13/17, *The Social Forum*, UN Doc. A/HRC/RES/13/17 (adopted without a vote on 25 March 2010).
29 HRC, *Report of the 2010 Social Forum (Geneva, 4–6 October 2010)*, UN Doc. A/HRC/16/62/Corr.1 (4 January 2011).
30 W. Kälin, *Report of the Representative of the Secretary-General on the Human Rights of Internally Displaced Persons*, UN Doc. A/HRC/13/21 (5 January 2010).
31 O. D. Schutter, *Report Submitted by the Special Rapporteur on the Right to Food*, UN Doc. A/HRC/16/49 (20 December 2010).

32 R. Rolnik, *Report on the Special Rapporteur on Adequate Housing as a Component of the Right to an Adequate Standard of Living*, UN Doc. A/65/261 (9 August 2010).
33 GA, *Report of the Independent Expert on Extreme Poverty and Human Rights*, UN Doc. A/65/259 (9 August 2010).
34 Special Rapporteur on the right to safe drinking water and sanitation, C. D. Albuquerque, *Position Paper: Climate Change and the Human Rights to Water and Sanitation* (2010), available online at www.ohchr.org/Documents/Issues/Water/Climate_Change_Right_Water_Sanitation.pdf.
35 S. Carmona, *Report of the Independent Expert on the Question of Human Rights and Extreme Poverty*, Addendum, Mission to Viet Nam (23 to 31 August 2010), Un Doc. A/HRC/17/34/Add.1 (9 May 2011).
36 *Ibid.*, at para 7.
37 *Ibid.*, at para 10 (d)(i).
38 O. D. Schutter, *Report of the Special Rapporteur on the Right to Food*, Addendum, Mission to the Syrian Arab Republic (29 August–7 September 2010), UN Doc. A/HRC/16/49/Add.2, 27 January 2011.
39 *Ibid.*, at para 21.
40 *Ibid.*, at para 23.
41 *Ibid.*, at para 66(a).
42 O. D. Schutter, *Report of the Special Rapporteur on the Right to Food*, Addendum, Preliminary note on the mission to China (15–23 December 2010), UN Doc. A/HRC/16/49/Add.3, 18 February 2011.
43 O. D. Schutter, *Report of the Special Rapporteur on the Right to Food*, Addendum, Mission to China (15–23 December 2010), UN Doc. A/HRC/19/59/Add.1, 20 January 2012.
44 *Ibid.*, at para 24.
45 *Ibid.*, at para 38.
46 T. Gelbspan, *Exposed – the Human Rights of the Poor in a Changing Global Climate* (March 2010), available at www.fes-globalization.org/geneva/documents/EXPOSED_web%20version.pdf (last visited 29 April 2017).
47 M.A. Orellana, M. Kothari and S. Chaudhry, *Climate Change in the Work of the Committee on Economic, Social and Cultural Rights* (May 2010), available at www.ciel.org/Publications/CESCR_CC_03May10.pdf.
48 HRC Res. 18/22, *Human Rights and Climate Change*, UN Doc. A/HRC/RES/18/22 (adopted without a vote on 30 September 2011).
49 This group of countries included also Costa Rica and Switzerland, later joined by Slovenia and Morocco.
50 HRC Res. 19/10, *Human Rights and the Environment*, UN Doc. A/HRC/RES/19/10 (adopted without a vote on 22 March 2012).
51 Expert meeting organized by the Center for International Environmental Law (CIEL), the Geneva office of the Friedrich-Ebert-Stiftung (FES) and Earthjustice (adding "environment" as a new related issue to the debate), Château de Bossey, Geneva, on 25–27 January 2012.
52 F. Hille and C. Hille, *Report – Expert Meeting on Human Rights, Environment and Climate Change* (2012), available online at www.fes-globalization.org/geneva/documents/FES%20Report_Expert%20Meeting%20on%20HR%20Env%20and%20CC%202012.pdf 9.
53 UNHCHR, *Report of the United Nations High Commissioner for Human Rights on the Outcome of the Seminar Addressing the Adverse Impacts of Climate Change on the Full Enjoyment of Human Rights*, UN Doc. A/HRC/20/7 (10 April 2012).
54 C. Beyani, *Report of the Special Rapporteur on the Human Rights of Internally Displaced Persons*, UN Doc. A/66/285 (9 August 2011).
55 *Ibid.*, at para 29.
56 *Ibid.*, at para 80.
57 C. Beyani, *Report of the Special Rapporteur on Internally Displaced Persons*, UN Doc. A/HRC/19/54 (30 January 2012).
58 *Ibid.*, at para 35.
59 F. Crépeau, *Report of the Special Rapporteur on the Human Rights of Migrants*, UN Doc. A/67/299 (13 August 2012).
60 *Ibid.*, at para 31.
61 Crépeau, *supra* note 59, at para 32.
62 *Ibid.*, at para 34.
63 *Ibid.*, at para 54.
64 *Ibid.*, at para 66.
65 *Ibid.*, at para 67.

66 C. D. Albuquerque, *Report of the Special Rapporteur on the Human Right to Safe Drinking Water and Sanitation*, Addendum, Mission to Tuvalu (17–19 July 2012), Un Doc. A/HRC/24/44/Add.2 (28 June 2013).
67 C. D. Albuquerque, *Report of the Special Rapporteur on the Human Right to Safe Drinking Water and Sanitation*, Addendum, Mission to Kiribati (23–26 July 2012), UN Doc. A/HRC/24/44/Add.1 (1 July 2013).
68 *Ibid.*, at para 46.
69 *Ibid.*, at para 63(k).
70 *Ibid.*, at para 42.
71 *Ibid.*, at para 39.
72 *Ibid.*, at para 44.
73 *Ibid.*, at para 45.
74 de Albuquerque, 28 June 2013, *supra* note 66, at para 53.
75 *Ibid.*, at para 54.
76 O. D. Schutter, *Report of the Special Rapporteur on the Right to Food*, Addendum, Mission to Canada (6 to 16 May 2012), UN Doc. A/HRC/22/50/Add.1, 24 December 2012.
77 *Ibid.*, at chapter VIII.
78 O. D. Schutter, *Report of the Special Rapporteur on the Right to Food*, Addendum, Mission to Cameroon (16–23 July 2012), UN Doc. A/HRC/22/50/Add.2, 18 December 2012.
79 *Ibid.*, at para 54.
80 *Ibid.*, at para 73 (e).
81 C. Beyani, *Report of the Special Rapporteur on Internally Displaced Persons*, Addendum, Mission to Maldives, UN Doc. A/HRC/19/54/Add.1 (30 January 2012).
82 C. Beyani, *Report of the Special Rapporteur on the Human Rights of Internally Displaced Persons*, UN Doc. A/HRC/23/44 (18 March 2013).
83 *Ibid.*, at para 19.
84 *Ibid.*, at para 52.
85 Beyani, 30 January 2012, *supra* note 81, at para 93.
86 HRC Res. 26/27, *Human Rights and Climate Change*, UN Doc. A/HRC/RES/26/27 (adopted without a vote on 27 June 2014).
87 OHCHR, *Summary Report of the Office of the United Nations High Commissioner for Human Rights on the Outcome of the Full-Day Discussion on Specific Themes Relating to Human Rights and Climate Change*, UN Doc. A/HRC/29/19 (1 May 2015).
88 HRC Res. 29/15, *Human Rights and Climate Change*, UN Doc. A/HRC/RES/29/15 (adopted without a vote on 2 July 2015).
89 OHCHR, *Analytical Study on the Relationship Between Climate Change and the Human Right of Everyone to the Enjoyment of the Highest Attainable Standard of Physical and Mental Health*, UN Doc. A/HRC/31/36, Geneva (4 December 2015).
90 OHCHR, *Mapping Report on Human Rights and Climate Change*, Special Rapporteur on Human Rights and the Environment, UN Doc. A/HRC/25/53 (30 December 2013).
91 J. H. Knox, *Report of the Independent Expert on the Issue of Human Rights Obligations Relating to the Enjoyment of a Safe, Clean, Healthy and Sustainable Environment*, 'Compilation of good practices', UN Doc. A/HRC/28/61 (3 February 2015).
92 OHCHR, *A New Climate Change Agreement Must Include Human Rights Protections for All*, An Open Letter from Special Procedures mandate holders of the Human Rights Council to the State Parties to the UN Framework Convention on Climate Change on the occasion of the meeting of the Ad Hoc Working Group on the Durban Platform for Enhanced Action in Bonn (20–25 October 2014), 17 October 2014.
93 OHCHR, C. D. Aguilar *et al.*, *Report on the Effects of Climate on the Full Enjoyment of Human Rights* (30 April 2015), available at www.thecvf.org/wp-content/uploads/2015/05/humanrightsSRHRE.pdf.
94 OHCHR, *Joint Statement by UN Special Procedures on the Occasion of World Environment Day (5 June 2015) on Climate Change and Human Rights* (2015), available at www.ohchr.org/EN/NewsEvents/Pages/DisplayNews.aspx?NewsID=16049&LangID=E.
95 The Universal Rights Group (URG) is an independent think tank based in Geneva dedicated to analyzing and strengthening global human rights policy.
96 Mary Robinson Foundation for Climate Justice, founded and chaired by Mary Robinson, former President of the Republic of Ireland and former United Nations High Commissioner for Human Rights.
97 Side event for the UN Human Rights Day on 'Climate Change Threatening Human Rights: Challenges and Actions' organized by the World Council of Churches, Religions for Peace, Quaker United

Nations Office, Center for International Environmental Law, Earthjustice and Friedrich-Ebert-Stiftung in Room Sipan, Army Headquarters Compound, Lima, Peru, Wednesday 10 December 2014.

98 G. Knies, *Report – Human Rights and Climate Change – NGO Information Meeting on the Coming Climate Change negotiations in Geneva and Their Relevance to Human Rights* (2015), available online at www.fes-globalization.org/geneva/documents/2015/2015_02_06_Ngo_meeting_report_Knies.pdf.

99 See Mary Robinson Foundation – Climate Justice, *Geneva Pledge on Human Rights and Climate Action Announced* (13 February 2015), available at www.mrfcj.org/resources/geneva-pledge-human-rights/ (last visited 29 April 2017); Human Rights & Climate Change Working Group, *Promoting the Geneva Pledge for Human Rights in Climate Action*, available online at http://climaterights.org/our-work/unfccc/geneva-pledge/; OHCHR, *Implementing the Geneva Pledge for Human Rights in Climate Action* (13 September 2016), available online at www.ohchr.org/FR/NewsEvents/Pages/DisplayNews.aspx?NewsID=20477&LangID=E.

100 OHCHR, *Understanding Human Rights and Climate Change*, Submission by OHCHR to the 21st Conference of Parties to the UNFCCC (27 November 2015), available at www.ohchr.org/Documents/Issues/ClimateChange/COP21.pdf.

101 Human Rights Watch, *'There Is No Time Left': Climate Change, Environmental Threats, and Human Rights in Turkana County, Kenya* (15 October 2015), available at www.hrw.org/report/2015/10/15/there-no-time-left/climate-change-environmental-threats-and-human-rights-turkana.

102 The round table was organized by the Human Rights & Climate Change Working Group, the Geneva Climate Change Consultation Group, CARE International, Center for International Environmental Law, Earthjustice, Franciscans International, Friedrich-Ebert-Stiftung and Human Rights Watch.

103 Paris Agreement, *supra* note 3, at Preamble: "*Acknowledging* that climate change is a common concern of humankind, Parties should, when taking action to address climate change, respect, promote and consider their respective obligations on human rights, the right to health, the rights of indigenous peoples, local communities, migrants, children, persons with disabilities and people in vulnerable situations and the right to development, as well as gender equality, empowerment of women and intergenerational equity".

104 Paris Agreement, *supra* note 3, at Preamble: "*Recognizing* the fundamental priority of safeguarding food security and ending hunger, and the particular vulnerabilities of food production systems to the adverse impacts of climate change," & "*Taking into account* the imperatives of a just transition of the workforce and the creation of decent work and quality jobs in accordance with nationally defined development priorities".

105 Paris Agreement, *supra* note 3, at Preamble: "*Recognizing* that Parties may be affected not only by climate change, but also by the impacts of the measures taken in response to it".

106 OHCHR, *Analytical Study on the Relationship Between Climate Change and the Human Right of Everyone to the Enjoyment of the Highest Attainable Standard of Physical and Mental Health*, UN Doc. A/HRC/32/23 (May 2016); HRC Res. 32/33, *Human Rights and Climate Change*, UN Doc. A/HRC/RES/32/33 (adopted without a vote on 1 July 2016).

Part III
Early lessons

11

Look before you jump

Assessing the potential influence of the human rights bandwagon on domestic climate policy

Sébastien Jodoin, Rosine Faucher, and Katherine Lofts

Introduction

Since the mid-2000s, numerous initiatives have drawn on human rights norms, arguments, and mechanisms to build legal and political support for combating climate change. NGOs have launched complaints before regional human rights bodies[1] and initiated litigation in domestic courts.[2] Indigenous Peoples have pressed their cases in multiple venues, such as the U.N. Permanent Forum on Indigenous Issues, the International Indigenous Peoples Forum on Climate Change, and the World Conservation Congress.[3] Governments, most notably those of small island states, have issued international declarations[4] and have requested that international bodies such as the United Nations Office of the High Commissioner for Human Rights initiate work on climate change issues.[5] Civil society actors and governments have also raised climate change matters in the context of the U.N. Human Rights Council (UNHRC), which has adopted several resolutions on the matter.[6] Finally, these civil society actors and states have pressed for the inclusion of human rights language in the international climate negotiations themselves,[7] securing such language in the preamble and operational provisions of several decisions, including the Paris Agreement adopted in December 2015.[8]

With few exceptions,[9] most scholars, including the contributors to this handbook, have been quite optimistic about the potential for human rights to make positive contributions to global efforts to combat climate change. Some argue that international human rights law provides a lens for understanding the human impacts of climate change, obliges states to undertake or support climate mitigation and adaptation measures, and provides, to varying degrees, potential remedies and recourses for affected states, communities, and individuals.[10] Others have focused instead on the normative case for climate justice provided by human rights and their implications for the development and implementation of international obligations in the climate regime.[11] Finally, some scholars have expressed enthusiasm about the role that human rights norms and arguments could play in spurring significant political progress in the fight against climate change.[12] Nicholson and Chong most clearly articulate this view, arguing that "jumping on the human rights bandwagon" can mobilize support for a new brand of climate politics that "properly recognizes, and that seeks to fully accommodate, the fact that traditionally marginalized nations and communities are suffering, and will continue to suffer from the worst impacts of climate change."[13]

While a body of literature that examines the strategies adopted by NGOs, Indigenous Peoples, and governments to build linkages between human rights and climate change at the international level has begun to grow,[14] whether and to what extent this "human rights bandwagon" can influence domestic climate policy efforts remains an open question. Our chapter draws on research in political science to assess the potential role that human rights norms, arguments, and mechanisms could play in leading states to adopt and implement effective climate mitigation policies at the domestic level.[15] To that end, we adopt a rationalist-constructivist framework that identifies four important causal mechanisms that can account for the influence of international norms on state behaviour: cost-benefit compliance, mobilization, persuasive argumentation, and acculturation. The first two of these causal mechanisms are anchored in a rationalist perspective that conceives of states as rational actors seeking to maximize their preferences and respond to changes in incentives and disincentives that affect their interests.[16] The latter two mechanisms are instead derived from a constructivist approach that views states as social actors influenced by norms that provide intersubjective understandings of appropriate behaviour, constitute their identities and interests, and shape their understanding of the world.[17] It is now well recognized that both types of explanation are necessary to give a full account of how international law influences state behaviour.[18]

Our chapter proceeds as follows. We begin by discussing the opportunities that the human rights bandwagon can provide for *pressuring* states to combat climate change through the mechanisms of cost-benefit compliance and mobilization. We then examine whether and to what extent the human rights bandwagon can *socialize* states into combating climate change because of persuasive argumentation and acculturation. We conclude by discussing how these different causal mechanisms may be combined to provide a complex assessment of the potential influence of human rights on global efforts to combat climate change as well as identify key lines of inquiry for future research in this field.

Does the human rights bandwagon increase opportunities for pressuring states to adopt and implement domestic climate mitigation policies?

The human rights bandwagon and cost-benefit compliance

Our assessment of the potential of the human rights bandwagon for climate politics begins with the following question: to what extent can drawing on human rights arguments and mechanisms alter the cost-benefit calculations that underlie state behaviour in the field of climate change? Given the generalized problem of getting states to implement existing international commitments to reduce their carbon emissions, we want to focus here on the comparative advantages that human rights may offer in relation to two common rationalist explanations of state compliance with international law. Indeed, as Nicholson and Chong argue, climate activists have turned to international human rights law, in part, because of the legal and political tools it can provide to pressure states into combatting climate change.[19]

Rationalist approaches in international relations generally emphasize that states may comply with international law in order to obtain or preserve reputational benefits vis-à-vis other states or their own domestic audience.[20] The existing literature evinces that the effectiveness of reputation as a motivation for compliance depends on the level of precision and obligation of international legal rules and the availability of information about compliance,[21] both of which increase the likelihood of the detection of non-compliance and the probability that reputational benefits will be lost or costs will be incurred as a result.[22]

Yet when it comes to state responsibility for human rights violations associated with climate change, a number of uncertainties and challenges diminish the effectiveness of reputation as a motivation for compliance. To begin with, it is not clear that the invocation of state responsibility for human rights violations arising from climate change finds strong support in existing international human rights law.[23] As the Office of the High Commissioner for Human Rights (OHCHR) has stated:

> Qualifying the effects of climate change as human rights violations poses a series of difficulties. First, it is virtually impossible to disentangle the complex causal relationships linking historical greenhouse gas emissions of a particular country with a specific climate change-related effect, let alone with the range of direct and indirect implications for human rights. Second, global warming is often one of several contributing factors to climate change-related effects, such as hurricanes, environmental degradation and water stress. Accordingly, it is often impossible to establish the extent to which a concrete climate change-related event with implications for human rights is attributable to global warming. Third, adverse effects of global warming are often projections about future impacts, whereas human rights violations are normally established after the harm has occurred.[24]

We do not meant to suggest here that states could never be held responsible under international human rights law for failing to act to reduce carbon emissions,[25] but merely that this an uncertain and unsettled legal matter. The ambiguity is all the more significant in the context of trying to hold developed states responsible for the human rights impacts resulting from their historic or current levels of carbon emissions, because the extent and scope of extra-territorial responsibility for human rights remains a controversial matter.[26] In the absence of clarity regarding state responsibility for human rights violations relating to climate change, it is therefore unlikely that the reputational risk for developed states will be high enough to alter their behaviour.

Another reason that states may decide to comply with international law according to a rationalist approach concerns their desire to avoid the sanctions that may be imposed upon them due to non-compliance[27] or, conversely, to gain access to material benefits (such as international aid or access to foreign markets) that provide rewards for compliance.[28] The latter option seems remote at this stage, since the developed states in a position to tie development aid or preferential trade agreements to human rights conditionalities do not themselves recognize the implications of climate change for human rights in their domestic climate policies and are not necessarily living up to their existing obligations under the UNFCCC.[29]

The key question, then, is whether the invocation of international human rights law can trigger the use of mechanisms to effectively sanction deviant behaviour.[30] A number of obstacles stand in the way of doing so under international human rights law. To begin with, admissibility can be a challenging hurdle for claimants to overcome. While admissibility requirements may differ depending on the tribunal in question, plaintiffs must have standing to bring a complaint, demonstrating that their rights are at stake. The issue in question must also be deemed judiciable; for example, it should be defined as legal, rather than political, in nature. The Inuit petition provides one example of the difficulties of admissibility in this context. In 2005, Inuit in the Canadian and Alaskan Arctic filed a complaint before the Inter-American Commission on Human Rights (IACHR) seeking compensation from the United States for alleged violations of their human rights resulting from climate change.[31] Although the complaint is generally considered to have played a critical role in raising awareness of the human rights implications of climate change,[32] it was ultimately deemed inadmissible by the IACHR on the grounds that the petition provided insufficient information on which to make a decision.[33]

In cases that are deemed admissible, the issue of causation may create a further obstacle. Two primary questions arise: firstly, "whether a clear and causal link exists between anthropogenic emissions and climate change," and second, "whether the particular damage suffered by one victim is effectively caused by CO_2 emissions from one particular source."[34] The first of these questions can be disposed of relatively easily, as scientific consensus clearly supports the fact that human activities are the main driver of climate change.[35] The second question is somewhat more challenging, however. As Faure and Nollkaemper note, under international law:

> the starting point is that a responsible state needs only to compensate for damage that is caused by the wrongful act. This requires a link between emissions, climate change, and harmful effects. Whether damage is "caused" by an act is primarily determined by the criteria of normality and predictability (or foreseeability). Under the criterion of normality, an injury is sufficiently linked to an unlawful act whenever the normal and natural course of events indicates that the injury is a logical consequence of the act. Under the criterion of predictability, an injury is linked to an unlawful act whenever the author of the unlawful act could have foreseen the damage it caused. The important question thus is whether a state emitting carbon dioxide could foresee the damage, or whether emissions would cause the harm in the "normal course of events."[36]

In the case of climate change, it may be difficult to link the actions or inaction of one particular state (or of a private entity within that state) to the harms experienced by an individual or community. This is particularly true given the temporal and geographical distances that often occur between the source of greenhouse gas emissions and their ultimate effects. Indeed, since virtually all states have enabled the generation of carbon emissions on their territories, the attribution of responsibility is a complex matter.[37] It may also be difficult to disentangle the impacts caused by climate change from underlying vulnerabilities that tend to be exacerbated by a changing climate, including poverty, discrimination and social exclusion, political disenfranchisement, and environmental degradation.[38]

Finally, even if a case brought before a tribunal is decided in favour of the claimants, it may be difficult to secure effective remedies and compliance with those remedies. In practical terms, the effectiveness of a remedy under international human rights law requires that a state have accepted the jurisdiction of the tribunal in question, and that enforcement mechanisms be in place to ensure compliance with the remedy granted. Yet when it comes to international human rights law, "[m]onitoring, compliance, and enforcement provisions are non-existent, voluntary, or weak or deficient."[39] Nor do non-legal factors tend to favour compliance, as "[p]owerful countries rarely employ sanctions – political, economic, military, or otherwise – to coerce other countries into improving their human rights record."[40] Thus, as Averill points out, "[h]uman rights law remains more aspirational than enforceable."[41] When it comes to compliance, scholars and activists may therefore overestimate the benefits that international human rights law offers.

Taken together, these challenges suggest that international human rights law is unlikely to effectively sanction deviant behaviour. As argued by Atapattu, applying the traditional model of state responsibility under international law to "global challenges such as climate change, where there are multiple perpetrators, multiple sources and multiple victims, is akin to fitting square pegs in round holes."[42]

The human rights bandwagon and mobilization

Mobilization is another potential advantage that the human rights bandwagon may offer for climate activism. Mobilization is defined here as the causal mechanism whereby domestic actors pressure governments into recognizing and implementing an international legal obligation.[43] As Checkel explains, the underlying causal logic of mobilization can be understood in rationalist terms to the extent that governments adjust their strategies when confronted by "pressure from below" and the "interests they pursue (re-election, political survival) remain the same."[44] In particular, international human rights law may provide domestic interest groups and coalitions with powerful rhetorical tools for pressuring state authorities to adopt these legal norms,[45] as well as enhanced opportunities for doing so, especially when state authorities have formally committed to them through the signature or ratification of an international treaty.[46]

The existing literature suggests that the effectiveness of mobilization depends on two important factors. The first factor has to do with the capacity of domestic interest groups to access and aggregate an array of ideational and material resources – generated by themselves or obtained from other actors – in order to pressure their governments.[47] The second factor is known as opportunity structure and refers to the set of institutional, ideational, and material conditions that may favour or constrain the emergence and mobilization of interest groups and coalitions in favour of change and reform.[48] According to Simmons, when domestic interest groups are successful in their mobilization efforts, this may result in a tactical concession on the part of state authorities to make superficial commitments to international human rights law.[49] In turn, the sustained or renewed pressure exerted by domestic interest groups may also explain why these state authorities may more fully implement this commitment at a later stage.[50]

Our assessment of the human rights bandwagon's potential for increasing the probability and effectiveness of domestic mobilization on climate issues distinguishes between two different forms of mobilization: pressure exerted by social movements on governments through political protest and advocacy[51] and pressure exerted through public interest litigation.[52] We will address these two forms of mobilization in turn.

To begin with, it is not clear that framing climate change as a human rights issue will lead to more political mobilization through protest and advocacy, due to a range of psychological and social barriers related to addressing climate change as a serious issue. First, the causes of climate change (namely, greenhouse gas emissions) are largely invisible, as are many of its impacts, which may be both temporally and geographically distant.[53] As a result, "[e]mitting greenhouse gases does not lead immediately to a noticeable, visible impact."[54] While catastrophic events such as flash floods or hurricanes do provide striking examples of the changing climate, it may be difficult to see an immediate connection between these events and greenhouse gas emitting activities. In addition, other impacts, such as shifting weather patterns or the increased prevalence of infectious disease, may be less immediately noticeable. It may also be difficult to see the linkages between mitigation actions and changes to the climate, due to "lags in the climate and social systems and the cumulative nature of emissions."[55]

One's personal experience of climate change can also impact individual behaviour.[56] For example, people are likely to give relatively less weight in decision-making to risks posed by statistically rare events or by those events of which one has no personal experience.[57] Indeed, in many regions of the world, "[t]he likelihood of seriously and noticeably adverse events as the result of global warming is bound to be small for the foreseeable future,"[58] leading to a low perception of risk and corresponding lack of urgency to act. In addition, people tend to "discount" future costs or risks, such as those associated with climate change.[59] Finally, people may experience denial when they perceive that they have no control over a situation such as climate change;[60] feeling overwhelmed by the uncertainty and complexity of the issue may lead to inaction. All in all, research in psychology and sociology suggests that the characteristics of climate change cause individuals to either underestimate or discount the risk it presents.[61] There is no reason to posit that presenting climate change as a human rights problem can overcome these challenges to citizen mobilization.

On the other hand, pressure exerted through domestic human rights litigation may offer greater opportunities for mobilization on climate issues. Indeed, recent developments at the national level highlight the promise of public interest litigation in getting domestic courts to enforce climate action. The *Urgenda* class action lawsuit in the Netherlands provides one such example.[62] In 2013, the Urgenda Foundation filed a class action in a Dutch court, claiming under both human rights and tort law that the government of the Netherlands had failed to adequately protect its citizens from climate change.[63] Urgenda's human rights claim rested on the jurisprudence of the European Court of Human Rights (ECHR), particularly its interpretation of the right to life as encompassing the right to a healthy environment.[64] The tort claim was based on the government's failure, under national law, to implement proper mitigation measures to avoid surpassing 2° Celsius warming, corresponding to a breach of duty of care in tort.[65] In 2015, the Hague District Court, ruling in favour of Urgenda, "adopted the European Court on Human Rights' view that 'human rights law and environmental law are mutually reinforcing.'"[66] The Court ordered the government to reduce the countries' greenhouse gas emissions by 25 per cent (from 1990 levels) by 2020. In justifying its decision, the Court referred to the fact that the country was a signatory to the UNFCCC as well as the Kyoto Protocol; the Netherlands had thereby expressly assented to "its responsibility for the national emission level and in this context accepted the obligation to reduce this emission level as much as needed to prevent dangerous climate change."[67] Moreover, the Court found that "a sufficient causal link can be assumed to exist between the Dutch greenhouse gas emissions, global climate change and the effects (now and in the future) on the Dutch living climate."[68]

In her analysis of the judgment, Bach highlights several interesting aspects of the Court's reasoning. To begin with, the Court accepted the Netherlands' ratification of the UNFCCC and the Kyoto Protocol as demonstrating knowledge of the serious threat to human populations caused by climate change.[69] The Court also noted the principles and commitments contained in Articles 3 and 4 of the UNFCCC, as well the 2°C threshold set out by the IPCC, taking these factors as evidence of the government's duty of care and subsequent breach.[70] Finally, the Court assessed the country's "voluntary pledge under the UNFCCC's Copenhagen Accord and Cancun Agreement . . . when determining insufficient action that would breach its duty of care to its citizens," thus drawing on "the Oslo Principle's vision of 'a network of intersecting sources' of climate change law."[71]

The groundbreaking American constitutional case of *Juliana v. U.S.* is another promising example of domestic human rights–based litigation being used to address climate change. While the case has not yet been heard on the merits, it has been deemed admissible and permitted to proceed to trial by a US District Court Judge, who rejected all arguments to dismiss raised by the federal government and representatives of the fossil fuel industry. Filed by 21 youth plaintiffs and climate scientist Dr. James Hansen on behalf of future generations, the lawsuit asserts that the US government's affirmative actions causing climate changed have violated children's rights to life, liberty, and property, and that the government has also failed to protect public trust resources. In determining the complaint to be valid, the Judge stated:

> This action is of a different order than the typical environmental case. It alleges that defendants' actions and inactions – whether or not they violate any specific statutory duty – have

> so profoundly damaged our home planet that they threaten plaintiffs' fundamental constitutional rights to life and liberty.[72]

These recent cases suggest that although international human rights mechanisms may, for a variety of reasons, be ill-suited to address the complexities of climate-related human rights complaints, domestic human rights litigation may present a more viable alternative. Domestic human rights litigation offers an important advantage for tackling climate change – the availability of a court that has the jurisdiction and powers to hear cases relating to the human rights implications of climate change. It should be noted, in this regard, that neither *Urgenda* nor *Juliana v. U.S.* advanced extra-territorial claims, but focused instead on seeking to protect the rights of Dutch and American citizens, respectively, from the impacts of climate change. A second advantage has to do with enforcement, at least in the context of well-functioning democracies operating under the rule of law.[73]

Of course, certain barriers must still be overcome. To begin with, access to justice remains a concern. For example, those currently most affected by climate change in their day-to-day lives tend to be from the most vulnerable segments of society.[74] Many will not have the resources necessary to bring a claim in court, particularly one whose outcome is uncertain.

The success of domestic climate change litigation will also depend on the willingness of courts to import international environmental law principles into municipal law or to inform their interpretation of municipal law with international legal principles. Linked to this issue is the fact that many of the harms caused by climate change are connected to rights, such as economic, social, and cultural rights, traditionally considered to be non-judiciable in both domestic and international tribunals (although in practice these rights have been adjudicated in a number of jurisdictions).[75] Moreover, governments enjoy broad discretion in relation to matters of public policy, a category into which climate-related decisions have frequently fallen. Claimants will therefore need to carefully consider the framing of their claims and weigh the costs and benefits of human rights–based litigation in order to address such challenges.[76]

Overall, the emerging trend of domestic litigation relating to climate change provides a strong example of the advantages of using this tool for the enforcement of international human rights law. This should not be surprising since domestic litigation is now understood as a key pathway for pressuring states into complying with their human rights obligations under international law more generally.[77]

Can the human rights bandwagon provide opportunities for socializing states into adopting and implementing domestic climate mitigation policies?

The human rights bandwagon and persuasive argumentation

A third way that the human rights bandwagon might affect domestic climate politics has to do with the causal mechanism of persuasive argumentation, whereby states internalize an international norm because it is consistent with the shared understandings they have developed with other actors (including other states, NGOs, and IGOs).[78] As many scholars argue, when actors engage in persuasive argumentation with one another, the truth claims inherent in their existing interests and identities are open to mutual contestation, and they may accordingly internalize new norms based on the normative consensus that they construct together.[79] According to Finnemore and Sikkink, when actors internalize a norm through persuasive argumentation, it *can* achieve a "taken-for-granted quality" that makes conformity "almost automatic."[80]

At the outset, it is important to note that there is little evidence to suggest states have thus far internalized a new international norm obliging them to combat climate change as a result of their responsibilities under international human rights law.[81] Many scholars have indeed highlighted the important role that human rights could play in shaping a new understanding of climate change focused on the ways that it will harm vulnerable groups and emphasizing the concordant obligations of states to respond under international law.[82] As Nicholson and Chong argue, "rights talk is an attempt to convert climate change from a dry and amorphous scientific problem into a tangible and actionable humanistic problem."[83] In this way, the human rights bandwagon, like the broader climate justice movement, seeks to challenge the dominant framing of climate governance as a domain that is primarily concerned with finding technological solutions to reduce carbon emissions that are compatible with patterns of production and consumption in industrialized countries and the pursuit of growth and the reduction of poverty in emerging economies and developing countries.[84]

The success of a coalition of NGOs, international organizations, and states in convincing a critical mass of states to prohibit landmines offers a number of lessons for thinking about the potential effectiveness of reframing climate change as a human rights issue. While landmines were considered a legitimate weapon and were widely used by militaries around the world for most of the twentieth century, activists and their allies succeeded in the late 1990s in reframing landmines as a humanitarian problem, rather than an arms control issue. Their efforts resulted in a comprehensive treaty to ban landmines that has achieved near universal ratification.[85] Price's analysis of these efforts reveals that two underlying strategies were key to the successful use of reframing in this context. First, the proponents of the landmine ban were able to stress the human tragedy associated with the use of landmines, point out the limitations of existing approaches to this problem, and assign blame in a simple and direct manner for the harms that they caused.[86] Second, they aligned their proposed norms with existing international norms internalized by states – something that Price describes as grafting.[87] In their efforts to ban landmines, activists relied on an important international norm – the protection of civilians from the indiscriminate effects of warfare – for which there existed an overwhelming legal and moral consensus.[88] Both strategies were critical to the emergence of a transnational coalition of NGOs, international organizations, and states that generated morally persuasive information about landmines, developed new norms delegitimizing the use of landmines, and obliged recalcitrant states to justify their continued use.[89]

To be sure, the proponents seeking to reframe climate change as a human rights issue have adopted both of the strategies used to great effect in the landmines case. Along with the broader climate justice movement, they have sought to generate morally persuasive messages about the impacts of climate change on the rights and lives of communities and individuals most vulnerable to the effects of climate change.[90] They have also called for states to build their responses to climate change based on existing rules and principles of international human rights law.[91] Indeed, there are reasons to believe that climate change may provide a particularly favourable field for the use of these strategies and more broadly for the generation and internalization of new international norms that reframe climate change as a human rights problem.

As was discussed above, psychological research suggests that the persuasiveness of framing climate change as a human rights issue may be constrained by how individuals generally perceive diffuse and complex forms of risk, causation, harm, and responsibility in the context of a phenomenon like climate change.[92] This is not an insurmountable obstacle, but it does require that climate activists be especially ingenious in developing innovative ways of communicating the human harms associated with climate change and attributing responsibility for these harms to the actions of particular actors. The advocacy accompanying the *Urgenda* case mentioned above may be worth emulating in this regard. Like the case itself, the key message of the *Urgenda* campaign is that the Dutch government is taking insufficient action to protect the rights of Dutch citizens from the impacts of climate change. While the Dutch government has yet to be persuaded by this argument and has appealed the judgement, the case has attracted significant attention from the media and civil society[93] and appears to have shifted the debate over climate politics in the Netherlands.[94]

The potential effectiveness of grafting as a strategy for seeking to reframe climate change as a human rights problem may also be limited. As seen in the section on the human rights bandwagon and cost-benefit compliance, climate change raises a number of complex legal questions, in terms of causation and the attribution of responsibility, that international human rights may not be particularly well-equipped to handle.[95] As well, the scope of application of economic, social, and cultural rights and the respective responsibilities of developed and developing countries in relation to their protection and realization continue to raise legal as well as political controversies.[96] The effort to generate a new international norm requiring states to take action to combat climate change because of their obligations to protect human rights may therefore be constrained by its inconsistency with existing international norms in relation to states' obligations under international human rights law. Here again, the *Urgenda* case may offer useful insights. It is important to note that *Urgenda* is grounded on the protection of the right to life (Article 2 of the ECHR) and the right to health and respect for private and family life (Article 8 of the ECHR) and that there is extensive case law in the European Court of Human Rights that provides that states have a responsibility, under both articles, to protect their citizens from environmental harms.[97] Although it remains to be seen whether grafting will ultimately be successful in the particular context of Dutch climate politics, it is possible that this technique may be more effective in certain jurisdictions or regions that possess favourable existing norms.

The human rights bandwagon and acculturation

The causal mechanism of acculturation offers one final way through which the human rights bandwagon might affect domestic climate politics. Acculturation is a form of socialization in which states conform with international norms embedded within a broader transnational reference group to which they belong. As argued by Goodman and Jinks, when actors are acculturated into an international norm, they alter their behaviour in ways that are consistent with that norm's prescriptions because of the internal cognitive pressures that drive social conformity, as well as the external social pressures exerted by exogenous actors that approve of conformity and sanction non-conformity.[98] Unlike persuasive argumentation, acculturation does not in and of itself lead states to be actively convinced of the validity of an international norm.[99] Work by Finnemore and Sikkink suggests that acculturation is often associated with a later stage in the diffusion of international norms across a population of actors, when the acceptance of legal norms within a reference group has reached a certain threshold such that conformity with a norm is perceived as critical to an actor's identity or role within this group.[100]

Our review of the evidence suggests that the human rights bandwagon is far from having succeeded in engendering a process of acculturation whereby the integration of human rights in domestic climate policies might have become central to the identity of states. We first looked at the Intended Nationally Determined Contributions (INDCs) submitted by states in the lead-up to the 21st Conference of the Parties to the UNFCCC held in Paris in December 2015.[101] In all, we found that only 26 states mentioned human rights in some way in their INDCs and that none of the industrialized countries, and no emerging economies, with the exceptions of Brazil and Mexico, included human rights in their INDCs.[102] In addition, most references to human rights in INDCs occur in the context of adaptation planning and capacity building[103] or the need to ensure that climate actions do not violate human rights,[104] or simply cite broader provisions or aspects of states' constitutions or sustainable development programmes.[105] We also performed searches for the key word "human rights" in two leading databases of climate laws and policies from around the world.[106] We failed to find any countries that included references to human rights in their domestic climate laws and policies.

Current state practice evinces that a growing number of developing countries recognize the linkages between human rights and climate change, but the way that this linkage is articulated is inconsistent. Most importantly, the recognition of human rights in domestic climate commitments articulated by states generally fails to capture the role that human rights could play in articulating a basis for the pursuit of climate mitigation. This finding is consistent with the Mary Robinson Foundation for Climate Justice's recent assessment of the use of human rights in the national communications submitted to the UNFCCC and national reports to the Human Rights Council under the Universal Periodic Review.[107]

It is hard to conclude that the diffusion of human rights norms in the context of climate policy has reached the sort of tipping point where acculturation might begin to exert causal influence on recalcitrant states opposed to, or wary of, adopting a rights-based approach to climate change. At the same time, current trends indicate that Latin America may eventually constitute a region in which the consideration of human rights issues in the context of climate change will be perceived by states as essential to their roles and identities, thereby reaching the threshold for acculturative processes to take hold in that reference group.[108]

Conclusion

As is discussed throughout this handbook, we agree that human rights provide an important conceptual, legal, and normative framework for understanding and addressing the implications of climate change for individuals and communities. At the same time, we are more sober about the prospects for the human rights bandwagon to pressure and socialize states into the adoption and implementation of climate mitigation policies at the domestic level.

Although there is little doubt that climate change will have significant implications for the enjoyment of a wide range of human rights, it is not clear that these implications can be defined in terms of violations giving rise to an established set of duties and remedies under international law. International human rights law may simply not be well equipped to grapple with a causally complex problem like climate change, and the invocation of human rights compliance mechanisms as a way of pressuring states to combat climate change risks diverting resources and focus away from more productive strategies for the pursuit of domestic policy change. Likewise, there is little evidence to suggest that human rights might alter the opportunity structure for successful domestic mobilization on climate change, given the range of psychological and sociological challenges that lead individuals to underestimate, discount, or deny the risk climate change presents. That said, as demonstrated by the *Urgenda* case, domestic litigation probably offers the best hope for holding governments accountable in line with their human rights obligations and pressuring them to adopt and implement effective domestic climate mitigation policies.

While it is true that the normative framing provided by human rights has significant transformative potential, it is not apparent that the underlying conditions for this potential to be fulfilled are present in the field of climate. There is little evidence to suggest that states have thus far internalized an international norm obliging them to combat climate change because of their responsibilities under international human rights law, meaning that persuasive argumentation is unlikely to be successful. Likewise, the efficacy of grafting as a strategy for reframing climate change as a human rights problem may also be limited due to complex legal issues of causation and attribution, and continuing political controversy related to the application of economic, social, and cultural rights. Finally, there is little evidence that the diffusion of the normative framing of climate change as a human rights issue has reached the necessary threshold for processes of acculturation to begin to take effect in the international community as a whole.

In sum, our chapter identifies the many obstacles that stand in the way of deploying the human rights bandwagon in a way that could influence climate mitigation policy-making at the domestic level. We nonetheless believe that there are reasons for optimism about rights-based strategies in the field of climate change. First, the transnational pursuit of change generally takes time, and several of the causal processes discussed above, particularly in terms of domestic legal mobilization, persuasive argumentation, and acculturation, may simply be in their earliest stages. Second, the results of more than 12 years of human rights advocacy in the context of climate change has laid the groundwork for additional and more effective strategies. For instance, the gains achieved in having human rights recognized in the preamble to the Cancun Agreements and the Paris Agreement open the door to their invocation in the context of domestic litigation.[109] Third, while the four causal mechanisms are treated here as distinct from one another, it is important to recognize that they may operate together, concurrently or as part of a sequence, in the transnational legal process through which international norms may emerge, evolve, and become effective.[110] Extensive literature in the field of human rights demonstrates the importance of considering how several causal mechanisms may work to close some of the initial gaps that may emerge between a state's decision to commit to an international norm and the extent to which they may actually implement this obligation in practice.[111]

Ultimately, our biggest reason for optimism has to do with our conviction, as exemplified by the *Urgenda* case, that it is time for the human rights bandwagon to shift its focus to the domestic level. That is not to say that ongoing advocacy in the UNFCCC and the U.N. human rights system should be abandoned altogether, but that it is time for activists to raise human rights arguments in the context of climate legal and policy debates taking place at the domestic level. Rather than simply jumping on the human rights bandwagon, we should look before we leap – ensuring that the tools of human rights are used in a way that will best effect positive change.

Notes

1 Inuit Circumpolar Conference, *Petition to the Inter-American Commission on Human Rights Seeking Relief From Violations Resulting from Global Warming Caused by Acts and Omissions of the United States, submitted by Sheila Watt-Cloutier* (7 December 2005), with the support of the Inuit Circumpolar Conference, on behalf of all Inuit of the Arctic Regions of the United States and Canada, available online at http://inuitcircumpolar.com/files/uploads/icc-files/FINALPetitionICC.pdf.

2 *Urgenda Foundation v. The State of the Netherlands (Ministry Of Infrastructure and the Environment)*, C/09/456689, 13–1396, Rechtbank Den Haag, The District Court of the Hague (24 May 2015), available online at https://uitspraken.rechtspraak.nl/inziendocument?id=ECLI:NL:RBDHA:2015:7196&keyword=urgenda, at paras 4.8 & 4.48 (last visited 9 April 2017).

3 A. Doolittle, 'The Politics of Indigeneity: Indigenous Strategies for Inclusion in Climate Change Negotiations', 8(4) *Conservation and Society* (2010) 256.

4 Center for International Environmental Law, *Male' Declaration on the Human Dimension of Global Climate Change* (14 November 2007), available online at www.ciel.org/Publications/Male_Declaration_Nov07.pdf.

5 J. H. Knox, 'Linking Human Rights and Climate Change at the United Nations', 33 *Harvard Environmental Law Review* (2009) 477 at 483–484.

6 *UNHRC Res 7/23 Human Rights and Climate Change, 7th Session* (14 July 2008), UN Doc. A/HRC/7/78; *UN Human Rights Council Res. 10/4 Human Rights and Climate Change, 10th session* (25 March 2009), UN Doc. A/HRC/10/L.11; *UNHRC Res 10/4 Human Rights and Climate Change, 41st*

Meeting (25 March 2009), UN Doc. A/HRC/10/L.11; *UNHRC Res 26/27 Human Rights and Climate Change, 40th Meeting* (27 June 2014), UN Doc. A/HRC/26/L.33/Rev.1; *UNHRC Res 29/15 Human Rights and Climate Change, 44th Meeting* (2 July 2015), UN Doc. A/HRC/29/L.21; Office of the High Commissioner for Human Rights, *Report of the Office of the United Nations High Commissioner for Human Rights on the Relationship Between Climate Change and Human Rights*, UN Doc. A/HRC/10/61 (15 January 2009), at para 70 ("OHCHR Report on Climate Change and Human Rights").

7 L. Rajamani, 'The Increasing Currency and Relevance of Rights-Based Perspectives in the International Negotiations on Climate Change', 22(3) *Journal of Environmental Law* (2010) 391; A. Schapper, 'Local Rights Claims in International Negotiations: Transnational Human Rights Networks at the Climate Conferences' (this volume).

8 B. Mayer, 'Human Rights in the Paris Agreement', 6 *Climate Law* (2016) 109–17.

9 See for example E. A. Posner, 'Climate Change and International Human Rights Litigation: A Critical Appraisal', 155 *University of Pennsylvania Law Review* (2007) 1925; S. Tully, 'Like Oil and Water: A Sceptical Appraisal of Climate Change and Human Rights', 15 *Australian International Law Journal* (2009) 213; S. Humphreys, *Competing Claims: Human Rights and Climate Harms* (2010) 37; O. W. Pedersen, 'Climate Change and Human Rights: Amicable or Arrested Development', 1(2) *Journal of Human Rights and the Environment* (2010) 236.

10 See for example J. H. Knox, 'Climate Change and Human Rights Law', 50(1) *Virginia Journal of International Law* (2009) 163; S. L. Kass, 'Integrated Justice: Human Rights, Climate Change, and Poverty', 18 *Transnational Law & Contemporary Problems* (2009) 115; M. Averill, 'Linking Climate Litigation and Human Rights', 18(2) *Review of European Community and International Environmental Law* (2009) 139; M. Limon, 'Human Rights Obligations and Accountability in the Face of Climate Change', 38 *Georgia Journal of International & Comparative Law* (2010) 543; M. Wewerinke and C. F. J. Doebbler, 'Exploring the Legal Basis of a Human Rights Approach to Climate Change', 10(1) *Chinese Journal of International Law* (2011) 141; M. Wewerinke-Singh, 'State Responsibility for Human Rights Violations Associated With Climate Change' (this volume); S. Harmeling, 'Climate Change Impacts: Human Rights in Climate Adaptation and Loss and Damage' (this volume).

11 See for example S. Caney, 'Climate Change, Human Rights and Moral Thresholds', in Humphreys (ed.), *supra* note 9, at 69; S. Caney, 'Human Rights, Responsibilities, and Climate Change', in C. Beitz and R. Goodin (eds.), *Global Basic Rights* (2009) 227; A. Sinden, 'Climate Change and Human Rights', 27(2) *Journal of Land, Resources & Environmental Law* (2007) 255; E. Cameron, 'Human Rights and Climate Change: Moving From an Intrinsic to an Instrumental Approach', 38 *Georgia Journal of International & Comparative Law* (2010) 673.

12 M. Limon, 'Human Rights and Climate Change: Constructing a Case for Political Action', 33 *Harvard Environmental Law Review* (2009) 439.

13 S. Nicholson and D. Chong, 'Jumping on the Human Rights Bandwagon: How Rights-Based Linkages Can Refocus Climate Politics', 11(3) *Global Environmental Politics* (2011) 121, at 123.

14 See A. Schapper and M. Lederer, 'Introduction: Human Rights and Climate Change: Mapping Institutional Inter-Linkages', 27(4) *Cambridge Review International Affairs* (2014) 666; Schapper, *supra* note 7.

15 As such, we do not focus on how human rights could influence state behaviour in relation to climate adoption policy.

16 D. Snidal, 'Rational Choice and International Relations', in W. Carlsnaes, T. Risse and B. A. Simmons (eds.), *Handbook of International Relations* (2013) 85, at 85.

17 M. Finnemore and K. Sikkink, 'International Norm Dynamics and Political Change', 52(4) *International Organization* (1998) 887, at 891.

18 H. H. Koh, 'Transnational Legal Process', 75 *Nebraska Law Review* (1996) 181, at 205.

19 Nicholson and Chong, *supra* note 13, at 129–133.

20 B. A. Simmons, 'Compliance With International Agreements', 1(1) *Annual Review of Political Science* (1998) 75, at 81.

21 K. W. Abbott *et al.*, 'The Concept of Legalization', 54(3) *International Organization* (2000) 401, at 408–415; K. W. Abbott and D. Snidal, 'Hard and Soft Law in International Governance', 54(3) *International Organization* (2009) 421, at 426–427; E. Hafner-Burton, *Forced to Be Good: Why Trade Agreements Boost Human Rights* (2009) 160.

22 A. T. Guzman, 'A Compliance-Based Theory of International Law', 90 *California Law Review* (2002) 1865, at 1861–1863.

23 S. Atapattu, *Human Rights Approaches to Climate Change: Challenges and Opportunities* (2016) 274–278.

24 OHCHR Report on Climate Change and Human Rights, *supra* note 6, at para 70.

25 For a perspective that argues that this is indeed possible, see Wewerinke-Singh, *supra* note 10.
26 Atapattu, *supra* note 23, at 89–91.
27 R. Goodman and D. Jinks, *Socializing States: Promoting Human Rights Through International Law* (2013) 31–32.
28 E. Hafner-Burton, 'Trading Human Rights: How Preferential Trade Agreements Influence Government Repression', 59 *International Organization* (2005) 593–629; F. Schimmelfennig and U. Sedelmeier, 'Governance by Conditionality: EU Rule Transfer to the Candidate Countries of Central and Eastern Europe', 11(4) *Journal of European Public Policy* (2004) 661–679.
29 UNEP, *Climate Change and Human Rights*, available online at http://web.law.columbia.edu/sites/default/files/microsites/climate-change/climate_change_and_human_rights.pdf, at 30–32.
30 T. C. Halliday and B. Carruthers, *Bankrupt: Global Lawmaking and Systematic Financial Crisis* (2009) 351–354.
31 Petition, *supra* note 1.
32 H. Osofsky, 'Inuit Petition as a Bridge? Beyond Dialectics of Climate Change and Indigenous Peoples Rights', 31 *American Indian Law Review* (2007) 675.
33 In a letter dated November 16, 2006, the IACHR informed the petitioners that it would not consider the petition because the information it provided was not sufficient for making a determination. In March 2007, however, the IACHR did hold hearings with the petitioners to address matters raised by the petition. See J. Gordon, 'Inter-American Commission on Human Rights to Hold Hearing After Rejecting Inuit Climate Change Petition', 7(2) *Sustainable Development Law and Policy* (2006–2007) 55.
34 M. Faure and A. Nollkaemper, 'International Liability as an Instrument to Prevent and Compensate for Climate Change', 26A *Stanford Environmental Law Journal* (2007) 123, at 158.
35 Intergovernmental Panel on Climate Change, *Climate Change 2014: Synthesis Report* (2014), in R. K. Pachauri and L. A. Meyer (eds.), *Contribution of Working Groups I, II and III to the Fifth Assessment Report*, available online at www.ipcc.ch/report/ar5/.
36 Faure and Nollkaemper, *supra* note 34.
37 Atapattu, *supra* note 23, at 277–278.
38 Averill, *supra* note 10, at 141.
39 E. Neumayer, 'Do International Human Rights Treaties Improve Respect for Human Rights?', 49(6) *Journal of Conflict Resolution* (2005) 925, at 926.
40 *Ibid.*
41 Averill, *supra* note 10, at 141.
42 Atapattu, *supra* note 23, at 277.
43 J. T. Checkel, 'Why Comply? Social Learning and European Identity Change', 55(3) *International Organization* (2001) 553, at 557–558; B. A. Simmons, *Mobilizing for Human Rights: International Law in Domestic Politics* (2010) 7; Goodman and Jinks, *supra* note 27, at 144–150.
44 J. T. Checkel, 'International Norms and Domestic Politics: Bridging the Rationalist—Constructivist Divide', 3(4) *European Journal of International Relations* (1997) 473, at 477.
45 Goodman and Jinks, *supra* note 27, at 144–150.
46 Simmons, *supra* note 43, at 127–129.
47 B. Edwards and P. F. Gillham, 'Resource Mobilization Theory', published on-line in *The Wiley-Blackwell Encyclopaedia of Social and Political Movements* (2013), available online at http://onlinelibrary.wiley.com/doi/10.1002/9780470674871.wbespm447/full (last visited 9 April 2017).
48 See D. McAdam, 'Conceptual Origins, Current Problems, Future Directions', in D. McAdam, J. D. McCarthy and M. N. Zald (eds.), *Comparative Perspectives on Social Movements: Political Opportunities, Mobilizing Structures and Cultural Framing* (1996) 23–40; L. Vanhala, 'Legal Opportunity Structures and the Paradox of Legal Mobilization by the Environmental Movement in the UK', 46(3) *Law & Society Review* (2012) 523.
49 B. A. Simmons, 'From Ratification to Compliance: Quantitative Evidence on the Spiral Model', in Risse, Ropp and Sikkink, *supra* note 16, 43, at 47–57.
50 Goodman and Jinks, *supra* note 27, at 144–150.
51 Simmons, *Mobilizing for Human* Rights, *supra* note 43, at 139; M. McCann, 'Law and Social Movements: Contemporary Perspectives', 2(1) *Annual Review of Law and Social Science* (2006) 17.
52 Simmons, *Mobilizing for Human Rights*, *supra* note 43, at 130–135; H. H. Koh, 'Transnational Public Law Litigation', 100 *Yale Law Journal* (1991) 2347.
53 S. C. Moser, 'Communicating Climate Change: History, Challenges, Process and Future Directions', 1 *WIREs Climate Change* (2010) 31, at 33.

54 *Ibid.*
55 *Ibid.*, at 34.
56 American Psychological Association's Task Force on the Interface Between Psychology and Global Climate Change, *Psychology and Global Climate Change: Addressing a Multi-Faceted Phenomenon and Set of Challenges* (2011), available online at www.apa.org/science/about/publications/climate-change-booklet.pdf, at 22.
57 *Ibid.*
58 *Ibid.*
59 *Ibid.*, at 24–25.
60 *Ibid.*, at 24.
61 *Ibid.*, at 27.
62 *Urgenda*, *supra* note 2, at para 4.66; T. Bach, 'Human Rights in a Climate Changed World: The Impact of COP21, Nationally Determined Contributions, and National Courts', 40 *Vermont Law Review* (2016) 561, at 582–585.
63 Bach, *supra* note 62, at 562.
64 *Ibid.*
65 *Ibid.*
66 *Ibid.*
67 *Urgenda*, *supra* note 2, at para 4.66.
68 *Ibid.*, at para 4.90.
69 Bach, *supra* note 62, at 583.
70 *Ibid.*, at 584.
71 *Ibid.*
72 *Kelsey Cascadia Rose Juliana v. United States of America et al.* (2016), Opinion and order 6:15-cv-01517-TC, United States District Court for the District of Oregon, Eugene Division, available online at https://static1.squarespace.com/static/571d109b04426270152febe0/t/5824e85e6a49638292ddd1c9/1478813795912/Order+MTD.Aiken.pdf, at 52.
73 The case of Shell in Nigeria provides a counter-example of the difficulties of securing remedies in a domestic context. In 2005, the Nigerian NGO Climate Justice sued Shell, an oil extracting company on the basis that its oil extraction practices caused pollution that violated the rights to life and dignity of local communities. Although the Nigerian high court ruled in favour of the NGO, Shell refused to comply with the court order to stop its problematic practices, and the government did not enforce the order. See Nicholson and Chong, *supra* note 15, at 130.
74 W. N. Adger *et al.*, 'Assessment of Adaptation Practices, Options, Constraints and Capacity', in M. L. Parry *et al.* (eds.), *Climate Change 2007: Impacts, Adaptation and Vulnerability*, Contribution of Working Group II to the Fourth Assessment Report of the Intergovernmental Panel on Climate Change (2007) 717, at 720.
75 International Commission of Jurists, *Courts and the Legal Enforcement of Economic, Social and Cultural Rights: Comparative Experiences of Judiciability* (2008) 99.
76 Averill, *supra* note 10, at 147.
77 See Simmons, *supra* note 10, at 129–135.
78 Goodman and Jinks, *supra* note 27, at 24–25; J. T. Checkel, 'International Institutions and Socialization in Europe: Introduction and Framework', 59(4) *International Organization* (2005) 801, at 812–813; T. Risse, '"Let's Argue !": Communicative Action in World Politics', 54(1) *International Organization* (2000) 1; J. T. Checkel, 'Why Comply? Social Learning and European Identity Change', 55(3) *International Organization* (2001) 553, at 562.
79 T. Risse, 'International Norms and Domestic Change: Arguing and Communicative Behavior in the Human Rights Area', 27(4) *Politics & Society* (1999) 529; A. I. Johnston, 'The Social Effects of International Institutions on Domestic (and Foreign Policy) Actors', in D. Drezner (ed.), *Locating the Proper Authorities: The Interaction of Domestic and International Institutions* (2003) 145.
80 Finnemore and Sikkink, *supra* note 17, at 904–905.
81 L. Wallbott and A. Schapper, 'Negotiating By Own Standards? The Use and Validity of Human Rights Norms in UN Climate Negotiations', 17(2) *International Environmental Agreements: Politics, Law, and Economics* (2017) 209; G. Beck *et al.*, 'Mind the Gap: The Discrepancy Between the Normative Debate and Actual Use of Human Rights Language in International Climate Negotiations', 14(2) *Consilience: The Journal of Sustainable Development* (2015) 25.

82 See Osofsky, *supra* note 32; L. Wallbott, 'Keeping Discourses Separate: Explaining the Non-Alignment of Climate Politics and Human Rights Norms by Small Island States in United Nations Climate Negotiations', 27(4) *Cambridge Review of International Affairs* (2014) 736, at 745–746.
83 Nicholson and Chong, *supra* note 13, at 131.
84 H. A. Smith, 'Disrupting the Global Discourse of Climate Change: The Case of Indigenous Voices', in M. E. Pettenger (ed.), *The Social Construction of Climate Change: Power, Knowledge, Norms, Discourses* (2007) 197; J. Goodman, R. Pearse and S. Rosewarne, *Climate Action Upsurge : The Ethnography of Climate Movement Politics* (2013).
85 See also R. Price, 'Reversing the Gun Sights: Transnational Civil Society Targets Land Mines', 52(3) *International Organization* (1998) 613.
86 *Ibid.*, at 617.
87 *Ibid.*, at 622–630.
88 *Ibid.*, at 627–631.
89 *Ibid.*, at 623–627 and 631–637.
90 Mary Robinson Foundation of Climate Justice, *Human Rights and Climate Change* (2015), available online at www.mrfcj.org/media/pdf/2015/BriefingNoteforClimateJusticeDialogue7Feb2015.pdf.
91 Center for International Environmental Law, *Climate Change and Human Rights: A Primer* (2011), available online at www.ciel.org/Publications/CC_HRE_23May11.pdf.
92 *Ibid.*
93 Urgenda, *Dutch Climate Case With Commentary of Dutch Politicians*, available online at www.youtube.com/watch?v=GQfa-htZVC0; see for example, Dutch Government Ordered to Cut Carbon Emissions in Landmark Ruling, *The Guardian* (24 June 2015), available online at www.theguardian.com/environment/2015/jun/24/dutch-government-ordered-cut-carbon-emissions-landmark-ruling.
94 A. Nelsen, Dutch Parliament Votes to Close Down Country's Coal Industry, *The Guardian* (23 September 2016), available online at www.theguardian.com/environment/2016/sep/23/dutch-parliament-votes-to-close-down-countrys-coal-industry.
95 *Ibid.*
96 Wallbott and Schapper, *supra* note 81, at 14.
97 R. H. J. Cox, 'The Liability of European States for Climate Change', 30(78) *Utrecht Journal of International and European Law* (2014) 125, at 130.
98 Goodman and Jinks, *supra* note 27, at 27–28.
99 Checkel, *supra* note 78, at 810–812.
100 Finnemore and Sikkink, *supra* note 17, at 902–904.
101 Intended Nationally Determined Contributions (INDCs) are documents which were prepared by UNFCCC parties in the eve of the Conference of the Parties in 2015 (COP21). INDCs outline what actions parties intended to take to put into practice their post-2020 goals, which were agreed during COP21. UNFCCC, *Intended Nationally Determined Contributions (INDCs)*, available online at http://unfccc.int/focus/indc_portal/items/8766.php.
102 UNFCCC, *INDCs as Communicated by Parties*, available online at http://www4.unfccc.int/submissions/indc/Submission%20Pages/submissions.aspx. These countries include Brazil, Cameroon, Central African Republic, Chad, Costa Rica, Cuba, Egypt, El Salvador, Georgia, Guatemala, Guyana, Honduras, Laos, Malawi, Marshall Islands, Mexico, Nepal, Paraguay, Peru, Philippines, South Africa, South Sudan, Yemen, Venezuela, Vietnam, and Zimbabwe.
103 See for example, UNFCCC, *Mexico's Intended Nationally Determined Contribution*, available online at http://www4.unfccc.int/Submissions/INDC/Published%20Documents/Mexico/1/MEXICO%20INDC%2003.30.2015.pdf, at 6: "The adaptation component of the INDC of Mexico was elaborated taking into account a gender equality and human rights approach." Here are the other countries studied which refer to human rights in the context of adaptation, capacity building and resilience: South Soudan, Yemen, Georgia, Marshall Islands, and Nepal.
104 See for example UNFCCC, *Federative Republic of Brazil's Intended Nationally Determined Contribution: Towards Achieving the Objective of the United Nations Framework Convention on Climate Change*, available online at http://www4.unfccc.int/Submissions/INDC/Published%20Documents/Brazil/1/BRAZIL%20iNDC%20english%20FINAL.pdf, at 1 (last visited 9 April 2017): "The Government of Brazil is committed to implementing its INDC with full respect to human rights, in particular rights of vulnerable communities, indigenous populations, traditional communities and workers in sectors affected by relevant policies and plans, while promoting gender-responsive measures." Mexico also

seems to use the human rights discourse similarly, i.e. in a way where climate action does not violate such rights.

105 See for example, UNFCCC, *Ecuador's Intended Nationally Determined Contribution (INDC)*, available online at http://www4.unfccc.int/Submissions/INDC/Published%20Documents/Ecuador/1/Ecuador%20INDC%2001-10-2015%20-%20english%20unofficial%20translation.pdf, at 2: "providing that Ecuador's national development model commits "the country to defend the right of its population to live in a healthy environment and respect the rights of nature"". Similarly, Zimbabwe makes use of the discourse of environmental rights within a constitutional framework and Egypt refers to a broader 'right to standard of living, UNFCCC, *Egyptian Intended Nationally Determined Contribution*, available online at http://www4.unfccc.int/Submissions/INDC/Published%20Documents/Egypt/1/Egyptian%20INDC.pdf, at 3.

106 The two databases that we searched are London School of Economics, *The Global Climate Legislation Database*, available online at www.lse.ac.uk/GranthamInstitute/legislation/the-global-climate-legislation-database/ and New Climate Institute, *New Climate Policy Database*, available online at http://climatepolicydatabase.org/index.php?title=Special:BrowsePolicy&rand=999&page=3&page=1.

107 Mary Robinson Foundation for Climate Justice, *Incorporating Human Rights Into Climate Action*, Version 2 (May 2016), available online at www.mrfcj.org/wp-content/uploads/2016/05/Incorporating-Human-Rights-into-Climate-Action-Version-2-May-2016.pdf.

108 Indeed, 10 of the 21 INDCs which mention human rights were published by Latin American countries, which are the following: El Salvador, Cuba, Guyana, Venezuela, Chile, Brazil, Ecuador, Honduras, and Mexico.

109 *Urgenda, supra* note 2, at paras 4.8 & 4.48.

110 S. Jodoin, *Forest Preservation in a Changing Climate: REDD+ and Indigenous and Community Rights in Indonesia and Tanzania* (2017).

111 T. Risse and S. C. Ropp, 'Introduction and Overview', in T. Risse, S. C. Ropp and K. Sikkink (eds.), *The Persistent Power of Human Rights. From Commitment to Compliance* (2013).

12

Rights, justice, and REDD+

Lessons from climate advocacy and early implementation in the Amazon Basin

Deborah Delgado Pugley

Introduction

Organizations that support forest communities are acutely aware of the global impacts that will ensue from climate change, on the one hand, and from related policy reforms, on the other. They have therefore worked to keep informed about – and actively participated in – international climate change talks. During the last 15 years, many of these organizations have avidly seized upon human rights developments in international law to ensure that positive impacts of environmental policies reach local communities. This chapter presents some early lessons on the integration of human rights into climate advocacy, as well as into reforms in forest governance driven by climate policies.

We will focus on the mechanism that has been under negotiation by the United Nations Framework Convention on Climate Change (UNFCCC) for the reduction of emissions from deforestation and forest degradation, known as REDD+. REDD+ encompasses a range of forest conservation and reforestation activities, as well as financing mechanisms to reduce deforestation and forest degradation and to improve forest carbon stocks in order to mitigate climate change.[1] In this chapter, lessons are taken mainly from (1) the international negotiations on REDD+ at the Conferences of the Parties of the United Nations Framework Convention on Climate Change and (2) the implementation of agreements from these Conferences on countries of the occidental Amazon Basin (which include Colombia, Ecuador, and Peru) by The United Nations Programme on Reducing Emissions from Deforestation and Forest Degradation (or UN-REDD Programme), the Forest Carbon Partnership Facility (FCPF) and states. We will therefore pay closest attention to the political arenas of planning and early implementation.

Recognition of the links between the enjoyment of human rights and environmental protection, broadly speaking, has risen for several decades within the United Nations, governmental agencies, and civil society. This process started with the 1972 Stockholm Declaration,[2] and its significant mainstreaming can be noticed by the adoption of the Sustainable Development Goals in 2015 and the Climate Agreement reached in Paris in 2016. Here we will analyze particularly the advocacy of two types of actors within civil society that played an active role in this process: (1) environmental NGOs working in nature conservation and (2) Indigenous Peoples' organizations, as their advocacy has been prominent, both during the agenda-setting process of

REDD+[3] and in the early implementation of policies on the national and subnational level.[4] As they come from different standpoints and have different power bases, these two civil-society actors have gained significant influence working in coalition. Yet their relationship was not always close. At the beginning of the 21st century, the conservation community was challenged to take stronger action to respect human rights and cultural diversity on the sites where they intervene.[5] The rapprochement of conservationist and Indigenous Peoples in the context of climate policies has been one of the responses to this challenge.

This chapter proceeds as follows. The first section analyzes the agenda-setting process of climate policies, paying particular attention to the influence on the design of REDD+ of political discourses framed in respect to human rights. The focus of the second section is the current implementation efforts taking place in the Occidental Amazon, which includes the subnational regions of Bolivia, Colombia, Ecuador, and Peru. The chapter will conclude with a set of lessons learned from these processes.

Methodologically, the analysis provided in this chapter is grounded in a combination of participant observation, semi-structured interviews, and an analysis of legal documents and reports issued by institutions such as the UNFCCC, the Green Climate Fund, the FCPF, and organizations working for nature conservation and Indigenous Peoples' representation between 2007 and 2016. Multi-sited ethnographic fieldwork[6] was done within the Conferences of Parties of the UNFCCC from 2010 to 2015, as well as during meetings regarding the implementation of REDD+ at the subnational level in the Amazon Basin of Ecuador and Peru.[7]

The advocacy for a rights-based approach at the outset of REDD+

The climate narrative has rekindled contention over who is politically and legally accountable for the past, present, and future environmental degradation, biodiversity loss, and hazards brought about by human-caused climate change. Discussion on several agenda items of the UNFCCC included complex considerations related to justice and equity. In the particular case of REDD+, national representatives and civil-society-accredited observers revealed different conceptions of justice from the outset.[8] Climate policies at a global level have favored technical solutions, tending to prefer market-based approaches.[9] In that sense, effectiveness, monitoring, and accounting have been at the core of technical policy debates, which intended to attain standardized procedures for REDD+ at a global level. At the same time, the protection of the traditional knowledge and land-management practices of inhabitants of forested lands has been one of the central topics in the policy debate in many of the countries that participate in REDD+. These topics were frequently framed within a rights-based approach.[10] As a result, negotiations on REDD+ have been the site of lively discussions on how to avoid perverse outcomes of market-based mechanisms and how to combine REDD+ activities and policies with the pursuit of co-benefits (or multiple benefits), such as improved forest governance, biodiversity conservation, and an array of social advantages that might be also delivered by climate actions.

In the academic debate, some authors have considered that REDD+ represents an example of a policy that reaches a balanced trade-off, since it covers both economic means and ecosystem benefits.[11] Others argue that REDD+ aims to confirm and legitimize particular discourses,[12] tools, and actors in order to achieve climate goals, while disregarding others.[13] For the latter authors, REDD+ has become a new technology of governance that reinforces an apolitical and mercantile vision of nature,[14] emphasizing the technical framing while tending to evade discussion of other inevitable but contentious topics (such as respect for human rights, and land tenure rights) and potentially fostering a perverse incentive for developed countries by creating the 'right to pollute'.[15]

Up until now, the most prominent actors of civil society that have being pushing for broader social and environmental considerations at the multilateral level are Indigenous Peoples' international organizations, as well as NGOs and think tanks that strive for a global agenda on conservation. I will argue in this section that 'environmental justice' as a political frame has not had much influence on the development of climate policy. However, a human rights–based approach has been effectively integrated into advocacy efforts and has had a limited but more meaningful impact. We can observe that it facilitated the representation of right bearers (such as Indigenous Peoples and representatives of people living in small island states) in various political arenas and helped bring to the fore sensitive regulation for respecting the rights of subaltern populations.

Lobbying for the integration of human rights in REDD+

REDD+ is first and foremost focused on reducing emissions from deforestation and forest degradation in developing countries. However, the means by which this objective is pursued have been established little by little over time. The item on reducing emissions from deforestation (RED) in developing countries was formally introduced into UNFCCC negotiations at COP-11 in 2005 by a submission of Papua New Guinea and Costa Rica. In 2006, after parties and accredited observers had submitted their views, the Subsidiary Body for Scientific and Technological Advice (SBSTA) expanded the possibilities technology could bring to the problem, considering relevant scientific, technical, and methodological issues and providing proposals on policy approaches and positive incentives for related mitigation.[16] In 2007, the Bali Action Plan, formulated at the 13th session of the Conference of the Parties (UNFCC COP-13) of the UNFCCC, stated that a comprehensive approach to mitigating climate change should include '[p]olicy approaches and positive incentives on issues relating to reducing emissions from deforestation and forest degradation in developing countries; and the role of conservation, sustainable management of forests and enhancement of forest carbon stocks in developing countries'.[17] A year later, the role of conservation, sustainable management of forests, and enhancement of forest carbon stocks was upgraded, to receive the same emphasis as avoided emissions from deforestation and forest degradation. In 2010, a 'plus' was added to the original designation REDD to reflect its new components.[18] Considerations on all of them have since been repeatedly confirmed, particularly in the Warsaw Framework for REDD+ issued by the COP-19 in 2013. So REDD+ includes (a) reducing emissions from deforestation; (b) reducing emissions from forest degradation; (c) conservation of forest carbon stocks; (d) sustainable management of forests; and (e) enhancement of forest carbon stocks.

Following the approach of Wallbot[19] and our recent work,[20] we view politics as shaped by institutions that, at the most general level, consist of clusters of rights, rules, and decision-making procedures that give rise to social practices, assign roles to participants, and govern interactions among occupants of these roles.[21] Hence, the design of institutions (including international 'sector-specific regimes', like the UNFCCC) includes foundational norms that can be defined as 'collectively held or "intersubjective" ideas and understandings on social life'.[22] Norms of justice and equity relate to the distribution of rights and responsibilities, of substantive rights (like the right to culture or water), and procedural rights (like the right to participate in decision-making processes). 'Agency', in turn, refers to the power and ability of actors to display 'conduct that possessed subjective meaning'.[23]

The advocacy for the inclusion of human rights in climate governance has struggled to make headway, gaining critical support from states at the lead-up to the Paris Conference when more than 30 states signed the 'Geneva Pledge for Human Rights in Climate Action'.[24] The 2015 Paris Agreement makes explicit preambular reference to overarching human rights obligations on the states. However, as Mayer notes, since it specifies no concrete measures, its direct impact on the protection of human rights in climate action might remain limited.[25]

In the case of REDD+, some more specific steps have been taken for the implementation of a rights-based approach. It was recognized early in the development of the mechanism that various human rights instruments, and the standards they put forth, provide a normative basis for establishing a REDD system based on human rights.[26] In such a system, people living in a forest would have continued access to their land and would participate in the decisions affecting how forests are used, who owns them, and how populations are compensated for any costs they bear and share any revenue they accrue. Let us now explore how advocacy by Indigenous Peoples' networks has worked to achieve this recognition.

The insider/outsider strategy of Indigenous Peoples: displaying influence as right bearers

Framing demands in human rights terms has been a way to aid common formulation of claims across diverse networks and to facilitate their insertion into global governance debates. Many movements have succeeded in making human rights relevant to their local and global struggles by pushing the boundaries of existing human rights.[27]

REDD+ is implemented in forest-rich countries of the global South, accounting for its weak governance systems and serious land tenure issues, which have raised alarms among multiple actors across scales of governance. These concerns especially resonated, since in the original proposal of RED in 2005 there was no mention of Indigenous Peoples (IP) or Indigenous Peoples' rights.[28] Without recognition of rights, weak tenure regimes could lead to displacement of communities if more powerful local actors[29] (e.g. ranchers, loggers, government, and corporations) acquire rights over their forest.[30]

Discussions on the limits of REDD+ have been impassioned, with indigenous organizations as key participants in them. Indigenous Peoples' networks have deployed an interesting inside/outside strategy of involvement in climate negotiations.[31] IP organizations have used their political engagement within the UNFCCC to gain respect for their territorial and human rights, as well as a means to exercise these rights. They strived to be recognized as efficient actors against deforestation, advancing the idea that respecting the rights of Indigenous Peoples were the best guarantee for forest resilience. As this chapter shows, IP networks pushed to change the framing of initiatives to control deforestation under the UNFCCC, in order to increase the participation of indigenous organizations and communities.

Indigenous Peoples advocacy within UNFCCC

Agency of indigenous movements was vital and, as I would like to demonstrate in this section, strongly connected with their advocacy in the UN for collective human rights. Their use of evidence from the ground, coupled with the advancement of international human rights law, brought them significant achievements in the climate regime and in the UNFCCC particularly.

IP advocacy within the UNFCCC process can be traced back to 1998, when the first indigenous participants from the northern hemisphere attended the COP[32] and issued a declaration demanding the inclusion of Indigenous Peoples' rights in the Convention (Indigenous Peoples of North America 1998). A global involvement followed. In order to articulate their action, the International Indigenous Peoples Forum on Climate Change (IIPFCC) begin working in 2000 as an open forum for all Indigenous Peoples who were interested in following the UNFCCC process.[33] One of the first public statements made by the IIPFCC was The Hague Declaration of the Second International Forum of Indigenous Peoples and Local Communities on Climate Change (2000). It consisted of a direct and simple request for inclusion in the political process and recommendations on how to include them as political actors:

> We are profoundly concerned that current discussions within the Framework Convention on Climate Change, as well as the practical implementation of the Kyoto Protocol do not recognize our right to adequate participation. These policies and mechanisms exclude us as participants, deny our contributions, and [marginalize] our Peoples.[34]

They recommended the establishment of an ad hoc, open-ended working group on Indigenous Peoples and climate change with the broad participation of Indigenous Peoples; the creation of a Division on Indigenous Peoples within the Convention's Secretariat, and the inclusion of a permanent agenda item on Indigenous Peoples in the permanent agenda of the COP.[35]

In 2001, Indigenous Peoples were recognized as a UNFCCC constituency and since then they have engaged in every meeting COP and subsidiary bodies' meeting with varying intensity. IP organizations came with well-prepared delegations to Bali meetings in 2006, as they felt that REDD+ compromised their lands and territories and that decision-making was done without proper consultation of their constituency.[36] What concerns Indigenous Peoples is that they own, or live within, much of the developing world's last standing forests. As Victoria Tauli-Corpuz, indigenous leader from the Philippines at the time, expressed it:

> We decided to engage actively in this process because we feel that with the role that forests will play in climate change, everybody is interested to go into our communities and be the ones who will be receiving such benefits to the detriment of indigenous peoples. And secondly, we also fear that governments will not recognize our rights to our territories and also to carbon.[37]

Indigenous organizations increasingly organized in order to have daily discussions and coordinate joint advocacy in the UNFCCC process. In 2008, the IIPFCC settled as the Caucus for IP participating in UNFCCC meetings.[38] This happened at the same time that in the 14th session of the UNFCCC (held in 2008 in Poznan, Poland), governments could not agree to include references to the rights of Indigenous Peoples and forest-dependent communities in the developing SBSTA guidance on REDD+. They referred instead to the need to 'promote the full and effective participation of indigenous people and local communities, taking into account national circumstances and noting relevant international agreements'. Most remarkably, this SBSTA text referred to 'Indigenous people' (singular) rather than 'Indigenous Peoples' (plural), thus avoiding even an implicit reference to the international human rights obligations owed to Indigenous Peoples by virtue of their status as peoples under international law.[39]

The IIPFCC continued its work in the UNFCCC, and organizations of Indigenous Peoples had deeper debates on the implications of REDD+ for the respect of their rights. Between 2008 and 2009, the discourse on Indigenous Peoples rights intensified but also became more visibly differentiated. Some IP networks expressed stark opposition to carbon markets for land-use activities,[40] while some networks considered these schemes as a way to regain control and autonomy in their territories and influence the regulation of other investments on their lands (such as oil extraction, cattle ranching, and palm plantations). As tensions exist between different organizational agendas, advocacy points are usually discussed before the COPs, and issues that are too contentious are not included in the IIPFCC objectives.[41] This way of working has enabled the IP constituency to speak with a single voice, while not preventing IP organizations from engaging in coalitions[42] with other social movements, NGOs, businesses, or donors.

One important process influenced the inclusion of human rights in their work influencing the climate negotiations. After decades of struggle in various international arenas, Indigenous Peoples succeeded in getting the United Nations Declaration of the Rights of Indigenous Peoples (UNDRIP) adopted by the UN General Assembly in 2007. The UNDRIP grants IP the right to self-determination, as well as collective rights to own, use, control, and manage their lands, territories, and resources.[43] Since 2007, the focus of IP has been on getting this new framework recognized and on seeing their new group rights implemented. The IIPFCC consistently insisted that a reference to the human rights of Indigenous Peoples should be included in the UNFCCC decisions.[44] In effect, one of the main goals of indigenous movements within the UNFCCC process has been to mainstream the UNDRIP.[45] Making human rights the common framework for advocacy represented a key point of consensus for the entire constituency and played a key role in the cohesion of IIPFCC.

As Jodoin notes, the Cancun Agreements provide that the implementation of REDD+ should 'promote' and 'support' a set of social and environmental safeguards, including the following two safeguards that are relevant to the rights of Indigenous Peoples and local communities:

(c) Respect for the knowledge and rights of Indigenous Peoples and members of local communities, by taking into account relevant international obligations, national circumstances and laws, and noting that the United Nations General Assembly has adopted the United Nations Declaration on the Rights of Indigenous Peoples;
(d) The full and effective participation of relevant stakeholders, in particular, Indigenous Peoples and local communities in [REDD+ activities].

It is not coincidental that the Cancun safeguards were pushed for after the approval of UNDRIP. Delegations of Indigenous Peoples, NGOs, and their allies lobbied in order to get an explicit reference to it. Support for safeguards came from different coalitions in the environmental NGOs. Conservation International declared right after the negotiations,

> From our development and documentary work with forest communities in 7 Latin American countries over 35 years, we believe that REDD+ should require that these resource statutory rights be made binding for all indigenous people and other forest peoples whose rights do not conflict with the rights of adjoining indigenous peoples. These rights should be a pre-requisite for the granting of REDD+ funds and funds should earmarked for the recognition, securing, resolution and enforcement of these rights.[46]

The following year, in Durban, countries confirmed that a 'Safeguards Information System' should be in place to receive results-based finance for REDD+.[47] At that moment, a coalition of NGOs and IPOs, called 'the REDD+ Safeguard Working Group', was created to ensure that safeguards would be monitored and respected. In any case, and from many regards, safeguards for REDD+ were perceived as necessary but not as strong as they should be. As Jodoin discusses, the expression 'shall respect' safeguards used in previous drafts was replaced by the expression 'promote and support'. Second, the safeguards do not specifically refer to the right of Indigenous Peoples to free prior and informed consent (FPIC). Third, the safeguards only 'note' the adoption of UNDRIP, a drafting term of negligible legal import within the formal practices of the UNFCCC decision-making process.[48]

From the indigenous activist, doubts over safeguards were explicitly advanced. Berenice Sanchez, of MesoAmerican Indigenous Womens BioDiversity Network from Mexico, said:

> The supposed safeguards are voluntary, weak and hidden in the annex. REDD+-type projects are already violating Indigenous Peoples' rights throughout the world. We are here to demand an immediate moratorium to stop REDD+-related land grabs and abuses because of REDD+.[49]

As a result, some IP organizations tried to influence the constitution of the REDD+ scheme, while others decided to boycott it or to ask for an international moratorium. IP representatives were united, however, in their demand to be recognized as actors in the territory, with local organizations holding tenure rights over land and forests. At the risk of losing the support of other actors who had come out strongly against REDD+, the IIPFCC strategically used REDD+ discussions to further its own agenda, demanding that REDD+ schemes respect human rights and include IP in the decision-making process and modalities of benefit sharing. The IIPFCC demanded the broad application of the principle of FPIC to REDD+ strategies at national and subnational levels and sought to demonstrate how to do this in a cost-effective manner (as in Indonesia, Thailand, or the Philippines).[50]

The fact that the global indigenous movement is not homogeneous is probably what has given it the capacity to make use of different frames and strategies without breaking solidarity.[51] As a result, Indigenous Peoples' organizations have used their political engagement within the UNFCCC to gain respect for their territorial and human rights and the means of exercising these rights. This insider strategy has proved effective to some extent, because the IIPFCC clearly succeeded in influencing the REDD+ schemes[52] by an explicit recognition of UNDRIP. How to balance participation in institutional processes and more confrontational types of activism have long been debated within the IP movement.[53] Until today, the consensus over the importance of autonomy and territory has not being broken. IP activists continue, overall, to see the international human rights framework as useful for asserting their rights in this regard.

Human rights compliance and REDD+

Human rights approaches provide benchmarks against which states' actions can be evaluated, and they offer the possibility of holding authorities to account. Human rights approaches may also offer additional criteria for the interpretation of applicable principles and obligations that states have to each other, to their own citizens, and to the citizens of other states in relation to climate change.[54] International safeguards are proposed to counteract potential negative social and environmental outcomes of prospective REDD+ projects. Currently, there are various safeguard initiatives for global REDD+ project implementation.[55] These include the UN-REDD Programme Social and Environmental Principles and Criteria (UN-REDD 2011), the Forest Carbon Partnership Facility (FCPF) Readiness Fund Common Approach to Environmental and Social Safeguards for Multiple Delivery Partners,[56] and the REDD+ Social and Environmental Standards.[57]

UN-REDD has adopted standards to support partner countries in developing national approaches to REDD+ safeguards, which are specified in the UN-REDD, 'Social and Environmental Principles and Criteria'.[58] In its Article 2, it specifies that the 'SEPC reflect the UN-REDD Programme's responsibility to apply a human rights-based approach to its programming, uphold UN conventions, treaties and declarations, and apply the UN agencies' policies and procedures'. The standards interpret the UNFCCC safeguards in light of human rights law and practice, applying a human rights–based approach to the work of the UN-REDD. It remains to be assessed comparatively how REDD+ safeguards have been interpreted in practice at national and subnational levels, as well as the weight of safeguard implementation on the result-based payments. It is worth noting that there are no specific sanctions attached to contraventions of UN-REDD standards. Similar concerns apply to the adopted GCF Environmental and Social Safeguards, which explicitly mention ensuring full respect of the human rights of Indigenous Peoples and their free, prior, and informed consent, at least in certain circumstances.[59]

Given these international efforts, only a few UNFCCC parties have pronounced themselves on the interpretation of REDD+ safeguards and their reporting as being closely linked with extant obligations under human rights law. As Savaresi[60] points out, views on the interpretation of safeguards' links with human rights law were expressed by El Salvador (on behalf of the Dominican Republic, El Salvador, Honduras, and Panama) and Switzerland.[61] For example, Switzerland suggested that 'information systems for safeguards embody and reinforce the guidance and rules of existing environmental and human rights treaties, particularly UNDRIP and FLEGT, when relevant'.[62] Likewise, Chad (on behalf of Burundi, Cameroon, Central African Republic, Chad, Congo, Democratic Republic of the Congo, Equatorial Guinea, Gabon, Rwanda, and Sao Tome and Principe) suggested that safeguard information systems guarantee consistency with international human rights law.[63]

Implementing rights-based tools for REDD+

REDD+ implementation is a multi-scalar and multi-actor challenge. As Peskett and Todd put it,[64] countries undertaking REDD+ activities need to develop country-level approaches that enable them to respond to requirements outlined in recent United Nations Framework Convention on Climate Change agreements. At the same time, they need to ground these country-level approaches into different regional contexts by efforts that involve different state sectors and stakeholders.

Besides governance complexities, it is necessary to step back and analyze forests as social-ecological systems,[65] and examine climate action with 'context-dependent, adaptive outcomes interlinked to nested ecological states'.[66] It is important to take into account that the institutional and governance elements represent only a piece of a larger puzzle that must match the scale of the forest.[67]

It is worth taking Eleanor Ostrom's theory of collective action for common pool resource (CPR) management[68] as an analytical framework to approach institutions governing forest conservation. The theory originally described eight principles that characterize the institutional governance structures in systems that have achieved sustainable management of CPRs, such as tropical forests. Ostrom divides the eight principles into three levels: operational, collective choice, and constitutional. Operational rules are those governing day-to-day activities that directly affect the physical world, such as resource appropriation, monitoring, and enforcement. For example, operational rules may describe when, where, and how resources may or may not be withdrawn from the common pool resources system. Constitutional rules define who is eligible to govern; who can make the rules. These 'working rules' may also be understood as the rules that are used to craft collective-choice guidelines; those actually used, monitored, and enforced.[69] We will focus here on the changes at the constitutional level.

Governance reforms on the ground

Communities that live in and around forested areas are diverse in their histories, internal organization, livelihood strategies, and networks. As there are usually limited opportunities for their involvement in politics, there are, of course, risks of resource appropriation and elite capture within communities.[70] Various channels have been created both globally and nationally to empower indigenous agency, such as giving them the position of stakeholders (with attached responsibilities) and providing them with a political status in the global meetings to make them part of the 'planned solution'. It remains necessary to analyze whether these participation mechanisms enhance their agency and collective decision-making or if they might represent predominantly an exercise of what some author call the biopower of indigenity.[71] But for now it certainly seems that strengthening supra-communal organizations and building autonomous indigenous REDD committees has been a powerful driver of agency at different levels of governance.

One important example of Indigenous Peoples achieving inclusion at the constitutional level is the Forest Investment Programme (FIP), which counts Indigenous Peoples' organizations among the participants of decision-making bodies, both globally and in some countries such as Peru and Colombia.[72] As a result, part of FIP investment and funding coming from bilateral agreements – such as the Norwegian and German Climate and Forest partnership, which allowed these countries to pledge support for the achievement of REDD+ milestones of $40M and, after 2017, for verified emission reductions up to $200M.[73] These actions foresee land titling and demarcation of indigenous territories.

Gaining influence at a constitutional level has been a key part of the agenda of the Indigenous Peoples' movement in the occidental side of the Amazon Basin. Although the impact and 'translation' of human rights at the international level is very important on the ground, national laws play a significant role in the mainstreaming of a rights-based approach.[74] To show this, we will use here the case of Peru. Following official data, Peru has the fourth largest surface of tropical forest coverage in the world. Almost 58% of its territory is covered by forest, and from this, 94% is clasified as tropical forest.[75] Peru's tropical forest has a significant role nationally and globally in tackling climate change. Since 2015, Peru has committed to align its development and environmental agenda in order to comply with the Sustainable Development Goals (approved in September 2015) and the commitments of the Paris Agreement (approved in December 2015). Peru's Readiness Preparation Proposal (a framework document which sets out a clear plan, budget, and schedule for a country to achieve REDD+ readiness) was presented in 2011 and has received comments from many actors,[76] but specifically from AIDESEP, the association that represents ethnic indigenous organizations of the Peruvian Amazon. In 2010, during the meeting on the 'Preparation Proposal' of REDD+, AIDESEP expressed its views on the process, especially how it excluded a discussion on the rights of indigenous people in Peru. Since that meeting, AIDESEP has been a key player in the readiness process.[77]

Significant progress has being achieved by indigenous organizations towards the inclusion of their rights on the projects implemented by the Forest Investment Programme and the whole REDD+ in the country. A total of 40 agreements were made from observations in the Plan of Investments of the FIP-Peru. These agreements were signed on August 2013 by the Committee of Directors of the FIP (CD-FIP), formed by the Ministries of Environment, Economy, Agriculture and Culture, and the Amazonian regional governments, together with national, regional, and local leaders of indigenous organizations. Across the 40 agreements, $50 million will be invested, of which $14.5 million will be granted to meet the demands of Amazonian

Indigenous Peoples, divided into three priorities: $7 million to meet the demands for land titles; $4 million to promote community-based forest management; and $3.5 million to support forest governance of indigenous communities and organizations.[78] The act also includes other points addressed by AIDESEP and CONAP to be included in CD-FIP and the National Forest Program, the indigenous participation in specific projects, and the project design of PTRT (a land titling project) with Inter-American Development Bank (IDB).[79]

Ensuring a dedicated grant mechanism (DGM) is a success in itself and can be a political tool. It is a platform for (a) direct funding allocation to Indigenous Peoples for solving enabling conditions and community forest management initiatives and (b) strengthening indigenous forest governance through the empowerment of its National Steering Committee (NSC) and its regional and local indigenous organizations. Finally, indigenous people and private banking cooperative agreements might constitute examples of 'in-house' climate funding for the implementation of indigenous economy initiatives. This will allow analysis of indigenous communities as economic actors within a framework of territorial sustainable development.

Facing incommensurability: Amazon Indigenous REDD+ (AIR)

Regional organizations in the Amazon Basin (including Colombia, Ecuador, and Peru) have agreed not to sign REDD+ contracts until the nature of REDD+ projects and programs have been clearly defined, IPs and local communities' rights are guaranteed, and due processes for FPIC is established. But IP movements have developed more ambitious proposals, such as the 'Amazonian Indigenous REDD+'.[80] The Indigenous REDD+ proposal is interesting because it places indigenous collective rights recognized by the UNDRIP at the core of REDD+, and goes beyond carbon capture by encouraging holistic management of the territories and by including non-carbon benefits.[81]

The Amazonian Indigenous REDD+ initiative deploys an instrument called 'community-based monitoring, reporting and verification system' (CB-MRV) as a local tool to monitor forest conservation over indigenous territories. AIR has the ambition to further develop a small-scale indigenous economic/development model that guarantees food security and supply of forest goods/services, and through this, to contribute to the national development and environmental agenda. Despite the limitations on capacity and resources, Indigenous Peoples' organizations have succeeded in scaling up the AIR narrative and incorporating it in some planning forestry and climate documents (e.g. the National Determined Contributions of countries such as Peru and Colombia).[82]

AIR has been embraced by national indigenous organizations with varying strengths, since the political opportunities differ in each country. Bolivian organizations could not keep the pace as the national pressure over avoiding REDD+ was high. The Amazonian Indigenous Peoples Association in Peru and Colombia's OPIAC have become active in debates about climate change at the national level, and have brought 'alternatives' to the fragmented vision of technical approaches to forest conservation and poverty alleviation. They argue for prioritizing indigenous 'life plans' as an overall natural resource-management strategy within indigenous territories, in which forest ecosystem services, such as carbon sequestration, are incorporated as one element of the holistic territorial management.

The limits of implementation

It is recognized that where funds could be mobilized through REDD for indigenous populations, there must be inquiry into what social transformation processes this may trigger.[83] Critical voices are concerned that the traditional relationship with the non-human will be eroded, generating a counter-productive impact. This underscores the need for Indigenous Peoples to actively shape REDD at the constitutional level in order to apply traditional knowledge right from the outset. The example of Panama can be illustrative: as Potvin and Mateo-Vega[84] argued, REDD+ started well in Panama. The country put the rights of Indigenous Peoples on the agenda of the United Nations Framework Convention on Climate Change, and REDD+ project promoters complied with the consent procedures of the Guna General Congress. Panama's National Coordinating Body of Indigenous Peoples (COONAPIP) drafted a plan in 2011 for comprehensive REDD+ capacity-building efforts in each indigenous territory. If the plan had been implemented, this would have stimulated debate among indigenous people over fears that REDD+ might threaten traditional land uses and rights, as well as over possible ways forward. In July 2013, one of Panama's leading traditional indigenous authorities, the Guna General Congress, banned a REDD+ project. As Potvin and Mateo-Vega The Congress, controls about 7% of Panama's primary forests, went further, forbidding organizations in the Guna Yala territory from engaging in REDD+ activities, and walked out of REDD+ discussions.

COONAPIP withdrew from the UN-REDD programme and called on Indigenous Peoples globally to proceed cautiously on REDD-related matters. This crisis stems from a failure to build REDD+ capacity for indigenous people at all levels. The plan failed to receive UN funding in a timely manner. It later became clear that knowledge transfer was the best antidote for the fear of REDD+.

Early lessons

It is crucial to build on lessons learned about multi-scale programmes such as REDD+ to provide broader ground for the development of impact strategies that ensure respect for human rights. We would therefore like to highlight some early lessons as concluding remarks of this chapter.

As Savaresi points out,[85] the complex debate on REDD+ safeguards boils down to concerns over striking the right balance between, on the one hand, avoiding perverse outcomes of climate actions while pursuing co-benefits, and, on the other, ensuring the feasibility of REDD+. So far, a cost-effective REDD+ framework has been implemented across multiple scales in a rather slow manner. Nevertheless, it has already abounded in political incentives for the public sector to listen to and engage communities living in forested areas in political processes, and to ensure respect for substantive and procedural human rights. Nevertheless, REDD+ implementation interacts with and affects socio-cultural dynamics and power in ways that are still to be apprehended.

Looking back at the dynamics of framing and counter-framing concerning the place of local and Indigenous Peoples during the last decade, it is striking how movements and (to a lesser extent) environmental NGOs have endeavored to change the perceptions among other actors in climate policy of communities living in forests. Changing the perceptions of local drivers of deforestation was one of the main challenges faced where big advancements have been achieved.[86] Framing representation and demands in terms of human rights has been a way to assist a common formulation of claims across diverse networks and to facilitate their insertion into global governance debates and climate politics.

The UN-REDD experience has confirmed that there is ample scope to build upon human rights to interpret REDD+ safeguards. The added value of making reference to human rights lies in avoiding duplications and exploiting the consensus underpinning existing human rights law. Building explicit links with extant human rights instruments and practice may be difficult, because not all state parties eligible to carry out REDD+ activities are parties to the same human rights treaties.[87] Taking this into account, insofar as several of these countries have since subscribed to treaties regarding Indigenous Peoples, reference to human rights has played a positive role so far.

After years of work on REDD+, land rights and clear local authority issues should still be considered as a priority at different scales of governance. Countries are making slow progress on land tenure and carbon rights reform. Absence of transparent and accountable processes, comprehensive carbon rights legislation, and dispute resolution mechanisms are still worrisome.[88] In this context, incorporating indigenous planning and decision-making processes requires additional time and resources, which is, nevertheless, fully rewarded as conflicts are prevented and the long-term sustainability of REDD can be secured. Funding should come in a timely manner to enable debates and support grounded visions on governance, conservation, and forest management. As an example, an interesting process is happening in this regard with the Climate Forest Partnership Agreement of Norway, Germany, and Peru for REDD+. Mainstreaming approaches that include multiple stakeholders for the success of new policies demand evolving knowledge. Capacity building and knowledge construction for multi-scale interventions seems to be a way forward.

As Jodoin highlights, the relationship between different sites of law in the complex legal framework for REDD+ is a key line of scholarly inquiry concerns. The field of REDD+ is now governed by multiple sites of law that are characterized by different forms and modes of law making.

Finally, it is a common stance to underestimate the agency of subaltern organizations. Rights over forests resources have historically been reserved to the state. Rural organizations, indigenous and peasant alike, face this context in many forest-rich countries in the world. Advances on policies related to REDD+ at the international level (either at the UNFCCC or by bilateral cooperation agreements) have increased available resources and incentives to define access and property rights to forest. This political opportunity is being used for the assertion of human rights recognized for collective actors. Alternative approaches coming from Indigenous Peoples can make relevant contributions to the overall forestry system with positive impacts for climate and environmental resilience as a whole.

Notes

1 Brown *et al.*, 'How Do We Achieve REDD Co-Benefits and Avoid Doing Harm?', in Center for International Forestry Research (ed.), *Moving Ahead With REDD: Issues, Options and Implications* (2008) 107.
2 Campese *et al.* (eds.), *Right Based Approaches: Exploring Issues and Opportunities for Conservation* (2009), available online at www.cifor.org/publications/pdf_files/Books/BSunderland0901.pdf.
3 Wallbot, 'Indigenous Peoples in UN REDD+ Negotiations: "Importing Power" and Lobbying for Rights Through Discursive Interplay Management', 19(1) *Ecology and Society (E&S)* (2014) 23, at 176.
4 Aguilar-Støen, 'Better Safe Than Sorry? Indigenous Peoples, Carbon Cowboys and the Governance of REDD in the Amazon', 44(1) *Forum for Development Studies (FDS)* (2017) 1891–1765; Aguilar-Støen, Toni and Hirsch, 'Forest Governance in Latin America: Strategies for Implementing REDD', in F. de Castro, B. Hogenboom and M. Baud (eds.), *Environmental Governance in Latin America* (2016) 205, at 233.
5 Chapin, 'A Challenge to Conservationists', 17(6) *World Watch* (2004) 17; Alcorn, Bristol and Royo, 'Conservation's Engagement With Human Rights: "Traction", "Slippage", or Avoidance?', 15 *Policy Matters* (2007) 115; Igoe and Brockington, 'Neoliberal Conservation: A Brief Introduction', 5(4) *Conservation and Society* (2007) 432.
6 For further references on this methodology, see Marcus, 'Multi-Sited Ethnography: Notes and Queries', in M-A. Falzon (ed.), *Theory, Praxis and Locality in Contemporary Research* (2009) 181–198.
7 As part of a broader research project, analyzing the attempts to reform land and resources management policies emanating from global environmental political regimes and concern the Upper Amazon region,

I completed 82 semi-structured elite interviews with individuals affiliated with international organizations, Andean governments, indigenous organizations, and NGOs actively working on REDD+ in South America. The majority of the interviews were conducted, in person, in Germany, Norway, South Africa, the United States during climate talks and other related gatherings.

8 Okereke and Dooley, 'Principles of Justice in Proposals and Policy Approaches to Avoided Deforestation: Towards a Post-Kyoto Climate Agreement', 20(1) *Global Environmental Change* (2010) 82, at 95.

9 Meckling, 'The Globalization of Carbon Trading: Transnational Business Coalitions in Climate Politics', 11(2) *Global Environmental Politics (GEP)* (2011) 26, at 50.

10 See Espinoza, Roberto and C. Feather, AIDESEP, *The Reality of REDD+ in Peru: Between Theory and Practice* (2011), available online at www.forestpeoples.org/sites/fpp/files/publication/2011/11/reality-redd-peru-between-theory-and-practice-november-2011.pdf, at 63. See also Griffiths, *Seeing 'REDD'? Forests, Climate Change Mitigation and the Rights of Indigenous Peoples* (3 December 2008), available online at www.forestpeoples.org/sites/fpp/files/publication/2010/08/seeingreddupdatedraft3dec08eng.pdf.

11 Malhi *et al.*, 'Climate Change, Deforestation, and the Fate of the Amazon', 319(5860) *Science* (2008) 169, at 172.

12 Luttrell *et al.*, 'Who Should Benefit From REDD+? Rationales and Realities', 18(4) *Ecology and Society (E&S)* (2013) 52, at 6.

13 Corbera, 'Problematizing REDD+ as an Experiment in Payments for Ecosystem Services', 4(6) *Current Opinion in Environmental Sustainability* (2012) 612, at 619; Thompson, Baruah and Carr, 'Seeing REDD+ as a Project of Environmental Governance', 14(2) *Environmental Science & Policy* (2011) 100, at 110; Shankland and Hasenclever, 'Indigenous Peoples and the Regulation of REDD+ in Brazil: Beyond the War of the Worlds?', 42 *Institute of Development Studies (IDS)* (2011) 80; Hiraldo and Tanner, UNRISD, *The Global Political Economy of REDD+ Engaging Social Dimensions in the Emerging Green Economy* (2011), available online at www.fes-globalization.org/geneva/documents/4%20Hiraldo-Tanner%20(with%20cover)%20Small.pdf.

14 Corbera, *supra* note 13.

15 Norgaard, 'Ecosystem Services: From Eye-Opening Metaphor to Complexity Blinder', 69(6) *Ecological Economics* (2010) 1219.

16 Wallbot, *supra* note 3.

17 UN Doc. FCCC/SBSTA/2006/5*, October 2006 http://unfccc.int/resource/docs/2006/sbsta/eng/05.pdf

18 At COP-16 (15) as set out in the Cancun Agreements.

19 Wallbot, *supra* note 3.

20 Claeys and Delgado Pugley, 'Peasant and Indigenous Transnational Social Movements Engaging With Climate Justice', 1 *Canadian Journal of Development Studies (CJDS)* (2016) 1, at 16.

21 Schroeder, 'Analysing Biosafety and Trade Through the Lens of Institutional Interplay', in O. R. Young *et al.* (eds.), *Institutional Interplay: Biosafety and Trade* (2008) 49, at 70.

22 Finnemore and Sikkink, 'International Norm Dynamics and Political Change', 52(4) *International Organization* (1998) 887.

23 Campbell, 'Distinguishing the Power of Agency From Agentic Power: A Note on Weber and the "Black Box" of Personal Agency', 27(4) *Sociological Theory* (2009) 407.

24 Geneva Pledge for Human Rights in Climate Action, adopted on 27 February 2015.

25 Benoit, 'Human Rights in the Paris Agreement', 6 *Climate Law* (2016) 109.

26 Lawlor and Huberman, 'Reduced Emissions From Deforestation and Forest Degradation (REDD) and Human Rights', in J. Campese *et al.* (eds.), *Rights-Based Approaches: Exploring Issues and Opportunities for Conservation* (2009) 269, at 286.

27 Claeys, 'Food Sovereignty and the Recognition of New Rights for Peasants at the UN: A Critical Overview of La Via Campesina's Rights Claims Over the Last 20 Years', 12(4) *Globalizations* (2015) 452; Brysk, 'Human Rights Movements', in D. A. Snow *et al.* (eds.), *The Wiley-Blackwell Encyclopedia of Social and Political Movements* (2013).

28 See Jodoin, 'The Rights of Forest-Dependent Communities in the Complex Legal Framework for REDD+', in C. Voigt (ed.), *Research Handbook on REDD-Plus and International Law* (2016), 157; Wallbot, *supra* note 3.

29 Claims for forests and forestlands may involve (a) wealthy farmers and agribusiness agents; (b) displaced peasants; (c) business consortia interested in extractive industries (logging, oil, mining); and (d) the expansion of protected areas. On top of these competing actors and claims, countries of the region have ongoing or projected state-supported infrastructure projects.

30 See Wollenberg and Springate-Baginski, *Incentives+: How Can REDD Improve Well-Being in Forest Communities?* (2009), No. CIFOR Infobrief no. 21 and Van Dam, 'Indigenous Territories and REDD in Latin America: Opportunity or Threat?' 2(1) *Forests* (2011) 394.
31 Claeys and Delgado Pugley, *supra* note 19.
32 Powless, 'An Indigenous Movement to Confront Climate Change', 9(3) *Globalizations* (2012) 411.
33 Doolittle, 'The Politics of Indigeneity: Indigenous Strategies for Inclusion in Climate Change Negotiations', 8(4) *Conservation and Society (C&S)* (2010), 286.
34 IIPFCC, *The Hague Declaration of the Second International Forum of Indigenous Peoples and Local Communities on Climate Change* (2000).
35 IIPFCC, *The Hague Declaration* (2000), *supra* note 35.
36 Personal interview realized in Durban 2012.
37 Interview, quoted in Claeys and Delgado, *supra* note 19.
38 The mandate of IIPFCC is to come to agreement on what IP representatives will be negotiating for, and to share information on the evolution of the negotiations (see webpage IIPFCC). It is composed of representatives from IP organizations from seven regions of the world, but any indigenous person present at a UNFCCC meeting has the right to participate in the IIPFCC. http://www.iipfcc.org/who-are-we/
39 See Jodoin, *supra* note 27 for more details on this process.
40 Personal Interview, realized in the climate talks in Cancun, 2010. See also the website of the campaign at www.ienearth.org/category/we-support/no-redd/.
41 Personal Interview, Warsaw 2013.
42 Meckling, 'The Globalization of Carbon Trading: Transnational Business Coalitions in Climate Politics', 11(2) *Global Environmental Politics (GEP)* (2011) 26, at 50.
43 E-I. A. Daes, Commission on Human Rights, *Final Report of the Special Rapporteur on the Prevention of Discrimination and Protection of Indigenous Peoples: 'Indigenous Peoples' Permanent Sovereignty Over Natural Resources'*, UN Doc. E/CN.4/Sub.2/2004/30 (13 July 2004).
44 TWN (Third World Network), Bangkok News Update, *Differences Over Indigenous Peoples' Rights and Forest Conversion in REDD-Plus* (9 October 2009), available online at www.twn.my.
45 Personal Interview with indigenous representatives of North America realized in the climate talks in Cancun in 2010.
46 Chacko, R. Painting Cancun REDD: A Win-Win? (December 10, 2010) [Blog post]. Retrieved from https://blog.conservation.org/2010/12/painting-cancun-redd-a-win-win/.
47 UNFCCC, Decision 1/CP.16, 15 March 2011, at para 71(d).
48 Jodoin, *supra* note 27.
49 See Global Alliance of Indigenous Peoples and Local Communities against REDD and for Life, *Indigenous Peoples Call for a Moratorium on REDD+* (6 December 2011), available online at www.carbontradewatch.org/articles/indigenous-peoples-call-for-a-moratorium-on-redd.html.
50 See Claeys and Delgado Pugley, *supra* note 19.
51 See Claeys and Delgado Pugley, *supra* note 19.
52 Martin, *The Globalization of Contentious Politics: The Amazonian Indigenous Rights Movement* (2014).
53 In my personal interviews and informal conversation, this was a big issue mainly until Durban climate talks.
54 Rajamani, 'The Increasing Currency and Relevance of Rights-Based Perspectives in the International Negotiations on Climate Change', 22(3) *Journal of Environmental Law (JEL)* (2010) 391.
55 Moss *et al.*, Forest Carbon Partnership and UN REDD, 'A Review of Three REDD+ Safeguard Initiatives' (2011), available online at www.cbd.int/forest/doc/analysis-redd-plus-safeguard-initiatives-2011-en.pdf; McDermott *et al.*, 'Operationalizing Social Safeguards in REDD+: Actors, Interests and Ideas', 21 *Environmental Science & Policy* (2012) 63.
56 Forest Carbon Partnership Facility (FCPF), *Readiness Fund Common Approach to Environmental and Social Safeguards for Multiple Delivery Partners* (2011), available online at www.forestcarbonpartnership.org/sites/forestcarbonpartnership.org/files/Documents/PDF/Nov2011/FCPF%20Readiness%20Fund%20Common%20Approach%20_Final_%2010-Aug-2011_Revised.pdf.
57 For further information, see www.redd-standards.org.
58 UN-REDD, *UN-REDD Programme Social and Environmental Principles and Criteria: UN-REDD Programme Eight Policy Board Meeting*, UN Doc. UNREDD/PB8/2012/V/1 (25–26 March 2012).
59 GCF, 'Guiding Framework and Procedures for Accrediting National, Regional and International Implementing Entities and Intermediaries, Including the Fund's Fiduciary Principles and Standards and Environmental and Social Safeguards, at 1.7.

60 Savaresi, 'The Legal Status and Role of REDD-Plus Safeguards', in C. Voigt (ed.), *Research Handbook on REDD+ and International Law* (2015) 126.
61 UNFCCC, *Views on Methodological Guidance for Activities Relating to Reducing Emissions from Deforestation and Forest Degradation and the Role of Conservation, Sustainable Management of Forests and Enhancement of Forest Carbon Stocks in Developing Countries*, UN Doc. FCCC/SBSTA/2011/MISC.7, 19 October 2011.
62 *Ibid.*, at 100.
63 UNFCCC, *Views on Experiences and Lessons Learned From the Development of Systems for Providing Information on How all the Safeguards Are Being Addressed and Respected and the Challenges Faced in Developing Such Systems*, UN Doc. FCCC/SBSTA/2014/MISC.6, 6 November 2014, at 17. See also Savaresi and Hartmann, *Human Rights in the 2015 Agreement*, Legal Response Initiative Briefing Paper, 2/15 (2015).
64 Peskett and Todd, 'Putting REDD+ Safeguards and Safeguard Information Systems Into Practice', 3 *Policy Brief* (2013).
65 S. B. Hecht, K. D. Morrison and C. Padoch (eds.), *The Social Lives of Forests: Past, Present, and Future of Woodland Resurgence* (2014); Nagendra and Ostrom, 'Polycentric Governance of Multifunctional Forested Landscapes', 6(2) *International Journal of the Commons (IJC)* (2012) 104.
66 Nagendra and Ostrom, *supra* note 62.
67 Nagendra and Ostrom, *supra* note 62.
68 Ostrom, Roy and Walker, 'The Nature of Common-Pool Resource Problems', 2(3) *Rationality and Society* (1990) 335, at 358.
69 Ostrom, *supra* note 65.
70 Peskett *et al.*, *Making REDD Work for the Poor*, A Poverty Environment Partnership (PEP) Report (September 2008), available online at www.odi.org/sites/odi.org.uk/files/odi-assets/publications-opinion-files/3451.pdf.
71 Lindroth and Sinevaara-Niskanen, 'Adapt or Die? The Biopolitics of Indigeneity – From the Civilising Mission to the Need for Adaptation', 28(2) *Global Society* (2014) 180, at 194.
72 See Climate Investment Funds, *Design Document for the Forest Investment Program, a Targeted Program Under the SCF Trust Fund* (7 July 2009), available at www.climateinvestmentfunds.org/cif/sites/climateinvestmentfunds.org/files/FIP_Design_Document_July_final.pdf.
73 See Joint Statement by Germany, Norway and the United Kingdom of Great Britain and Northern Ireland, *Unlocking the Potential of Forests and Land Use Paris*, COP21 (30 November 2015), available online at www.bmub.bund.de/fileadmin/Daten_BMU/Download_PDF/Klimaschutz/joint_statement_redd_cop21_en_bf.pdf.
74 Gready, 'Rights-Based Approaches to Development: What Is the Value-Added?', 18(6) *Development in Practice* (2008) 735.
75 See W. A. L. León, *MINAM – National Forest Monitoring in Peru*, available online at http://claslite.ciw.edu/en/success-stories/minam-national-forest-monitoring-in-peru.html.
76 Romijn *et al.*, 'Assessing Capacities of Non-Annex I Countries for National Forest Monitoring in the Context of REDD+, 19 *Environmental Science & Policy* (2012) 33, at 48.
77 For a coverage of this process, see C. Lang, *AIDESEP Critique of Peru's Readiness Preparation Proposal* (8 March 2011), available at www.redd-monitor.org/2011/03/08/aidesep-critique-of-perus-readiness-preparation-proposal/.
78 For information about the funds, see World Bank, document about *Saweto Dedicated Grant Mechanism for Indigenous Peoples and Local Communities in Peru* (19 August 2015), available at https://static1.squarespace.com/static/550abd2ce4b0c5557aa4f772/t/574328c7ab48de0aeb435567/1464019144570/DGM_Peru_Saweto_PAD_18.8.2015_EN.pdf.
79 For a coverage of the process, see Rights and Resources, *Indigenous Participation in REDD+ Increases in Peru* (13 August 2013), available online at http://rightsandresources.org/en/blog/indigenous-participation-in-redd-increases-in-peru/#.WNCd1GTytsM.
80 An ambitious project to implement pilot cases of the 'Amazon Indigenous REDD+' in Madre de Dios (Peru) and Inirida (Colombia) and a test case in Ecuador was launched last December 9 at UNFCCC COP 20. See http://wwf.panda.org/wwf_news/?235956/Amazon-Indigenous-REDD-launched-at-UNFCCC-COP-20.
81 For an extensive report on Peru, see REDD+ INDÍGENA EN EL PERÚ: Perspectivas, avances, negociaciones y desafíos desde la mirada de los actores involucrados. www.proyecto-cbc.org.pe/admin/recursos/publicaciones/c4073-REDD_Indigena_En_El_Peru.pdf.

82 This information was recollected in personal interviews with national officials and also NGO allies, who are working with indigenous peoples' proposals that are currently ongoing in Colombia and Peru. No public documents are available at the moment of writing this chapter.
83 Sheil, Douglas and Wunder, 'The Value of Tropical Forest to Local Communities: Complications, Caveats, and Cautions', 6(2) *Conservation Ecology* (2002) 9; P. West, *Conservation Is Our Government Now: The Politics of Ecology in Papua New Guinea* (2006).
84 Potvin and Mateo-Vega, 'Panama: Curb Indigenous Fears of REDD+', 500(7463) *Nature* (2013) 400.
85 See Savaresi, *supra* note 57.
86 For a case study, see Wehkamp *et al.*, 'Analyzing the Perception of Deforestation Drivers by African Policy Makers in Light of Possible REDD+ Policy Responses', 59 *Forest Policy and Economics (FPE)* (2014) 7.
87 Savaresi, *supra* note 57; Savaresi and Hartmann, *supra* note 60.
88 Dunlop and Corbera, 'Incentivizing REDD+: How Developing Countries Are Laying the Groundwork for Benefit-Sharing', 63 *Environmental Science and Policy* (2016) 44.

13

Protecting Indigenous Peoples' land rights in global climate governance

Ademola Oluborode Jegede

Introduction

The global climate change governance is a top-down approach characterized by the development under the aegis of United Nations Framework Convention on Climate Change (UNFCCC),[1] the Kyoto Protocol[2] and the recent Paris Agreement[3] aimed at addressing the challenge posed by climate change. In itself, the top-down approach is not problematic considering that climate change is a global challenge.[4] Action is necessary at other levels; however, issues such as the differentiation of responsibilities between developed and developing states, unequal capacity to adapt and mitigate adverse climate effects, and the allocation and transfer of resources make a global governance inevitable and distinct from other levels of climate governance.

The development at the global governance level encompasses the interaction of hard and soft instruments along with an array of institutions. This chapter demonstrates that although the protection of Indigenous Peoples' land rights has considerably featured in the climate global governance, it accommodates certain notions which may undermine the protection of Indigenous Peoples' land rights at the national level. The key notions are 'sovereignty', 'country-driven' and 'national legislation'. Arguably, these notions limit the importance of the development at that level of governance in addressing the adverse impacts of climate change on Indigenous Peoples' land rights. Following this introduction, the second section briefly explains the global climate governance, emphasising relevant development at that level, while the third section considers the notions which may undermine its activities. The fourth section is the conclusion.

Global climate governance and Indigenous Peoples' land rights

Scholars use the word 'governance' in relation to the environmental field interchangeably with phrases such as 'architecture' and 'regime'. For instance, according to Le Preste, 'governance' connotes either 'architecture' or 'regime' and refers to

> A set of interrelated norms, rules and procedures that structure the behaviour and relations of international actors so as to reduce the uncertainties that they face and facilitate the pursuit of a common interest in a given area of issue.[5]

Deere-Birkbeck defines climate change governance as referring to the processes, traditions, institutional arrangements and legal regimes through which authority is exercised, and decisions taken at the global level for implementation.[6] In agreeing with this description, Thompson et al. note that governance connotes structures, arguably institutional and policy, through which decisions are made and resources are managed.[7] This structure, in the view of den Besten et al., may be shaped by various actors and groups with which it interacts in negotiation.[8] What is certain is that all the definitions agree that governance is made up of rules and institutions. Against this backdrop, this section demonstrates the extent of Indigenous Peoples' lands protection under the emerging set of instruments and established institutions that form global climate governance.

Emerging set of instruments and Indigenous Peoples' lands

Although the UNFCCC does not mention 'Indigenous Peoples', the decisions reached under the framework identify 'Indigenous Peoples' as the focus of attention. For example, the decision of UNFCCC Conference of Parties (COP) 13 requires that the needs of 'local and indigenous communities' should be addressed in the context of initiatives aimed at reducing emissions from deforestation and forest degradation in developing countries.[9] Subsequently, at COP 16 in Cancun, the Conference of the Parties to the UNFCCC affirmed that developing countries implementing such initiatives should, among other safeguards, promote and support:

> (c) Respect for the knowledge and rights of Indigenous Peoples and members of local communities, by taking into account relevant international obligations, national circumstances and laws, and noting that the United Nations General Assembly has adopted the United Nations Declaration on the Rights of Indigenous Peoples.[10]

The recognition of UNDRIP in the above provision proves that respect of Indigenous Peoples including their land rights is a standard requirement at the international level for the implementation of REDD+. This position can be reinforced by the Paris Agreement, which urges all parties to 'respect, promote and consider their respective obligations' towards vulnerable groups including Indigenous Peoples while implementing all climate actions. Besides the foregoing normative trend, there is evidence that windows exist within the institutional framework of global climate governance that engage with Indigenous Peoples' land rights.

Institutional windows for protecting Indigenous Peoples' lands

Key institutions at the global climate governance level relevant to the discussion are the Conference of Parties (COP), Meeting of the Parties (MOP), the Intergovernmental Panel of Climate Change (IPCC), Subsidiary Body for Scientific and Technological Advice (SBSTA), Subsidiary Body for Implementation (SBI), Ad Hoc Working Group on Long-Term Cooperative Action Under the Convention (AWG-LA), and Ad Hoc Working Group on Further Commitment for Annex 1 Parties Under the Kyoto Protocol (AWG-KP).[11] Arguably, these institutions offer windows of opportunity, at least for the discussion and mobilization of attention around issues pertaining to Indigenous Peoples' lands.

Conference of parties/meeting of the parties

Established pursuant to Article 7 of the UNFCCC, the COP is the highest decision-making body under the UNFCCC,[12] and functions as the MOP or the CMP (Conference of the Parties serving as the meeting of the Parties to the Kyoto Protocol) by virtue of Article 13 of that protocol. There is evidence that Indigenous Peoples–based NGOs have observer status, which qualifies them to attend debates at the forum and therefore indirectly influence the agenda of Indigenous Peoples at that fora. There is evidence that Indigenous Peoples–focused organisations, such as Forest Peoples Programme[13] and the International Alliance of Indigenous and Tribal Peoples of the Tropical Forests, enjoy observer status and contribute to climate discussions through their submissions. Of importance to Africa is the Indigenous Peoples of Africa Co-ordinating Committee (IPACC), which has made substantial submissions at COP on a range of issues affecting Indigenous Peoples in Africa. There is evidence on how this forum has been used in advancing the cause of Indigenous Peoples' lands. At COP 17, for instance, IPACC recommended to the African Group of Negotiators the need to integrate land tenure systems, particularly of the nomadic tribe in Africa, into climate discussions.[14]

International panel on climate change

The Intergovernmental Panel on Climate Change was jointly established by the United Nations Environment Programme (UNEP) and the World Meteorological Organisation (WMO and subsequently endorsed by the United Nations General Assembly (UNGA) in 1988.[15] The primary mandate of the IPCC is to offer 'a clear scientific view on the current state of knowledge with regard to climate change and its potential environmental and socio-economic impacts.'[16] Though this attention has remained minimal, through its reports over the years, the IPCC has considered to some extent the issues of Indigenous Peoples in its research outcomes. The IPCC Working Group II First Assessment Report (FAR) alludes to the protection of Indigenous Peoples in connection with the value to be placed on forest produce, an issue at the heart of Indigenous Peoples' land regime.[17] The IPCC Working Group III FAR on response strategies to climate change points out the practice of Indigenous Peoples in the context of those in the boreal region.[18] The IPCC Working Group II Second Assessment Report (SAR) refers to the impact of climate change on the ecosystem of Indigenous Peoples,[19] while the IPCC Working Group II Third Assessment Report (TAR) on vulnerability discusses Indigenous Peoples in the Arctic and Americas.[20]

Indigenous Peoples' vulnerability,[21] health,[22] and related risks[23] are described in the IPCC Working Group II Fourth Assessment Report (AR4) on impact and vulnerability, although it was largely in the context of the Americas and Arctic. A brief reference to property rights[24] and pastoralist coping strategy[25] is discernible at least in relation to Indigenous Peoples' land rights largely within the Americas and Arctic. Reference is made to Indigenous Peoples' land rights, in the IPCC Working Group III AR4, as a structural challenge which must be addressed in forest management.[26] In its report, the IPCC Working Group II AR5 devotes a section to Indigenous Peoples, acknowledging that vulnerability to climate change impact is high among these peoples and that considerable challenges will be witnessed in terms of their culture, livelihoods and food security as a result of the adverse impacts of climate change.[27]

Subsidiary body for scientific and technological advice

The Subsidiary Body for Scientific and Technological Advice is established pursuant to Article 9 of the UNFCCC. It is largely composed of government experts who provide assessments of scientific knowledge and evaluations of scientific/technical aspects of national reports and the effects of implementation measures.[28] In the main, the SBSTA serves a 'multi-disciplinary' purpose in that it provides expeditious information and advises on scientific and technological matters relating to the UNFCCC.[29] The SBSTA has contributed significantly to the discussion of a range of issues, such as the impact of climate change as well as the vulnerability of different regions and potential response measures.[30]

In its deliberations, the SBSTA offers an important platform for showcasing the pertinent questions relating to land tenure and use by Indigenous Peoples. For instance, in response to its invitation for submissions by parties to the SBSTA made at the 11th session of the COP in 2006 regarding forest activities, Bolivia emphasised the need for the protection of Indigenous Peoples.[31] In particular, it stressed that REDD should 'dignify the living conditions'[32] and promote the participation of relevant stakeholders, including Indigenous Peoples in the forest.[33] Similarly, on behalf of the African countries of the Congo Basin, Gabon argued that sustainable management of the forests cannot be achieved without the participation of Indigenous Peoples.[34]

On the status of Indigenous Peoples and local communities in the formulation of an appropriate approach to forest emission reduction, the contribution of parties was specifically invited by the SBSTA. These contributions were considered at the 13th session of the SBSTA.[35] In its submission, the Czech Republic contends that for any REDD to be effective, there is a need to respect the rights of Indigenous Peoples as guaranteed under UNDRIP.[36] It also advised that local communities and Indigenous Peoples should be involved in the monitoring activities of the status of forest carbon stocks.[37] Similarly, Ecuador submitted that the development and implementation of methodologies for REDD should safeguard the rights of Indigenous Peoples, incorporate a prior consultation clause and assure benefit sharing which accommodates incentives for Indigenous Peoples and local communities.[38]

Subsidiary body for implementation

Article 10 of the UNFCCC establishes the Subsidiary Body for Implementation (SBI), which is composed of government experts that review policy aspects of national reports and help the COP in evaluating summative effects of implementation measures.[39] Compared with the SBSTA, the mandate of the SBI is narrower in nature as it is restricted to matters of implementation, including the determination of timetables and ensuring that targets are being achieved.[40] In performing its role, the SBI scrutinizes the information submitted by state parties in documentation, such as the national communications and emission inventories.[41]

Indigenous Peoples' rights, arguably including their land tenure and use, have gained considerable space in the SBI role. For instance, at its eighth session held in Doha, 2012, the invitation was extended to parties and admitted observer organisations to submit to the secretariat (by 25 March 2013) their positions on possible changes to the modalities and procedures for the CDM.[42] To this end, the session requested the secretariat to organise a workshop and compile submissions for consideration by the SBI at its thirty-eighth session,[43] which can then make recommendations on possible changes to the modalities and procedures for the CDM.[44] This process is required to be carried out in preparation for a review by the COP serving as the MOP to the Kyoto Protocol at its ninth session in 2013.[45]

At its 38th session and workshop held by the SBI, a range of submissions were made, including the need to respect substantive and procedural human rights.[46] After reviewing the submissions of participants, the SBI prepared a report which documents some of the recommendations highlighted by participants for key sections of CDM modalities.[47] Notably, recommendations of significance to Indigenous Peoples include the necessity to ensure in the modalities procedures that processes are made more transparent,[48] compensation for deficiencies in validation, verification and certification reports,[49] and respect for human rights.[50]

Ad Hoc Working Group on Long-Term Cooperative Action Under the Convention

The Ad Hoc Working Group on Long-Term Cooperative Action Under the Convention (AWGLCA) was established as a subsidiary body under the UNFCCC at COP 13 as part of the Bali Action Plan[51] to conduct a wide-ranging process to enable the full, effective and sustained implementation of the instrument through long-term cooperative action, up to and beyond 2012.[52] One of the main purposes of the AWGLCA is to negotiate the issue of non-Annex 1 contributions to reducing greenhouse gas emissions over time.[53] Noteworthy achievements of the AWGLCA include the Cancun Agreements and the resultant implementing decisions, including the Cancun Adaptation Framework.[54]

In relation to the consideration of Indigenous Peoples' issues, in February 2008 a contribution over the need to promote additional information, views and a proposal in relation to paragraph 1 of the Bali Action Plan was jointly made by Kenya, Tanzania and Uganda at the fourth session of the AWGLCA.[55] In that submission, it was highlighted that the rights and roles of local communities and Indigenous Peoples as well as their social, environmental and economic development should not be undermined by REDD.[56]

On a similar matter, at a later session in the same year, intergovernmental organisations enjoying accredited status of the UNFCCC made submissions which highlight the importance of safeguarding the rights of Indigenous Peoples, particularly in relation to their land.[57] The International Labour Organisation (ILO) advised that the success of REDD will depend on availing the forest dwellers and communities of access to sustainable forest and land use within the mechanism and providing them with sufficient employment and income opportunities. Any policy for REDD, in its view, should channel incentives to and respect the rights of indigenous and tribal peoples in the conservation of forests as carbon sinks, in line with the provisions of ILO Convention 169.[58] It also notes that local communities and Indigenous Peoples should participate and be included in the measurement, reporting and verification of the impact of REDD activities 'with respect to income, employment, migration and cultural identity'.[59] Other intergovernmental organisations, including IPACC, noted that the protection of natural areas are of cultural and religious significance to Indigenous Peoples.[60]

The foregoing trend shows that the protection of Indigenous Peoples' lands and related issues are not strange to the discussions and decisions at the institutional global climate governance level. However, key notions have developed along this line which can undermine the development.

Subordinating notions in the international climate regulatory framework

The emerging international climate change regulatory framework reflects certain notions which may legitimise states' inadequate formulation of the domestic regulatory framework in addressing the adverse impacts of climate change on Indigenous Peoples' land tenure and use. The key notions are 'sovereignty', 'country-driven', and 'national legislation'.

Notion of 'Sovereignty'

The concept of 'sovereignty' is the keystone of international law.[61] There are various ways in which the concept has been discussed.[62] The traditional concept of international law considers sovereignty as a status in which each state is co-equal and has final authority within the limits of its territory.[63] This meaning of sovereignty under international law aptly reflects the definition by Max Huber in *Island of Palmas* case (*Netherlands v USA*).[64] In that matter, Hubber notes:

> Sovereignty in the relations between States signifies independence. Independence in regard to a portion of the globe is the right to exercise therein, to the exclusion of any other State, the functions of a State.[65]

Sharing the above position, in *Corfu Channel* (*UK v Albania*),[66] Alvarez J considered sovereignty as 'the whole body of rights and attributes which a state possesses in its territory, to the exclusion of all other states, and also its relations with other states'.[67] As Cassese argues, one of the sweeping powers and rights of sovereignty includes the power to assume authority over the populations in a given territory and the power to freely use and dispose of the territory under the state's jurisdiction and to do all activities considered essential for the benefit of the population.[68] The concept of 'sovereignty' has always been a major statement of defence in a world system largely considered by some as unequal. According to Keck and Sikkink, although the claims by third world leaders to sovereignty are viewed as the self-interested argument of authoritarian leaders, states' attachment to the concept is not without basis:

> The doctrines of sovereignty and non-intervention remain the main line of defence against foreign efforts to limit domestic and international choices that this world affairs (and their citizens) can make. Self-determination, because it has so rarely been practised in a satisfactory manner, remains a desired, if fading, utopia. Sovereignty over resources, as fundamental part of the discussions about a new international economic order, appears particularly to be threatened by international action on the environment. Even where third world activists may oppose the policies of their own governments, they have no reason to believe that international actors would do better, and considerable reason to suspect the contrary. In developing countries, it is much the idea of the state, and it is the state itself, that warrants loyalty.[69]

The possibility that the issue of 'sovereignty' is controversial and can shape the approach of states in relation to the protection of Indigenous Peoples is evidenced in the response of parties and accredited observers to the invitation extended by the SBSTA at its twenty-ninth meeting.[70] This invitation sought their views on issues relating to Indigenous Peoples and local communities for the development and application of methodologies for REDD+.[71] The Czech Republic on behalf of the European Community and its member states, Ecuador, Guatemala, Panama, Costa Rica, Bolivia and Tuvalu argued that states reserve to themselves a large measure of discretion on certain issues pertaining to Indigenous Peoples,[72] others generally prefer the principle of 'consultation', instead of 'consent', in dealing with climate-related actions affecting Indigenous Peoples.[73]

At the fifteenth session of the Ad Hoc Working Group on Long-Term Cooperative Action Under the Convention in 2012, which was convened to discuss the idea of creating a REDD+ market mechanism, nations belonging to the Commission of Central African Forests (COMIFAC), that is Burundi, Cameroon, the Central African Republic, Chad, Congo, the DRC, Equatorial Guinea, Gabon, Rwanda, Sao Tome and Principe, emphasised that to fully respect the notion of 'sovereignty', parties involved in REDD+ activities should have the discretion to decide the approach they deem most appropriate.[74] Hence, it is not surprising that the Readiness Preparation Proposal (R-PP) template used at the national level to initiate the process incorporates safeguard principles as listed under Appendix 1 to the Cancun Agreements, which include respect for sovereignty and national legislation, confirming their centrality to the implementation of REDD+ activities.[75] Indeed, the fact that nations place sovereignty above the climate change mitigation safeguards may well have informed the provision that compliance with the decision of the COP requesting state parties to describe activities on safeguards is voluntary.[76]

For countries in Africa where basic instruments that specifically aim to safeguard the land rights of Indigenous Peoples are largely not ratified, this is worrying. For instance, only one African state has ratified the ILO Convention 169.[77] In all, it can be summed up that the foregoing discussion reflects the possibility that the notion of 'sovereignty' has the potential to inform a domestic climate change regulatory regime which essentially does not include normative content that recognises the protection of Indigenous Peoples' land use and tenure. It further signifies that a state may justifiably hide under the concept of sovereignty to do as it wishes, including the exclusion of specific instruments dealing with Indigenous Peoples.

Notion of 'Country-Driven'

Related to the notion of 'sovereignty' is the concept of 'country-driven', which implies state ownership of implementation process and attracts significant mention in the climate change regulatory framework on adaptation and mitigation. The notion is perhaps justified considering when decisions are taken at that level, at least there is the presumption that it is taken for the purpose of implementation on behalf of the entire population, which includes Indigenous Peoples. In relation to adaptation, state ownership of the concept is discernible from the documentation process for adaptation. Article 4(1)(b) of the UNFCCC enjoins all parties to 'formulate, implement, publish and regularly update national programmes on adequate adaptation and mitigation to climate change'. Also, Article 4(1)(e) requires parties to the UNFCCC to cooperate 'in preparing for adaptation to the impacts of climate change', as well as plans for 'coastal zone management, water resources and agriculture, and for the protection and rehabilitation of areas, particularly affected by drought and desertification, as well as floods' in Africa. In the decisions of the COP, there is heavy focus on the state government for the facilitation of adaptation process. This features prominently in the guidelines formulated for the preparation of National Adaptation Plan of Actions (NAPA Guidelines).[78] The NAPA Guidelines, in paragraphs 6(a) and (c), affirm that the programme will be 'action-oriented and country-driven' and that NAPA will set out 'clear priorities for urgent and immediate adaptation activities in relation to the countries'. Paragraph 7(f) of the NAPA Guidelines reiterates that it is 'a country-driven approach'. In paragraph 7(a), it is pointed out that NAPA is 'a participatory process involving stakeholders, particularly local communities', while paragraph 7(j) declares that the process will ensure 'flexibility of procedures based on individual country circumstances'.

The COP 7 largely lays the ground which signifies that adaptation should be country-driven and that policy measures at the national level are required in attending to adaptation needs. Subsequent COP meetings, namely COP 8[79] and COP 9[80] respectively, endorsed the NAPA Guidelines. At COP 10,[81] it was decided that actions in relation to adaptation and mitigation should reflect the needs and information indicated in national communications, thus tacitly highlighting the role of national communication on adaptation issues. The developing countries in both the LDC and non-LDC are enjoined to file a national communication to document their adaptive concerns and need for funds. The basis for this is Article 12, paragraphs 1 and 4 of the UNFCCC. The combined reading of these paragraphs enjoins parties to the Convention to communicate to the COP measures being taken in response to climate change. This angle to the formulation of adaptation actions was projected at the Cancun meeting of COP 16, which emphasised country-driven 'enhanced action on adaptation' and invites parties to take actions in NAPA and national communications toward its achievement.[82]

Also, paragraph 1 of Appendix 1 (c) to the Cancun Agreements provides that the activities of REDD+ should follow a 'country-driven' approach and consider 'options available to parties'. Although stakeholders' participation in the REDD+ process is key, this is generally intended to take place within 'country-specific interpretation of safeguards for REDD+ and in the development of the elements of the safeguards system'.[83] In a decision reached at COP 17, titled 'Guidance on systems for providing information on how safeguards are addressed and respected and modalities relating to forest reference emission levels and forest reference levels as referred to in decision 1/CP.16', the COP agrees that the system for providing information on compliance with safeguards must be 'country-driven and implemented at the national level'.[84] The notion is further reinforced by the template of the UN-REDD and FCPF for the Readiness Preparation Proposal which is state-centred.[85] For instance, funding or support for REDD+ activities is commenced by the formulation of a Readiness Proposal Idea Note (R-PIN), through which a country expresses its interest in participating in the FCPF and presents early ideas for how it might organise itself to get ready for REDD+. If successful, the country is then asked to formulate a Readiness Preparation Proposal, with funding assistance subsequently made available to the country to carry out the activities laid out in the R-PP.[86]

In all, the possibility exists that a country-specific interpretation of safeguards for REDD+ may fall below the standard of protection afforded to Indigenous Peoples, particularly in relation to their lands under global climate change governance. The implication of this for Indigenous Peoples is that they may be excluded from climate response actions. It is difficult to imagine an effective engagement with peculiar issues relating to Indigenous Peoples' lands when the state is the only recognised host of the project under the REDD+ activities. For Indigenous Peoples, who are often marginalised or not recognised at all by the states, it is uncertain such activities will be as beneficial to them, if at all, as would be the case if they could directly formulate proposals and participate in the initiative. In the absence of clear guidance, the idea of 'country-driven' may signify that states can elect to act as they please, adopting a different position on issues relating to Indigenous Peoples land tenure and use.

Deference to 'National Legislation'

Deference to national legislation is a concern in that it generally places emphasis on compliance with national legislation without insistence on the need for such legislation to be in conformity with an international framework on the implementation of programmes. This emphasis is more pronounced and can be illustrated, particularly in relation to REDD+. An exception is a proposition found in the submission of Tuvalu in response to the invitation by SBSTA at its twenty-ninth session to seek the views of parties and accredited observers on issues relating to indigenous people and local communities for the development and application of methodologies for REDD+.[87] The submission made by Tuvalu on a model legal framework for REDD+ that safeguards Indigenous Peoples is most instructive. According to its submission, a legal framework for REDD+ should include the principles:

> [A]cknowledge and recognise the rights enshrined in the UN Declaration on the Rights of Indigenous Peoples; It should establish similar rights and provisions to those found within the UN Declaration on the Rights of Indigenous Peoples so that all UNFCCC Parties are able to apply these rights concurrently whether or not they are signatories to this

> Declaration and require that all Parties undertaking REDD activities to establish legal systems to recognise and put into place these rights; A framework should be established whereby indigenous peoples from all UN regions are fully represented on any decision-making body associated with REDD; it should establish a legal basis whereby no REDD legal regime is able to displace indigenous peoples or local communities from their land or expropriate their right to the use of their land; it should establish appropriate prior informed consent decision-making processes at the national and sub-national level to ensure that the rights of indigenous peoples and local communities are properly recognised.[88]

In order to achieve the foregoing, Tuvalu suggested a national legislation framework that protects the rights of Indigenous Peoples and local communities.[89] At the same session, Mexico, however, affirmed:

> We believe that indigenous peoples and local communities, rights, visions and experiences should be taken into account in the discussions of any topic regarding REDD. Furthermore, there should be enough flexibility in the discussion to allow for the consideration of parties, circumstances and legislation regarding consultation processes and property rights of these communities.[90]

The consequence of this tension is a range of COP decisions and initiatives on safeguards stressing national legislation as a context for the implementation of REDD+. Evidence is found in paragraph 2 of the Appendix 1 to the Cancun Agreements: although it requires respect for the knowledge and rights of Indigenous Peoples and local communities, it only urges parties to note that the United Nations General Assembly has adopted UNDRIP.[91] Mainly, in respecting the knowledge and rights of Indigenous Peoples and local communities, it calls on parties to take into account relevant international obligations along with national circumstances and laws.[92] Also, parties are required to ensure that actions taken in connection with REDD+ are consistent with objectives of their national forest programmes along with applicable international conventions and agreements.[93]

Similarly, the preamble to the COP 17 decision regarding the systems for providing information on the safeguards for REDD+, provided under paragraph 1 of Appendix 1 to the Cancun Agreements, states that such systems should be consistent with national legislation and circumstances.[94] Although in contrast with the provisions that follow, a preamble is not a source of law; however, it has a significant legal effect.[95] It is useful in identifying the purpose of a statute and serves as an important aid in construing unclear legislative language.[96] In *Reference re Remuneration of Judges*, Chief Justice Lamer explained that 'the preamble articulates the political theory which the Act embodies'.[97] On this authority, it can be argued that in indicating in the preamble to this decision that reporting about REDD+ safeguards will be consistent with 'national legislation and circumstances', the instrument offers the necessary context in which the provisions following the preamble should be understood. Further reinforcing this position, the decision calling for the collection of information at the domestic level indicates, along with related international obligations and agreements, that there is the need to take into account the 'national circumstances and respective capabilities' as well as national legislation.[98]

At the thirty-sixth SBSTA meeting, suggestions were made on the elements to describe when giving information on how safeguards are being addressed. It underscored the need for parties to provide information on national forest governance structures, taking into account national legislation and indicating the applicable and relevant administrative bodies, laws, policies, regulations and law enforcement mechanisms, the nature of land tenure and/or land rights for REDD+ activities, and arrangements on how to transfer the rights and incentives of carbon.[99] This requirement becomes problematic where national forest governance in terms of law and institutions does not recognize Indigenous Peoples' rights to forest resources. It means that approval of projects and initiatives is possible as long as states can defend their approach as in line with their national legislation and circumstances.

In all, the foregoing notions set the ground for the legitimacy of a domestic climate change regulatory regime that may undermine the protection of Indigenous Peoples' lands. In states where the identity of Indigenous Peoples and the use of their territories are disputed, deferring to national legislation is capable of being interpreted as indirectly endorsing approaches which do not recognise or respect the existence of Indigenous Peoples and their right to the use of land.

Conclusion

The protection of Indigenous Peoples is progressively featuring in the global climate governance. It is particularly discernible in the normative arrangement under the framework and the functioning of institutions at that level. Through their representation and presentations, issues around Indigenous Peoples' lands can feature in the activities of the key institutions in the governance. However, along with the developments at that level, the recognition has emerged of the notions of 'sovereignty', 'country-driven' and 'national legislation'. In granting states the space to implement measures according to their sovereignty, approach and domestic laws, without qualification, these notions provide the basis for a global climate governance which may undermine Indigenous Peoples' lands.

Accordingly, it is proposed that the notion of 'sovereignty' should conceptually shift toward a 'human-centred' perspective which protects vulnerable groups such as Indigenous Peoples, instead of the notion of state-centred sovereignty which continues to retain its presence in the negotiation of international climate change instruments. This proposed approach should emphasise the rights of Indigenous Peoples as human rights and regard the protection of their lands as critical in the formulation and implementation of all climate-related actions. Rather than deferring to state-driven and national circumstances, implementation of initiatives should require that legislative and practical steps be taken to ensure the enjoyment by Indigenous Peoples of the rights on a non-discriminatory basis. Guidelines on adaptation and mitigation measures should require states to indicate legislative reforms carried out in relation to property rights that accommodate Indigenous Peoples' land rights in the light of global climate change challenge. Only then can the development within the global climate governance have an enduring effect at the national level on the protection of Indigenous Peoples' land rights.

Notes

** In developing this chapter, I draw from and build on the content of my book *The Climate Change Regulatory Framework and Indigenous Peoples' Lands in Africa: Human Rights Implications* (Pretoria: Pretoria University Law Press, 2016).

1 *United Nations Framework Convention on Climate Change (UNFCCC)* (1992), 1771 UNTS 107.

2 UNFCCC, *United Nations Kyoto Protocol to the United Nations Framework Convention on Climate Change* (1997), UN Doc. FCCC/CP/1997/7/Add.1, 11 December 1997.

3 UNFCCC, *Paris Agreement Under the United Nations Framework Convention on Climate Change 2015, Adopted by Conference of the Parties, 21st Session Paris* (30 November to 11 December 2015), UN Doc. FCCC/CP/2015/L.9/Rev.1, 12 December 2015.

4 J. Dunnof, 'Levels of Environmental Governance', in D. Bodansky, J. Brunnee and E. Hey (eds.), *The Oxford Handbook of International Environmental Law* (2007) 85 at 87.

5 Cited in Smouts, 'The Issue of an International Forest Regime', 10 *International Forestry Review (IFR)* (2008) 429, at 432.
6 C. Deere-Birkbeck, 'Global Governance in the Context of Climate Change: The Challenges of Increasingly Complex Risk Parameters', 85 *International Affairs* (2009) 1173, at 1194.
7 M.C. Thompson, M. Baruah and E.A. Carr, 'Seeing REDD+ as a Project of Environmental Governance', 14 *Environmental Science & Policy (ESP)* (2011) 100, at 102.
8 W.J. Den Besten, B. Arts and P. Verkooijen, 'The Evolution of REDD+: An Analysis of Discursive Institutional Dynamics', 35 *Environmental Science & Policy (ESP)* (2014) 40, at 40.
9 UNFCCC, *Reducing Emissions From Deforestation in Developing Countries: Approaches to Stimulate Action' (Decision 2/CP.13)*, UN Doc. FCCC/CP/2007/6/Add.1 (14 March 2008).
10 UNFCCC, *The Cancun Agreements: Outcome of the Work of the Ad Hoc Working Group on Long-Term Cooperative Action Under the Convention* (Decision 1/CP.16), UN Doc. FCCC/CP/2010/7/Add.1 (15 March 2011).
11 F. Gale, 'A Cooling Climate for Negotiations: Intergovernmentalism and Its Limits', in T. Cadman (ed.), *Climate Change and Global Policy Regime: Towards Institutional Legitimacy* (2013) 32, at 32; D. Bodansky, 'International Law and the Design of a Climate Change Regime', in U. Luterbacher and D. F. Sprinz (eds.), *International Relations and Global Climate Change* (2001) 201, at 206. In 2012, the CMP, at its eighth session, adopted the Doha Amendment which effectively decided that the AWG-KP had fulfilled the mandate set out in decision 1/CMP.1, and that its work was finished.
12 Bodansky, *supra* note 11, at 213–214.
13 UNFCCC, *Admitted NGOs*, available online at http://unfccc.int/parties_and_observers/ngo/items/9411.php.
14 IPACC, *IPACC Recommendations to UNFCCC COP 17, Durban, South Africa* (2011), available online at http://ipacc.org.za/en/2011/40-ipacc-recommendation-to-unfccc-cop17.html?path=.
15 IPCC, *Organisation*, available online at www.ipcc.ch/organisation/organisation.shtml.
16 *Ibid.*
17 R. S. de Groot *et al.*, 'Natural Terrestial Ecosystems', in W. J. McG. Tegart, G. W. Sheldon and D. C. Griffiths (eds.), *Climate Change: The IPCC Impact Assessment Report Prepared for IPCC by Working Group II FAR* (1990) Chapter 3, at 3.
18 D. Kupfer and R. Karimanzira, 'Agriculture, Forestry and other Human Activities', in World Meteorological Organisations/ United Nations Environment Programme (eds.), *Working Group III The IPCC Response Strategies IPCC FAR* (1990) 90 at 113.
19 R. T. Watson, M. C. Zinyowera and R. H. Moss, *Impacts, Adaptations and Mitigation of Climate Change: Scientific-Technical Analyses Contribution of Working Group II to IPCC SAR* (1996) 7, at 30, 99, 257.
20 A. Allali *et al.*, 'Africa', in J. J. McCarthy *et al.* (eds.), *Impacts, Adaptation and Vulnerability Contribution of Working Group II to IPCC TAR* (2001) 487.
21 A. Fischlin *et al.*, 'Ecosystems, Their Properties, Goods, and Services', in M. L. Parry *et al.* (eds.), *Impacts, Adaptation and Vulnerability Contribution of Working Group II to IPCC AR4* (2007) 211, at 248.
22 U. Confalonieri *et al.*, 'Human Health', in M. L Parry *et al.* (eds.), *Impacts, Adaptation and Vulnerability Contribution of Working Group II to IPCC AR4* (2007) 391, at 395.
23 S. H. Schneider *et al.*, 'Assessing Key Vulnerabilities and the Risk From Climate Change', in M. L. Parry *et al.* (eds.), *Impacts, Adaptation and Vulnerability Contribution of Working Group II to IPCC AR4* (2007) 779 at 791.
24 Fischlin *et al.*, *supra* note 23, at 247.
25 M. L. Parry *et al.*, 'Technical Summary', in M. L. Parry *et al.* (eds.), *Impacts, Adaptation and Vulnerability Contribution of Working Group II to IPCC AR4* (2007) 23, at 39.
26 *Ibid.*
27 J. Barnett *et al.*, 'Human Security', in *IPCC WGII AR5* (2013), available at http://ipcc-wg2.gov/AR5/images/uploads/WGIIAR5-Chap12_FGDall.pdf, at para 12.3.2.
28 Bodansky, *supra* note 11, at 214.
29 Gale, *supra* note 11, at 36.
30 *Ibid.*
31 UNFCCC Subsidiary Body for Scientific and Technological Advice (SBSTA), *Paper No. 3: Bolivia Agenda Item 6: Reducing Emissions from Deforestation in Developing Countries: Approaches to Stimulate Action, 24th session Bonn, 18–26 May 2006*, Item 6 of the Provisional Agenda, UN Doc. FCCC/SBSTA/2006/MISC.5, 11 April 2006, at 10 (*Bolivia* paper).
32 *Bolivia* paper, *supra* note 31, at 10.

33 *Ibid.*, at 11.
34 UNFCCC SBSTA, *Paper No. 8: Gabon on behalf of Cameroon, Central African Republic, Chad, Congo, Democratic Republic of the Congo, Equatorial Guinea and Gabon*, UN Doc. FCCC/SBSTA/2006/MISC.5 (11 April 2006), at 75.
35 *Ibid.*
36 UNFCCC SBSTA, *Paper No. 1: Czech Republic on Behalf of the European Community and its Member States, 13th Session Bonn*, 1–10 June 2009, UN Doc. FCCC/SBSTA/2009/MISC.1 (10 March 2009), at 3 (*Czech Republic* presentation).
37 *Czech Republic* presentation, *supra* note 36, at 4.
38 Ibid., at 5.
39 Bodansky, *supra* note 11, at 214.
40 Gale, *supra* note 11, at 37.
41 *Ibid.*
42 UNFCCC, Guidance relating to the clean development mechanism (Decision 5/CMP.8), UN Doc. FCCC/KP/CMP/2012/13/Add.2 (28 February 2013), at para 10 (Decision 5/CMP.8).
43 Decision 5/CMP.8, *supra* note 42, at para 11.
44 *Ibid.*, at para 14.
45 *Ibid.*, at para 10.
46 'Submission on views regarding the revision of the CDM Modalities and Procedures' by Asociación Interamericana para la Defensa del Ambiente, Center for International Environmental Law, Earthjustice, and International Rivers, 26 March 2013, available at http://unfccc.int/resource/docs/2013/smsn/ngo/352.pdf (last visited 12 April 2017), at 6.
47 UNFCCC SBI, *Report on the Workshop on the Review of the Modalities and Procedures of the Clean Development Mechanism*, UN Doc. FCCC/SBI/2013/INF.6 (10 June 2013) (SBI Report).
48 SBI Report, *supra* note 47, at paras 20–22.
49 *Ibid.*, at para 23.
50 *Ibid.*
51 UNFCCC CP, *Bali Action Plan* (Decision 1/CP.13), UN Doc. FCCC/CP/2007/6/Add.1 (14 March 2008).
52 UNFCC, *Ad Hoc Working Group on Long-Term Cooperative Action Under the Convention*, available at http://unfccc.int/bodies/body/6431.php.
53 Gale, *supra* note 11, at 38.
54 Decision 1/CP.16, *supra* note 10.
55 UNFCCC, *Bali Action Plan*, *supra* note 51, at para 1 of which launches 'a comprehensive process to enable the full, effective and sustained implementation' of the UNFCCC 'through long-term cooperative action, now, up to and beyond 2012'. See also UNFCCC AWGLCA, *Paper No. 1: Belize, Central African, Costa Rica, Dominician Republic, Democratic Republic of the Congo, Ecuador, Equatorial Guinea, Honduras, Ghana, Guyana, Kenya, Madagascar, Nepal, Nicaragua, Panama, Papua New Guinea, Singapore, Solomon Islands, Thailand, Uganda, United Republic of Tanzania, Vanuatu and Vietnam*, Ad Hoc Working Group on Long-Term Cooperative Action Under the Convention, 4th Session, February 2008, UN Doc. FCCC/AWGLCA/2009/MISC.1/Add.4 (7 April 2009), 3, at 11 (*Belize* paper).
56 *Belize* paper, *supra* note 55.
57 UNFCCC AWGLCA, *Ideas and Proposals on the Elements Contained in Paragraph 1 of the Bali Action Plan, Submissions From Intergovernmental Organisations*, 4th Session, Poznan, 1–10 December 2008, UN Doc. FCCC/AWGLCA/2008/MISC.6/Add.2 (10 December 2008), at 1.
58 UNFCCC AWGLCA, *Paper No. 2: International Labour Organisation Submission to Be Considered in the Update of the Assembly Document (Bali Action Plan) to the AWG-LCA*, Poznan, 6 December 2008, UN Doc. FCCC/AWGLCA/2008/MISC.6/Add.2 (10 December 2008), at 32–33 (International Labour Organisation Submission).
59 *Ibid.*, at 33.
60 UNFCCC AWGLCA, 'Paper No. 4: International Union for the Conservation of Nature on behalf of the International Union for the Conservation of Nature, The Nature Conservancy, WWF, Conservation International, Birdlife International, Indigenous People of Africa Co-ordinating Committee, Practical Action, Wild Foundation, Wildlife Conservation Society, Fauna and Flora International and Wetlands International Ecosystem-based adaptation: An approach for building resilience and reducing risk for local communities and ecosystems', UN Doc. FCCC/AWGLCA/2008/MISC.6/Add.2 (10 December 2008), at 65.

61 R. C. Gardner, 'Respecting Sovereignty', 8 *Fordham Environmental Law Review (FELR)* (2011) 133 at 133.
62 Four ways in which the term is used are 'domestic sovereignty' to refer to political authority and the level of control enjoyed by a state; 'interdependence sovereignty' dealing with the ability of a state to control movements across its border; 'international legal sovereignty' which treats the state as a subject of international law in the same way that an individual is considered as a citizen at national level; and 'Westphalian sovereignty' which construes the concept in two terms, namely, territorially and the exclusion of external actors from domestic structures of authority. See SD Krasner, *Sovereignty: Organised Hypocrisy* (1999) 73–90.
63 See J. Dugard, *International Law: A South African Perspective* (2012) 125, on the notion of co-equality. See however, A. Cassese, *International Law in a Divided World* (1986) 129, who contends that it is not valid to maintain that the United Nations is based on the full equality of its members, considering that Article 27(3) of its Charter grants the right of veto to the permanent members of the Security Council only. Hence, at best, the principle of equality laid down in Article 2(1) can only be interpreted as merely a general guideline, which is weakened by the exceptions particularly laid down in law.
64 *Island of Palmas* case (*Netherlands, USA*), 4 April 1928, vol II 829–871 (*Island of Palmas* case).
65 *Ibid.*, at 838.
66 *Corfu Channel (UK v Albania)*, Advisory Opinion (Alvarez J), 9 April 1949, ICJ Reports (1949) 39, at 43 (*Corfu Channel* case).
67 *Ibid.*
68 Cassese, *supra* note 63, at 49–52.
69 M. E. Keck and K. Sikkink, *Activists Beyond Borders: Advocacy Networks in International Politics* (1998) 215.
70 UNFCCC SBSTA, *Report of the Subsidiary Body for Scientific and Technological Advice on its 29th session, held in Poznan from 1–10 December 2008*, UN Doc. FCCC/SBSTA/2008/13 (17 February 2009), at para 45.
71 *Ibid.*
72 UNFCCC SBSTA, *Paper No. 1: Czech Republic on Behalf of the European Community and Its Member States, Submission Supported by Bosnia and Herzegovina, Croatia, Montenegro*, UN Doc. FCCC/SBSTA/2009/MISC.1, 10 March 2009 at 3, 4 (Czech Submission); UNFCCC SBSTA, Paper No. 4, Panama Submission, UN Doc. FCCC/SBSTA/2009/MISC.1, 10 March 2009, at 9 (Panama Submission).
73 UNFCCC SBSTA, Paper No. 2: Ecuador, UN Doc. FCCC/SBSTA/2009/MISC.1, at 5; Czech Submission, *supra* note 72, at 4; Panama Submission, *supra* note 72, at 9.
74 UNFCCC AWGLCA, *Submission from Burundi, Cameroon, Central African Republic, Chad, Congo, Democratic Republic of the Congo, Equatorial Guinea, Gabon, Rwanda, Sao Tome and Principe*, 15th session Bonn, 15–24 May 2012, UN Doc. FCCC/AWGLCA/2012/MISC.3/Add.2 (21 May 2012).
75 R-PP Template Version 6 Working Draft' April 4, 2012, which replaces Version 5 of December 22.
76 UNFCCC CP, *The Timing and the Frequency of Presentations of the Summary of Information on How All the Safeguards Referred to in Decision 1/CP.16, Appendix I, Are Being Addressed and Respected* (Decision 12/CP.19), UN Doc. FCCC/CP/2013/10/Add.1 (31 January 2014), at para 5.
77 Only Central African Republic has ratified ILO Convention 169. It did so on 30 August 2010. The Convention is available at www.ilo.org/dyn/normlex/en/f?p=1000:11300:0::NO:11300:P11300_INSTRUMENT_ID:312314.
78 UNFCCC CP, *Guidelines for the Preparation of National Adaptation Programmes of Action* (Decision 28/CP.7/2001), UN Doc. FCCC/CP/2001/13/Add.4 (21 January 2002).
79 UNFCCC CP, *Review of the Guidelines for the Preparation of National Adaptation Programmes of Action* (Decision 9/CP.8/2002:1), UN Doc. FCCC/CP/2002/7/Add.1 (28 March 2003).
80 UNFCCC CP, *Review of the Guidelines for the Preparation of National Adaptation Programmes of Action* (Decision 8/CP.9/ 2003:1), UN Doc. FCCC/CP/2003/6/Add.1 (22 April 2004).
81 UNFCCC CP, *Buenos Aires Programme of Work on Adaptation and Response Measures* 1 (Decision 1/CP.10/2004), UN Doc. FCCC/CP/2004/10/Add.1 (19 April 2005), at 4.
82 Decision 1/CP.16, *supra* note 10, at 11–14.
83 *Ibid.*
84 UNFCCC CP, *Guidance on Systems for Providing Information on How Safeguards Are Addressed and Respected and Modalities Relating to Forest Reference Emission Levels and Forest Reference Levels as Referred to in Decision 1/CP.16* (Decision 12/CP.17), UN Doc. FCCC/CP/2011/9/Add.2, 15 March 2012.
85 Forest Carbon Partnership Facility (FCPF) and the United Nations Collaborative Programme on Reducing Emissions from Deforestation and Forest Degradation in Developing Countries (UN-REDD),

Readiness Preparation Proposal (R-PP), Version 6 Working Draft (4 April 2012), available at www.unredd.net/index.php?option=com_docman&task=doc_download&gid=6953&Itemid=53.

86 *Ibid.*

87 UNFCCC SBSTA, 'Report of the Subsidiary', *supra* note 70, at para 45.

88 UNFCCC SBSTA, Paper No. 3 Tuvalu, *Reducing Emissions From Deforestation in Developing Countries: Approaches to Stimulate Action, Issues Relating to Indigenous People and Local Communities for the Development and Application of Methodologies*, 13th session Bonn, UN Doc. FCCC/SBSTA/2009/MISC.1,17 April 2009, available at http://unfccc.int/resource/docs/2009/sbsta/eng/misc01a01.pdf, at 7.

89 *Ibid.*

90 UNFCCC SBSTA, Paper No. 2: Mexico Submission, *Reducing Emissions From Deforestation in Developing Countries: Approaches to Stimulate Action, Issues Relating to Indigenous People and Local Communities for the Development and Application of Methodologies*, 13th session Bonn, UN Doc. FCCC/SBSTA/2009/MISC.1 (9 April 2009) http://unfccc.int/resource/docs/2009/sbsta/eng/misc01a01.pdf, at 5–6.

91 Decision 1/CP.16, *supra* note 10, at 2 (c).

92 *Ibid.*

93 Decision 1/CP.16, *supra* note 10, at 2 (a).

94 Decision 12/CP.17, *supra* note 84, preamble.

95 *Ibid.*, at para 216.

96 *Ibid.*

97 See Lamer CJ in *Reference re Remuneration of Judges* (1998), 1 S.C.R., at para 95.

98 Decision 12/CP.17, *supra* note 84, at para 2.

99 UNFCCC SBSTA, *Submission From the United States of America: Potential Additional Guidance on-Informing How All Safeguards Are Being Addressed and Respected*, 36th session Bonn, 14–25 May 2012, UN Doc. FCCC/SBSTA/2012/MISC.9 (12 April 2012), at 4.

14

The indigenous rights framework and climate change

Ben Powless

Introduction

Indigenous Peoples worldwide find themselves in the unenviable position of being amongst the most vulnerable to climate change, and amongst the groups least powerful to thwart it. Many are already being impacted, particularly in northern and coastal regions, and the impacts will only be exacerbated as more changes to the global climate accumulate. While not even present at the inception of the UN Framework Convention on Climate Change (UNFCCC), Indigenous Peoples have rapidly emerged as one of the most engaged non-state actors. Apart from solely engaging with the UN system, Indigenous Peoples have also offered devastating critiques of the dominant social, economic, and development models held responsible for this crisis.

As Indigenous Peoples have navigated these spaces, they've seen and partially been responsible for the transition of the UNFCCC system and parallel domestic processes away from understanding climate change in a strictly technical and economic issue, to one that is seen as fundamentally tied to all aspects of personal and political life. This nascent identity-based Indigenous climate movement also parallels, and is to a degree born out of, the broader global Indigenous movement, which has primarily been concerned with human rights issues and land- or territory-based struggles. In fact, we can see a strong mirroring of the rise of engagement, as Indigenous Peoples were also absent – or excluded – from international treaty negotiations before emerging as a notable actor.

The central argument of this piece is that Indigenous Peoples, coming together from around the globe, have formed a unique, identity-based social movement that works both inside and outside of official spaces to advance both a positive vision of social and economic systems, while also contesting and engaging with dominant understandings of climate change and their hegemonic and neocolonial causes. The movement is both a challenge to official decision-making processes, while also demanding to be included, however, on its own terms. The movement also seeks to work with like-minded allies, building alternative sites of collective power both within and outside of the formal UN and state systems. The movement is grounded in an understanding that the roots of the climate crisis and environmental injustice are generally to be found in colonialism and capitalism. Most significantly, the movement is responsible for developing an Indigenous rights framework and related vision that can be an effective tool for tackling climate change, and can serve as a framework for the larger environmental movement going forward.

In this chapter, we will first look at a brief history of the global Indigenous movement to understand its unique relation to and shared discourses with the Indigenous climate movement. Then we will survey the Indigenous climate movement's own history, with an overview of its primary concerns, agenda, and methods of self-organization that have developed within official climate negotiations. This leads to a discussion of the autonomous, parallel, and alternative spaces that have been employed by the movement, while also engaging and building power with allied movements. We will then look at the progress achieved in the recent Paris round of negotiations, as well as a review of the related commitments that Canada has made in its own climate plan, in relation to Indigenous Peoples. Lastly, we will turn to a reflection on how Indigenous movements continue to challenge state and corporate power as it pertains to climate change across North America, and what we can expect to see of this movement going forward.

A brief note: this work draws on both my scholarly research on Indigenous Peoples' rights and struggles in relation to environmental conflicts, as well as my own participation and engagement as an activist who works in this field. I've participated in formal climate negotiations and alternative Indigenous spaces, as well as related domestic activism. Therefore, many of the reflections are drawn from direct, first-hand observations to enhance the existing, yet scarce, literature on international Indigenous movements.

The emergence of a global Indigenous movement

The global Indigenous movement (GIM) can be said to have emerged from the engagement of various Indigenous governments, organizations, networks, professionals, and individuals who discovered each other over decades of engagement with the United Nations human rights systems. Members of the movement gathered under a shared identity and ideology, with similar goals and interests, particularly concerning the promotion of Indigenous rights and defense of Indigenous lands and ways of life.[1]

Indigenous Peoples, despite their obvious diversity, have a shared history of both oppression and general dispossession that forms the basis for collective demands and mobilization that are not within the boundaries of traditional political systems or confined to state borders. Indigenous demands have often sought not to reform or accept the dominant political systems and terms of engagement, but have instead sought to challenge them directly, and reshape them on their own terms.[2] This has led Indigenous Peoples to not just engage with formalized, official (inter)governmental processes, but also to engage with 'extraordinary politics' up to and including active resistance efforts in many situations.[3]

The first antecedents for a GIM came from the 1920s, as representatives of the Haudenosaunee (Iroquois) people of North America and Maori people of Aotearoa (New Zealand) went abroad to seek redress with the League of Nations in 1922 and British Commonwealth in 1923, respectively. Even though their respective missions would end in failure, decades later more Indigenous Peoples would take up the mantle, after social movements had begun to coalesce nationally. These movements began to take hold in Canada, the United States, Australia, and New Zealand after and related to the rise of the civil rights movement in the United States. For many, this also marked the first awareness that Indigenous Peoples in each of these nations would have had of each other's shared histories and experiences. This was not as true with other movements, which only rarely concerned themselves with the unique situation and struggles of Indigenous Peoples, instead choosing to focus on their identity-based or broader economic and social justice demands to the state.[4] After coming to regard national governments as unresponsive and ineffective during this time, many of these groups would engage in a scale shift[5] that would take their concerns and issues to the international level.

In Latin America, governments were less responsive, and often outwardly hostile, to Indigenous demands, particularly broad mobilization and protest.[6] The social movements born out of this particular situation were entirely dissuaded from engaging with state apparatuses, leading them to create political visions that were unique, highly based on their local situation, and reflective of a tremendous cultural diversity.[7] Indigenous movements became increasingly effective at bringing these unique visions and struggles, fostered in relative autonomy, to the forefront across Latin America. Their demands often were a direct challenge to state sovereignty and coloniality, seeking a differentiated version of citizenship and human rights, more democratic states that would be open to their visions, and even Indigenous visions of pluri-national states. States responded by increasingly allowing or ignoring corporate incursions into Indigenous territories for resource extraction, aided by laws and trade agreements that sought to erode any remaining Indigenous control of land. Their movements having been forged in a repressive environment, and increasingly recognizing the arbitrary borders dividing their peoples, Indigenous movements throughout Latin America developed transnational relationships from their inception, leading to their natural participation in international forums like the UN.[8]

The UN working group on Indigenous populations

Indigenous Peoples began to see the United Nations and its many bodies as a unique venue bestowed with a tremendous amount of moral and legal authority, a kind of soft power that could be harnessed[9] to not just set international norms, but also to define *peoplehood* and the rights bestowed by it.[10] An incipient transnational movement thus began to form around the United Nations in the 1970s, seeing it as also a more open venue than the closed doors they encountered nationally. Indigenous petitions saw their first success with the creation of the Working Group on Indigenous Populations in 1982, the first international venue where Indigenous Peoples found themselves with any substantial voice.[11]

As a result, the venue became one of the most highly attended UN human rights bodies, with special permissions given for Indigenous participation that didn't exist in other parts of the international system. The effect of this allowed Indigenous Peoples to challenge states from 'below' by creating a requirement that decisions of the Working Group had to be reached by consensus with Indigenous Peoples – effectively giving non-state actors a veto over state actions within the UN system. This was unprecedented. These efforts were coordinated by the creation of an Indigenous caucus, a tactic that would continue in not just climate negotiations, but also in many other UN bodies. Within the Indigenous caucus, delegates would meet separately from the states to strive to come to consensus positions of their own, from which they could then return to negotiate with states from a position of unity and strength.[12]

After years of listening to a litany of human rights abuses suffered by Indigenous Peoples from around the world, the Working Group was given the historic task in 1987 to develop a draft declaration to protect and promote Indigenous rights. It was a laborious task, the details of which are omitted here, but suffice it to say that this document was only finalized in 2007 and presented to the United Nations General Assembly for a vote. There, only four states voted against it: Canada, the United States, Australia, and New Zealand – all former British colonies with sizeable Indigenous populations. The document had taken on the name of the UN Declaration on the Rights of Indigenous Peoples (UNDRIP). Its passing created a new set of rights standards for Indigenous Peoples that would be used by Indigenous movements the world over as a framework against which to compare national and international laws and policies.[13]

Within a few years, all would recant their initial opposition after concerted national and international pressure. Still, this initial opposition showed Indigenous Peoples that powerful states were still willing to step out and oppose them and the recognition of their rights. This adversarial relationship is another dynamic that would spill over into the climate talks.

The passing of the UNDRIP ushered in a new group of rights previously unrecognized in international law and norms. Of importance for us are three classes of rights recognized that would later become prominently utilized by the yet-unborn international Indigenous climate movement (IICM). The first was a recognition of Indigenous Peoples as collective rights-holders, a very strong break from previous human rights norms, which recognized the rights of individuals who happened to belong to groups. This was one of the most arduous conflicts while drafting the UNDRIP – essentially an argument over whether the Declaration would apply to 'people' or to 'peoples.' Indigenous activists saw themselves as members of stateless nations, and they were fighting for a type of sovereignty, as both a positive right to self-determination and a means to stave off state and corporate rights violations. In the end, states acquiesced and acknowledged a right to self-determination for Indigenous *Peoples*, while excluding the right to secession, which they feared would be the natural consequence of a right to self-determination.[14]

The second right recognized of interest is the right to free, prior, and informed consent (FPIC). The right appears multiple times within the Declaration, applicable to policies and laws that would impact Indigenous Peoples, as well as development projects that would affect their lands and territories. As its name suggests, the right requires consent to be granted by Indigenous Peoples, in a non-coerced way, before any law or project is approved, and with all requisite information freely available. This would become a large part of the demands that the IICM would urge, especially as related to energy projects such as oil drilling, biofuel plantations, or dams. Lastly, there was a right for Indigenous Peoples to both maintain their traditional knowledge[15] and for states to take appropriate action to protect that right.

Indigenous Peoples that came together to form the GIM have therefore been part of two powerful challenges to Western conceptions. The first was a challenge to the normative values based on Eurocentric values considered inherent to the international system. Indigenous Peoples, by insisting on speaking in their own voices without conforming to decorative norms, forced international actors to translate their radically different linguistic and social constructs into a language that could be understood. This was reinforced by Indigenous caucus meetings and counter-summits, creating their own forms of language and norms. The GIM was therefore able to not just successfully demand inclusion on their own terms, but in doing so, were also challenging and helping to deconstruct dominant meanings and concepts.[16]

Secondly, Indigenous Peoples were continuing the legacy to constrain the rights and sovereignty of states that began with the first postwar human rights instruments, while also challenging some of the foundational concepts of a Western, liberal, individualist human rights regime and discourse. In articulating an alternative construction of sovereignty and self-determination, not necessarily tied to a state, they were of necessity challenging the statist and dominant understanding of sovereignty and self-determination based on territorial control and citizenship. This was intimately tied to the assertion that Indigenous rights are collective rights. This contributed to an attack on the theoretical foundations of the modern nation-state and how states exercise jurisdiction over their societies, peoples, territories, and even species. We should also note that this stands in considerable contrast to the demands of most nongovernmental actors in the international system, whose efforts seek to reinforce state sovereignty and its philosophical, moral, and legal justifications.[17] This was just as true within international climate change negotiations, to which we shall now turn.

The rise of an international Indigenous climate movement

While the UN Framework Convention on Climate Change began its negotiations in 1995, it wouldn't be until 1998 that Indigenous Peoples began to participate in any form, by issuing an Indigenous declaration. It was only in 2001 that Indigenous Peoples were recognized as a constituency within the UNFCCC and granted the same rights as other stakeholder groups, such as the ability to hold meetings on site and to make statements to the parties gathered. Building on experience gained negotiating within the UN Convention on Biological Diversity, a sister convention to the UNFCC, Indigenous Peoples began to demand an ad hoc working group on Indigenous issues as a means to have some of the same rights as states within the forum, though without success until now.[18]

The year 2006 would mark the first time that Indigenous issues were addressed in any substantial way within the UNFCCC. At that year's negotiations in Nairobi, the Nairobi Work Programme was adopted, which encouraged states to consider "local and Indigenous knowledge" when considering adaptation programs to climate change. This was the first instance of Indigenous traditional knowledge being recognized, and the first concrete achievement for the IICM. Against the backdrop of this singular success were repeated failures to secure protections for Indigenous rights in other key parts of adopted texts, often opposed by a number of colonial states. At the time, many Indigenous Peoples were becoming aware that they were not only impacted by many of the drivers of climate change, such as oil extraction, but that they were doubly at risk because of the proposed solutions to climate change. Indigenous representatives sought assurances that they wouldn't, for instance, be displaced by governments keen to sell carbon credits for forests they inhabited, or have their lands flooded to make way for dams, or any other of the so-called solutions being considered.[19]

The 15th meeting of the Conference of Parties (COP 15) to the UNFCCC in 2009 was touted as the opportunity for a major climate agreement breakthrough. Ahead of that meeting, Indigenous Peoples met for the first time ahead and distinct from the talks in Alaska, with observers from the UN system also invited and attending. Hundreds of Indigenous delegates – more than typically attended the formal UNFCC meetings – attended a four-day gathering and came up with a declaration by consensus stating what was to be considered as a position of Indigenous Peoples worldwide. This was significant for allowing Indigenous Peoples to identify what they thought were the key causes of climate change and to delineate their role in resolving it. This allowed for many Indigenous Peoples to have the opportunity to fully express themselves in ways that would not be possible within the UN. While there were some divisions over the role of limiting fossil fuel developments – since some communities depended upon them for their prosperity – there was a widespread sense of conciliation about a path forward for the upcoming deliberations. The process was also useful for building solidarity with other movements, who were provided with the final document and asked for their support.

Unfortunately, the 2009 Copenhagen negotiations ended in failure. Indigenous Peoples were widely excluded from the final deliberations, and there was no consensus over the final language, leaving the talks in shambles. The only outcome of note was that Bolivia decided to convene a gathering in 2010 under Indigenous president Evo Morales to try to salvage a moral authority. Over 35,000 people ended up gathering, the majority Indigenous from the Andean regions of South America. Many of the discussions were conducted in Indigenous languages, leading to the final declaration proclaiming Indigenous Peoples and their critiques of Western capitalism and colonialism to be at the forefront of the climate movement. This gathering was also significant for bringing non-Indigenous movements into a space of building solidarity and cohesion with Indigenous movements, often for the first time. The resulting declaration was submitted to the UNFCCC and supported by a few global South countries, for the first time bringing the Indigenous movement to the front of international climate politics. A number of other groups endorsed and advocated for the "Cochabamba Declaration," granting a newfound legitimacy and maturity to the Indigenous movement.

It is worth mentioning briefly some of the substances of these declarations, as well as statements and other documents produced by the IIPFCC and broader IICM. As mentioned, there is a strong focus on Indigenous rights and traditional knowledge in the documents. What is significant here is that these are both seen as ways to prevent further climate change. Indigenous rights and traditional knowledge, particularly collective rights over the use and control of lands, territories, and resources, are seen as one of the best defenses for some of the planet's remaining areas of biodiversity in the face of resource extraction and expanding urban areas. There is likewise a strong emphasis placed on protecting and enhancing a spiritual, non-materialist, and non-commercial relationship with the planet. This is complemented by often strong critiques of the dominant economic and social models that have led to the commodification and desolation of the planet.[20]

Outside of the UNFCCC process, there had been other attempts to tie climate change to Indigenous rights. In 2005, the Inuit Circumpolar Committee filed a petition against the United States in the Inter-American Commission on Human Rights. There, they claimed that the US was violating their human right "to be cold" – essentially an assertion that their rights to their culture and way of life depended upon certain conditions, and that the US was violating these rights by not acting to stop climate change. In doing so, they relied not just on scientific arguments, but also included traditional knowledge in their claims. Even though the case was unsuccessful, it did allow for further petitions to be heard, making it the first case of international law recognizing the human rights and Indigenous rights implications of climate change. This was one of the first attempts to treat climate change under a human rights law, an approach that would be used by many later groups.[21] Indigenous Peoples had finally broken into the mainstream ideations of climate change and human rights. Indigenous groups would also use the venues of the UN Permanent Forum on Indigenous Issues and the UN High Commissioner for Human Rights to bring up issues of climate change under a human rights and Indigenous rights frame.

It is also worth noting that the way that Indigenous Peoples organized themselves within the confines of the UNFCCC was unique. Similarly to the Working Group on Indigenous Populations, Indigenous Peoples gathered in a type of Indigenous caucus that became formally known as the International Indigenous Peoples' Forum on Climate Change (IIPFCC). The IIPFCC behaved in a similar way to other stakeholder groups, in terms of hosting meetings and lobbying government officials, but also broke a number of unspoken norms by participating in unsanctioned protests and direct action events to push the ideas of Indigenous vulnerability and sovereignty.[22]

Outside of the officially sanctioned events, the IICM is typically heavily involved with alternative forums, side events, and even demonstrations. These spaces are typically hosted by like-minded civil society groups seeking a venue where non-governmental groups can have the chance to share information and perspectives (particularly ones that do not find such an open reception by states), as well as coordinate positions and tactics. Indigenous groups have often taken prominent roles in protests and other demonstrations, and have often fought to take the lead in many of these demonstrations. The use of unconventional tactics, appropriation of public space, and distinct forms of organization within the halls of the UN mean that Indigenous Peoples more closely resemble a social movement than a typical constituency group.[23]

The IICM, directly or indirectly, has ended up challenging many of the largest colonial states, such as Canada, the US, and Australia – who also happen to be some of the biggest polluters, some of the most opposed to Indigenous rights, and at times some of the most intransigent actors in the negotiations.[24] In doing so, Indigenous groups are also making space for both action and discourse for other groups with similar demands. Grossman[25] argues that Indigenous groups are particularly well placed to enact such a platform, being able to assert similar rights to states in the negotiations, and leveraging their claims to sovereignty and self-determination. By also being amongst the first to forcefully apply an Indigenous rights – and therefore human rights – approach to climate change, they also help to enable and build a foundation for other movements with similar positions.[26]

By the time of the Paris COP 21 in 2015, states had thawed somewhat to the inclusion of Indigenous concerns and rights within their negotiations. Significantly, the final text of the Paris Agreement includes a stipulation in the preamble:

> Acknowledging that climate change is a common concern of humankind, Parties should, when taking action to address climate change, respect, promote and consider their respective obligations on human rights, the right to health, the rights of indigenous peoples, local communities, migrants, children, persons with disabilities and people in vulnerable situations and the right to development, as well as gender equality, empowerment of women and intergenerational equity.

The Agreement also recognized that approaches to adaptation should take into consideration "traditional knowledge, knowledge of indigenous peoples, and local knowledge systems" where appropriate.[27]

Not only had Indigenous rights been recognized, it was recognized as part of a broader set of human rights. While the Indigenous caucus stated their preference for including this provision within the operative section of the text (as opposed to the preamble), other Indigenous commentators pointed out that the preamble in effect served as a framework for how the entire text was to be interpreted and implemented. This was welcomed as an undoubted success for advancing the protection of Indigenous rights with respect to climate change, and in international law generally. A number of countries were also thanked for their work in stepping up to support Indigenous issues in the negotiations.[28]

As a follow-up to the Paris Agreement, the government of Canada held a gathering of federal and provincial ministers, including the prime minister, as well as Indigenous leaders, to develop a national climate change plan. The result was titled the Pan-Canadian Framework on Clean Growth and Climate Change. Notably, the document contained repeated references to Indigenous issues and communities as a priority, while also providing an explicit endorsement of the UNDRIP, specifically naming the right to free, prior, and informed consent. There are also a few references to the use of Indigenous traditional knowledge. Indigenous leaders also managed to get a commitment to work on an Indigenous-specific climate change plan.[29]

Less than a year after the release of the climate framework, the Canadian government also announced it would be reviewing its laws and policies related to Indigenous Peoples, with the aim of modernizing many, and bringing them in line with the UNDRIP, a large shift in federal policy from less than 10 years earlier, when Canada had voted against the UNDRIP at the UN General Assembly. While the details are still unclear at press time, Canada's largest representative Indigenous organization was scheduled to work with the government on the review.

There remain lasting doubts about the government's commitment to both of these proclamations because of its support for the development of the Alberta tar sands (oil) and related infrastructure. The tar sands represent one of the largest single sources of emissions; they have been the fastest growing source of emissions in Canada and a constant target of environmental groups nationally. They also happen to be primarily situated in and around Indigenous territories, who are unduly impacted by the emissions and by the release of oil and other contaminants into nearby waters and ecosystems. Indigenous communities in the area maintain that their rights are continuously being violated, including the initial act of land appropriation and subsequent granting of the lands to oil companies.[30]

As the oil industry has increasingly sought further ways to bring the tar sands to international markets, they've begun planning and constructing a number of pipelines across North America. Incidentally, most of the proposed pipelines also cross through Indigenous territories and watersheds, which has led to a recent high level of activism and legal challenges founded on Indigenous rights. Both Canada and the United States have seen high-profile fights against these pipelines, which include the Keystone XL and Dakota Access pipelines in the US, and the Northern Gateway, the Kinder Morgan Trans Mountain, and the Energy East pipelines in Canada. While the Keystone XL pipeline was initially rejected after overwhelming public pressure, the Northern Gateway pipeline was approved in Canada, only to lose a court challenge based on Indigenous rights to consultation. The government subsequently revoked that permit. As more companies plan expansions to fossil fuel–related infrastructure across North America, they're likely to increasingly face off against Indigenous communities, who are aware of their rights and willing to challenge these projects in the courts and locally.

The global Indigenous movement was birthed out of a response to widespread human rights abuses around the world. The movement was responsible for developing a framework of Indigenous rights that has proven to be useful from the level of Indigenous communities to international negotiations. The international Indigenous climate movement took that framework and applied it to the issue of climate change. Here, it provides us with a rubric for understanding and remedying the human rights impacts that climactic changes are expected to bring. It also allows for a political vision to be presented, one where Indigenous traditional knowledge and relationships with the earth are respected, with Indigenous self-determination replacing state sovereignty. Lastly, it provides a strong set of legal tools to actually achieve that vision – including most prominently the right to free, prior, and informed consent over laws, policies, and projects that would affect Indigenous Peoples. Taken as a whole, the framework of Indigenous rights is a critical challenge to statist and even corporatist power and worldviews. Its further adoption and implementation, both at the international and national levels, may be of crucial importance not just for the protection of Indigenous Peoples, but also for the very climate itself.

Notes

1 S. E. Álvarez, 'Latin American Feminisms "Go Global": Trends of the 1990s and Challenges for the New Millennium', in S. E. Álvarez, E. Dagnino and A. Escobar (eds.), *Cultures of Politics, Politics of Cultures: Re-Visioning Latin American Social Movements* (1998) 293; A. Brysk, *From Tribal Village to Global Village: Indian Rights and International Relations in Latin America* (2000); S. Escárcega, 'Authenticating Strategic Essentialisms: the Politics of Indigenousness at the United Nations', 22(1) *Cultural Dynamics* (2010) 3; S. Escárcega, 'The Global Indigenous Movement and Paradigm Wars: International Activism, Network Building, and Transformative Politics', in J. S. Juris and A. Khasnabish (eds.), *Insurgent Encounters: Transnational Activism, Ethnography, and the Political* (2013) 129; J. Y. Henderson, *Indigenous Diplomacy and the Rights of Peoples: Achieving UN Recognition* (2008); M. C. Lâm, *At the Edge of the State: Indigenous Peoples and Self-Determination* (2000); R. Niezen, *The Origins of Indigenism: Human Rights and the Politics of Identity* (2003); F. Wilmer, *The Indigenous Voice in World Politics: Since Time Immemorial* (1993).

2 Brysk, *supra* note 1, at 33.

3 Wilmer, *supra* note 1, at 135.

4 S. Escárcega, *Internationalization of the Politics of Indigenousness: A Case Study of Mexican Indigenous Intellectuals and Activists at the United Nations*, Ph.D. Dissertation, University of California – Davis (2003); Jelin, 'Toward a Culture of Participation and Citizenship: Challenges for a More Equitable World', in S. E. Alvares, E. Dagnino and A. Escobar (eds.), *Cultures of Politics, Politics of Cultures: Re-Visioning Latin American Social Movements* (1998) 305; Lâm, *supra* note 1; Niezen, *supra* note 1; L.T. Smith, *Decolonizing Methodologies: Research and Indigenous Peoples* (2005).
5 R. Reitan, *Global Activism* (2007).
6 Álvarez, *supra* note 1, at 9.
7 S. Varese, *Witness to Sovereignty: Essays on the Indian Movement in Latin America* (2006) 217.
8 N. G. Postero and L. Zamosc, 'Indigenous Movements and the Indian Question in Latin America', in N. G. Postero and L. Zamosc (eds.), *The Struggle for Indigenous Rights in Latin America* (2004) 1, at 2–3; Smith, *supra* note 4, at 108; Varese, *supra* note 7, at 3, 228; K. B. Warren and J. E. Jackson, 'Introduction: Studying Indigenous Activism in Latin America', in K. B. Warren and J. E. Jackson (eds.), *Indigenous Movements, Self-Representation, and the State in Latin America* (2003) 1.
9 J. Nye, *Soft Power: The Means to Success in World Politics* (2005).
10 J. Anaya, *Indigenous Peoples in International Law* (2004); Escárcega 2003, *supra* note 4, at 5; Lâm, *supra* note 1.
11 Brysk, *supra* note 1, at 130; Escárcega, *supra* note 4, at 135; Muehlebach, '"Making Place" at the United Nations: Indigenous Cultural Politics at the U.N. Working Group on Indigenous Populations', 16(3) *Cultural Anthropology* (2001) 415.
12 Escárcega, *supra* note 4; Henderson, *supra* note 1; Lâm, *supra* note 1.
13 Henderson, *supra* note 1, at 51, 67.
14 Escárcega, *supra* note 4, at 74; Lâm, *supra* note 1, at 64, 172; Henderson, *supra* note 1, at 67–70.
15 For more on Indigenous traditional knowledge, see H. Fourmile (1999) and K.-A.S. Kassam (2009): Fourmile, 'Indigenous Peoples, the Conservation of Traditional Ecological Knowledge, and Global Governance', in N. Low (ed.), *Global Ethics and Environment* (1999) 215; K.-A.S. Kassam, *Biocultural Diversity and Indigenous Ways of Knowing: Human Ecology in the Arctic* (2009).
16 Henderson, *supra* note 1; Lâm, *supra* note 1, at 211; Postero and Zamosc, *supra* note 8, at 7; Warren and Jackson, *supra* note 8, at 16; Wilmer, *supra* note 1, at 1993.
17 Escárcega, *supra* note 4, at 34–35; Jelin, *supra* note 4, at 407.
18 Grossman, 'International Indigenous Responses', in Parker *et al.* (eds.), *Climate Change and Pacific Rim Indigenous Nations* (2006) 39, at 39, 43; M. Macchi *et al.*, *Indigenous and Traditional Peoples and Climate Change* (2008), available online at http://cmsdata.iucn.org/downloads/indigenous_peoples_climate_change.pdf, at 9, 11.
19 Macchi *et al.*, *ibid*, 12; M. Abhainn *et al.* (eds.), *Indigenous Peoples and Climate Change: Vulnerabilities, Adaptation, and Responses to the Mechanisms of the Kyoto Protocol – a Collection of Case Studies* (2007).
20 Warren and Jackson, *supra* note 8, at 13.
21 Grossman, *supra* note 18, at 47–48.
22 Abhainn *et al.*, *supra* note 19, at 134–135.
23 Álvarez 1998, *supra* note 1, at 19.
24 Grossman, *supra* note 18, at 42, 49–50.
25 *Ibid.*, at 50.
26 R. Reitan and S. Gibson, 'Climate Change or Social Change? Environmental and Leftist Praxis and Participatory Action Research', 9(3) *Globalizations* (2012) 395.
27 United Nations, *Paris Agreement* (2015).
28 International Indigenous Peoples Forum on Climate Change, *Statement at Closing Plenary of UNFCCC COP21* (2015), International Indian Treaty Council, *Press Release: The Paris Agreement: An "Incremental Advance" for International Recognition of the Rights of Indigenous Peoples* (2015), available online at http://hosted.verticalresponse.com/1383891/b821424379/545546365/eebd23fa9a/.
29 Government of Canada, *Pan-Canadian Framework on Clean Growth and Climate Change* (2016).
30 J. Huseman and D. Short, 'A Slow Industrial Genocide': Tar Sands and the Indigenous Peoples of Northern Alberta', 16(1) *The International Journal of Human Rights (IJHR)* (2012) 216.

15

Using the Paris Agreement's ambition ratcheting mechanisms to expose insufficient protection of human rights in formulating national climate policies

Donald A. Brown

Introduction – links between protecting human rights and climate change

This chapter focuses on how human rights standards should be considered in the Paris Agreement's transparency mechanisms that have been created to assure that governments adopt policies and measures that are sufficiently ambitious and fair to avoid dangerous climate change and thus prevent human rights violations. Although this chapter will focus on national obligations to reduce greenhouse gas (GHG) to prevent human rights violations, much of what is discussed in this chapter also applies to national obligations to support adaptation, and compensate for losses and damages, associated with climate change impacts, when these are necessary to protect human rights.

There is widespread agreement among scholars that climate change interferes with numerous human rights. For instance, Caney explains that climate change violates many human rights, including three of the most fundamental and least controversial rights, namely (1) right to life, (2) right to health, and (3) right to subsistence.[1] Climate change violates the right to life because a changing climate is already killing people through more intense storms, floods, droughts, and killer heat waves. Climate change violates the right to health by increasing the number of people suffering from disease, death, and injury form heat waves, floods, storms, fires, and droughts; the range of malaria and the burden of diarrheal diseases, cardio-respiratory morbidity associated with ground-level ozone; and the number of people at risk from dengue fever. Climate change violates the right to subsistence by increasing droughts which will undermine food security, water shortages, sea level rise which will put some agricultural areas under water, and flooding which will lead to crop failure.[2]

At COP 21 in Paris, 195 nations adopted the Paris Agreement.[3] Under Articles 2–4 of the Agreement, each Party is obliged to prepare, communicate, and maintain successive Nationally Determined Contributions (NDCs) that it intends to pursue through domestic mitigation measures, with the aim of reducing greenhouse gas emissions to hold the increase in the global average temperature to well below 2°C above pre-industrial levels and to pursue efforts to limit the temperature increase to 1.5°C above pre-industrial levels.[4]

Because climate change is already harming millions of people around the world and thereby often violating their human rights, it is not sufficient to limit warming to between 1.5°C and 2°C to prevent human rights violations. According to one report, climate change was already responsible for 400,000 deaths a year by 2012.[5] A case can therefore be made that nations should reduce future warming to the maximum extent possible to prevent human rights violations, that is to prevent any increase in warming, not just increases of 1.5°C or greater. Yet, there is growing evidence that it may be too late to prevent warming of 1.5°C.[6] Because it may not be possible to limit all future warming to levels that will not cause some additional human rights violations, I argue that national climate change policies should focus on preventing human rights violations to the maximum extent feasible.

This chapter next describes the Paris Agreement's transparency mechanisms that have been created to provide periodic critical review of the NDCs submitted by nations to determine whether they are sufficiently ambitious and fair to achieve the Agreement's warming limit goal. Following this, the chapter next explains why experience with national GHG emissions reduction commitments made before Paris demonstrates that nations will need to submit specific information with their NDCs, to allow citizens and governments to effectively evaluate whether the national commitments are adequately ambitious and fair. Finally, the chapter describes a five-step process that a nation should follow in formulating its NDC to assure that the nation's GHG emissions reduction commitment constitutes the nation's fair share of a global GHG emissions that will achieve the Agreement's warming limit goal.

The Paris Agreement's transparency mechanisms

The effectiveness of the Paris Agreement concerning reducing national GHG emissions to safe levels that protect human rights to the maximum degree feasible depends on finding ways of assuring that nations adequately strengthen their NDCs to levels that will achieve the Paris Agreement's global warming goals and protect human rights. The mechanism in the Paris Agreement that has been created to achieve this strengthening is through successive five-year cycles of review, referred to as the "stocktake" mechanism, which aims to assure that Parties strengthen their contributions.[7] Before the first official stocktaking event in 2023, the Parties agreed in Paris to have a "facilitative dialogue among parties in 2018 to take stock of the collective actions."[8] The purpose of this 2018 dialog is to evaluate the effectiveness of what has been committed to so far to set the stage for the regular five-year "stocktake," when nations are expected to upgrade prior GHG emissions reductions commitments in their NDCs. Also in 2018, the Intergovernmental Panel on Climate Change (IPCC) has been invited to submit a special report on the impacts of warming of 1.5°C and related global emissions pathways.[9]

Although the Paris Agreement urges nations to upgrade their prior commitments before 2020, there is no obligation under the Agreement to strengthen prior GHG reduction commitments until before the first stocktake in 2023 and then every five years thereafter.[10] For the post-2020 commitments, nations are required to submit commitments that "represent a progression beyond the Parties then current NDCs."[11]

In addition to the regular stocktakes, the Paris Agreement established an "enhanced transparency framework" for action and support.[12] The aim of the framework is "to provide a clear understanding of climate action, and to track progress."[13] Under the transparency framework, all nations, except Small Island Developing States (SIDS) and the Least Developed States (LDS), must submit biennial reports which include, among other things:

a Information necessary to track progress made in implementing and achieving its NDC under Article 4,[14] and
b For developed countries, information on financial support for mitigation, adaptation, and technology transfer.[15]

Thus, the transparency framework should generate information that will be of use in the periodic stocktakes on how nations are complying with their obligations under the Paris Agreement, including the extent to which they have enacted policies to reduce GHG emissions consistent with their NDCs and whether the policies have been successful in reducing GHG emissions.

Instead of negotiating "top-down" legally binding GHG emissions reduction commitments sufficient to meet the agreed goal of well below 2°C while pursuing efforts to achieve a warming limit of 1.5°C, the Paris Agreement relies solely upon a "bottom-up" nationally determined approach, presumably in the hope that a transparent review process will invoke diplomatic and civil society pressure strong enough to achieve needed increases in the mitigation ambition of Parties whose GHGs targets are inadequate. Thus, one observer characterized the Paris Agreement's approach to achieving greater ambition of inadequate national GHG emissions reduction commitments as a process which relies on "naming and shaming."[16]

For the transparency instrument and the periodic stocktakes to work effectively to increase national ambition on GHG emissions reductions to levels that achieve agreed upon warming limits and protect human rights, it will be necessary to make sure that nations submit sufficient information in support of their NDCs to allow critical reflection on how the nation responded to its obligations under the UNFCCC to adopt policies to achieve the warming limit goal "on the basis of equity and common but differentiated responsibilities and respective capabilities."[17]

Armed with this information, other governments and NGOs will be in a position to evaluate the actual factors considered by a nation in formulating its NDC relevant to the nation's fair share of maximum global GHG emissions that must not be exceeded to assure that the warming limit goal is achieved. This information will allow civil society, during the regular "stocktake" events created by the Paris Agreement or during the formulation of NDCs at the national level, to engage in dialogues with the government about the adequacy of its NDC in regard to its ethical and legal obligations to adopt policies that are consistent with the nation's fair share of safe global GHG emissions.

To ensure that the Paris Agreement reaches its goals, nations will need to submit information on how nations responded to their international responsibilities to prevent dangerous climate change. Recent research project of Widener University Commonwealth Law School and the University of Auckland (hereinafter Widener/Auckland research) examined on what basis 23 countries set their national GHG emissions reduction targets before the 21st Conference of the Parties (COP 21) under the United Nations Framework Convention on Climate Change in Paris in December, 2015.[18] These targets were formulated to fulfill national responsibilities agreed to in Warsaw at COP 19 in 2013 to submit to the UNFCCC Secretariat their Intended Nationally Determined Contributions (INDCs) before COP 21. This research concluded that most nations had based their international emissions reduction commitments on national economic self-interest rather than their global responsibility under the UNFCCC, and therefore the national targets did not represent the nation's fair share of safe global emissions.[19] It follows that most national INDCs were not formulated to prevent human rights violations.

The Widener/Auckland research project also concluded that in the discussions that took place during the formulation of the national targets, even those NGOs that favored stronger national reduction targets very frequently failed to frame their objections on the basis of the sufficiency of the target to reach the then-applicable warming limit goal of 2°C agreed to by the international community in Cancun in 2010,[20] or the equity and justice considerations that nations agreed under the UNFCCC should guide the formulation of their national climate policy,[21] matters which the process for setting national targets described below requires national governments to expressly consider. Thus, an understanding of the steps that nations should follow when setting national GHG emissions reduction targets described below will facilitate more sophisticated critical reflection on the adequacy of national climate policy, including national responsibility to protect and provide human rights.

Furthermore, the Paris Agreement makes clear that NDCs, in addition to reflecting the highest possible ambition, must also reflect the principles of common but differentiated responsibilities and respective capabilities, in the light of different national circumstances, on the basis of equity, and in the context of sustainable development and efforts to eradicate poverty.[22] It follows that NDCs must be fair and sufficiently ambitious to prevent dangerous climate change, including impacts leading to violations of human rights.

The Widener/Auckland research project also revealed that when Parties submitted their INDCs before Paris, the actual basis for a nation's INDC was not discernable from the INDC submitted.[23] Although nations sometimes claimed that their GHG emissions reductions commitments were sufficiently ambitious and fair, almost all failed to provide the information needed to evaluate the actual considerations that had affected their INDCs, including what warming limit or equity considerations determined the basis of the final INDC. In fact, none of the 23 INDCs examined explained how equity considerations specifically affected their INDC, despite their obligation as a Party to the UNFCCC to adopt policies and measures to prevent dangerous climate change based on "equity and common but differentiated responsibilities and respective capabilities."[24]

For this reason, the success of the Paris Agreement in achieving adequate GHG emissions reductions commitments sufficient to meet the 1.5°C–2°C warming limit goal and prevent human rights violations likely depends on assuring that nations submit sufficient information in support of their initial and future NDCs. This will facilitate an evaluation of whether the commitment is ambitious enough to achieve the agreed warming limit or levels that cause human rights violations with a reasonable level of probability and what considerations were taken into account in determining the nation's fair share of global GHG emissions reductions necessary to achieve this warming limit. Unless nations increase the ambition of their NDCs in response to civil society pressure at the national level during formulation of their GHG emission reduction commitments, or in response to NGO or other government pressure during proceedings created by the Paris Agreement, there is little hope that nations will raise their level of ambition to reduce their GHG emissions to levels that will achieve the warming limit goal on the basis of the equity requirement.

Five key steps in formulating and communicating a national GHG emissions reduction target

There are five steps that nations should follow in formulating a national GHG emissions reduction target in response to their obligations to formulate an NDC. Even if nations do not formally follow these steps, any national GHG emissions reduction target is implicitly a position on each of these steps; therefore, understanding how nations made decisions on these steps is essential to permit critical evaluation of the adequacy of national GHG emissions targets.

Any national GHG reduction target should be consistent with a target formulated by following these steps:

1 Select a global warming limit to be achieved by the GHG emissions reduction target.
2 Identify a global carbon budget that is consistent with achieving the global warming limit at an acceptable level of probability.
3 Calculate the annual rates by which global GHG emissions will be reduced on the pathway to zero emissions.
4 Determine the national fair share of the global carbon budget based upon equity and common but differentiated responsibilities and respective capabilities.
5 Specify the annual rates of its national GHGs reductions on the pathway to zero emissions.

All these steps require making judgements on issues that affect the interests of people around the world, including their rights to enjoy human rights, and for this reason nations should explain how it arrived at judgements on these issues so that the Paris Agreement's transparency mechanisms can work. The following describes these steps in more detail to identify sub-issues in some of these globally significant steps, so that a description of the minimum information that nations should include with their NDCs can be identified.

Determine a warming limit not to be exceeded

The first step in setting a national GHG emissions reduction is to identify a warming limit that the target seeks to achieve. Every national GHG target is implicitly a position on how much warming is acceptable, because it condones elevation of atmospheric GHG concentrations to levels entailed by the GHG target. As mentioned earlier, the international community agreed in Paris to limit the increase in global average temperatures to well below 2°C and to pursue efforts to limit temperature increases to 1.5°C, in light of more recent scientific evidence that a 2°C warming limit may not prevent dangerous climate change. The lower 1.5°C goal was agreed to because of intense pressure from Small Island Developing States and the most vulnerable Least Developed States that a warming limit of 2°C would leave them exposed to dire climate impacts, including in some cases existential threats to their sovereignty.[25] Yet, any additional warming from already elevated current global temperatures is likely to cause some human rights violations. For this reason, nations should be expected to explain what warming limit the nation's NDC is designed to achieve and, given that any elevation of current global temperatures is likely to cause some human rights violations, the nation should explain on what basis it can justify not preventing additional human rights violations.

If a nation claims that a 1.5°C warming limit or lower would impose unfair costs on the nation, the nation should make this claim when it identifies the equity criteria it applied to allocating its fair share of a global carbon budget in Step 4, not initially in selecting a warming limit. In this regard, although many developing countries will very likely be able to claim that fairness justifies that their NDC should allow them to emit GHGs at higher levels than most developed countries, the fairness considerations that lead to this conclusion should be taken into account in Step 4 when it determines its fair allocation of acceptable global emissions based on the Paris equity requirement.

Because any national GHG emissions target is implicitly a position of the nation on how much warming the nation deems to be acceptable to impose on others, the nation should be required to expressly identify the warming limit goal the NDC is designed to achieve. Because any warming above current increased warming levels is likely to cause additional human rights violation, any Party that commits to a warming limit that is higher than 1.5°C should explain why it is appropriate for it to cause additional human rights violations entailed by the additional warming level. In this regard, the nation should be expected to explain why its discretion in setting a GHG target is not limited by the following implications of a human rights approach identified by Caney.

- Because human rights are violated, costs to those causing climate change entailed by policies to reduce the threat of climate change are not relevant for policy. That is, if a person is violating human rights, he or she should desist even if it is costly. The abolition of slavery was immensely costly to slave owners, yet because basic human rights were violated, costs to the slave owner of abolishing slavery were not relevant to the slave owners' duty to free the slaves.[26]
- If climate change is a human rights problem, compensation is due to those whose rights have been violated. The human rights approach generates both duties for mitigation and adaptation. It also generates duties of compensation for harm.[27]
- Human rights apply to each and every human being, as they are based on the idea that all human beings are born free and entitled to certain rights.[28]
- Human rights usually take priority over other human values, such as efficiency and promoting happiness.[29]

If the warming limit entailed by a nation's GHG emissions reduction target authorizes additional human rights violations, the nation should explain on what basis it can ignore the above prohibitions against causing human rights violations.

Identify a global carbon budget which will achieve the global warming limit at an acceptable level of probability

To operationalize a warming limit, it is necessary to identify a carbon budget or the total amount of CO_2 equivalent gases (CO_2-e) that can be emitted into the atmosphere before atmospheric concentrations of GHGs exceed a level that will cause the temperature warming limit goal that the target is designed to achieve. Therefore, for instance, to limit warming to 1.5°C, a global carbon budget can be calculated that will limit atmospheric CO_2-e to a concentration that is assumed will limit warming to 1.5°C.

A scientifically respected way of determining a carbon budget for different warming limits is to rely on budgets generated from global climate change models reported by the IPCC's Fifth Assessment.

For instance, the IPCC's Fifth Assessment reported on the following budgets:

> Limiting the warming caused by anthropogenic CO_2 emissions alone with a probability of >33%, >50%, and >66% to less than 2°C since the period 1861–1880 will require cumulative CO_2 emissions from all anthropogenic sources to stay between 0 and about 1570 GtC (5760 $GtCO_2$), 0 and about 1210 GtC (4440 $GtCO_2$), and 0 and about 1000 GtC (3670 $GtCO_2$) since that period, respectively. These upper amounts are reduced to about 900 GtC (3300 $GtCO_2$), 820 GtC (3010 $GtCO_2$), and 790 GtC (2900 $GtCO_2$), respectively, when accounting for non-CO_2 forcings as in RCP2.6.[30]

These numbers must be adjusted for GHG emissions to include the 515 [445 to 585] GtC (gigatonnes of carbon) emitted by 2011[31] and additional amounts since 2012, which can be determined by examination of data maintained by the Global Carbon Budget project.[32]

IPCC has not yet developed budgets for a 1.5°C warming limit, although it has been asked to report on a 1.5°C warming limit budget by 2018. In the meantime, the United Nations Environment Programme (UNEP) has identified a warming limit budget for 1.5°C of between 200 to 415 GtC.[33] Recently, Carbon Brief has also identified carbon budgets for a 1.5°C warming limit in 2016 of 204.62 GtC for a 66% probability, 354.62 GtC for a 50% probability, and 652.62 GtC for a 33% probability, which at current GHG emissions levels provide 5.15, 8.93, and 16.49 years left, respectively, before the budgets are exhausted at 2015 emissions level rates of 39.70 GtC per year.[34]

Given the scientific uncertainty about what temperatures will be caused by different concentrations of GHGs, an issue usually discussed under the topic of "climate sensitivity," any carbon budget will need to make an assumption about the probability of limiting warming to the desired warming limit given different atmospheric concentrations of GHGs. Thus, any identification of a carbon budget will need to make an assumption about the "equilibrium climate sensitivity (ECS)," Any carbon budget, therefore, will need to take a position on the probability that the temperature limit will be achieved. And so any NDC will need to acknowledge that the carbon budget relied on to determine the emissions reductions necessary to achieve the warming limit has made an assumption about the probability that an atmospheric concentration will limit the warming to the warming limit goal. The probability of achieving any warming limit has ethical significance, because those who are most vulnerable to climate change have a right not to be threatened by the actions of others. Therefore, the probability of achieving a warming limit chosen by a nation in determining a carbon budget on which it calculates its NDC will affect the risks the nation is imposing on others.

The IPCC projections from the global climate change models have an assigned probability, with the general rule of thumb being that the higher the probability the smaller the carbon budget. Budgets most widely discussed by the international community that will limit warming to 2°C are ones that will provide a 50% or 66% probability that the 2°C warming limit will not be exceeded, probabilities that most vulnerable nations find unacceptable. Because different carbon budgets are understood to have different probabilities of limiting warming to desired warming limits, different budgets have different probabilities of preventing human rights violations. Therefore, each nation should be expected to justify why it has chosen any budget that increases the probability of increasing human rights violations.

A nation should be expected to justify the carbon budget it assumed in formulating its GHG emission reduction target, in regard to the probability that the budget will achieve the warming limit assumed, and why the budget selected is consistent with the nation's duty to prevent human rights violations. Because the carbon budget relied upon by a nation in formulating its NDC will be a position on the probability of preventing human rights violations, the nation should explain its justification for selecting a carbon budget that fails to prevent human rights violations at acceptable levels of probability.

Calculate the rate that global GHG emissions will be reduced to define the contraction pathway to zero emissions

The third step in setting a national GHG reduction target is to determine the rate or variable rates at which GHG emissions must be reduced by the entire international community before a carbon budget of zero is reached. This determination must be mindful of the fact that living within the constraints of a carbon budget is sensitive to achieving rapid rates of reduction early on to prevent creating a situation where rates of reduction required later are unachievable. It is necessary to determine the rate of reduction necessary to live within a carbon budget, because waiting to reduce GHG emissions to a given percentage in the future will deplete a dwindling carbon budget faster than if GHG emissions are reduced earlier in a given period.

Determine the national fair share of the carbon budget based upon equity and common but differentiated responsibilities and respective capabilities

All nations have agreed that their GHG emissions reduction commitments must prevent dangerous climate change and represent the nation's equitable share of necessary GHG emissions reductions. The Parties to the UNFCCC agreed to:

> protect the climate system for the benefit of present and future generations of humankind, on the basis of equity and in accordance with their common but differentiated responsibilities and respective capabilities.[35]

These principles were recommitted to in the Paris Agreement.[36] Therefore, a nation is obliged to take a position on what the concepts "equity" and "common but differentiated responsibilities" and "respective capabilities" mean and apply this understanding when determining its fair share of the global carbon budget specified in Step 2. These terms are usually discussed in relevant climate literature under the term "equity."

The principle of common but differentiated responsibilities and respective capabilities (CBDRC) captures the idea that it is the common responsibility of states to protect and restore the environment, but that the levels and forms of states' individual responsibilities may be differentiated according to their own national circumstances.[37] Under the 1997 Kyoto Protocol, national responsibilities for making GHG emissions reductions commitments were differentiated into two groups. First, 38 industrialized countries known as Annex I countries had specific targets to be achieved between 2008 to 2012, and second, non-Annex I developing countries had no target.[38] In the second commitment period of the Kyoto Protocol, 37 nations accepted targets that were to be achieved between 2013 and 2020, while other countries had no target.[39] Under the Paris Agreement, this stark distinction between the obligations of developed and developing countries to make commitments to reduce GHG emissions has been softened to allow differentiation not on the basis of the absence of any obligations to set a target but to allow delay in making "economy-wide" commitments if justifiable on the basis of "national circumstances."[40]

The Paris Agreement says:

> Each Party shall prepare, communicate and maintain successive nationally determined contributions that it intends to achieve [and] shall pursue domestic mitigation measures with the aim of achieving the objectives of such contributions.[41]

Thus, all nations must prepare an NDC without regard to whether it is a developed or developing nation. Yet, the Paris Agreement allows the following differentiation between developed and developing countries:

> Developed country Parties shall continue taking the lead by undertaking economy-wide absolute emission reduction targets. Developing country Parties should continue enhancing

> their mitigation efforts, and are encouraged to move over time towards economy-wide emission reduction or limitation targets in the light of different national circumstances.[42]

Although the Paris Agreement still differentiates the obligations of developed from developing countries in setting GHG emissions targets, by allowing developing countries to transition to economy-wide targets as conditions permit, all countries, including developing countries, are required to prepare and communicate NDCs that achieve the objectives of the Paris Agreement. Yet the clause "in light of different national circumstances" in the Paris Agreement introduces an element of ambiguity about the obligations of developing countries as to when they must make economy-wide emissions reductions commitments. For this reason, developing countries that do not identify economy-wide NDCs because of alleged national circumstances should explain the "national circumstances" on which they have formulated an NDC that is not economy wide.

In addition to allowing nations to make distinctions about the stringency of their NDCs based on "common but differentiated responsibilities and national circumstances," the stringency of any national GHG emissions reduction target must be based on "equity."

The Intergovernmental Panel on Climate Change in its Fifth Assessment IPCC in Chapter 4 explained that equity was a fair burden-sharing concept.[43]As such, it should be understood as a synonym of "distributive" justice. Furthermore, IPCC said that despite a lack of unambiguous meaning of what equity means:

> there is a basic set of shared ethical premises and precedents that apply to the climate problem that can facilitate impartial reasoning that can help put bounds on the plausible interpretations of 'equity' in the burden sharing context. Even in the absence of a formal, globally agreed burden sharing framework, such principles are important in establishing expectations of what may be reasonably required of different actors.[44]

The IPCC went on to say that these equity principles are given along four key dimensions: responsibility, capacity, equality, and the right to sustainable development.[45]

And so, each nation must determine its allocation of the global carbon budget after consideration of:

1. Responsibility, which is usually understood to be historical responsibility for current elevated GHG atmospheric concentrations;
2. Equality, or the right of each citizen to an equal right to use the atmosphere as a sink for its GHGs;
3. The right of poor countries to pursue economically sustainable development; and
4. Capacity, or the economic ability of a nation to reduce its GHG emissions.

Each of these criteria for determining what fairness requires raises questions acknowledged in the relevant literature about how to interpret them so that they can be applied to the determination of national obligations to prevent dangerous climate change. A brief description of each of these equity principles, along with an overview of some of the questions each principle raises, follows.

Historical responsibility

According to the IPCC, the principle of responsibility can be understood as follows:

> In the climate context, responsibility is widely taken as a fundamental principle relating responsibility for contributing to climate change (via emissions of GHGs) to the

> responsibility for solving the problem. The literature extensively discusses it, distinguishing moral responsibility from causal responsibility, and considering the moral significance of knowledge of harmful effects. Common sense ethics (and legal practice) hold persons responsible for harms or risks they knowingly impose or could have reasonably foreseen, and, in certain cases, regardless of whether they could have been foreseen.[46]

Like all the potential equity criteria, historical responsibility raises interpretive questions about such matters as when should a nation be responsible for its GHG emissions – for all historical GHG emissions from the nation, or when it became aware of the harm GHG emissions could cause others, or when it ratified the United Nations Framework Convention on Climate Change? If a nation chooses to modify its GHG emissions reduction obligations based on historical responsibility, it should explain the justification for its interpretation of when historical responsibility is triggered.

Equal per capita emissions

In regard to equality, IPCC said,

> Equality means many things, but a common understanding in international law is that each human being has equal moral worth and thus should have equal rights. Some argue this applies to access to common global resources, expressed in the perspective that each person should have an equal right to emit.[47]

Although IPCC acknowledges several interpretations of equality, a strong case can be made that in determining what equity requires, nations should acknowledge an equal right of all humans to use the atmosphere as a sink for GHGs, a consideration that leads eventually to an entitlement to equal per capita shares. Based on this interpretation, applying the "equality" principle, nations could determine their fair share of a remaining carbon budget by allocating the budget based on the nation's percentage of global population. As was the case with the equity criteria of historical responsibility, equity shares based on equal per capita rights raises questions about when people should be entitled to equal per capita allocations of GHG emissions reductions responsibilities. Nations basing their NDC on equal per capita entitlements should explain the basis for their interpretations of this concept.

The right of developing countries to sustainable development

The equitable principle based on the right to development is derived from the 1986 United Nations "Declaration on the Right to Development."[48] The UNFCCC also acknowledges a right to promote sustainable development, and "the legitimate priority needs of developing countries for the achievement of sustained economic growth and the eradication of poverty."[49]

The right of developing countries to pursue sustainable economic development also raises many interpretative questions, such as how to determine (a) what level of poverty should justify GHG emissions rates lower than rate reductions required of the entire world needed to prevent dangerous warming; (b) what limitations on GHG emissions, if any, should be imposed on developing countries who are legitimately pursuing their right to development; and (c) whether developing nations should be expected to limit GHG emissions generated by some human activities, such as unnecessary deforestation, indiscriminate flaring of gases, high emitting transportation systems, etc. In other words, if one acknowledges the right of poor nations to pursue sustainable economic development, questions remain about what limits on GHGs emission rates should be placed on their development activities. Nations who rely on the right of poor nations to develop to determine their equitable share of safe global emissions should explain the basis for their interpretation of these equity criteria. Any poor nation which seeks to justify its future GHG emissions based on its right to sustainable economic development should be expected to explain the basis for exempting emissions reduction commitments from activities such as wasteful gas flaring or economically unnecessary deforestation, which could produce GHG emissions reductions without impeding economic development needs.

Economic capacity to reduce GHG emissions

As we have seen, under the UNFCCC, nations agreed to reduce their GHG emissions to prevent dangerous climate change on the basis of equity and in accordance with their common but differentiated responsibilities and *respective capabilities* (emphasis added).[50] Therefore, a valid consideration for determining a nation's responsibility to reduce its GHG emissions is the economic capability of a nation to reduce its climate-causing GHG emissions. Yet this principle also raises numerous interpretive questions very similar to the issues raised by the principle of the right to sustainable development, such as (a) What levels of poverty or GDP per capita should be considered in determining a nation's responsibility to reduce GHG? (b) Are all GHG-emitting activities excused from reductions by a nation's poverty level, including, for instance, deforestation, gas flaring, and wasteful use of electricity generated by fossil fuels? and (c) Should wealthy people's GHG-emitting activities in poor countries be excused from GHG emissions restrictions? For this reason, those nations who seek to justify their GHG emissions reductions obligations on low levels of economic capacity should explain the basis for the distinctions they make about economic capacity.

In summary, the need to apply an equity step to determine a nation's NDC commitment to reduce GHG emissions is a classic problem of distributive justice. There is a sizeable literature on how principles of distributive justice should guide allocation of burdens to nations to reduce the threat of climate change.[51] Some of this literature focuses on ethical problems, with some frequently made arguments in support of proposed allocations of national responsibility to reduce the threat of climate change.

A frequent argument made for national allocations is the claim that those countries that have emitted the highest level of GHG emissions should have higher emissions targets, because status quo levels should entitle emitters to status quo rights. However, Meyer points out that other reasons why the argument of status quo fails to pass ethical scrutiny is because humans have no prescriptive right to assign common pool resources to themselves.[52]

Another common argument in support of allocations of national targets that fails to pass ethical scrutiny is that national allocations should be based on economic efficiency or cost-benefit analyses. Ekart and others argue that economic efficiency is an inadequate distributive principle, because economic rationality is based on maximizing individual preferences and preference satisfaction has no moral status.[53]

Distributive justice holds that all people should be treated as equals in any allocation of public goods unless some other distribution can be justified on morally supportable grounds. Yet, distributive justice does not require that all shares of public goods be equal but puts the burden on those who want to move away from equal shares to demonstrate that their justification for their requested entitlement to non-equal shares is based on morally relevant grounds. Therefore, someone cannot justify his or her desire to use a greater share of public resources on the fact that he or she has blue eyes or that he or she will maximize his or her economic self-interest through greater use of public goods. Such justifications fail to pass the test of morally supportable justifications for being treated differently. As IPCC concluded in its 2014 chapter on equity, which synthesized the peer-reviewed literature on climate change equity and justice, equity principles can be understood to comprise the four dimensions identified above, namely responsibility, capacity, equality, and the right to sustainable development.[54]

So although reasonable people can disagree on what fairness requires of nations when allocating national shares of a carbon budget, not all claims made by nations about the fairness of their GHG emissions targets are entitled to respect or are of equal merit. Following Amartya Sen, people do not have to reach agreement on what perfect justice requires to obtain agreement on injustice.[55] For this reason, there is hope that some progress can be made on the issue of what fairness and equity require of nations in the "stocktake" meetings that will take place under the Paris Agreement in the years ahead, provided nations are required to specifically explain how equity considerations and the other steps that should be followed in formulating the GHG emission reduction target affected the calculation of their NDC.

Conclusion

In determining an NDC, nations must justify any target that will increase human rights violations. This should be done for any step in formulating an NDC, which can affect whether the target will harm human rights. These justifications should expressly be included with the submission of NDCs under the Paris Agreement.

As this chapter explains, the setting of a national GHG emissions reduction target requires a nation to take a position on five steps that must be considered in formulating an emissions reduction target following the Paris Agreement. Decisions on each of these steps can affect the extent to which people around the world will be harmed by climate change impacts and deprived of their human rights. Because the success of the Paris Agreement depends on the ability of the Agreement's transparency mechanisms to expose inadequate national GHG emissions reduction commitments, nations should be required to explain and justify their decisions on these five steps in a transparent manner, so that civil society can effectively evaluate nations' obligations to prevent dangerous climate change and assure that human rights are enjoyed to the maximum extent feasible by people around the world.

Notes

1 S. Caney, 'Climate Change, Human Rights and Moral Thresholds', in S. Gardiner *et al.* (eds.), *Climate Ethics, Essential Readings* (2010) 163.
2 For one of many reports on the connection between human rights and climate change, see United Nations Environment Program (UNEP), *Climate Change and Human Rights* (2015), available online at www.unep.org/newscentre/Default.aspx?DocumentID=26856&ArticleID=35630 (last visited 3 November 2016). See also Caney, *ibid.*, for a discussion of why climate change violates the most basic human rights and therefore the least controversial human rights.
3 United Nations Framework Convention on Climate Change (UNFCCC), *Paris Agreement*, UN Doc. FCCC/CP/2015/L.9/Rev.1 (12 December 2015), available online at https://unfccc.int/resource/docs/2015/cop21/eng/109r01.pdf.
4 *Ibid.*, Arts 2–4.
5 DAR International, *Climate Vulnerability Monitor* (2012), available online at http://daraint.org/wp-content/uploads/2012/09/CVM2ndEd-FrontMatter.pdf, at 24.
6 R. McKie, Guardian, *Scientists Warn World Will Miss Key Climate Target* (6 August 2016), available online at www.theguardian.com/science/2016/aug/06/global-warming-target-miss-scientists-warn.
7 UNFCCC, *Paris Agreement*, *supra* note 3, Art. 14.

8 *Ibid.*, decision para. 20.
9 *Ibid.*, decision para. 21.
10 *Ibid.*, decision paras. 23–24.
11 *Ibid.*, Art. 13.
12 *Ibid.*
13 *Ibid.*
14 *Ibid.*, para. 7.
15 *Ibid*, para. 9.
16 W. Obergassel *et al.*, *Phoenix From the Ashes – an Analysis of the Paris Agreement to the UNFCCC* (1 March 2016), available online at http://wupperinst.org/uploads/tx_wupperinst/Paris_Results.pdf.
17 *United Nations Framework Convention on Climate Change*, UN Doc. UNFCCC/INFORMAL/84 (1992), Art. 3.1.
18 Widener Commonwealth University School of Law & the University of Auckland (Widener/Auckland Research Project), *Research Project on Ethics and Justice in Formulating National Climate Policies* (2015), available online at http://nationalclimatejustice.org.
19 *Ibid.*
20 UNFCCC Executive Secretary, *The Cancun Agreements* (2010), available online at http://cancun.unfccc.int/cancun-agreements/significance-of-the-key-agreements-reached-at-cancun/.
21 UNFCCC 1992, *supra* note 17, Art. 3.1.
22 UNFCCC, *Paris Agreement*, *supra* note 3, Art. 4.
23 Widener/Auckland Research Project, *supra* note 18, at "Lessons Learned".
24 UNFCCC 1992, *supra* note 17, Art. 3.1.
25 Obergassel *et al.*, *supra* note 16.
26 Caney, *supra* note 1, at 171.
27 *Ibid.*
28 *Ibid.*, at 165.
29 *Ibid.*
30 Intergovernmental Panel on Climate Change (IPCC), *Climate Change 2013: The Physical Science Basis, Summary for Policymakers, Contribution of Working Group I to the Fifth Assessment Report of the Intergovernmental Panel on Climate Change* (2013), available online at www.ipcc.ch/report/ar5/wg1/.
31 *Ibid.*
32 Carbon Brief, *Carbon Budget Update* (2016), available online at www.globalcarbonproject.org/carbonbudget/.
33 UNEP, *The Emissions Gap Report 2015* (2015), available online at http://uneplive.unep.org/media/docs/theme/13/EGR_2015_301115_lores.pdf.
34 Carbon Brief, *supra* note 32.
35 UNFCCC 1992, *supra* note 17.
36 UNFCCC, Paris Agreement, *supra* note 3, Art. 2.
37 J. Brunnée and C. Streck, 'The UNFCCC as a Negotiation Forum: Towards Common But More Differentiated Responsibilities', 13(5) *Climate Policy* (2013) 589. For a synthesis of much of the literature on common but differentiated responsibilities, see Pauw *et al.*, German Development Institute, *Different Perspectives on Differentiated Responsibilities: A State-of-the-Art Review of the Notion of Common But Differentiated Responsibilities in International Negotiations* (2014), available online at www.die-gdi.de/uploads/media/DP_6.2014.pdf.
38 For a discussion of obligations of Annex I countries under the Kyoto Protocol and how differentiation has changed since then, see Deleuil, 'The Common But Differentiated Responsibilities Principle: Changes in Continuity After the Durban Conference of the Parties', 21(3) *Review of European, Comparative & International Environmental Law (RECIEL)* (2012) 271. For a discussion of the how the Paris Agreement created a more nuanced approach to differentiation than did the Kyoto Protocol, see Voight and Ferreira, 'Differentiation in the Paris Agreement', 6 *Climate Law* (2016) 58.
39 UNFCCC, *Kyoto Protocol.*
40 UNFCCC, *Paris Agreement*, *supra* note 3, Art. 4.2.
41 *Ibid.*
42 *Ibid.*, Art. 4.
43 IPCC, 'Chapter 4: Sustainable Development and Equity', in IPCC Contributors (eds.), *Climate Change 2014: Mitigation of Climate Change. Contribution of Working Group III to the Fifth Assessment Report of the Intergovernmental Panel on Climate Change* (2014), at s. 4.6.2.1.

44 *Ibid.*, at 48.
45 *Ibid.*, at 4.
46 *Ibid.*, at 49.
47 *Ibid.*, at 51.
48 GA Res. 41/128, 4 December 1986.
49 UNFCCC 1992, *supra* note 17, at Preamble.
50 *Ibid.*, Art. 3.1.
51 See, for instance, L. Meyer and D. Roser, 'Distributive Justice and Climate Change: The Allocation of Emission Rights', 28 *Analyse and Kritik* (2006) 223; M. Patterson, 'Principles of Justice in the Context of Global Climate Change', in U. Luterbacher and D. F. Sprinz (eds.), *International Relations and Global Climate Change* (2001) 119; J. Moeller, *Distributive Justice and Climate Change: The What, How, and Who of Climate Change Policy*, Theses, Dissertations, Professional Papers (2016); F. Ekardt, *Climate Change and Social Distributive Justice* (31 May 2010), available at www.kas.de/china/en/publications/19733/.
52 Meyer and Roser, *ibid.*
53 Ekardt, *supra* note 51, s. 4.4.
54 IPCC, Chapter 4, *supra* note 43.
55 A. Sen, *The Idea of Justice* (2009).

Part IV
Stakeholder perspectives

16

From Marrakesh to Marrakesh

The rise of gender equality in the global climate governance and climate action

Anne Barre, Irene Dankelman, Anke Stock, Eleanor Blomstrom, and Bridget Burns

Global progress towards gender equality in all spheres of our societies is still too slow to realize the full potential of women within our lifetime.[1] *Existing inequalities between men and women, such as access to natural and productive resources, social status and decision-making, can be exacerbated by climate change impacts. At the same time, climate mitigation and adaptation measures that do not take gender dimensions and women's empowerment into account could increase existing inequalities, therewith endangering human rights. As of 2014, 143 out of the 195 countries guarantee equality between women and men in their constitution, yet discrimination between women and men persists in many areas, directly and indirectly through laws and policies, gender-based stereotypes and social norms and practices.*[2] *Initially limited to considerations related to women's participation and representation in technical climate change bodies, gender equality has come a long way under the UNFCCC and is now recognized among the human rights' obligations to consider when undertaking climate action.*

Focusing first on the journey from COP7 (2001) to COP22 (2016), both held in Marrakesh, this chapter will then provide examples of methods and tools, based on illustrations from the field, which effectively contribute to the promotion of gender equality and women's empowerment in the fight against climate change. Concluding remarks will highlight the crucial challenges that remain to be addressed for implementing gender-just climate action.

History of gender in the climate negotiations, the journey from COP7 to COP22

Gender-responsive action on climate change: global policy

Climate change magnifies existing inequalities, and in particular, gender inequalities, but women and men together are vital to climate solutions. In recent years, global consensus has recognized that the integration of women's rights and gender equality into the mitigation of and adaptation to climate change is not only essential but maximizes the efficacy of interventions, programs and resources. This is coherent with the normative frameworks establishing the linkages between gender equality, women's human rights and environment, to which Governments have already agreed, including the Convention on the Elimination of all Forms of Discrimination against Women (CEDAW), Hyogo Framework for Action, Rio+20, Agenda21, and the Beijing Platform for Action.[3]

The links between gender equality and climate change are supported by undeniable data in terms of differentiated impacts and contributions to action. A gender perspective frames the enabling conditions needed for solutions to be effective and combats the potential of climate impacts to further exacerbate inequalities.

Signed in 1992, the United Nations Framework Convention on Climate Change (UNFCCC)[4] was the sole Rio Convention (out of three, including on biodiversity and desertification) that did not recognize gender equality or women's socially constructed roles and realities in relation to the management of natural resources.

In 2001, Parties to the UNFCCC agreed on the first text on gender equality and women's participation, adopting two decisions at the seventh Conference of Parties in Marrakesh. Nine years later in the 'Shared Vision' of the Cancun Agreements adopted at COP16 in 2010,[5] the Parties to the UNFCCC stated that gender equality and women's participation are necessary for effective action on all aspects of climate change, agreeing to several decisions mainstreaming gender aspects across finance, adaptation and capacity building.[6]

Since then, UNFCCC Parties have recognized gender equality as a concept by integrating it in decisions on nearly every UNFCCC thematic area,[7] including the 2012 Decision 23/CP.18 on gender balance and women's participation and most notably through the 2014 launch of the Lima Work Programme on Gender (LWPG), which was extended for a term of three years at COP22 in Morocco, Marrakesh.

Some examples[8] of how gender has been mainstreamed across decisions of the UNFCCC include:

- Under the topic of *adaptation*, where gender issues have long been recognized, Decision 5/CP.17 (COP17 in Durban) on Guidelines for National Adaptation Plans (NAPs) emphasizes that adaptation should follow a country-driven, gender-sensitive, participatory and fully transparent approach and should be based on and guided by gender-sensitive approaches.[9] Additionally, the guidelines for the formulation of NAPs state that in developing NAPs, consideration should be given to the effective and continued promotion of participatory and gender-sensitive approaches. However, no such guidance exists for the Nationally Appropriate Mitigation Actions (NAMAs).[10] Significantly, the Paris Agreement adopted in 2015 foresees gender-responsive adaptation and capacity-building activities.
- In relation to *technology*, in 2011 when countries decided to establish a Climate and Technology Center and Network (CTCN) as the operational arm of the UNFCCC Technology Mechanism, the Terms of Reference for the network stated as part of its mission, 'to facilitate the preparation and implementation of technology projects and strategies taking into account gender considerations to support action on mitigation and adaptation and enhance low emissions and climate-resilient development'.[11]
- As a final example, under the topic of *climate finance*, in the 2011 decision to establish the Green Climate Fund, Parties agreed as part of the Fund's Governing Instrument that 'The Fund will strive to maximize the impact of its funding for adaptation and mitigation, and seek a balance between the two, while promoting environmental, social, economic and development co-benefits and taking a gender-sensitive approach.'[12]

The 2016 extension of the Lima Work Programme on gender[13] provides a new opportunity to advance the progress made towards integrating gender in UN climate policies by helping to ensure their implementation in an effective and cohesive manner. The decision adopted in Morocco takes forward many of the activities that were part of the initial work programme, including (a) training for delegates on gender and climate change; (b) capacity building/negotiation skills for women delegates; (c) a set of in-session workshops in 2018 and 2019 to further explore the topic; and (d) technical guidance on entry points related to gender across other bodies of the UNFCCC. However, it also advances the work through newer actions, such as (e) requesting both technical bodies and Parties, as well as the financial mechanism, of the UNFCCC to enhance communications and reporting on progress implementing gender-responsive climate policy; (f) requesting that a gender perspective be considered in the organization of the technical expert meetings on mitigation and adaptation; and (g) inviting Parties to appoint and provide support for a national gender focal point for climate negotiations, implementation and monitoring. Importantly, Parties requested that a concrete action plan, a 'Gender Action Plan (GAP)', be developed in 2017 in order to set clear timelines and responsibilities for undertaking and monitoring these activities.

There are several factors, which influenced Parties in their decision-making on UN climate policies, to shift from nominally 'gender-blind' to 'gender-responsive', including:

- Shifts in climate policy discussions and literature that allowed for a more socially focused, society-wide debate on climate action around the launch of the Bali Action Plan (BAP) in 2007.
- Sustained investment in the advocacy, engagement and knowledge-sharing efforts of gender experts, practitioners and women's rights organizations in the climate change debate. For example, in 2007 at COP13 in Bali, the Global Gender and Climate Alliance (GGCA)[14] was launched, bringing together UN agencies, intergovernmental organizations and civil society, with the goal of ensuring that climate change decision-making, policies and programmes, at all levels, would be gender-responsive. In addition, an active and organized UNFCCC Women and Gender Constituency[15] was formalized in 2010, leveraging the space for women and gender experts and activists to deliver interventions, hold press conferences and engage directly with negotiators.
- Even with some opposition from a handful of Parties that challenged the relevance of gender to climate change policy, there remained strong political will from heads of state, ministers, key government negotiators and political leaders in the UN, particularly champions within the UNFCCC Secretariat.

In addition, it is important to note that recognition of the links between gender equality, women's rights and climate change exist outside of the UNFCCC. The 1979 Convention on the Elimination of All Forms of Discrimination Against Women (CEDAW), which is fundamental to advancing gender equality and regarded as the first international bill of women's rights, has direct implications for climate change policy making. CEDAW obliges Parties to take 'all appropriate measures to eliminate discrimination against women in rural areas in order to ensure, on a basis of equality of men and women, that they participate in and benefit from rural development' and participate in all levels of development planning.[16] It further addresses issues of resources, credit, family planning, education and the right to work, to participate in forming and implementing government policies and to represent the country at international level – all of which impact a woman's capacity to adapt to impacts of climate change and to participate in planning and implementation to address climate change.

At its seventh session in March 2008, the UN Human Rights Council adopted by consensus Resolution 7/23 on Human Rights and Climate Change, and as a result, the Office of the

High Commissioner released a follow-up report in January 2009. The report recognizes the need for more country-specific and gender-disaggregated data to effectively assess and address gender-differentiated effects of climate change. It simultaneously reports that women have high exposure to climate-related risks exacerbated by unequal rights, and that women's empowerment and the reduction of discriminatory practices have been crucial to successful community adaptation and coping capacity.[17]

The Commission on the Status of Women (CSW) meets annually to follow up on implementation of the Beijing Platform for Action (1995), to ensure the mainstreaming of a gender perspective into UN work and to identify emerging issues and trends important to gender equality. In 2011, at its 55th session, Parties adopted a resolution to mainstream gender equality and promote the empowerment of women in climate change policies and strategies. The resolution is the first resolution by the CSW to address the link between gender equality and climate change.[18] In 2014, the 58th session of the CSW passed a resolution entitled 'Gender equality and the empowerment of women in natural disasters', which outlined the link between women, gender equality and disasters, and referred to the climate resolution from CSW55.[19]

Finally, the 2015 Sustainable Development Goals include goals on both climate change and gender equality and women's empowerment, explicitly recognizing the intersection between the two. For example, one of the targets under SDG13 on Climate Action says, 'Promote mechanisms for raising capacity for effective climate change–related planning and management in least developed countries and small island developing States, including focusing on women, youth and local and marginalized communities.'[20]

Methods and tools to promote gender equality in climate policies

The promotion of gender equality and women's empowerment in the fight against climate change has to be pushed on all levels – internationally, nationally and locally, and should be reflected in policies and practices. Gender mainstreaming processes benefit from the application of diverse approaches and tools.

Gender mainstreaming[21] is the overarching strategic approach for achieving gender equality and women's empowerment at all levels, as mandated by the Beijing Platform for Action. Results show that a consistent use of mainstreaming gender into other sectors leads to creating more effective policies, reaching the right target groups, increasing ownership of women and men, and bringing about more effective and sustainable implementation and results.[22] Even though the method has been known and practiced for over 20 years, it has not been applied systematically and therefore responses and impacts are still limited.[23] A major challenge is the context-specific approach requiring teams with diverse expertise and the willingness to cooperate across disciplines.

Gender mainstreaming and human rights principles

Deriving from the human right of gender equality (see the history section of this chapter), all gender-responsive policies and programmatic interventions should be consistent with human rights principles.[24] The principles include *equality and equity, non-discrimination, participation, empowerment* and *accountability*. The last three principles will be elaborated below, as they can enhance the mainstreaming of gender in policies and practices.

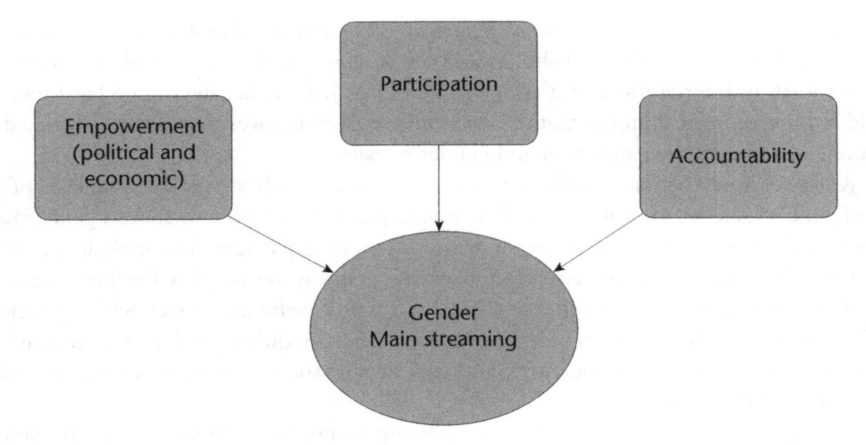

Figure 16.1 Principles of gender mainstreaming

Effective *participation* is crucial for gender mainstreaming and includes two main aspects:

1 Focusing on ways of working that enable women, men, girls and boys to be actively involved and encouraging the participation of marginalized, disempowered and discriminated against groups of women and men in decisions.
2 Focusing on processes of decision-making, referring to issues like *how* decisions are taken, *at what stage* of programming and project cycles, and *how views* of men and women are *taken into consideration.*

Empowerment has both political and economic dimensions. Empowering women in climate change mitigation and adaptation programmes and policies has an additional significance, since in particular the mitigation sector with its economic orientation and a centralized technical infrastructure is deemed to be a traditionally 'hard' sector. This category of centralized technical systems attracts a lot of funds. However, this area is often male dominated and lacks gender-responsive approaches. Empowerment of women in these sectors is essential.

Accountability is crucial to achieve long-term results in mainstreaming gender. This includes the adoption of gender indicators, which should be the basis of any monitoring and evaluation system. The principle of mutual accountability supports the development or improvement of accountability mechanisms to hold all stakeholders, such as donors and governments, to account for their work to reduce gender gaps and empower women.

Mainstreaming gender in national climate change policies and programmes

The following part describes tools and methods for the integration of gender equality issues in existing or planned climate change laws and policies at the national level.

A prerequisite for gender mainstreaming is an *enabling environment* within the respective country's political system encompassing the legal framework (including international obligations as well as national and regional legislation), the policy/ies, the governmental institutions, public and private utilities, and finances.

In general, the *national and regional legislation* (laws and regulations) need to address discrimination based on gender and should aim for gender equality (in accordance with obligations set forth in international treaties, such as CEDAW). A further developed legal framework would require mainstreaming so that the laws and regulations governing climate change directly address gender-based discrimination and gender equality.

A *gender policy* or a national gender action plan serves as an indicator of a certain level of awareness of gender issues at the national level, as well as guidance for how to address gender issues at national and sub-national levels. However, without proper implementation, including a sufficient allocation of budget for gender-related activities, any policy is meaningless. Furthermore, the collection of sex-disaggregated data goes hand-in-hand with developing proper gender policies.

The effective implementation of legislation and policies requires gender awareness and qualified as well as committed management and staff in relevant *governmental agencies*, as well as in public and private utilities.

Financing and budgeting policies not only referring to public revenues but also pertaining to economic policy – in particular to climate change mitigation and adaptation – should be gender-responsive and just.

An active civil society with NGOs – that have sufficient resources to work effectively and efficiently – are a determining factor of a balanced political system. This includes the existence of organizations working on gender equality and the empowerment of women.

Which tools and which steps?

Gender mainstreaming is a strategy that requires the use of various tools at different stages, singular or cumulatively. Some key tools are described below, including examples from WECF's (Women Engage for a Common Future) fieldwork implementing the same.

Gender analysis

A 'gender analysis helps to gain an understanding of the different patterns of participation, involvement, behaviour and activities that women and men in their diversity have in economic, social and legal structures and the implications of these differences'.[25] Sex-disaggregated data are the basis for the analysis, which can be conducted quantitatively and qualitatively, including through the use of anecdotal evidence (by desktop research, participatory means, etc.). Additional issues will need to be addressed in the context of climate laws and policies; however, the following issues provide a good starting point for a gender analysis:[26]

- Division of Labour between Women and Men

  Who does what kind of labour (unpaid and paid work, work within the household, work for wages outside the household)? How much time do women and men spend on these different tasks? How is it done and when? Why is it done? How do people perceive these differences?

- Access, Power and Control over Resources

  Who has access, power and control over natural and productive resources (e.g. land), income, information, time, technologies and services? How are access, power and control (legally) granted? Why is this so? Is it fair (i.e. how do existing power relations factor in to access, power and control)?

- Recognition of Differentiated Needs and Interests

  Is there broad understanding of different needs of women and men (e.g. sanitation, menstrual hygiene)?

- Decision-Making Ability

  Who has the ability to make decisions in the household and in the public sphere (i.e. how are benefits being shared)? How is the decision-making power granted? Why is this so? Is it fair?

- Status of Men and Women before the Law

  Who has which rights in accordance with law? How are men and women differently treated by customary and formal legal codes and the judicial system (e.g. inheritance, employment, legal representation)? Why is this so? Is it fair? What measures are taken to ensure that these laws are adhered to?

- Most Vulnerable

  What are the most vulnerable groups and who are the most vulnerable people within those groups? How is this reflected (e.g. financially, exposure to contamination)? Why is this so (e.g. single-headed household, rural, disability, migrant, widow, minority, sexual orientation, age)?

Below an example of a gender-responsive energy program conducted in Georgia.

Since 2010, Women Engage for a Common Future has been developing a gender-responsive program in Georgia, providing access to affordable renewable energy and energy efficiency (RE/EE) solutions to low-income communities in rural areas. Based on six years of technical capacity building for 200 men and women, this program was presented to the Nationally Appropriate Mitigation Action (NAMA) facility[27] for financial support.[28] It aims to promote a clean energy transition from carbon-intensive wood burning in seven regions of Georgia and foresees the construction and installation of 15,000 solar water heaters, 15,000 fuel-efficient stoves and 15,000 thermal insulations, thus improving the lives of 180,000 people. The social and gender co-benefits of this program are two-fold; it will substantially reduce the burden of unpaid domestic work for women (collecting wood, heating water, cooking), and will reduce 30–50% of energy costs, which has a direct impact on women as they are responsible for the energy budget in their homes. To increase the accessibility of the RE/EE solutions, a gender-sensitive financing mechanism with 0% interest rate has been specifically developed for low-income, women-headed households.

The program relies on the creation of gender-responsive renewable energy cooperatives to ensure the production and distribution of the technologies (four have already been established in four different areas). The cooperatives reflect gender equality in their statutes so that their governance structure (including the board, management and members) consists of at least 40% women. Through the cooperatives, the program creates local employment for women who participate in dedicated training programmes.[29] The participation of women in decision-making processes is strengthened not only via their representation in the cooperatives' governance structure, but also as promoters of the EE/RE technologies. More specifically, special programs train women to lead as 'renewable energy ambassadors' with a rewarding scheme. Finally, additional trainings foster women's leadership and participation in communities and local policy processes.

Gender-differentiated data and gender statistics

Gender statistics are defined as statistics that adequately reflect differences and inequalities in the situation of women and men in all areas of life.[30] While this data serves as a means to collect sex-disaggregated information, it must also:

- reflect gender issues;
- be based on concepts and definitions that adequately reflect the diversity of women and men and capture all aspects of their lives; and
- be collected according to methods taking into account stereotypes and social and cultural factors that may induce gender bias in the data.[31]

Gender statistics are the basis for all gender mainstreaming tools; they are also relevant for the development of gender indicators and for gender monitoring and evaluation. These tools are not discussed within this chapter.

Capacity building and awareness raising/information

Information on gender issues in climate change must be shared via formal climate training, mainstream and social media, as well as through outreach and meetings with key groups of stakeholders. Generally, the messaging should emphasize the relevance of climate change to men and women, and how women and men can be equal actors and agents of change in adapting to and mitigating climate change. The value of both men's and women's contributions in submitting information and taking part in future discussions and consultations needs to be highlighted.

Campaigns to raise awareness must reflect gender-specific needs and may require capacity building. In particular, information must be easily accessible and presented in an accessible manner. Language and literacy issues as well as access to information via the internet and other sources must be taken into account.

Gender-responsive financing and budgeting (GRB)

This tool focuses on an equality-oriented evaluation of the distribution of resources. The aim of GRB is to guarantee equal opportunities and equal benefits for women and men. Governmental budgets have to undergo a gender analysis looking at the taxation policies and the use of public funds and their impacts on women and men.[32] Among other things, the analysis must take into account whether the budget reflects the differing participation rates of women and men in the care economy. Gender procurement is another method that falls within the scope of GRB.

Gender-responsive budgeting in Morocco

An excellent example of gender-responsive budgeting is the Kingdom of Morocco. In 2014, Morocco received the UN Public Service Award for its results-based and gender-responsive public finance management. Initiated by the Ministry of Economy and Finance in 2003,[33] this effort led to the adoption of a new finance law, legally institutionalizing gender equality through all budget processes in the country. This law requires budgets to take gender equality into account at all levels, in its objectives, results and indicators. Morocco also produces an annual Gender Report,[34] as required by the finance law, which gives information on the work conducted by each sector with gender-disaggregated data. In 2015, a total of 31 ministerial departments – representing over 80% of the nation's budget – were included in the report. This reporting process makes ministries accountable to the parliament and citizens, and provides an analysis of budgets and programmes benefiting women. For example, the Gender Report 2004 highlighted that women constitute 9% of the beneficiaries of agricultural extension services, although they are responsible for 39% of the economic activity in rural areas. Because of this finding, Morocco significantly increased the budget for women's productive activities in agriculture.

In addition, the Ministry of Economy and Finance (MEF) invested in capacity building on gender budgeting for public staff and published a Gender Manual[35] and a Guideline for Budget Reform[36] that focuses on GRB. Then in 2013, the MEF established a Centre for Excellence on GRB, in partnership with UN-Women Morocco. This effort is now being implemented at the regional and local levels through a participative approach. The results have been overwhelmingly positive: in the 'Idmaj' programme for access to employment, women received 48% of the 70,000 jobs created in 2015, and women represented 73% of the beneficiaries of training programmes in 2016. We should recall, however, that women's employment rate in Morocco is still below 25%. Another notable effort is the creation of the guaranty fund 'Ilayki' dedicated to feminine enterprises: it mobilized 39 million dirhams (3,8 M €) in 2015.

Stakeholder mapping and consultation

When developing any climate change legislation, policies, programmes and/or plan, a critical first step involves the engagement of relevant stakeholders. The mapping of stakeholders helps to identify the groups, institutions, governmental bodies and NGOs that will be affected by and/or can influence climate change outcomes.[37]

This analysis aims to identify who can contribute to the discussion of a law and/or policy and how they can facilitate the development of such action. Criteria should include the stakeholder's potential contribution to each sector and topic area, and his/her knowledge of gender issues.

Consultation needs to respect certain principles to guarantee fair and equal participation of women and men in the decision-making process. Possible questions are provided below:

- Is the consultation inclusive of women and men at local and national levels across sectors and topic areas?
- Is gender balance an objective when setting up consultation meetings?
- Are processes defined that guarantee the early and effective participation of male and female stakeholders focusing especially on poor and marginalized groups in the design of the law, policy or plan?
- Have factors affecting the participation of women and men been taken into account (e.g. time constraints, women's limited mobility, childcare times, seasons of high labour demand, capacity to speak out, cultural constraints)?
- Have specific measures been taken to involve women (e.g. separate meetings for 'women-only' groups)?
- Is capacity building for women available (either related to making their voices heard or to sectoral issues)?
- Are there lessons learned from related processes to demonstrate gender dynamics by sector?

The process of engaging stakeholders requires more than one consultation in the initial phase; the objective is to build long-lasting partnerships to facilitate sustainable cross-sector cooperation throughout the life of the law, policy, programme or plan. Effective co-operation is critical to the success of any gender-equitable climate change solution.

Concluding remarks

As this book is showing, it is essential to identify climate change as a major human rights' challenge. Global legal frameworks underline the goal of gender equality and women's empowerment as an essential step in attaining sustainable and just development. This chapter has shown how over the past two decades the relevance of a gender perspective in climate change policies and actions has been recognized, and is gradually being implemented.

Women's organizations and gender experts have played an important role in integrating such a perspective at all levels and have designed specific approaches and tools, such as the ones described above. However, there is still a long road to go to make gender equality a central element of climate policies and actions at all levels. For example, there is still a gender gap in the political participation in climate change fora – including the UN climate negotiations.

Overall, the climate change and energy sector, including ecological modernization, are male-dominated and follow a masculine discourse.[38] Therefore, a more profound reform of the economy and society is needed, one that reflects basic human needs, environmental security and the 'ethics of care'. As Ingólfsdóttir (2016) underlines, this approach and moral theory is concerned with justice and rights, challenges dominant approaches and sees people as interdependent and relational. 'The ethics of care . . . demands not only equality with women, it calls for equal consideration for the experience that reveals the values, importance, and moral significance of caring.'[39]

This also has major implications for our climate change policies and practices, or as Robinson states: 'Our ability to care for each other depends fundamentally upon our ability to maintain a healthy environment.'[40]

Notes

1 World Economic Forum, *The Global Gender Gap Report 2016* (Cologny/Geneva: WEF, 2016).
2 UN Women (2016), available online at http://UNWomen.org/en/news/in-focus/women-and the SDGs/sdg-5-gender-equality; I. Dankelman (ed.), *Gender and Climate Change: An Introduction* (2010) 41–42.
3 B. Burns and J. Patouris, *United Nations Framework Convention on Climate Change (UNFCC) Decisions and Conclusions: Existing Mandates and Entry Points for Gender Equality* (2014), available online at www.wedo.org/wp-content/uploads/GE-Publication-ENG-Interactive.pdf, at 3.
4 UN Secretary General, *1992 United Nations Framework Convention on Climate Change*, 1771 UNTS 107 / [1994] ATS 2 / 31 ILM 849 (1992).
5 UNFCCC, *Report of the Conference of the Parties on Its Sixteenth Session, Held in Cancun From 29 November to 10 December 2010. Addendum. Part Two: Action Taken by the Conference of the Parties at Its Sixteenth Session*, FCCC/CP/2010/7/Add.1, 15 March 2011.
6 *Ibid.*, at para 7.
7 More in L. Aguilar, M. Granat and K. Owen, *Roots for the Future* (2015), available online at https://portals.iucn.org/library/sites/library/files/documents/2015-039.pdf, at 67 seqq.
8 A full compilation of gender mandates under the UNFCCC is available online at genderclimatetracker.org.
9 UNFCCC, *Report of the Conference of the Parties on Its Seventeenth Session, Held in Durban From 28 November to 11 December 2011. Addendum. Part Two: Action Taken by the Conference of the Parties at Its Seventeenth Session*, FCCC/CP/2011/9/Add.1 (2011).
10 Aguilar, Granat and Owen, *supra* note 7, at 223.
11 UNFCCC, *Report of the Conference of the Parties on Its Seventeenth Session, Held in Durban From 28 November to 11 December 2011*, FCCC/CP/2011/9/Add.1. Decision 2/CP.17 (15 March 2012) 47.

12 *Ibid.*, Decision 3/CP.17, at 58.
13 The extension is for a period of three years, to be reviewed at COP25 in 2019.
14 Global Gender and Climate Alliance (GGCA), available online at http://gender-climate.org.
15 UNFCCC Women and Gender Constituency, available online at www.womengenderclimate.org.
16 Article 14(2), Convention on the Elimination of All Forms of Discrimination (1979) 1249 UNTS 13.
17 UN Human Rights Council, *Report of the Office of the United Nations High Commissioner for Human Rights on the Relationship Between Climate Change and Human Rights*, A/HRC/10/61, 15 January 2009.
18 CSW Res E/CN.6/2011/L.1.
19 CSW Res E/CN.6/2014/L.4.
20 UN Women, *SDG 13, Take Urgent Action to Combat Climate Change and Its Impacts*, available online at www.unwomen.org/en/news/in-focus/women-and-the-sdgs/sdg-13-climate-action#sthash.tY0BIsan.dpuf.
21 Definition by ECOSOC Agreed Conclusion 1997/2: "The process of assessing the implications for women and men of any planned action, including legislation, policies or programmes, in all areas and at all levels. It is a strategy for making women's as well as men's concerns and experiences an integral dimension of the design, implementation, monitoring and evaluation of policies and programmes in all political, economic and societal spheres so that women and men benefit equally and inequality is not perpetrated. The ultimate goal is to achieve gender equality."
22 UNDP, *Gender-Responsive National Communications Toolkit* (2015), available online at www.undp.org/content/dam/undp/library/gender/UNDP%20Gender%20Responsive%20National%20Communications%20Toolkit.pdf?download, at 19.
23 E.g. CSW, *Challenges and Achievements in the Implementation of the Millennium Development Goals for Women and Girls, Agreed Conclusions*, E/CN.6/2014/L.7 2014 (2014), at para 37: "progress on the MDGs for women and girls has been limited owing to the lack of systematic gender mainstreaming and integration of a gender perspective in the design, implementation, monitoring and evaluation of the Goals."
24 UN Women, *Guidance Note: Gender Mainstreaming in Development Programming* (2014) 15.
25 European Institute for Gender Equality, *Gender Analysis*, available online at http://eige.europa.eu/gender-mainstreaming/tools-and-methods/gender-analysis.
26 Based on the following frameworks: Harvard Analytical Framework (designed to demonstrate that there is an economic case for allocation resources to m/w); Moser Framework (gender planning as type of planning in its own rights – transformative); Gender Analysis Matrix (community-based technique to identify difference between w/m; aims to determine different impact development interventions have on w and m). For more information on various approaches for gender analysis see C. March, I. Smyth and M. Mkhopadhyay, *A Guide to Gender-Analysis Frameworks* (1999).
27 *NAMA Facility*, available online at www.nama-facility.org/.
28 For project description see: WCF International, *Building Local Capacity for Domestic Solar Heating, Hot Water and Insulation for Rural and Remote Areas in the EEC Region*, available online at www.wecf.eu/english/about-wecf/issues-projects/projects/localcapacity-eecregion.php.
29 WECF International, *WECF and Partners Developed Training Module on Construction, Use, Monitoring and Maintenance of Solar Water Heaters*, available online at www.wecf.eu/english/articles/2016/03/solarwaterheater-modules.php.
30 UN Statistics Division, *Gender Statistics Manual* (2015), available online at https://unstats.un.org/unsd/genderstatmanual/What-are-gender-stats.ashx.
31 *Ibid.*
32 UN Women, *supra* note 24, at 28.
33 *Kingdom of Morocco – Ministry of Economy and Finance*, available online at www.finances.gov.ma.
34 See for example Kingdom of Morocco, Ministry of Economy and Finance, *Rapport sur le Budget Genre* (2014), available online at www.ogfp.ma/uploads/files/Rapport%20budget%20genre%202014.pdf.
35 Kingdom of Morocco – Ministry of Economy and Finance, and UNIFEM, *Intégration de la dimension genre dans l'élaboration et la planification du budget: Manuel de formation* (2006), available online at www.finances.gov.ma/depf/SitePages/dpeg_action/genre/MANUEL%20BSG.pdf.
36 Kingdom of Morroco and UNIFEM, *Guide de la réforme budgétaire* (2005).
37 The following stakeholders could be relevant: government agencies (national to local if relevant)/ministry responsible for gender equality; bilateral and multilateral development partners; national and local NGOs and initiative groups, in particular those working on gender equality and women's empowerment (special attention has to be paid to those representing poor and/or marginalized groups), but also others interested and involved; business and private service providers; universities/research institutes

concerned with gender studies; religious leaders; stakeholders who are working on men's and women's issues related to the specific area of the planned programme/activity, e.g. vulnerability, adaptation, mitigation and emissions reduction; and women and men representing specific sector interests, e.g. agriculture, fisheries and coastal resources use, energy, forestry, manufacturing, service, economic development and business.

38 S. MacGregor, 'Gender and Climate Change: From Impacts to Discourses', *Journal of the Indian Ocean Region* 6(2) (2010) 223.

39 A. H. Ingólfsdóttir, 'Climate Change and Security in the Artic', 338 *Acta Universitatis Lapponiensis* (2016) 55–56.

40 F. Robinson, *The Ethics of Care: A Feminist Approach to Human Security* (2011).

17

Energy justice

The intersection of human rights and climate justice

Allison Silverman

Introduction: there are real problems with the global energy system

Almost two decades into the twenty-first century and we still live in an age of severe energy poverty and pervasive energy injustice. Although the established global energy system has catalyzed economic wealth and social development for some, it has significantly contributed to the climate crisis and has adversely affected the most vulnerable, who have not benefited from the energy boom. Considering that the energy sector is the largest source of greenhouse gas emissions globally,[1] "climate change makes the most persuasive case for a justice framing."[2] Thus, global energy security and energy access are some of the central environmental justice issues of our times, with implications for numerous fundamental human rights.

There are significant inequalities associated with the lack of access to electricity and how and where energy is produced, distributed and used. There are also profound environmental, social and economic impacts embedded in the global energy system. Environmentally, the processes used in generating and distributing energy produce water contamination, solid waste generation and the extraction of limited natural resources. Socially, the global energy system causes widespread social inequity, placing the environmental health hazards and associated burdens on those who do not stand to profit but rather suffer the consequences of energy production. Economically, "in modern times, no country has substantially reduced poverty without greatly increasing the use of energy."[3] The power and resources involved in the global energy system are thus consolidated in the hands of only a few countries,[4] some of which commit human rights abuses in the process.

Furthermore, there is an intergenerational equity issue at play. The global energy system's reliance on fossil fuels and the inequitable distribution of resources are directly tied to adverse impacts of climate change, causing harm to current and future generations. Although the UN Framework Convention on Climate Change (UNFCCC) provides that Parties' actions should be consistent with the principles of equity and common but differentiated responsibilities "for the benefit of present and future generations of humankind,"[5] the global energy system is not responsive to these principles. While many negative consequences of climate change are already manifesting themselves – food insecurity, humanitarian disasters, habitat destruction and biodiversity loss, to mention a few – others, predicted to be even more severe, will be experienced by future generations that have not contributed to the problem. Hence, energy justice is essential both for current and future generations; it is a key component of a climate justice agenda.

Rather than continuing to espouse a harmful approach to generating and distributing energy that leads to human rights abuses as well as social instability, any efforts to reform the energy system should involve a rights-based approach. Within international legal and policy frameworks, there is a strong foundation for integrating human rights in efforts to combat climate change and to support sustainable development. However, the global energy system has not taken rights or justice into account, and as a result, it may create more problems than it solves. In this chapter, we discuss the issues associated with energy access – specifically how, where and what type of energy is produced and distributed – highlighting the need to advance energy justice as part of the climate justice movement. In order to upend the current energy system and manage the climate crisis that threatens the world as we know it, a rights-based approach to energy production and distribution grounded in principles of equity, justice and fairness[6] is essential.

Inadequate, inequitable access to energy is the norm

Energy is fundamental to modern life,[7] yet the current global energy system does not provide energy for all. This inequitable distribution of energy resources leaves 1.2 million people worldwide without basic access to electricity,[8] 80 percent of whom live in rural areas in Asia and Africa.[9] Lack of energy access causes real harm, as it impacts people's health and the environment, as well as educational and economic development opportunities. Without electricity, many people rely on biomass as their primary fuel source for cooking and on fossil fuels for indoor lighting. These practices result in burn injuries, house fires and exposure to smoke and serious air pollution, which then contribute to respiratory illnesses and 3.5 million premature deaths every year.[10] Other health issues stemming from the inequitable access to energy include the reduced ability to secure medical treatment at night and the inability to store vaccines, other drugs and blood needed to fight preventable diseases that require refrigeration. By contrast, energy access provides clean water and sanitation. Water can be pumped and purified for drinking, irrigation and agricultural productivity,[11] as well as used to operate water treatment and sanitation facilities.[12]

Lack of energy access also limits educational and other economic development[13] opportunities. First, without adequate classroom lighting and fans, the learning environment for students and teachers is severely impacted. Second, the number of hours available each day for studying both in and out of the classroom is cut short when the sun sets. Third, without electricity, students are unable to access educational technologies, such as radios, televisions, computers, the internet and projectors, as well as other tools that have been demonstrated to improve learning and thus job prospects.[14] If schools had electricity, then the students could gain from an enhanced educational experience yielding broader social and economic co-benefits, such as better-quality sanitation, health, women's empowerment and resilience.[15] In sum, access to energy is a key component of promoting economic advancement.

With respect to economic development, the benefits of energy access extend far beyond increased educational opportunities. Numerous studies demonstrate that access to modern energy enables the global poor to engage in new income-generating activities[16] that lead to improved living conditions.[17] Access to energy access helps to facilitate increased communication and connectivity; it also prompts the development of small, often informal enterprises and leads to increased worker productivity.[18] The financial benefits of switching to cleaner, more efficient and cost-effective fuel sources include cost savings, given that candles, charcoal, kerosene and batteries are more expensive than light bulbs. Even the least efficient light bulb lasts longer and thus minimizes the time and effort involved in frequently purchasing fuel-based lighting.[19] Therefore, the inequity is striking when millions of households without electricity pay 60 to 80 times more for energy-related products than do those living in two of the wealthiest cities in the world, namely New York and London.[20]

Not surprisingly, lack of energy access has some of the most significant adverse impacts on women in developing countries. Women are primarily responsible for cooking and cleaning, among other household tasks, and therefore suffer disproportionate health impacts associated with poor energy choices, as previously discussed. Given these responsibilities, most of a woman's "free" time for school, studying and developing her own business is extremely limited because she spends her day searching for water, firewood and other fuel sources.[21] Also, women entrepreneurs confront discrimination and even greater barriers in accessing electricity,[22] which further limit their economic opportunities and ability to succeed independently.[23] The role of women in the energy sector is often overlooked; for this reason, it is critical that efforts to expand access to modern, clean energy must integrate women's rights as well. In fact, studies demonstrate that electrification greatly enhances women's lives in the form of improved health and increased leisure time, among other benefits.[24]

In sum, the roughly one quarter of the global population that still lacks access to basic, modern energy services[25] are those most marginalized – politically and socio-economically – and most vulnerable to climate change. This climate injustice must be remedied by committing to a clean energy future that is distributed equitably to all. The transformative effects that even small amounts of clean energy can have on people's lives have been documented throughout the world and by environmental, human rights and development organizations alike. Yet somehow, the numbers of those lacking modern energy services are still staggering. Another strategy is needed to support increased energy access; taking a right-based approach, and even framing energy access as a human right, would offer an effective way forward.

Energy access as a human right

Access to energy is not explicitly recognized as a fundamental human right; however, given that energy access is essential to fulfilling basic human needs and serves as a key pillar in achieving internationally agreed upon climate and sustainability goals, the need to provide energy access to all has never been clearer. As discussed, energy access supports health, education and other means of economic development and is considered a "quintessential intermediate good."[26] In addition to the importance of energy access on its own, it is critical for enabling and safeguarding other human rights, including the rights to food, water, shelter and a healthy environment,[27] among other rights that enable physical security. Lack of energy access also implicates the right to life, when it produces unhealthy conditions that create greater risk of mortality. All of these substantive rights are threatened by climate change, yet could be addressed through investments in renewable energy and the promotion of energy justice.

It could also be argued that a number of procedural rights, such as access to information, the right to participation, due process and the right to free prior and informed consent (FPIC) are so intertwined with energy access that energy access should be its own right too. Especially as it relates to the siting of energy projects, key stakeholders such as communities located near a proposed energy project often do not receive adequate or timely information about the project's development, leaving them unable to, or even excluded from, participating in key decisions that directly impact their lives, livelihoods and other fundamental rights. Upholding procedural access rights is tied to the idea of "energy sovereignty," enabling communities to choose their energy sources.[28] Thus, energy access is implicitly conferred by various international treaties and conventions.

Framing energy access as a right in the context of promoting energy justice provides individuals and communities a more sustainable, promising future. Without recognizing energy access as a right, the production, consumption and distribution of energy will continue to lead to great harm and suffering, including human rights violations. Under the current global energy system, companies and governments get away with planning and developing energy projects without providing those impacted with adequate information or allowing for their participation in relevant discussions or decisions. As a result of denying basic procedural rights, these vulnerable communities suffer from negative environmental, social and health impacts of the project without benefiting from the energy generated. Compounding these preventable problems, many communities are forced to relocate for an energy project without anywhere to go and without the means or the contacts to successfully start a new life elsewhere. Therefore, establishing an explicit right to energy access would not only encourage better distribution of much-needed modern energy resources, but also result in other social, economic and environmental benefits that facilitate the protection of other fundamental rights.

Acknowledging a fundamental right to energy access would help to ensure that everyone has access to clean, reliable, affordable energy as a baseline. Furthermore, it would impose additional obligations on States and companies, and empower civil society and community groups fighting for rights to include energy access as an important tool in their efforts to achieve climate justice.

Discussions focused on whether energy access should be a human right in and of itself have been ongoing. When the United Nations established their Millennium Development Goals (MDG), intended to clarify the basic human rights to which all people are entitled and to reduce extreme poverty using targets with a deadline of 2015, energy was not a major focus. However, energy access has since been recognized as key to sustainable development. In the post-2015 Agenda for Sustainable Development, there are explicit linkages between energy services and each of the MDGs, including poverty, gender equality and health.[29] Energy access is now its own stated Sustainable Development Goal (SDG), which seeks to "ensure access to affordable, reliable, sustainable, and modern energy for all."[30] The sustainable development framework also provides a clear target by when this goal must be achieved: "By 2030, ensure universal access to affordable, reliable and modern energy services."[31]

In addition, the World Bank and the United Nations have launched the Sustainable Energy for All campaign (SEforAll), which places energy access at the heart of the SDGs as well as the Paris Climate Agreement. SEforAll calls on governments, businesses and civil society to take action to ensure universal access to sustainable energy as a means to eradicate poverty and address climate change. This integration of energy access with efforts to fight poverty and climate change demands a rights-based approach to climate change and energy justice.

Unsustainable global energy systems: energy at what costs?

At current consumption levels, it is evident that global energy systems are having devastating impacts on the environment, human health and human rights (both individual and collective). Indigenous communities and other marginalized populations are most impacted by these dirty energy investments, despite the fact that they are least responsible for the climate crisis and are on the frontlines of fights to protect clean air, water, land and natural resources. As the demand for energy grows and becomes more essential for securing rights to life, liberty and property, among others, issues of how, where and what types of energy are produced and distributed become front and center. Considering just solutions to these energy issues in the context of climate change is critical.

First, all forms of energy are not equal. At present, the global energy system is dependent on fossil fuels. This reliance is greatly contributing to the climate crisis, disrupting ecological systems worldwide and accelerating socio-political disruption.[32] There is broad consensus that the world must shift to low- or no-carbon energy sources to mitigate the direct and indirect impacts associated with climate change.

Where energy is produced impacts certain people and communities more than others. There are many factors that go into the siting of both fossil fuel–based and renewable energy projects, including the location of the actual energy source, the potential impacts to human health and the environment, among others. Yet somehow, the majority of energy projects are concentrated in the most socio-economically impoverished and vulnerable communities, leading to forced relocation, livelihood deprivation or land grabs. Moreover, the communities that are most burdened do not often benefit from these projects – they do not gain access to the energy produced, enjoy employment opportunities or receive other economic rewards.

In addition, *how energy is produced* implicates civil, political, economic, social and cultural rights, as well as issues of property, economic development, human health, safety and natural resources.[33] Not surprisingly, it is far too common for energy projects to be developed without any community engagement, including the dissemination of accessible and timely information as well as the ability to participate in the decision-making process. Human rights violations range from failure to consult or seek consent to physical and bodily harms (e.g. rape, torture, murder) incurred simply for opposing an energy project. Thus, where and how energy is generated and distributed results in rights violations, often without recourse or remedy.

The current global energy system must be changed to reduce greenhouse gases as well as to promote more equitable energy access and limit human rights abuses. Fossil fuel-based energy is being phased out in favor of renewable energy sources, which are critical for reducing greenhouse gas emissions and thwarting the speed at which the climate is changing. However, while promoting renewable energy is critical for a sustainable future, the way in which renewable energy projects are sited and constructed needs to be managed in a way that enables energy justice. Even renewable forms of energy can cause social and environmental harms.[34]

In building a more sustainable and just energy future, it is important to limit the negative impacts involved in the production, consumption and distribution of energy to meet the growing needs. Recognizing that there are inextricable links between today's energy systems and the climate crisis is a first step in promoting energy justice. Next, the climate justice movement must press for clean energy access for all to support the most vulnerable and marginalized communities who are already disproportionately burdened with the negative impacts from climate change, yet who have contributed the least to the problem.

Climate justice requires energy justice

At a time when the world is confronted with the overwhelming challenges of promoting socio-economic development, eradicating poverty and limiting environmental degradation, it may seem at odds with efforts to expand energy access to all. However, linking these challenges creates an opportunity to encourage true sustainable development and climate justice. As UN Secretary-General Ban Ki-moon has expressed, "Energy is the golden thread that connects economic growth, increased social equity, and an environment that allows the world to thrive."[35]

Nevertheless, advancing clean energy access equitably remains a complex challenge. Despite the recognition that providing clean energy services enhances economic development, public health, education, water and sanitation, gender equality and a range of other benefits,[36] it is still expected that one billion people will still live in the dark by 2030.[37] Many policies, initiatives and actions are needed to secure energy justice and advance climate justice. Promoting a right to energy access will be important to enable each strategic solution. Transforming the global energy supply from one based on fossil fuels to one that is grounded in human rights and in clean forms of energy is essential.

Re-envisioning the global energy system in response to the current climate reality will require providing everyone access to clean energy that has been produced, distributed and consumed in a way that respects human rights and promotes sustainable development. The results of an energy-just world would include the promotion of welfare, freedom and happiness. It would guarantee due process both for the energy producers as well as the consumers, ensuring that all stakeholders have access to information and participation in energy decision-making.[38]

In his Papal Encyclical, Pope Francis envisions an energy-just world as one that "equitably shares the benefits and burdens involved in the production and consumption of energy services, as well as one that is fair in how it treats people and communities in energy decision-making."[39] The Pope is not alone in this vision. Around the world, individuals, communities and civil society organizations have been promoting this future and resisting a range of carbon-intensive energy schemes. Thanks to their hard-fought efforts, there is an improved understanding of how sustainable, reliable and affordable energy for all supports socio-economic development and climate justice.

As the planet reaches new tipping points, the imperative for climate justice has never been greater. A rights-based approach that is driven by climate and energy justice is essential to protect both current and future generations, as well as the planet.

Notes

1 Intergovernmental Panel on Climate Change (IPCC), 'Summary for Policymakers', in O. Edenhofer *et al.* (eds.), *Climate Change 2014: Mitigation of Climate Change. Contribution of Working Group III to the Fifth Assessment Report of the Intergovernmental Panel on Climate Change* (2014) 1, available online at www.ipcc.ch/pdf/assessment-report/ar5/wg3/ipcc_wg3_ar5_summary-for-policymakers.pdf.

2 G. Walker, *Environmental Justice: Concepts, Evidence, and Politics* (2012) 179.

3 J. Saghir, 'Energy and Poverty: Myths, Links, and Policy Issues', 4 *Energy Working Notes: Energy and Mining Sector Board* (2005) 5.

4 The top countries who produce the most energy are the U.S., Saudi Arabia and Russia. China, Canada and Iraq follow in their energy production; and China, the U.S., Russia and Japan are among the top countries in their energy consumption. See The U.S. Energy Information Administration's independent statistics and analysis here: www.eia.gov/beta/international/rankings.

5 Article. 3(1), UN General Assembly, *United Nations Framework Convention on Climate Change: Resolution / Adopted by the General Assembly*, 20 January 1994, A/RES/48/189.

6 For an excellent review of other theories of justice and the problems and principles around energy justice, see B. Savacool and M. Dworkin, *Global Energy Justice: Problems, Principles, and Practices* (2014).

7 Energy Access Targets Working Group, *More Than a Lightbulb: Five Recommendations to Make Modern Energy Access Meaningful for People and Prosperity* (April 2016), available online at www.cgdev.org/app/reader/3124016.

8 International Energy Agency, *World Energy Outlook Electricity Access Database* (2016) available at online at www.worldenergyoutlook.org/media/weowebsite/2015/WEO2016Electricity.xlsx.

9 Organisation for Economic Co-Operation and Development/International Energy Agency, *Energy for All: Financing Access for the Poor* (2011), available online at www.worldenergyoutlook.org/media/weowebsite/energydevelopment/weo2011_energy_for_all.pdf.

10 S. S. Lim *et al.*, 'A Comparative Risk Assessment of Burden of Disease and Injury Attributable to 67 Risk Factors and Risk Factor Clusters in 21 Regions, 1990–2010: A Systematic Analysis for the Global

Burden of Disease Study 2010', 380 *The Lancet* (2010) 2224, available online at www.thelancet.com/journals/lancet/article/PIIS0140-6736(12)61766-8/fulltext.

11 D. F. Barnes and H. P. Binswanger, 'Impact of Rural Electrification and Infrastructure on Agricultural Changes', 21 *Economic and Political Weekly* (1986) 26, available online at www.jstor.org/stable/4375175?seq=1#page_scan_tab_contents.

12 United Nations, *UN Information Brief: Securing Access to Water and Energy* (2014), available online at www.un.org/waterforlifedecade/pdf/01_2014_securing_access_eng.pdf.

13 World Bank Group, *Enterprise Surveys: Infrastructure*, available online at www.enterprisesurveys.org/data/exploretopics/infrastructure#sub-saharan-africa.

14 M. Millinger, T. Mårlind and E. O. Ahlgren, 'Evaluation of Indian Rural Solar Electrification: A Case Study in Chhattisgarh', 16 *Energy for Sustainable Development* (2012) 486, available online at http://dx.doi.org/10.1016/j.esd.2012.08.005; The World Bank, *Rural Electrification and Development in the Philippines: Measuring the Social and Economic Benefits* (2002), available online at http://documents.worldbank.org/curated/en/2002/05/2087958/rural-electrificationdevelopment-philippines-measuring-social-economic-benefits.

15 United Nations Department of Economic and Social Affairs, *Electricity and Education: The Benefits, Barriers, and Recommendations for Achieving the Electrification of Primary and Secondary Schools* (2014).

16 B. Attigah and L. Mayer-Tasch, *The Impact of Electricity Access on Economic Development: A Literature Review* (2013), available online at www.produse.org/imglib/downloads/PRODUSE_study/PRODUSE%20Study_Literature%20Review.pdf.

17 UN Development Program/World Health Organization, *The Energy Access Situation in Developing Countries: A Review Focusing on the Least Development Countries and Sub-Sahara Africa* (2009), available online at http://content-ext.undp.org/aplaws_assets/2205620/2205620.pdf.

18 Charles Kirubi *et al.*, 'Community-Based Electric Micro-Grids Can Contribute to Rural Development: Evidence From Kenya', 37(7) *World Development* (2009) 1208.

19 T. Walters *et al.*, *Policies to Spur Energy Access: Volume 1* (2015), available online at www.nrel.gov/docs/fy15osti/64460-1.pdf.

20 International Renewable Energy Agency, *Solar PV in Africa: Costs and Markets* (2016), available online at www.irena.org/DocumentDownloads/Publications/IRENA_Solar_PV_Costs_Africa_2016.pdf.

21 UN Development Program, *Gender and Energy – Gender and Climate Change: Asia and the Pacific, Policy Brief 4*, available online at www.undp.org/content/dam/undp/library/gender/Gender%20and%20Environment/PB4-AP-Gender-and-Energy.pdf.

22 There are structural barriers and different challenges that women face over men in the business environment, which negatively impact women entrepreneurs' opportunities in benefiting from increased energy access. Recent research has found that efforts to encourage greater energy access must ensure equal benefits for both women and men entrepreneurs. See Y. Glemarec, F. Bayat-Renoux and O. Waissbein, *Removing Barriers to Women Entrepreneurs' Engagement in Decentralized Sustainable Energy Solutions for the Poor* (2016), available online at www.wocan.org/sites/default/files/Women%20entrepreneurts-energy-04-00136.pdf.

23 P. Alstone *et al.*, *Expanding Women's Role in Africa's Modern Off-Grid Lighting Market*, available online at www.esmap.org/sites/esmap.org/files/gender_lighting_highres_LOW%20RES.pdf.

24 D. Barnes, The Impact of Electrification on Women's Lives in Rural India, 5 *ENERGIA News* (2004).

25 International Energy Agency, *Energy Access Database* (2016), available online at www.worldenergyoutlook.org/resources/energydevelopment/energyaccessdatabase/.

26 Attigah and Mayer-Tasch, *supra* note 16.

27 The right to a clean and healthy environment has been integrated into a number of countries' national constitutions, and its recognition as a human right at the international level continues to grow.

28 See *Solar Electric Light Fund's Energy Is a Human Right™ Campaign*, which provides a voice to the billions of people around the world living without access to modern energy services, available online at http://self.org/energy-is-a-human-right/.

29 V. Modi *et al.*, *Energy Services for the Millennium Development Goals* (2005), available online at http://unmillenniumproject.org/documents/MP_Energy_Low_Res.pdf.

30 UN, *UN Sustainable Development Goal* 7, available online at www.un.org/sustainabledevelopment/energy/.

31 *Ibid.*, Sustainable Development Target 7.1.

32 The impacts of our reliance on fossil fuels and its destruction are clearly illustrated in the most recent and fifth IPCC assessment report: Core Writing Team, R. K. Pachauri and L. A. Meyer (eds.), *Climate*

Change 2014: Synthesis Report. Contribution of Working Groups I, II and III to the Fifth Assessment Report of the Intergovernmental Panel on Climate Change 2014, 151, available online at www.ipcc.ch/report/ar5/syr/.

33 C. Ballard and G. Banks, 'Resource Wars; the Anthropology of Mining', 32 *Annual Review of Anthropology* (2003) 287.

34 There are numerous examples of poor planning and improper implementation of renewable energy projects resulting in human rights abuses. As a start, review the Business and Human Rights Resource Centre's case studies available online at https://business-humanrights.org/en/case-studies-renewable-energy.

35 UN, *Sustainable Energy 'Golden Thread' Connecting Economic Growth, Increased Social Equity, Secretary-General Tells Ministerial Meeting* (May 2014), available online at www.un.org/press/en/2014/sgsm15839.doc.htm.

36 T. Walters *et al.*, 'Policies to Spur Energy Access: Volume 1: Engaging the Private Sector in Expanding Access to Electricity', *National Renewable Energy Lab* (2015) 2–4.

37 International Energy Agency, 'Energy Access Projections' (2014), available online at www.worldenergyoutlook.org/resources/energydevelopment/energyaccessprojections/.

38 B. Savacool and M. Dworkin, *Global Energy Justice* (2014) 13.

39 *Ibid.*, at 5.

18

Overlooked and undermined

Child rights and climate change

Joni Pegram

The call to tackle climate change for the sake of children and future generations has by now become a well-worn phrase – a declaration commonly made by heads of state, but too rarely translated in the policies and laws over which they preside, despite near universal ratification of the UN Convention on the Rights of the Child (CRC).[1] While the UN Human Rights Council has emphasised that children are among the most vulnerable to climate change,[2] the interface between child rights and climate change remains one of the least understood and least represented in the work of global, regional and national actors and advocates concerned with child rights. Conversely, or perhaps reflecting this, children's rights have been largely overlooked in international negotiations under the UN Framework Convention on Climate Change (UNFCCC) and in the multitude of regional and national policies and processes that derive from these more broadly.

This represents a fundamental injustice on multiple fronts. Firstly, children make up one of the largest groups affected by climate change, as many of the countries that are most vulnerable to its impacts – due to their location and relatively weak adaptive capacity – are also those in which children account for the greatest share of the total population. Projected demographic trends are expected to further accentuate this reality. Recent UNICEF data suggests that more than half a billion children currently live in areas with extremely high risk of flooding, 115 million are at high or extremely high risk from tropical cyclones, and almost 160 million are exposed to high or extremely high drought severity.[3]

Secondly, children and future generations will bear the heaviest burden of our failure to decisively act on climate change today, despite being least responsible for its causes. More intense and frequent weather-related events, and slow-onset changes such as rising sea levels, changing rain fall patterns, salinization, desertification and dwindling resources, are already undermining a whole raft of children's rights, but as these impacts escalate over time, this harm will become more profound and widespread.

Finally, and perhaps most significantly, children experience distinct and more acute risks resulting from climate-induced changes than do adults, due to their unique stage of physiological and mental development. These risks expose them to potentially lifelong harm and are likely to fall hardest on the most disadvantaged children.

This chapter outlines some of the multiple ways in which climate change disproportionately affects children and their rights. It reviews the current policy landscape and the opportunities presented by the 2030 Sustainable Development Agenda in particular to facilitate greater coherence between States' obligations under the CRC on the one hand, and their efforts to tackle climate change on the other. It concludes with some concrete recommendations on the steps required to ensure that the best interests of the child firmly underpin decision-making in this area.

Climate change and the rights of the child

The UN Committee on the Rights of the Child has recognized climate change as "one of the biggest threats to children's health",[4] as well as its adverse impacts on a range of other rights, including the rights to education, adequate housing, safe drinking water and sanitation.[5] Indeed, children's vulnerability to climate change impacts poses an immediate and pervasive threat to the enjoyment of nearly all of the rights enshrined in the CRC, and notably the rights to life, survival and development.

Health

Rising temperatures, drought and flooding create breeding grounds for vector-borne diseases such as malaria, dengue and zika, which already disproportionately affect children. In 2014, children under five accounted for nearly 80 per cent of all deaths from malaria.[6] By 2030, the World Health Organisation (WHO) projects that the impacts of climate change on the incidence of this disease could lead to an additional 60,000 deaths per year among children under the age of 15.[7] Similarly, drought, excessive heat, flooding and variable rainfall patterns affect children's access to clean water, increasing the incidence of diarrhoeal disease, which represents another major cause of mortality for children, responsible for the deaths of approximately 760,000 children aged under five each year.[8]

The disproportionate impacts of climate change on children are placed into particularly sharp relief in the context of increasing food insecurity. Malnutrition is responsible for almost half of deaths among children under the age of five globally,[9] while undernutrition in the first two years of life can lead to irreversible stunting, affecting physical and cognitive development with consequences for a child's health, educational outcomes and livelihoods in later life.[10] The implications are of course far-reaching for societies more widely, a scenario acknowledged by the president of the World Bank in observing that inequality and poverty are currently being "baked into the brains" of 25 per cent of children before the age of five due to stunting.[11] Climate change is projected to increase severe child stunting by 23 per cent in central sub-Saharan Africa and by 62 per cent in South Asia by 2050, reversing decades of development gains.[12]

The root causes of climate change also undermine children's right to health. As temperatures rise and countries continue to industrialise, pollutants resulting from vehicle emissions and fossil fuel combustion, smog and dust mix with stagnant air, creating toxic conditions for children. Air pollution now represents a leading cause of child death, responsible for over half a billion mortalities in children under the age of five.[13] Health complications include respiratory conditions such as pneumonia, bronchitis and asthma, harm to children's physical and cognitive development, loss of education, and potentially life-long consequences for their health.[14] Currently, two billion children live in areas that exceed international air quality guidelines, while approximately 300 million are exposed to outdoor air pollution that exceeds WHO guidelines by at least six times.[15]

Protection

In the context of severe weather events, children may suffer from physical and psychological trauma, both because of injuries incurred during the event and in the aftermath, when they may be displaced from their homes and separated from their family, or forced into circumstances where they face heightened risk of abuse, child labour, exploitation and violence.[16] In Bangladesh, for example, Human Rights Watch has documented how extreme poverty caused by weather-related disasters has led to arranged child marriages.[17] In Assam, India, protracted internal displacement resulting from flooding and land erosion, linked to unprecedented glacial melt from the Himalayas, has helped to create conditions for child trafficking to thrive.[18] Furthermore, there is increasing consensus that the role of climate change in compounding other complex political and socio-economic stressors can contribute to instability and the outbreak of conflict, with catastrophic consequences for children.[19]

Education

Globally, approximately 120 million children are not enrolled in primary or secondary school, and 140 million adolescents do not attend upper secondary school.[20] Children are prevented from attending school when family livelihoods and resources are disrupted by the impacts of climate change. Families may no longer be able to afford school fees and children may be required to take on an increasing burden of domestic tasks or to engage in labour. During climate-related disasters, school infrastructure is frequently destroyed, schools may be taken over as shelters for communities, or children may be displaced to areas that are too far away to attend. The impacts of climate change on physical and psychological health also undercut children's education, for example through loss of school days due to diarrhoea and other diseases,[21] or impacts on their concentration and performance due to undernutrition or psychosocial trauma. Research suggests that three years after Hurricane Katrina in the United States, more than one third of children displaced or seriously affected by the hurricane were one year or more behind in school – double the pre-storm rate,[22] while negative impacts on attendance, suspension, expulsion and drop-out rates were also observed.[23,24]

Loss of education also impairs children's access to the very knowledge and skills they require to increase their own and their communities' resilience to climate-related impacts, and to adopt and promote more sustainable low-carbon lifestyles, including through their meaningful participation in climate-related decision-making, advocacy and action. Indeed, the CRC is one of the very few universal human rights instruments that explicitly requires States to take steps to protect the environment, including through education. Article 29(1) states, "State Parties agree that the education of the child shall be directed to: . . . (e) The development of respect for the natural environment."[25] In this way, the Convention acknowledges that learning to protect the environment is intrinsically linked to the realisation of children's rights more broadly.

Participation and access to justice and effective remedy

Upholding children's right to climate change and environmental education and to participation in decision-making strengthens the impact of both adaptation and mitigation interventions, as well as the quality and scope of data.[26] Fulfilling these rights also provides important tools for children's access to effective and timely remedy for climate-related harms. This extends to seeking redress for child rights violations that may occur in the context of poorly conceived climate action. Incorporating strong child and human rights safeguards in development projects, such as the construction of renewable energy infrastructure or changes in land use, can reduce the risk of tragedies, such as the murder of two indigenous children following their community's protests against the construction of the Santa Rita hydrodam in Guatemala.[27]

Increasingly, there are promising signs that strategic climate change litigation can also be effective in the context of harm that is expected to occur in the future, as a powerful means for demanding accountability and securing more ambitious mitigation action, particularly as climate attribution models evolve. Cases such as the landmark lawsuit against the U.S. government by 21 youth plaintiffs (*Juliana v United States*) claims that the federal government is violating these young people's constitutional rights to life, liberty and property, and failing to protect essential public trust resources. In Pakistan, 7-year-old Rabab Ali is suing the government for violating her rights, and the rights of her generation, to a healthy life.[28] Such cases, regardless of their outcome, represent an influential advocacy tool for bringing pressure to bear on both State and non-State actors, as well as raising public awareness of children's claims.

Not all children are affected equally

Climate change exacerbates inequality and falls hardest on those already suffering from poverty, discrimination and marginalisation. Families afflicted by poverty have fewer resources available to cope with the impacts of climate change and less access to essential services, such as water and sanitation, than their wealthier counterparts have. Girls must walk farther to find safe water in the context of drought and desertification, forcing them to miss out on education, play and leisure, and exposing them to increased risk of sexual violence on their journeys.[29] Indigenous children's close relationship to the environment and its resources means that they are particularly vulnerable to the impacts of climate change on traditional species and the land on which they depend. Environmental degradation and climate-induced displacement and migration carries profound implications for these children's specific rights to learn about and enjoy their culture, language and beliefs, and to preserve their collective identity.[30] Children with disabilities also face specific and heightened challenges in the context of climate change, including higher exposure to climate impacts, lower adaptive capacity, and a lack of access to information and adequate and inclusive social protection policies.

Falling through the cracks

In addition to the immediate redress that this situation demands from the perspective of climate justice and intergenerational equity, the practical implications of children's near total omission from climate change policies and action are equally profound. These include a glaring lack of information and disaggregated data on children most at risk from the impacts of climate change, and an absence of mechanisms to support children's full and meaningful participation in climate-related discussions and initiatives, hampering the formulation of effective and child-sensitive responses to one of the gravest dangers they face. Similarly, policies that address children's rights in many climate-relevant sectors – including disaster risk reduction, water, sanitation and hygiene, health and education – are often informed superficially at best by climate and disaster risk assessments, undermining the long-term resilience of these interventions.

The most recent round of communications submitted by Parties to the UNFCCC clearly illustrates this gap. Just 27 of the Nationally Determined Contributions put forward by countries mention children explicitly – less than one in six – rising to 47 when references to 'youth' and 'young people' are considered as well. The vast majority of these references are superficial and tend to portray children as passive victims, rather than as rights holders.[31] Amplifying the voices of children will be essential to filling this gap. A recent UNICEF poll conducted among more than 5,000 children across over 60 countries found that 77 per cent considered climate change to be one of the most pressing issues facing young people today, while 98 per cent thought that governments needed to tackle this through urgent action.[32]

Prospects for asserting child rights in an evolving policy context

As the Committee on the Rights of the Child has stated, a child rights approach to climate change is overdue,[33] and there is an urgent need for the best interests of the child to be systematically applied in shaping local, national and international responses to this growing crisis. In particular, evolving national laws and policies flowing from the adoption of key international frameworks in 2015 under the 2030 Sustainable Development Agenda provide a significant opportunity for States to finally connect the dots between their CRC commitments and climate action. Notably, the Paris Agreement recognises that children's rights and intergenerational equity should guide States' action to address climate change.[34] The Sustainable Development Goals (SDGs) and Sendai Framework for Disaster Risk Reduction also contain important provisions for upholding child rights in the context of climate change.[35]

In parallel, important processes are underway to further clarify the child rights obligations that apply in the context of climate change, and to provide guidance on how these can be promoted and implemented by duty bearers. Through its General Comment on the Right to Health, and various Concluding Observations – principally to developing countries – the Committee on the Rights of the Child has played a progressive role in beginning to elucidate these norms, including through emphasis on protecting children's rights in national climate and disaster risk management policies, ensuring that their views are taken into account, and increasing their awareness and preparedness through education.[36] And in a new development, the Committee issued its first recommendation to a developed country on the issue of climate change during its 2016 periodic review of the United Kingdom, citing concern about the impact of air pollution on children's health in the UK, and its contribution to climate change affecting various rights of the child in the UK and in other countries.[37] The Committee called on the UK to "place children's rights at the centre of national and international climate change adaptation and mitigation strategies".[38]

Further guidance is expected following the Committee's 2016 Day of General Discussion on Child Rights and the Environment, while the UN Human Rights Council resolution on child rights and climate change adopted in 2017, and a forthcoming report on child rights from the UN Special Rapporteur on Human Rights and the Environment in 2018, will further consolidate this body of work.

Possible actions

- Fundamentally, respecting children's rights in the context of climate change will entail urgent action to limit temperature rises through a rapid and just transition to renewable energy, helping to reduce both air pollution and greenhouse gases. Significant investment in resilient healthcare facilities, schools and water and sanitation systems will also be required, underpinned by a child-sensitive analysis of climate and disaster risk.
- Increasing children's meaningful participation in decision-making and their access to judicial processes, including through investment in climate change education, is also essential.
- These child-targeted measures should be informed by the collection of robust and disaggregated data and child-centred research methods, which capture their unique perspectives and experiences.

- Consideration of children's rights should be mainstreamed in countries' Nationally Determined Contributions, National Adaptation Plans, national and local Disaster Risk Reduction strategies and SDG implementation plans. In addition, systematic reporting on action taken to safeguard these rights – for example in national communications to the UNFCCC, and in the context of the Global Stocktake – would facilitate the exchange of best practice and enhance accountability.
- In parallel, climate change should be incorporated in States' periodic reporting to the Committee on the Rights of the Child and other human rights monitoring mechanisms, including its impact on child rights, and steps taken to prevent harm.
- Dedicated funding streams for children should be established in bilateral and multilateral climate financing mechanisms.
- A standing agenda item on human rights should be introduced in the UNFCCC's Conference of Parties to address child rights and intergenerational equity – as well as the rights of other vulnerable groups – in a cross-cutting manner.
- Consideration of future generations could be elevated in decision-making at the national level, for example through establishing a Commissioner for Future Generations, or expanding the remit of existing institutions responsible for children.

Notes

1 In addition to the rights that are available to all humans, there are rights that are distinct to children, in recognition of their unique needs and the additional protections that they require. These are set out in the UN Convention on the Rights of the Child, the most widely and rapidly ratified international human rights treaty.

2 Human rights and climate change, A/HRC/32/L.34, 28 June 2016.

3 UNICEF, *Unless We Act Now: The Impact of Climate Change on Children* (2015).

4 UN Committee on the Rights of the Child, General Comment No. 15 (2013) on the right of the child to the enjoyment of the highest attainable standard of health (art. 24), at paragraph 50. All UNCRC decisions are available online at www.ohchr.org/EN/HRBodies/CRC/Pages/CRCIndex.aspx.

5 See e.g. UN Committee on the Rights of the Child, *Concluding Observations on Jamaica* (2015); *Saint Lucia* (2014); *Tuvalu* (2013).

6 World Health Organisation, *World Malaria Report* (2014).

7 World Health Organisation, *Quantitative Risk Assessment of the Effects of Climate Change on Selected Causes of Death, 2030s and 2050s* (2014).

8 World Health Organisation, *Climate Change and Health* (June 2016), available online at www.who.int/mediacentre/factsheets/fs266/en/.

9 UNICEF, *supra* note 3.

10 *Ibid.*

11 World Bank, *Human Capital Summit Highlights Need to Invest in the Youngest Children* (September 2016), available online at http://blogs.worldbank.org/voices/human-capital-summit-highlights-need-invest-youngest-children.

12 S. J. Lloyd, 'Climate Change, Crop Yields, and Undernutrition: Development of a Model to Quantify the Impact of Climate Scenarios on Child Undernutrition', 119 *Environmental Health Perspectives* (2011) 1817.

13 World Health Organization, *Burden of Disease – Data by Region* (September 2016), available online at http://apps.who.int/gho/data/node.main.156?lang=en.

14 D.V. Bates, 'The Effects of Air Pollution on Children', 103 *Environmental Health Perspectives* (1995) 49, at 49–53; Calderon-Guarciduenas *et al.*, 'Mental Effects on Children's Brain: The Need for a Multi-Disciplinary Approach to the Issue Complexity and Challenges', 8 *Frontiers in Human Neuroscience* (2014) 1.

15 UNICEF, *Clear the Air for Children* (2016).

16 UNICEF, *supra* note 3.

17 Human Rights Watch, *Marry Before Your House Is Swept Away: Child Marriage in Bangladesh* (2015).
18 UK Committee for UNICEF, *No Place to Call Home: Protecting Children's Rights When the Changing Climate Forces Them to Flee* (2017).
19 *Ibid.*
20 UNESCO Institute for Statistics and the Global Education Monitoring Report, *Leaving No One Behind: How Far on the Way to Universal Primary and Secondary Education?* Policy Paper 27 (July 2016), available online at http://unesdoc.unesco.org/images/0024/002452/245238E.pdf.
21 UNICEF, *The Benefits of a Child-Centred Approach to Climate Change Adaptation* (2011).
22 D. M. Abramson *et al.*, 'Children as Bellwethers of Recovery: Dysfunctional Systems and the Effects of Parents, Households, and Neighborhoods on Serious Emotional Disturbance in Children After Hurricane Katrina', 4(Suppl 1) *Disaster Medicine and Public Health Preparedness* (2010) S17.
23 J. H. Pane *et al.*, *Student Displacement in Louisiana After the Hurricanes of 2005: Experiences of Public Schools and Their Students*, Santa Monica (2006), available online at www.rand.org/pubs/technical_reports/TR430.html.
24 M. E. Ward *et al.*, 'Hurricane Katrina: A Longitudinal Study of the Achievement and Behavior of Displaced Students', 13 *Journal of Education for Students Placed at Risk* (2008) 297.
25 UN General Assembly, *Convention on the Rights of the Child*, 20 November 1989, Art 29(1) e.
26 UNICEF, *Written Submission to the Study of the UN Office of the High Commissioner for Human Rights on Climate Change and the Full and Effective Enjoyment of the Rights of the Child* (2016), available online at www.ohchr.org/EN/Issues/HRAndClimateChange/Pages/RightsChild.aspx.
27 The Guardian, *"Green" Dam Linked to Killings of Six Indigenous People in Guatemala* (26 March 2015), available online at www.theguardian.com/environment/2015/mar/26/santa-rita-green-dam-killings-indigenous-people-guatemala.
28 Climate Justice Programme and Heinrich Böll Foundation, *Climate Justice: The International Momentum Towards Climate Litigation* (2016).
29 UK Committee for UNICEF, *Children and the Changing Climate* (2015).
30 UNICEF Innocenti Research Centre, *Ensuring the Rights of Indigenous Children* (2004).
31 Based on a search for words 'child', 'infant', 'girl', 'boy', 'young people' and 'youth' across 160 NDCs. Three NDCs were excluded due to language or unavailability of search function (Iraq, Indonesia, Timor Leste). All NDCs are available online at http://www4.unfccc.int/submissions/indc/Submission%20Pages/submissions.aspx.
32 U-Report poll conducted in the UK and more than 60 countries globally, September 2016, available online at http://uk.ureport.in/poll/106/ and www.ureport.in/poll/1474/.
33 K. Sandberg, *UN Human Rights Council Panel Discussion on Climate Change and the Rights of the Child* (2 March 2017), available online at www.ohchr.org/EN/Issues/HRAndClimateChange/Pages/RightsChild.aspx.
34 The preamble to the Agreement states: Parties should, when taking action to address climate change, respect, promote and consider their respective obligations on human rights, the right to health, the rights of indigenous peoples, local communities, migrants, children, persons with disabilities and people in vulnerable situations and the right to development, as well as gender equality, empowerment of women and inter-generational equity.
35 For a comprehensive mapping of references to child rights in the 2030 Agenda for Sustainable Development, see Children in a Changing Climate Coalition, *A View From 2016: Child-Centred Disaster Risk Reduction and Climate Change Adaptation in the 2030 Agenda for Sustainable Development* (2016). In particular, SDG 13 b "promote[s] mechanisms for raising capacity for effective climate change-related planning and management in least developed countries and small island developing states, including focusing on women, youth and local and marginalized communities". Sendai Framework on Disaster Risk Reduction: "Children and youth are agents of change and should be given the space and modalities to contribute to disaster risk reduction, in accordance with legislation, national practice and educational curricula" (36a(ii)/p. 23).
36 UN Committee on the Rights of the Child, *supra* note 5.
37 UN Committee on the Rights of the Child, *Concluding Observations to the United Kingdom* (2016)
38 *Ibid.*

19

Human rights, differentiated responsibilities?

Advancing equity and human rights in the climate change regime

Gita Parihar and Kate Dooley

Introduction

The concepts of human rights and equity are both rooted in the goal of achieving societal fairness and human flourishing. This chapter examines how legal obligations relating to equity and human rights interact with each other in the context of the climate negotiations, the complementarities between the two and the extent to which each is necessary for realising the other.[1]

Climate change as a phenomenon has clear moral and ethical dimensions, because those who contributed least to the problem will be the most impacted. Both human rights and equity are well placed to speak to these dimensions to ensure a climate response that actively protects those most impacted. The examination below illustrates that the principles are complementary; an awareness of their respective spheres of operation is also important to ensure that they are mutually reinforcing and thereby improve human and environmental well-being overall.

The core role of equity within the international climate framework is evident from its placing in the core objective (Article 2) of the Paris Agreement. However, while undisputably a principle of international law, its manner of application within the climate regime remains the subject of controversy. For their part, human rights legal norms and applications have been established for many decades and yet while they have been referred to in the preamble of the Paris Agreement, the guiding interpretation of the treaty, they were not incorporated into the operational text of the agreement.

This chapter begins by providing a brief introduction to equitable and human rights principles. It then looks at how they have interacted within the regime. From this point, it considers the application of human rights and equitable principles to particular thematic areas within the climate change regime, with a particular focus on loss and damage. It demonstrates that differentiated obligations in the UNFCCC context can serve to enforce human rights obligations as well as restrict the possibility of solutions that are both inequitable and undermining of international human rights standards and obligations. It concludes that it is not possible to fully implement human rights obligations in the climate change context without also implementing obligations relating to equity and vice versa. While it will of necessity examine relevant legal obligations, this chapter is not intended primarily as a legal contribution to the debate but rather highlights the interaction of human rights and equitable principles from a conceptual and practical standpoint.

The principle of equity and its application in the climate regime framework

Equity in international environmental law

Equity is an established principle within the international climate change framework and international environmental law generally,[2] with references to it made in international treaties such as the United Nations Convention on the Law of the Sea.[3] It is a principle of international law that addresses fairness and is particularly relevant to climate governance, because the perception of equity in an international regime is key to ensuring that agreement is reached.[4]

While issues of equity were discussed at the Brundtland conference on the human environment in Stockholm in 1972, the Rio "Earth Summit" of 1992 was a significant step forward for the application of equitable principles in international environmental law: the title of the conference specifically linked the environment and development agendas.[5]

One of its achievements was the Rio Declaration, a statement of soft law principles, many of which are now recognised as formal legal principles of international law. Principle 7 of the Rio Declaration is a key principle outlining the basis for sharing responsibilities between countries and sets out the concept of common but differentiated responsibilities (CBDR) in the form below:[6]

Principle 7 of the Rio Declaration states:

> In view of the different contributions to environmental degradation, States have common but differentiated responsibilities. The developed countries acknowledge the responsibility that they bear in the international pursuit of sustainable development in view of the pressures their societies place on the global environment and of the technologies and financial resources they command.[7]

Therefore, CBDR incorporates two distinct concepts: the common responsibility of states to protect the environment and the need to take into account historical contributions to environmental pollution when determining responsibility.[8] Principle 7 links CBDR directly to contributions to global environmental harm.[9] However, negotiations around this principle in Rio were fraught and 'the obligation of the industrialised North to bear their proportionate share of the responsibility for the environmental crisis . . . were areas of deep North-South conflict'.[10]

The application of equitable principles in the UNFCCC

The United Nations Framework Convention on Climate Change (UNFCCC) refers to both equity and the differential contribution of states to climate change in its first governing principle in Article 3.1:

> The Parties should protect the climate system for the benefit of present and future generations of humankind, on the basis of equity and in accordance with their common but differentiated responsibilities and respective capabilities. Accordingly, the developed country Parties should take the lead in combating climate change and the adverse effects thereof.

Article 3.2 of the UNFCCC refers to the

> specific needs and special circumstances of developing country Parties, especially those that are particularly vulnerable to the adverse effects of climate change and of those parties, especially developing country Parties that would have to bear a disproportionate or abnormal burden under the convention.

Article 4 (8) adds that all parties are to consider what actions, including funding, insurance and transfer of technology, may be necessary to meet the specific needs of specially affected states.

Further, Annex 1 (developed country) parties agree to take into account the need for 'equitable and appropriate contributions' by each of them to the global effort regarding the achievement of the objective of the Convention.[11]

Equity is also given a central placing in Article 2 of the Paris Agreement. This provision sets out a long-term temperature goal which commits parties to limit global temperature rise to 'well below' 2°C above pre-industrial levels and to 'pursue efforts' towards 1.5°C while reflecting equity and CBDR in implementation. The mitigation goal described in Article 4.1 is to be achieved 'on the basis of equity' and the global stocktake is to assess implementation of the Agreement 'in light of equity'.

CBDR and historical responsibility

As we have seen above, Principle 7 of the Rio Declaration states that CBDR is based on contribution to environmental problems. Rajamani argues that the notion of differentiated responsibility derives from both the differing contributions of states to climate change and the difference in capacity to address the problem,[12] and this is widely perceived to be the case. However, the extent to which CBDR should be based on *historical* responsibility remains contested. There are those who argue that, given climate change is caused by the cumulative concentration of greenhouse gasses (GHG) in the atmosphere over centuries, responsibility should be based on historical contributions to cumulative emissions. Others contend that holding states accountable for emissions prior to knowledge of the damage they were causing is unreasonable.

The Paris Agreement did little to resolve these controversies. The linked question of legal liability for climate harm based on cumulative emissions surfaced in the negotiations leading up to the Paris Agreement when the issue arose in the context of negotiations around loss and damage. Developed countries strenuously objected to any suggestion of such liability in this context.[13]

Discussions relating to equity and CBDR are so politicised that there is even a debate about the role of scholarship relating to these principles, with a number of academics writing an article to support the role of equity within the UNFCCC framework.[14] When viewed from the context of justice and fairness, there are compelling arguments for CBDR to be understood as including historical responsibility, as we will see when considering the relationship between human rights and equity in the following discussion.

Human rights and their application in the climate regime framework

The role of human rights in the international sphere

Human rights are the basic rights inherent to all human beings, whatever their nationality, sex, national or ethnic origin, colour, religion, language or other status. They are universal, inalienable, interdependent, indivisible, equal and non-discriminatory. Human rights have been recognised as "the most successful normative project of all time".[15]

Following on from the 1948 UN Declaration of Human Rights, they were given international legal recognition through the International Covenant on Civil and Political Rights (ICCPR) and the International Covenant on Economic, Social and Cultural Rights (ICESCR) respectively.

The ICESCR addresses environmental protection directly in the context of its importance to realising economic, social and cultural rights rather than as a 'good' in itself. The Committee on Economic, Social and Cultural Rights (CESCR) recognises the relevance of the environment for current enjoyment of human rights, as well as 'the need for sustainability in the means chosen to realize human rights'.[16] In its statement on the green economy at Rio+20, the Committee stressed that many provisions of the ICESCR are linked with the environment and sustainable development and that 'without an adequate life support system, the human rights upheld by the ICESCR . . . cannot plausibly be attained on a universal scale.[17] The Convention also places an obligation on states to cooperate to progressively realise the rights enshrined within it,[18] overlaying the traditional state-individual paradigm of human rights. In addition, the UN Charter refers to the need for states to work collectively to realise human rights.[19]

In the environmental context, this collective dimension is vital to ensure that states in the developing world are able to uphold the rights of individuals in their territory impacted by climate change.

The right to a healthy environment can be regarded as a 'third generation' human right. It is protected in a large number of national constitutions across the world. It is also referred to in international environmental documents; Principle 1 of the Stockholm declaration made a linkage between a healthy environment and the protection of human rights, and Principle 1 of the Rio Declaration stated that human beings were at the centre of concerns for sustainable development and are entitled to a 'healthy and productive life in harmony with nature'.

As will be seen below, in order to truly protect human rights and prevent dangerous climate change, it is necessary to develop and acknowledge the collective aspect of enforcing such rights in terms of the duty to cooperate, extra-territorial obligations and the realisation of global conditions that will enable such obligations to be upheld.

The application of human rights obligations within the international climate framework

A human rights–centred approach to climate change draws attention to the impact of climate harms on individuals and their rights.[20] As recognised by the Office of the UN High Commissioner for Human Rights,

> looking at climate change vulnerability and adaptive capacity in human rights terms highlights the importance of analysing power relationships, addressing underlying causes of inequality and discrimination, and gives particular attention to marginalized and vulnerable members of society.[21]

The relationship between the goals of the UNFCCC and human rights has drawn much academic commentary. Advocates for inclusion have highlighted the centrality of human rights to achieving the goals of the Climate Convention (stabilisation of atmospheric concentrations of greenhouse gasses at levels to prevent dangerous climate change), in terms of both the impact a warming climate will have on human rights and the impact on human rights from policies and actions to address climate change.[22]

Human rights were not directly referenced in the treaty text of the UNFCCC, and indeed arguments for their inclusion have only gained currency in recent years. There was NGO lobbying (and some state support) for the inclusion of human rights language in the run-up to the Copenhagen climate summit of 2009, and the first explicit reference to human rights was in the 2010 Cancun Agreements (which are decisions under the UNFCCC rather than protocols or treaties). This reference states that countries should fully respect human rights in all climate-related actions.[23]

The lead-up to the Paris climate summit in 2015 saw an increase in advocacy efforts for human rights language to be included in a new post-2020 climate agreement. In 2014, 72 special procedures of the Human Rights Council issued an open letter to parties to the UNFCCC, calling for human rights protections in the Paris Agreement.[24] The following year over 200 organisations made a submission to the UNFCCC with the same demand.[25] Following the increasing recognition of the impact of climate change on human rights and calls for greater attention to be paid to human rights in developing the next phase of the climate regime, initial versions of the draft Paris Agreement included reference to human rights obligations in the overall objective. Ultimately, human rights language was included in the new climate treaty, albeit limited to the preamble:

> Acknowledging that climate change is a common concern of humankind, Parties should, when taking action to address climate change, respect, promote and consider their respective obligations on human rights, the right to health, the rights of indigenous peoples local communities, migrants, children, persons with disabilities and people in vulnerable situations and the right to development, as well as gender equality, empowerment of women and intergenerational equity.[26]

This language recognises that parties have existing human rights obligations and points to the need to take these into account when acting on climate change. It explicitly draws existing human rights obligations into the climate change framework by making a direct link between the two sets of obligations, with consequences for how parties take human rights into account when developing or implementing climate policies under the future climate regime.

Compatibility of human rights and equity frameworks – differential obligations in human rights law

We have seen that there is broad congruence between human rights and equity as norms promoting justice and fairness-related outcomes in the UNFCCC. We can also note that after some political battles, equity-related concepts were accepted in the Rio Declaration and form a central component of the UNFCCC, including the Paris Agreement. At the same time, human rights, as arguably the most successful international normative framework, are now directly referenced in the Paris Agreement. These are significant developments.

Despite this, the concept of historical responsibility continues to play a limited role in determining what CBDR means because of lack of broad acceptance as to its relevance, despite morally compelling arguments. It was not defined or referred to as a component of CBDR in the Paris Agreement. This in turn has implications for how effectively equity is promoted within the climate regime in practice.

Likewise, the reference to human rights in the Paris Agreement remains limited to the preamble. While this placement does not alter the legal obligations of states in this area, it does give a sense that human rights obligations were not viewed as an area of political emphasis.

We will see below that equity and human rights play an important role in reinforcing each other in this context. Indeed, there appears to be an awareness that, in the context of climate change, neither can be achieved if the other is not also achieved. In his most recent report to the Human Rights Council, the Special Rapporteur on human rights and the environment, John Knox, referred to climate change as a global problem requiring a global response as an application of the duty of international cooperation.[27] His report made direct reference to the language relating to CBDR in the UNFCCC text and added:

> To be clear, the duty of international cooperation does not require each state to take exactly the same actions in response to climate change . . . all states have a duty to work together to address climate change but the particular responsibilities necessary and appropriate for each state will depend in part on its situation.

This recognition of the different positions of states, of the need for progressive realisation of human rights, and of the division of responsibility among states to ensure these rights are fulfilled dovetails strongly with the concepts of equity and differentiation. However, the duty to cooperate in the human rights context is rather vague and not as developed as, for example, the clear obligations in relation to climate finance and technology transfer in the international climate framework.[28]

Poverty and development are related matters which fall within the purview of both human rights and equity concerns. This was acknowledged as far back as the Rio Conference, where UN High Commissioner on Human Rights Navi Pillay described the Rio Declaration as having 'integrated human rights in its approach to sustainable development', and being 'thoroughly infused with human rights considerations essential to sustainable development'.[29]

Shelton states that poverty has come to be seen both as a major source of environmental degradation and as a human rights issue, because it means that individuals lack an adequate standard of living – including the food, shelter, medical care and education that are guaranteed by international human rights law.[30] The Paris Agreement also acknowledges poverty eradication and sustainable development as the context in which both the temperature goal and the long-term mitigation goal are to be achieved.[31]

Poverty and development concerns have a direct bearing on considerations relating to historical responsibility for climate harm. As Shelton notes:

> Fairness and a morally coherent response suggest that these states, which attained their current developed status through imposing non-internalized costs on the environment, take the major abatement actions, rather than demanding that everyone equally mitigate the externalities, including those not responsible for initially creating the problem. Equity, in this sense, is justified as a means of corrective justice, requiring remedial conduct to correct past wrongs.[32]

For this reason, Shelton makes an ethical argument for historical responsibility with compensatory or reparative justice as a basis for differential treatment of developing countries, especially where they were colonies whose resources were exploited as the means for development by industrialized countries. A counter-argument is that it is unfair to penalise current generations for harm caused by their forebears.[33] Henry Shue contends, however, that although present and future generations in developed countries did not request or consent to the carbon emissions of their ancestors, they benefit by living amidst national affluence produced by means of those emissions.[34]

As we have seen, in the international environmental context, equity focuses on differing contributions to environmental harm at the global level and is a guiding value in the development of the international climate legal framework. But human rights instruments are also not blind to the importance of international frameworks in the achievement of human rights. For example, Article 28 of the Universal Declaration of Human Rights (UDHR) states that 'Everyone is entitled to a social and international order in which the rights and freedoms set forth in this Declaration can be fully realised.'

The role played by the international order is key both to the protection of the environment and to the attainment of human rights. Nevertheless, it is often overlooked in human rights discourse, even though problems, such as global poverty, can become entrenched through a particular international order.[35]

The international economic order has a significant impact on both the attainment of human rights and equitable environmental protection. In the run-up to Rio+20, Commissioner Navi Pillay recognised the links between human rights and equity, stating that

> strategies based on the narrow pursuit of economic growth without due regard for equity and related environmental, social and human rights considerations will both fail in their environmental objectives, and risk damaging the planet and the fundamental rights of the people who live here.[36]

Also worthy of note in highlighting concerns shared in both the human rights and development arenas is the UN Declaration on the Right to Development, which has a specific entitlement – the right to participate in, contribute to, and enjoy economic, social, cultural and political development. The Declaration highlights the 'fair distribution of the benefits of development'.[37] Indeed, Pillay goes so far as to state that 'equity is codified in the right to development'.[38] It is therefore clear that equity forms a component in the collective dimension of human rights, which, as we have seen above, is particularly important in the context of climate change. Collective efforts by the international community to realise these rights will become increasingly crucial to maintain a social and international order where states can uphold their wider obligations, including those pertaining to civil and political rights.

There are other potential bases for highlighting the synergies between climate change and human rights. For example, while non-discrimination obligations are generally considered to apply to the relationship between individual and state, in reality, there is also a clear international aspect to inequality. As Shelton explains:

> Non-discrimination is clearly a predominant obligation in international human rights law. Yet equal treatment is required only for those equally situated . . . Requiring all states to reduce GHGs in an identical fashion would make many developing countries, or groups in those countries, worse off . . . From the perspective of equity toward the most vulnerable or least well-off, environmental protection should not result in further deterioration of their well-being. In order to address this problem, non-equal or differential obligations can, and are, being imposed as an equitable means of fostering substantive equality in the long term.[39]

Based on the above, it can be argued that differential obligations are a determinative element of the human rights regime so that inequitable development, or distribution of benefits and burdens, would breach human rights obligations. While the focus in the climate regime is on states rather than individuals, it is clear that individuals within states that are already suffering from climate harm for which they are not responsible will be disadvantaged further through an inequitable outcome. As Shelton puts it, 'any allocation of benefits and burdens that makes vulnerable populations worse off, even if harm is felt outside the boundaries of the state, cannot be regarded as equitable or in conformity with international human rights law'.[40]

Further work by the human rights treaty bodies and the Human Rights Council as well as academic commentators would be helpful in order to advance thinking on integrating the collective dimension of human rights obligations and links with equity into climate change responses. There are also calls for scholarship on equity to include work that explores how human rights could be differentially impacted by climate change and climate policy.[41]

Addressing climate change: applying human rights and equity in practice

Having established that human rights and equity are complementary frameworks and explored some of their key aspects and areas of convergence, we now turn to what this means for implementation of duties and obligations in the context of the international climate framework.

Mitigation and the long-term goal

As discussed above, the Paris Agreement achieved agreement on a temperature goal, committing parties to limit global temperature rise to 'well below' 2°C above pre-industrial levels, and 'pursue efforts' towards 1.5°C, while reflecting equity and CBDR in implementation.[42] The preambular reference to human rights also indicates that any actions to mitigate climate change that adversely impact human rights would be contrary to the spirit of the treaty.

The mitigation objective requires parties to achieve a 'balance between anthropogenic emissions by sources and removals by sinks . . . in the second half of this century' in order to meet the temperature limit.[43] Removing CO_2 from the atmosphere would rely on large areas of land to sequester carbon in forests, soils and other biomass.[44] Estimates of the area of land required to keep within the 1.5°C or 2°C temperature limit ranges from one-third to four times current global agricultural land.[45] Land use on this scale would likely undermine food security, poverty eradication and other development goals, indicating that large-scale removal of emissions from the atmosphere through this method is incompatible with either human rights or the equitable approach to climate mitigation referenced in the treaty text.[46]

In response, civil society groups proposed an international 'fair shares' approach – a burden-sharing framework based on the equitable principles of historical responsibility, respective capacity and a right to development.[47] Under this approach, developed countries (who have already used their 'fair share' of the carbon budget) would take on emission reduction targets of well over 100%; larger developing countries would need to reduce emissions to zero quickly and the poorest countries have a few decades before emissions must decrease. This is a much faster and higher rate of carbon reduction than that currently envisaged in the Paris Agreement. However, it would avoid the human rights consequences mentioned above and minimise the need to find a human rights– and equity-compliant method of removing (as opposed to reducing) emissions from the atmosphere to meet the temperature goal in the Paris Agreement.

Failure to deliver an equitable burden-sharing framework could result in a breach of human rights obligations, but for poorer countries, this breach would only be in respect of unsupported actions (their own fair share of the climate burden), rather than situations where countries are unable to carry out actions beyond their fair share due to failure of richer countries to provide adequate support.

Finance and technology transfer

Article 4.7 of the UNFCCC confers obligations for developed countries relating to finance and technology transfer. The Paris Agreement builds on this by requiring developed countries to provide financial resources to developing countries to assist with mitigation and adaptation.[48] While intended to apply at the national level, human rights discourse concerning the duty to take affirmative action to address discrimination could provide a relevant analogy here, reinforcing the need to provide finance and technology transfer to address the unfair position in which developing countries find themselves.[49]

The Human Rights Council also recognises the need for international dialogue and cooperation concerning the adverse impacts of climate change on the enjoyment of human rights in developing countries, and calls for the implementation of practical steps to promote capacity-building, financial resources and technology transfer in accordance with the UNFCCC.[50] As pointed out by Stephen Humphreys, 'without robust and detailed policies of technology transfer and adaptation, [the poorest countries'] development and policy options will shrink with deleterious effects on basic rights'.[51]

The need for an energy transition is a global one, requiring financial support. As Shue states:

> If we are to escape from the fossil-fuel energy regime into an alternative-energy regime, the costs of the transition must be borne by someone. So we must have an allocation of the burdens (and benefits), and it ought to be fair.[52]

We can see therefore that both human rights and equitable approaches point to the need for finance and technology transfer to assist developing countries and prevent further breach of the human rights of the most vulnerable. Adopting the fair shares approach outlined above, the difference between developed countries' domestic emissions reductions and their international fair share could be met through the transfer of financial and technological resources, to allow developing countries to undertake more than their fair share of emission reductions, but be supported in doing so.[53]

Adaptation

The UNDP has noted that 'the world's poor cannot adapt their way out of dangerous climate change', but 'the impacts of global warming can be diminished through good policies'.[54]

The capacity to adapt to climate change, from the level of an individual to that of a global region, often correlates with general capacity in other areas. As a result, climate change has particularly intense consequences for the human health, security and stability of developing countries and their residents.[55] Consequently, adaptation efforts bring both human rights and equity considerations into play.

Equity points to the provision of funding and technology transfer by developed countries to assist developing countries in dealing with the harm they are already experiencing because of climate change. Adaptation is also necessary to protect basic human rights. For example, adapting to climate change can require people to move from one geographical location to another; this is discussed in more detail in the section related to loss and damage below.[56] Equally, the human rights framework provides a method to ensure that climate finance received within country and other adaptation measures are directed to those who are most likely to experience harm from climate change. Human rights concerns can help identify priorities for adaptation funds at the international and national levels, ensure that local people are able to participate in decisions about what appropriate adaptation for them might look like and contribute to discussions concerning how proactive (as opposed to reactive) adaptation might take place.[57]

Loss and damage[58]

The intersection of human rights and equity concerns is particularly evident in the area of loss and damage. An equity perspective requires the resolution of climate impacts that cannot be adapted to in countries that bear either no or little responsibility for climate harm. As mentioned above, considering equity in this context brings the issue of historical responsibility for causing climate change, as a basis for compensation or distribution of responsibilities, to the fore. NGOs have specifically recognised and campaigned on the human rights and equity-related impacts of loss and damage.[59]

The political fault lines concerning loss and damage parallel many of those relating to responsibility for climate change discussed above. Unsurprisingly, therefore, it was one of the most difficult areas to negotiate in the run-up to the Paris Agreement. Some developed states argued that it should be addressed within the Hyogo framework for action (now replaced by the Sendai framework, both of which focus on natural disasters), thus side-stepping discussions on responsibility and causality.[60] The eventual resolution of the matter has been described as being achieved through a process of 'constructive ambiguity'.[61]

The result of these tense negotiations is encapsulated in Article 8 of the Paris Agreement and the accompanying decision text. Article 8 establishes a loss and damage mechanism.[62] However, the COP decision text produced alongside it states that Article 8 'does not involve or provide a basis for any liability or compensation'.[63] It remains to be seen how this will be interpreted over time and what discussions develop in practice.

John Knox urged parties to incorporate a human rights perspective in identifying the types of loss and damage to be addressed.[64] The discussion on links between human rights and equity above indicates that a human rights approach would also favour an approach to loss and damage that takes into account historical responsibility when considering an appropriate response in this area. It is important to note that loss and damage encompasses non-economic losses, such as cultural identity and loss of place, and not only financial loss. The human rights framework may well assist in developing proposals that address such kinds of loss.

On the entry into force of the Paris Climate Change Agreement on 4 November 2016, UN High Commissioner for Human Rights, Zeid Ra'ad Al Hussein, called on states to prioritize the protection of the rights of people disproportionately affected by climate change.[65] Amongst those who fall into this category and are currently with limited legal protection are those displaced a result of climate change. The question of burden sharing and international responsibility to manage such population flows again arises in this context. Legal protections for those who are forced to flee borders are necessary, as is protection of those displaced internally.

However, states remain reluctant to enlarge the category of people entitled to formal legal protection; proposals for a climate displacement facility in Paris were also shelved in favour of a task force to discuss approaches to displacement, with the involvement of bodies such as the International Organization for Migrations (IOM) and UN Human Rights Council (UNHCR).

Both human rights– and equity-based approaches to addressing climate change highlight the need for support for the countries and communities most affected by loss and damage, including through (but not limited to) monetary compensation and the creation of legal and other frameworks to protect climate refugees. Where relocation of communities and populations is necessary, a human rights–oriented approach would favour planned (and consensual) relocation over reactive/forced relocation, as the latter is likely to lead to a much greater breach of human rights.

It is clear that loss and damage is a context in which human rights and equity norms clearly intersect. As such, they mutually strengthen calls for a governance framework that protects the most vulnerable and works to both promote the fulfilment of human rights and provide redress when human rights are breached.

Conclusion

As this chapter demonstrates, human rights and equity are complementary legal frameworks. Beyond that, differentiated obligations under the UNFCCC can be viewed as the 'carrier' through which human rights obligations operate when addressing climate change, a context in which poverty reduction, environmental protection and the attainment of human rights are inextricably interwoven.

There is general agreement that equity means that developed countries should take the lead. Compliance with human rights obligations also requires differential obligations to be addressed in a manner that protects the most vulnerable. In addition, human rights considerations weigh in favour of arguments that historical responsibility is a factor to be considered when determining what differentiation means in practice, although this remains a controversial area.

Having gained a fuller understanding of the shared values and aims of the two normative frameworks we can see that both human rights and equity have the same end goal – an effective climate regime that delivers climate justice. The different perspectives brought to the task by the two frameworks broaden their effectiveness by appealing to a wider audience overall. In addition, it is clear that there are difficulties with their application across the thematic areas of the international climate regime, and this is particularly evident in the context of loss and damage. Further consideration is needed of ways to enhance their combined impact to effectively advocate for a fair and just climate regime that protects and promotes human rights and improves the position of the most vulnerable.

Notes

1 This chapter builds upon a previous joint publication by the authors in a volume edited by C. Sampford, T. Cadman and R. Maguire and which drew from discussions in the NGO context in the run-up to the Paris Agreement. See K. Dooley and G. Parihar, 'Human Rights and Equity: Governing Values for the International Climate Regime', in C. Sampford, T. Cadman and R. Maguire (eds.), *Governing the Climate Change Regime: Institutional Integrity and Integrity Systems* (2017). The authors are very grateful to the editors of the volume mentioned herein for their valuable comments and guidance.

2 See the discussion in P. Sands, *Principles of International Environmental Law* (2003) 152. Sands discusses the *North Sea Continental Shelf* cases where the ICJ described equity as a 'direct emanation of the idea of justice' and a 'general principle directly applicable as law', which should be applied as part of international law 'to balance up the various considerations which it regards as relevant in order to produce an equitable result'.

3 Article 59 of the latter states that when conflicts arise with regard to the attribution of rights and jurisdictions in the EEZ, 'the conflict should be resolved on the basis of equity'.

4 H. Winkler and L. Rajamani, 'CBDR&RC in a Regime Applicable to All', 14(1) *Climate Policy* (2014) 102.

5 The G77 stressed the need to do this to address poverty eradication, North-South equity issues and causes and responsibilities for environmental problems. As a result, discussions dealt with the recognition of differentiated obligations among states, depending on their relative contribution to the problem and differing capacities to act. C.Y. Ling, *The Rio Declaration on Environment and Development: An Assessment*, Penang: Third World Network (2012).

6 Equity was also referred to in Principle 23 of the Stockholm Declaration of 1972.

7 The specific link between the differentiation of obligations and equity is noted by academic commentators such as Sands, *supra* note 2, who states that 'in many respects UNCED (ie the Rio conference)

was about equity: how to allocate future responsibilities for environmental protection between states which are at different levels of economic development, have contributed in different degrees to particular problems, and have different environmental and developmental needs and priorities. This is reflected in each UNCED instrument.'

8 L. Rajamani, 'The Principle of Common But Differentiated Responsibilities and the Balance of Commitments Under the Climate Regime', 9 *Review of European, Comparative & International Environmental Law* (2000) 120.

9 Winkler and Raja, *supra* note 4.

10 Ling, *supra* note 5.

11 Arts 3.1 and 4.2(a), *United Nations Framework Convention on Climate Change*, FCCC/INFORMAL/84 (1992). Article 2 defines its ultimate objective as 'stabilisation of greenhouse gas concentrations in the atmosphere at a level that would prevent dangerous anthropogenic interference in the climate system'.

12 Rajamani, *supra* note 8.

13 See discussion in relation to loss and damage below, and L. Vanhala and C. Hestbaek, 'Framing Climate Change Loss and Damage in Climate Negotiations', 16(4) *Global Environmental Politics* (2016) 111.

14 See S. Klinsky, et al., 'Why equity is fundamental in climate change policy research', 44 Global Environmental Change (2017) 170–173.

15 R. Maguire *et al.*, 'Ethical Values and Global Carbon Integrity System', in H. Breakey *et al.* (eds.), *Ethical Values and the Integrity of the Climate Change Regime* (2015).

16 G. Baruchello and R. L. Johnstone, 'Rights and Value: Construing the International Covenant on Economic, Social and Cultural Rights as Civil Commons', 5(1) *Studies in Social Justice* (2011) 94, at 109; UN Committee on Economic, Social and Cultural Rights (CESCR), General Comment No. 19: The right to social security (Art. 9 of the Covenant), E/C.12/GC/19, 4 February 2008, at para 11, recognising the need for social security schemes to be sustainable 'to ensure that the right can be realized for present and future generations'.

17 G. Baruchello and R. L. Johnstone, 'Comment on Rights and Value: The Committee on Economic, Social and Cultural Rights Addresses the Environment', 7(1) *Studies in Social Justice* (2013) 175, at 176.

18 Art 2(1): 'Each State Party to the present Covenant undertakes to take steps, individually and through international assistance and co-operation, especially economic and technical, to the maximum of its available resources, with a view to achieving progressively the full realization of the rights recognized in the present Covenant by all appropriate means.' In Art 15(4): 'States parties recognize the benefits to be derived from the encouragement and development of international contacts and cooperation in the scientific and cultural fields.'

19 See Arts 55 and 56.

20 L. Rajamani, 'The Increasing Currency and Relevance of Rights Based Perspectives in the International Climate Negotiations', 22(3) *Journal of Environmental Law* (2010) 391.

21 Human Rights Council, *Report of the Office of the United Nations High Commissioner for Human Rights on the Relationship Between Climate Change and Human Rights*, A/HRC/10/61, 15 January 2009, at para 81.

22 Environmental Investigation Agency and Centre for International Environmental Law, *A Rights-Based Approach to Land-Use in a Future Climate Agreement: Policy and Implementation Framework* (2015).

23 United Nations Framework Convention on Climate Change (UNFCCC), *The Cancun Agreements*, U.N. Doc. FCCC/CP/2010/7/Add.1, December 2010.

24 Available online at www.ohchr.org/Documents/HRBodies/SP/SP_To_UNFCCC.pdf.

25 B. Lewis, 'The Contribution of Human Rights to the Effectiveness and Integrity of the Global Carbon Regime', in H. Breakey *et al.* (eds.), *supra* note 15.

26 UNFCCC, *Adoption of the Paris Agreement*, U.N. Doc. FCCC/CP/2015/L.9/Rev. 1, 12 December 2015, at preamble para 11.

27 Human Rights Council, *Report of the Special Rapporteur on the Issue of Human Rights Obligations Relating to the Enjoyment of a Safe, Clean, Healthy and Sustainable Environment*, A/HRC/31/52 (2016), at paras 42 and 46.

28 See for example: Arts 4 and 11, *United Nations Framework Convention on Climate Change*, *supra* note 11.

29 N. Pillay, *Pillay Urges States to Inject Human Rights Into Rio+20, Geneva* (18 April 2012), available online at www.ohchr.org/EN/NewsEvents/Pages/DisplayNews.aspx?NewsID=12070&LangID=E.

30 Dinah Shelton, "Equitable Utilization of the Atmosphere: A Rights-based Approach to Climate Change", in *Human Rights and Climate Change*, ed. Stephen Humphries (Cambridge University Press, 2010).

31 Arts 2.1 and 4.1, UNFCCC, *The Paris Agreement*, *supra* note 26.

32 Shelton, *supra* note 30, at 121.

33 Maguire, *supra* note 15.
34 H. Shue, 'Historical Responsibility, Harm Prohibition and Preservation Requirement: Core Practical Convergence on Climate Change', 2(1) *Moral Philosophy and Politics* (2015) 7.
35 S. Humphries, 'Conceiving Justice: Articulating Common Causes in Distinct Regimes', in S. Humphries (ed.), *supra* note 32, at 299, states: 'There is an at least plausible case . . . that the persistence and exacerbation of global poverty over recent decades is itself a human rights violation, largely sustained by international actors and systematised through international law.' See also Shelton, "Equitable Utilization of the Atmosphere," at 92 of the same volume, who points out that this model can generally address violations of civil and political rights but 'harm to economic, social and cultural rights, as well as to the underlying environmental conditions necessary to the enjoyment of all rights, often originates outside the jurisdiction where the harm is felt'.
36 OHCHR, *Remarks of High Commissioner for Human Rights Navi Pillay at Rio+Social*, available online at www.ohchr.org/en/NewsEvents/Pages/DisplayNews.aspx?NewsID=12254+LangID=E.
37 United Nations General Assembly, *Declaration on the Right to Development*, A/RES/41/128, (1986).
38 OHCHR, *Introduction Statement by the High Commissioner: Development Is a Human Right for All*, available online at www.ohchr.org/EN/Issues/Development/Pages/IntroductionStatement.aspx.
39 Shelton, *supra* note *32*, at 125.
40 Shelton, *supra* note *32*, at 125.
41 Klinsky *et al.*, *supra* note 14. While it is evident that theories of justice can be effectively employed to argue for the principle of CBDR to incorporate historical responsibility, as we have demonstrated above, Breakey demonstrates that human rights have achieved moral authority based on appealing to widely accepted norms, with a variety of different actors (for different reasons) endorsing the same norms. H. Breakey et al . (eds.), *supra* note 15.
42 Art 2, UNFCCC, *The Paris Agreement*, *supra* note 26.
43 *Ibid.*
44 P. Smith *et al.*, 'Biophysical and Economic Limits to Negative CO_2 Emissions', 6 *Nature Climate Change* (2016) 42.
45 P. Smith *et al.*, 'Agriculture, Forestry and Other Land Use (AFOLU)', O. R. Edenhofer *et al.* (eds.), *Climate Change 2014: Mitigation of Climate Change. Contribution of Working Group III to the Fifth Assessment Report of the Intergovernmental Panel on Climate Change* (2014).
46 Arts 2.1 and 4, UNFCCC, *The Paris Agreement*, *supra* note 26.
47 ActionAid International, *Fair Shares: A Civil Society Equity Review of INDCs 2015*, available online at www.oxfam.org/en/research/fair-shares-civil-society-equity-review-indcs.
48 Art 9.1, UNFCCC, *The Paris Agreement*, *supra* note 26.
49 Noting that discrimination duties are state obligations towards their citizens in the human rights context, as mentioned earlier.
50 Human Rights Council, *Resolution 26/27: Human Rights and Climate Change*, A/HRC/26/L.33/Rev.1, (2014), at para. 5.
51 Humphries, *supra* note 35, at 307.
52 H. Shue, *Climate Justice: Vulnerability and Protection* (2014) 23–24.
53 T. Athanasiou, S. Kartha and P. Baer, *National Fair Shares* (EcoEquity and Stockholm Environment Institute, 2014).
54 UNDP, *Fighting Climate Change: Human Solidarity in a Divided World*, Human Development Report 2007/8, at 47–48.
55 M. J. Hall and D. C. Weiss, 'Avoiding Adaptation Apartheid, Climate Change Adaptation and Human Rights Law', 2(37) *Yale Journal of International Law* (2012) 309.
56 Issues of human displacement and migration are considered under both the adaptation and loss and damage thematic areas in the UNFCCC.
57 Hall and Weiss, *supra* note 56.
58 With additional thanks to Olivia Serdeczny for her input.
59 'Tackling loss and damage is about climate justice. It is about protecting people, their livelihoods and, most importantly, their human rights and dignity. It is time for those who are mainly responsible for climate change to act here in Warsaw', excerpt from ECO, an NGO newsletter on November 19, 2013, prior to the setting of the Warsaw Mechanism on Loss and Damage in the UNFCCC framework.
60 Vanhala and Hestbaek, *supra* note 13, at 125.
61 *Ibid.*
62 Art 8.3, UNFCCC, *The Paris Agreement*, *supra* note 26.

63 *Ibid.*, at para 52.
64 Human Rights Council, *Report of the Special Rapporteur on the Issue of Human Rights Obligations Relating to the Enjoyment of a Safe, Clean, Healthy and Sustainable Environment*, A/HRC/31/52 (2016), at para 64.
65 *Spotlight on Indigenous Rights at COP22 Climate Talks*, available online at www.ohchr.org/EN/NewsEvents/Pages/COP22.aspx. He also added, 'I urge all the parties that will be at the COP22 in Marrakech to ensure that the meeting is about States taking action in accordance with their international human rights obligations.'

20

Climate justice and human rights

Doreen Stabinsky

Introduction

> *As global citizens we commit to build a climate movement, whose beating heart is justice, that can break out of its silo and create a broad based progressive movement alongside Black Lives Matter, Indigenous movements, women's movements, student movements, LGBTQI communities, migrants movements, labour movements, and local movements against corporate power and the fossil fuel industry that work together to address the inequalities and injustices that blight our world.*[1]

Climate justice – the fight for climate justice – is a meeting point, a political space for many different movements for action against the injustices of climate change. One community that comfortably arrives and actively engages in that political space are those fighting for human rights in the context of climate change. Indeed, some might consider a rights-based approach and demands for the protection of human rights in the face of climate change to be the core strategy and approach to climate justice, although the quote above indicates that perhaps there are many more paths. As we shall see, the climate justice movement is necessarily diverse in its membership, strategies, and tactics as its members seek to address a wide array of injustices and inequality associated with climate change and its impacts.

The roots of the climate justice movement may be found the soil of a number of social justice movements. It is clearly tied to the movement for environmental justice, which grew up in marginalized communities in the United States that were victims of environmental racism, and their counterparts in communities around the world.[2] Movements against neoliberal globalization; movements to call for the repayment of ecological debt; and struggles of Indigenous Peoples, women, and workers against unjust distribution of resources, lack of recognition, and exclusion from decision-making – all feed into the climate justice movement when people become aware of how climate change exacerbates their existing struggles – for recognition, rights, and justice.

Many of these social and climate justice struggles manifest as struggles for rights, and the intersections between a human rights agenda and a climate justice agenda are numerous. So there are certainly compelling strategic and other reasons to pursue a rights-based approach in the pursuit of climate justice. As Limon notes, a human rights framework "helps amplify the voices of those who are disproportionably [sic] affected by climate change – the poor, marginalized, and vulnerable people."[3] Still, the call for climate justice goes beyond an articulation of rights-based demands – it is broader and deeper.

In this essay, we explore the broader landscapes and terrain of struggles for climate justice and intersections with a human rights and climate change agenda. We start by trying to define climate justice – through story, academic, and activist accounts. We look at the intersections between human rights and climate justice agendas, with the intent to identify continuities and discontinuities, gaps, and synergies. We conclude with some thoughts on a path forward to enhance the conversation and possibilities of convergence to together bring about a more climate-just world.

Defining climate justice

Stories and a definition

In December 2016, climate justice activists in Bangladesh sent an email call to communities around the world to ask for their support with a global day of action on January 7, 2017. In the Sundarbans region of the Ganges delta, local communities were protesting the construction of a new coal-fired power plant. The Rampal Thermal Power Plant is a Bangladesh-India joint venture, with construction by an Indian company and financing from Indian and Bangladeshi state power companies. It would be the country's largest coal plant, requiring annual imports of coal estimated at almost 5 million tons per year – 13,000 tons per day – imports that would travel upriver through the mangroves, and the burning of which would release close to 8 million tons of carbon dioxide into the atmosphere each year.

The Sundarbans are a UNESCO World Heritage Site, the largest mangrove forest on the planet, so large it can be identified from space. It is home to the Bengal tiger, a complex and valuable ecosystem where millions of people depend on livelihoods based in and around the mangroves – agriculture, shrimp farming, fishing. The mangroves are critically important in protecting *tens* of millions from storm and tidal surges that accompany the frequent cyclones and storms that regularly hit the delta region, and which are growing in frequency and intensity due to climate change.

The biggest concern being voiced by those opposing the plant is about the possible impacts on local water supplies from the discharge of waste into the river, pollution that would harm mangroves, marine animals, and local fishing and farming communities. Activists also charge that the land on which the plant is being built was acquired by force. All these are elements in what the organizers themselves understand as a fight for climate justice.

On the other side of the globe, also in the final days of 2016, hundreds of "water protectors" were gathered at the Standing Rock reservation in the U.S. state of North Dakota. There, thousands of indigenous Sioux and many non-Sioux – from other indigenous nations and non-indigenous joining them in solidarity – gathered to prevent the construction of the Dakota Access Pipeline (DAPL). The pipeline would run from North Dakota to Illinois, carrying up to 570,000 barrels of oil per day from oil fields near the Canadian border and crossing under the Missouri River just north of the reservation. The Standing Rock Sioux were fighting to protect the river that runs across their territory and the sacred ancestral burial sites on their land that would be damaged by the construction – and more broadly to defend Native American sovereignty over their territories.

As with the fight in the Sundarbans to protect mangroves, the Standing Rock "water protectors" also understand their struggle as one of climate justice. "Stopping DAPL is a matter of climate justice and decolonization for indigenous peoples. It may not always be apparent to people outside these communities, but standing up for water quality and heritage are intrinsically tied to these larger issues."[4] The protectors are

> saying no to the continuation of the development and maintenance of fossil fuel industries, for the sake of indigenous survival. That statement reverberates for a lot of indigenous people, especially those such as people in the Arctic, in the Pacific, and in the Gulf of Mexico who are experiencing climate change impacts right now that pose threats of the highest severity.[5]

These stories illustrate two important consequences of climate change and current approaches to address climate change:

- Those most marginalized and vulnerable are also more exposed to the impacts of climate change, while they are least responsible for the problem;
- Those most affected are not included in decision-making processes about their future and the future of the resources upon which their lives and livelihoods depend.

Political action to address these consequences leads directly to movement demands for recognition, participation, a just distribution of resources, and a just apportionment of responsibility – demands we address in a later section of this chapter.

Before we move on, let us consider a provisional definition for climate justice. There are many that can be found in the academic literature and online, and it is not our intent to either review them here or craft our own. But as so much of climate justice is defined by action at the grassroots, by multiple actors and movements, it might not be too awkward or inappropriate to rely on a grassroots, crowd-sourced effort for a working definition: from Wikipedia. Indeed, it is actually a pretty good definition, at least as it appears at the time of writing of this chapter. According to Wikipedia, climate justice is:

> a term used for framing global warming as an ethical and political issue, rather than one that is purely environmental or physical in nature. This is done by relating the effects of climate change to concepts of justice, particularly environmental justice and social justice and by examining issues such as equality, human rights, collective rights, and the historical responsibilities for climate change. A fundamental proposition of climate justice is that those who are least responsible for climate change suffer its gravest consequences.[6]

Further elaborating elements of climate justice

As noted above, a wide range of actors have sought to define or otherwise utilize the term "climate justice" – from social movements engaged in struggles against the fossil fuel industry, to legal and policy academics engaged in the question of designing a new climate treaty,[7] to ethicists and environmental justice scholars.[8]

All these authors provide valuable contributions not just to a working definition, but to a longer list of elements of climate justice, which include, *inter alia*: recognition, rights, responsibility, procedures, distribution, in/equality, human rights, collective rights, historical responsibility, decolonization, development, just transition, equity, impacts of climate change on indigenous communities and ways of life, finance, and repaying ecological and climate debt. Our intent in this section is neither to be exhaustive, nor to engage with any of these elements in detail, but to provide a brief flyover of the different manifestations of "climate justice" in order to eventually engage productively in a conversation between climate justice and a human rights approach.

From the academy

TWO RESPONSIBILITIES OF CLIMATE JUSTICE: BURDEN SHARING AND HARM AVOIDANCE

Simon Caney, one of the most prominent and early theorists of climate justice, names two kinds of climate justice. First, there is burden-sharing justice: "An agent's responsibility . . . is to do her fair share." Second is harm-avoidance justice. "Its focus is primarily on ensuring that the catastrophe is averted . . . This perspective is concerned with the potential victims . . . and it ascribes responsibilities to others to uphold these entitlements."[9]

Both these responsibilities are found at the center of many climate justice framings at the international level. They play out in developing country and activist calls for paying the climate debt according to the historical responsibilities of developed country state actors.[10] Climate debt is conceptualized in terms of both emission reduction obligations on the part of developed countries – including the obligation that they should act first and urgently – and payment for an adaptation debt. In paying the adaptation debt, historical developed country emitters provide support and compensation for both *the cost* of adapting to impacts that developing countries did not cause and for *lost development prospects* that are the consequence of redirecting resources from sustainable development efforts towards adaptation and disaster risk reduction.[11] This second element of climate debt maps onto what Caney calls harm avoidance justice.

BEYOND RIGHTS AND RESPONSIBILITIES: RECOGNITION AND JUSTICE WITHIN NATIONS

Bulkeley and colleagues, in their work on climate justice and cities, "engage with the emerging work which regards justice as a multivalent concept encompassing notions of recognition."[12] They argue for a more expansive framing of the concept of climate justice than the current two-dimensional approach, which they say frames justice "in terms of rights and responsibilities . . . within a predominantly distributive mindset, albeit one engaged with procedural justice."[13] They see considerable potential with "a post-distributive framing of justice around the notion of 'recognition'."[14] Their conception of climate justice includes five core elements: recognition, rights, responsibilities, distributions, and procedures, visualizing these elements as a three-dimensional pyramid, with recognition as its base.

They call particular attention to what they see as a fundamental weakness of this two-dimensional emphasis on rights and responsibilities: the assumption of the intergovernmental arena as the relevant locus for addressing these justice claims. An important consequence for them of such an approach is that, "structural patterns of inequality within nations have tended to be overlooked."[15] Their attention is drawn to "the complex geographies of inequality which are compounded by the costs and benefits of climate change action,"[16] reflecting that

> notions of climate justice cannot be spatially agnostic . . . urban responses to climate change need to consider . . . the ways in which they serve to entrench or address questions of injustice within urban arenas, economies and communities.[17]

FROM ENVIRONMENTAL JUSTICE TO CLIMATE JUSTICE

Environmental justice scholars prominently note the historical links between environmental justice (EJ) and climate justice (CJ) movements: "climate justice developed directly out of the history and conceptualization of the EJ discourse."[18] As they make those links, that sharing of knowledge of the history, experiences, vocabulary, methodologies, and theories of environmental justice is quite helpful in better understanding the breadth and depth of climate justice claims.

For Schlosberg and Collins, the environmental justice movement has always been concerned about more than just equity – that "environmental justice has always focused on *how* injustice is constructed"[19] (emphasis added), echoing the conclusion by Bulkeley et al. that attention should be paid to structural inequities found within countries.[20] In this work on the climate justice movement(s), these EJ scholars find that "overall, the definitions of justice used . . . address distributive inequity, lack of recognition, disenfranchisement and exclusion, and, more broadly, an undermining of the basic needs, capabilities, and functioning of individuals and communities."[21] Schlosberg and Collins suggest that climate justice "focuses on local impacts and experience, inequitable vulnerabilities, the importance of community voice, and demands for community sovereignty and functioning."[22]

Echoing the direction of Bulkeley et al. towards inclusion of recognition as a fundamental element of climate justice, EJ scholars have called for expanding EJ/CJ methodology and theory

> beyond the unequal distributions of impacts and/or responsibilities to include the processes of disrespect, devaluation, degradation, or insult of some people versus others; inclusion or exclusion in participation and procedure; and the provision and protection of the basic capabilities or needs of everyday life.[23]

CLIMATE JUSTICE AND THREATS TO DEVELOPMENT

Writing in the space between academics and activists, Adams and Luchsinger, in a 2009 brief on climate justice for the UN Non-Governmental Liaison Service, emphasize the importance of sustainable development and the significant impacts on development posed by climate change that will set countries back years of effort and progress. They note that climate justice "builds on a platform of equitable development, human rights and political voice," with an agenda aimed at "reducing disparities in development and power that drive climate change and continued injustice."[24]

From the streets

The elements of climate justice identified by actors in the climate justice movement do not stray so far from those of the academics we quote in the previous section. In his work looking across scales at local and global climate justice, Evans identifies four elements of climate justice as articulated by campaigning organizations and social movements:

- developed countries and their corporations are historically responsible for the bulk of the climate problem;
- there is a need for recognition of the "inequitable and disproportionate" impacts of climate change on indigenous communities and poor and vulnerable communities;
- developed countries that are historically responsible for emissions and impacts should repay the resulting ecological/climate debt that is owed to these communities; and
- rejection of the idea that market solutions can solve the climate change problem.[25]

These elements are found in the communication materials of a wide spectrum of organizations, located in the North and the South. For example:

- The Pan African Climate Justice Alliance calls for an equitable sharing of the atmosphere and for historical polluters to repay climate debt and compensate for climate harms.[26]
- Christian Aid (UK) asks developed countries to "fund and support poor nations to cut their own emissions and develop cleanly, without adopting a highly polluting approach to economic development as rich countries have in the past" and "help poor countries to adapt to the impact of climate change."[27]
- CIDSE (a network of Catholic development organizations) says, "in recognition of their ecological debt to the international community, industrialized countries must assume significant responsibility for taking the lead in making absolute reduction of GHG emissions."[28]

The grassroots Global Campaign to Demand Climate Justice articulates a more expansive understanding of critical elements of climate justice in its platform, Fight for Climate Justice![29]

Fight for climate justice!

1. Fight for the transformation of energy systems
2. Fight for food sovereignty, for peoples' rights to sufficient, healthy and appropriate food and sustainable food systems
3. Fight for peoples' rights to sufficient, affordable, clean, quality water
4. Fight for just transitions for all workers
5. Fight for people's safety and security of homes and livelihoods from climate disasters
6. Fight for the social, political, economic, cultural and reproductive rights and empowerment of all of our people and communities
7. Fight for reparations for climate debt owed by those most responsible
8. Fight for the mobilization and delivery of climate finance
9. Fight for the end to deception and false solutions
10. Fight for the end to policies, decisions and measures by governments, elites, institutions and corporations (domestic, regional and global) that increase the vulnerabilities of people and planet to impacts of climate change
11. Fight to stop the commodification and financialization of nature and nature's functions
12. Fight for an international climate agreement that is rooted in science, equity and justice

Wings of the movement. Athanasiou[30] identifies three wings to the climate justice movement: grassroots resistance, or what Klein[31] has called "Blockadia"; "just transitions," linked with labor movement framings;[32] and "fair shares." In his argument about the way forward *After the catastrophe* (the results of the U.S. presidential election in 2016), he challenges the climate justice movement to "understand and acknowledge all sides of the challenge that it's taking on" – along all three of the wings – to embrace a definition of climate justice that explicitly includes both just transitions and fair shares.[33]

A just transition. "A just transition . . . focuses on jobs, livelihoods and ensuring that no one is left behind as we race to reduce emissions, protect the climate and advance social and economic justice."[34] For Athanasiou, "an expansive approach to the just transitions problems is a big part of the answer . . . The climate transition, if we really intend it to occur, has to stand for a world in which there's a place for everyone, a real place, one with at least a modicum of dignity."[35]

Fair shares. Consideration of "fair shares" (fair shares in relation to climate debt, where developed countries do their fair share of reducing emissions and addressing adaptation and loss and damage based on their historical responsibility and respective capabilities) requires the justice conversation to include equity and address inequality in both its national and international dimensions.[36] Who must take action? How is that action to happen? What are the national responsibilities and capabilities to address climate change, "in ways that take inequality within countries into explicit account"?[37] Given the substantial costs of action, this equity element also assumes "substantial and predictable channels of finance and technology support."[38]

Climate justice in conversation with the human rights and climate change agenda

Recall the long list of elements of what might comprise climate justice: recognition, rights, responsibility, procedures, distribution, in/equality, human rights, collective rights, historical responsibility, decolonization, development, just transition, equity, addressing impacts of climate change on indigenous communities and ways of life, finance, and repayment of ecological and climate debt.

Human rights are indeed central to any framing of climate justice. Indeed, the rights to life, adequate food, the enjoyment of the highest attainable standard of physical and mental health, safe drinking water and sanitation, adequate housing, self-determination, and development are all inarguably necessary elements of a climate-just world. The struggle to address climate change is of course a struggle both for justice and for human rights as core components of that justice. Ahmed Khaleel, Permanent Representative of the Republic of the Maldives to the United Nations, noted during general debate at the UN General Assembly in 2007 that, for the Maldives, climate change

> is not solely a development issue, but also a moral issue, an ethical issue, a political issue, a legal issue, a human rights issue as well as a grave security issue . . . Addressing the injustices of climate change is . . . the moral and ethical responsibility of the entire international community. It is time that we put people back on the heart of the climate change debate. We believe a comprehensive rights based approach to sustainable and just development, anchored in the concept of common but differentiated responsibility, is now an imperative.[39]

No doubt, the human rights framework and approach are powerful tools in the fight for climate justice, not least because they "supply not only legal imperatives, but also a set of internationally agreed values around which common action can be negotiated and motivated."[40] That climate justice encompasses a broader agenda for environmental and social justice beyond human rights is clear, but so what? How does that shape our actions both as climate justice activists and as human rights advocates?

Finding convergences and synergies

How do we use both a climate justice and a human rights agenda in the struggle against climate change? We might start by considering some of the most important unresolved challenges faced by those following a climate justice agenda or human rights–based approaches. In examining challenges, the hope is that we also can identify opportunities – for synergies that might help us to surmount individual and shared obstacles.

Gaps in achieving climate justice

In the lead-up to UNFCCC COP15 in Copenhagen, Adams and Luchsinger identified four main gaps in addressing climate justice in the negotiations.[41] These gaps remain relevant:

Disparities in development. "Justice comes from recognizing that people who are poor have the right to pursue opportunities to seek well-being – just as wealthier people have already done – rather than being further penalized by poverty."[42] Climate change affects most those who are most vulnerable in any society. It steals dreams of better lives when government priorities and resources must be shifted away from proactive sustainable development efforts to addressing impacts of climate change. Development disparities become exacerbated when developing countries are pushed to contribute more than their fair share of efforts – for mitigation or adaptation or assuming the costs of loss and damage. As noted by Nicholas Stern:

> in the case of the . . . "contents of the atmosphere," it is hard to think of an argument as to why rich people should have more of this shared resource than poor people. They are not exchanging their labor for somebody else's and they are not consuming the proceeds of their own land, or some natural resource that lies beneath it.[43]

Some groups face more threats: women, children, Indigenous Peoples, migrants. "The effects of climate change will fall hardest on the rights of those people who are already in vulnerable situations 'owing to factors such as geography, poverty, gender, age, indigenous or minority status and disability.'"[44]

Meaningful participation. Those who are most affected have little or no say on who takes action and how mitigation or adaptation is achieved. Intergovernmental mechanisms such as the CDM have imposed "solutions" without consultation. As the climate crisis deepens and action becomes even more urgent than it is today, observers fear geoengineering schemes will be imposed that further threaten lives and livelihoods.

Who should pay? "There is no doubt: large quantities of resources will be required to combat climate change." Yet, historical responsibility for climate debt is not recognized by developed countries, who see the clock starting in 1990 for responsibility and claims on atmospheric space.

Limitations of a human rights approach

> *More than most other issues, climate change throws into relief the inadequacies of the international justice system, given the scale and intimacy of global interdependence that drives the problem and must also drive its solutions.*[45]

The U.S. government, in its submission to the Office of the High Commissioner for Human Rights (OHCHR) report on human rights and climate change, minced few words to challenge the human rights space as a forum for addressing climate change. The submission notes that climate change is a highly complex issue, a global phenomenon, and a long-term challenge, inappropriate to being addressed in a human rights framework that requires identifiable violations and harms. In commenting on the submission, Limon notes that such a framework "provides no useful kind of accountability or redress framework for situations arising from phenomena such as climate change, where responsibility and harm are largely trans-national."[46]

Limon comments that "indeed, for someone from the Maldives to prove that his or her rights have been violated as a result of climate change and to hold those responsible (wherever they may be) accountable, would require a wholesale reconceptualization and reconfiguration of international human rights law as it is now understood."[47]

A crucial challenge is thus that, as reflected in the Mali government submission to the OHCHR report on human rights and climate change:

> laws and institutions for the defense of human rights [must] rapidly evolve to adapt to the new reality of climate change. When vulnerable communities have tried to use human rights laws to defend their rights and seek climate justice, important weaknesses have been revealed. It is almost impossible for populations in poor countries to identify and pursue channels of justice, to have their cases heard, or to prove responsibility.[48]

Conclusion: the road ahead together

Let me end by identifying four intersecting and intersectional challenges for the road ahead:

Equity and equality. At the intergovernmental level, resolving issues of equity in apportioning responsibility is crucial. At the national and sub-national levels, structural inequalities threaten access to justice – in terms of distribution, procedure, and recognition. Inequalities underlie vulnerability, exposure to and consequences of harm. Where are intersections here in terms of moving forward? Limon asks whether the principle of common but differentiated responsibilities might "also be helpful in reshaping international human rights law to make it more reflective of and more responsive to the needs of a globalized world?"[49] In the climate negotiations, many developing countries have asserted that equity is critical to ambition, a point reiterated by a large group of academics in a recent contribution to the scholarly debate on equity in climate change policy.[50] Equity is not visibly a part of the human rights agenda, but could it be? And how can the human rights agenda be framed to help address systematic inequalities within the nation-state?

Responsibility. One of the main obstacles to achieving equity between nation-states in the climate regime, and to effectively addressing the impacts of climate change on countries and communities, is the recognition and apportionment of responsibility. A first step to remedying harm is to identify responsible parties. A human-rights approach has much to contribute here, in that

> the right to a remedy is at the heart of a rights-based agenda. To operationalize this right, a mechanism must be established to provide a remedy to individuals, communities or indigenous peoples whose rights are adversely affected by climate impacts and response measures. Ensuring the right to a remedy also requires that developed countries responsibly repay their climate debts and release resources for financial compensation to climate victims in developing countries.[51]

Recognition. We agree with Bulkeley and colleagues – recognition is foundational to both the exercise of all human rights as well as to the broader demands for climate justice. Struggles for recognition are most often at the community level and require efforts to give voice to communities and empower sovereignty over their lives and livelihoods. There are strong links here to efforts towards the right to development that are not merely distributive or procedural in nature, but fundamentally about who gets to speak and define their own future.

Solidarity. And finally a word, idea, goal to strive for over the coming decades – solidarity. The president of the Maldives has observed that "the world has failed to humanize climate change."[52] Beyond rights is the necessity of solidarity – of taking care of our fellow humans – including recognizing our (Western) history and our overwhelming responsibility both for how we arrived at this moment and for how we move forward. Climate justice (equity) acknowledges that those of us most privileged, those living highly consumptive lifestyles in countries that have greatly benefited from past emissions, have a collective responsibility not just to protect human rights, as they exist now and in the future, but to right climate wrongs.

Writing the next climate justice and human rights story

Recently, the issue of climate refugees has resurfaced onto the global climate negotiations agenda – though for most small island states, their demands on this issue never really went away. As an outcome of the Paris climate conference, Parties agreed to set up a task force on climate displacement under the Warsaw International Mechanism for Loss and Damage. This step is timely, was urgently needed, and now pushes us to begin to answer some very difficult questions at the intersections of human rights and climate justice, such as, but not limited to, How will we take care of climate refugees now and the numbers that will only continue to grow as temperatures rise and productive land and livelihoods are lost? Those who are internally displaced? Who move regionally on, say, the African continent? Who arrive on the shores of Europe alive?

Answers to these difficult questions to date have been sadly few and insufficient – still, both human rights and climate justice agendas require that we now begin to demand answers. But in so doing, we must start from a foundation of the rights-based approach *and go further.* Harvard professor and justice scholar Michael Sandel points us in that direction. His critique of liberal justice is that it only: "requires that we respect people's rights . . . not that we advance their good."[53] Solidarity takes us beyond rights in this very direction, towards the good. Filling this particular gap between climate justice and a rights-based approach will steer us on a truer course on the road to justice.

Notes

1 Global Campaign to Demand Climate Justice, Global Climate Justice Movements Refuse to Be Overshadowed by Election of Climate Change Denier to U.S. Presidency, Declaration to COP22 in Marrakech, Morocco, November 2016, available online at http://climatejusticecampaign.org/sign-now?highlight=WyJtYXJyYWtlY2giXQ==.

2 Schlosberg and Collins, 'From Environmental to Climate Justice: Climate Change and the Discourse of Environmental Justice', 5(3) *WIREs Climate Change* (WCC) (2014); J. Martinez-Alier, *The Environmentalism of the Poor* (2002).

3 Limon, 'Human Rights and Climate Change: Constructing a Case for Political Action', 33 *Harvard Environmental Law Review* (HELR) (2009) 439.

4 K. P. Whyte, *Why the Native American Pipeline Resistance in North Dakota Is About Climate Justice* (16 September 2016), available online at https://theconversation.com/why-the-native-american-pipeline-resistance-in-north-dakota-is-about-climate-justice-64714.

5 Whyte as quoted in Bagley. See K. Bagley, *At Standing Rock, a Battle Over Fossil Fuels and Land* (10 November 2016), available online at http://e360.yale.edu/features/at_standing_rock_battle_over_fossil_fuels_and_land.

6 Wikipedia, *Climate Justice*, available online at https://en.wikipedia.org/wiki/Climate_justice.

7 S. Vanderheiden, 'Atmospheric Justice: A Political Theory of Climate Change' (2008); Posner and Weisbach, 'Climate Change Justice' (2007); Okereke, 'Climate Justice and the International Regime',

1 *WIREs Climate Change (WCC)* (2010) 462; Roberts and Parks, 'Ecologically Unequal Exchange, Ecological Debt, and Climate Justice: The History and Implications of Three Related Ideas for a New Social Movement', 50 *International Journal of Comparative Sociology (IJCS)* (2009) 385.
8 Hayward, 'Human Rights Versus Emissions Rights: Climate Justice and the Equitable Distribution of Ecological Space', 21(4) *Ethics & International Affairs (EIA)* (2007) 431; Agyeman *et al.*, 'Trends and Directions in Environmental Justice: From Inequity to Everyday Life, Community, and Just Sustainabilities', 41 *Annual Review of Environment and Resources* (2016) 321.
9 Caney, 'Two Kinds of Climate Justice: Avoiding Harm and Sharing Burdens', 22 *The Journal of Political Philosophy (JPP)* (2014) 125.
10 See Climate Fairshares, available online at http://climatefairshares.org.
11 Stilwell, 'Climate Debt – a Primer', 61 *Development Dialogue* (2012) 41.
12 Bulkeley *et al.*, 'Contesting Climate Justice in the City: Examining Politics and Practice in Urban Climate Change Experiments', 25 *Global Environmental Change (GEC)* (2014) 31.
13 *Ibid.*
14 See Bulkeley *et al.*, 'Climate Justice and Global Cities: Mapping the Emerging Discourses', 23 *Global Environmental Change (GEC)* (2013) 914, citing Fraser: N. Fraser, *Justice Interruptus: Critical Reflections on the "Post-Socialist" Condition* (1997).
15 See Bulkeley *et al.*, *supra* note 12.
16 *Ibid.*
17 See Bulkeley *et al.*, *supra* note 14.
18 For example, see Agyeman *et al.*, *supra* note 8.
19 Schlosberg and Collins, *supra* note 2.
20 Bulkeley *et al.*, *supra* note 12; Bulkeley *et al.*, *supra* note 14.
21 Schlosberg and Collins, *supra* note 2.
22 Schlosberg and Collins, *supra* note 2.
23 Agyeman *et al.*, *supra* note 8.
24 Adams and Luchsinger, UNCTAD, *Climate Justice for a Changing Planet: A Primer for Policy Makers and NGOs*, UN Doc. UNCTAD/NGLS/2009/2, November 2009.
25 Evans, 'A Rising Tide: Linking Local and Global Climate Justice', 66 *Journal of Australian Political Economy (JAPE)* (2011) 199.
26 Pan African Climate Justice Alliance, Statement to the Conference of African Heads of States and Governments on Climate Change (2009), document on file with author.
27 Christian Aid, Time for climate justice (2009), document on file with author. Flyer for Countdown to CO_2penhagen campaign.
28 CIDSE, *Development and Climate Justice*, Policy Paper (November 2008).
29 *Global Campaign to Demand Climate Justice, Fight for Climate Justice!* (2015), available online at http://climatejusticecampaign.org/resources/25-fight-for-climate-justice.
30 T. Athanasiou, *After the Catastrophe: Climate Justice as the Post-Trump Slingshot* (18 January 2017), available online at www.ecoequity.org/2017/01/after-the-catastrophe/.
31 N. Klein, *This Changes Everything: Capitalism vs. the Climate* (2014).
32 ITUC, Just Transition Center, *Welcome Page*, available online at www.ituc-csi.org/just-transition-centre.
33 Athanasiou, *supra* note 29.
34 ITUC, *supra* note 31.
35 Athanasiou, *supra* note 29.
36 See also Klinsky *et al.*, 'Why Equity Is Fundamental in Climate Change Policy Research', *Global Environmental Change* (25 August 2016), DOI: 10.1016/j.gloenvcha.2016.08.002, for a discussion about including equity in discussions.
37 Athanasiou, *supra* note 29.
38 *Ibid.*
39 Permanent Mission of the Republic of Maldives to the United Nations, Statement by His Excellency Mr. Ahmed Khaleel, Permanent Representative of the Republic of Maldives to the United Nations at the General Debate of the Sixty Third Session of the United Nations General Assembly (29 September 2007), available online at www.un.org/ga/63/generaldebate/pdf/maldives_en.pdf.
40 International Council on Human Rights Policy (ICHRP), *Climate Change and Human Rights: A Rough Guide* (2008), available online at www.ichrp.org/files/reports/45/136_report.pdf, at 8.
41 Adams and Luchsinger, *supra* note 24.
42 *Ibid.*

43 Stern, 'The Economics of Climate Change', 98 *American Economic Review: Papers & Proceedings* (2008) 1.
44 Adams and Luchsinger, *supra* note 24.
45 ICHRP, *supra* note 40, at 64.
46 Limon, *supra* note 3.
47 *Ibid.*
48 République du Mali, *Ministère de l'Environnement et de l'assainissement, Droits de l'homme et changements climatiques au Mali* (September 2008), available online at http://www2.ohchr.org/English/issues/climatechange/docs/Mali.pdf.
49 Limon, *supra* note 3.
50 Klinsky *et al.*, *supra* note 36.
51 Climate Justice Briefs, *Human Rights and Climate Justice*, brief 12, November 2010.
52 Limon, *supra* note 3.
53 Sandel, 'Democracy's Discontent', in M. J. Sandel (ed.), *Justice: A Reader* (2007) 328, at 332.

21

Securing workers' rights in the transition to a low-carbon world

The just transition concept and its evolution

Edouard Morena

Introduction

"This COP is not just a COP of decisions; it is a COP of solutions." When he pronounced these words, Laurent Fabius, French foreign minister and acting president of COP21, was referring to an important dimension of the Paris climate conference and resulting agreement. Through its ambitious long-term goal and system of five-year stocktakes to evaluate progress and set future targets, the Paris Agreement marks for its architects the start of a systemic shift towards a decarbonised economy; a shift that has to take place over the coming decades if we are to limit the rise of global temperatures to 1.5–2°C above pre-Industrial Revolution levels by the end of the century. The Paris 'moment' was intended to provide the framework and generate the political impetus for ambitious climate action. Paris also marks a clear shift away from a top-down, multilateral and legally binding approach to climate governance and a move towards a more bottom-up approach through voluntary actions at the national level.

These shifts in international climate governance have led to growing attention towards the ways of encouraging key economic and social actors to contribute to climate change mitigation and adaptation. The Paris conference was as much about negotiating a new agreement as it was about showcasing existing and future national, corporate and civil society solutions to the climate crisis. Multilateral, state and non-state initiatives were presented during the two-week Paris COP through side events, stands, press conferences and seminars both inside and outside the official Conference center in Le Bourget. They offered a chance to determine how economic sectors – agriculture, industry, transport, energy – local governments and cities, and NGOs can collectively contribute to achieving the long-term goals laid down in the agreement.

Among the various terms and concepts that were referred to, one in particular appears to have made its mark in Paris: the 'just transition' (JT). The concept, whose origins lie in the trade union community, draws on the social – and particularly employment – implications of climate change and climate action. In addition to being widely referred to and discussed in the run-up to and during the Paris conference, it was also included in the final text of the agreement. In its preamble, the Paris Agreement refers to the need to "[take] into account the imperatives of a just transition of the workforce and the creation of decent work and quality jobs in accordance with nationally defined development priorities."[1] The JT concept's inclusion in the Paris Agreement is the outcome of years of efforts by a group of committed trade union representatives and sympathisers. As we will see, the JT concept's originality and strength not only resides in its focus area but also in its mobilizing effect in trade union circles. The concept's widespread re-appropriation and re-interpretation by non-union groups risks undermining this important dimension of the JT concept.

Origins and spread of the just transition concept within the trade union movement

At the international level, JT is generally associated with the international trade union movement – and in particular the International Trade Union Confederation (ITUC). By 'just transition,' the ITUC refers to a "conceptual tool that the trade union movement shares with the international community aiming to ensure a soft passage towards a more sustainable society." While the term has been traced back to the 1970s and 1980s in the United States, it was during the early 1990s that US and Canadian unions began to systematically use it in response "to new regulations to prevent air and water pollution, which resulted in the closure of offending industries."[2] On the back of the 1992 UN Conference on Environment and Development (Rio92), JT was progressively taken up by the international trade union movement – through the International Confederation of Free Trade Unions (ICFTU) which became the ITUC in 2006 – and included in position papers and statements at international environment and development conferences.[3] In 1997, for example, the JT concept was included in the ICFTU's official statement at the Kyoto climate conference (COP3). We would have to wait until the second half of the following decade to see more active and coordinated efforts to mainstream JT inside the trade union community and to lobby for the inclusion of JT in UN processes and agreements (United Nations Commission on Sustainable Development [UNCSD], United Nations Framework Convention on Climate Change [UNFCCC], International Labour Organisation [ILO], United Nations Environment Program [UNEP]).[4] This coincides with the foundation of the ITUC in 2006 – replacing the ICFTU – whose original mandate included environmental issues and the launch of the Sustainlabour Foundation in 2004, whose function was to engage the trade union community on sustainable development issues.

Given its growing importance, the UN climate process – through the UNFCCC – logically became a privileged space for the ITUC to promote its JT agenda. Consequently, and within the international climate community, JT was increasingly framed and recognized as the trade union movement's contribution to the international climate debate. It reflected a growing preoccupation within the trade union movement that social – and in particular employment – concerns were not appropriately addressed within the UNFCCC process. Given the scale of the climate challenge and measures required to mitigate and adapt to it, there was a growing fear amongst unions that workers would lose out in terms of job security and quality. Corporate interests – especially coal – in trying to undermine the international climate process further exacerbated this fear. By engaging in the climate process and ultimately pushing for the inclusion of JT into the Paris Agreement, the goal was to counter denialist influences within the movement by highlighting the benefits for workers and their communities of decisive climate action.

By shedding light on the social implications of climate change, the JT concept filled an important gap in the international climate debate. Up to recently, equity and justice issues were essentially framed along a North-South axis. The priority for many climate justice activists involved in and around the UNFCCC was to get northern countries to recognize their historical responsibilities and to act upon them – both through more ambitious national mitigation efforts and through financial and technological assistance to developing countries (that are also the hardest affected by climate change). While climate justice groups referred to climate change's inequitable social impacts, they tended to focus on geographical differences.[5] In other words, fairly limited attention was paid to the differentiated social implications of both climate change and climate policies on the world of work – in the global North and South. The JT concept addressed this shortfall.

Instead of adopting a defensive 'corporatist' stance – an approach that is still favoured by some sector-based unions (especially coalminers) – the idea was to show how a better inclusion of world of work concerns into international development and environmental processes could facilitate the transition towards a low-carbon world. As Anabella Rosemberg, the ITUC's environmental officer explains, the 'just transition' seeks to "strengthen the idea that environmental and social policies are not contradictory but, on the contrary, can reinforce each other."[6] In particular, the ITUC regularly stresses the tremendous opportunities in terms of job creation of a just transition towards a green economy. In this sense, the ITUC's framing of JT has frequently been associated with the 'green growth' paradigm and associated narrative that presently dominates the international development and environmental agenda.[7] This has led some observers, such as Stefania Barca, to criticize the ITUC for being out of touch with reality and for blindly accompanying the dominant growth and techno-centred economic model.[8]

By focusing on its ideological underpinnings, existing analyses tend to neglect the JT concept's other functions for the international trade union movement. For many JT advocates, the priority is less of coming up with a refined and ready-to-use roadmap for socially inclusive climate action than of (a) making sure that world of work concerns are taken up in international environment and development processes; and (b) getting the wider trade union community to engage in the climate change debate – and environmental debate more broadly. As with other concepts circulated in international arenas – including 'green growth' and 'sustainable development' – JT acts as a policy instrument to galvanize action on the social implications of environmental change.

In addition to raising the trade union movement's profile in international environmental processes (see above), the JT concept was instrumental in spurring greater trade union engagement with environmental concerns.[9] The development of the JT concept and its mainstreaming formed part of a wider strategy developed by the ITUC and its allies (especially Sustainlabour) that consisted in framing environmental issues in a 'union friendly' manner.[10] The fact that JT was a product of the trade union movement was key, given the climate of mistrust that has traditionally plagued union-environmentalist relations. JT was a core element in various ITUC climate-related activities. These included the organisation of training sessions for regional, national and sub-national union representatives, the drafting and diffusion of training manuals and information materials. These types of activities were, for example, at the heart of the UNEP-Sustainlabour "Project on Strengthening Trade Union Participation in International Environmental Processes" (2008–2010). The Project's three main objectives were to:

> increase participation of workers and trade unions in international environmental processes; increase workers and trade unions' capacities to replicated/adapt case studies on environmental issues in their workplaces and communities; and increase awareness of the environmental issues among workers and trade unions and how they can potentially affect their workplaces and worklife.[11]

More recently, the ITUC – in collaboration with the Friedrich-Ebert-Stiftung (FES) – organised a series of training and awareness-raising sessions in countries such as Ghana, Argentina and more recently Morocco. The ITUC also contributed to the preparation and delivery of training courses on JT at the ILO's International Training Center (ITC) in Turin, Italy.[12] In addition to serving as a lobbying space, COP meetings also offer opportunities to inform and engage with local unions about environmental and climate-related issues. At the COP22 in Marrakech, for instance, the ITUC used the event to engage with local Moroccan unions and familiarize them with the JT concept.

By repeatedly participating in international negotiations and conferences, and actively pushing for the JT concept's official recognition and inclusion in the negotiations, the trade union movement was able to build up internal capacities, while at the same time deepening relations with other stakeholders – NGOs, business representatives, country delegates and intergovernmental organisations (ILO, UNFCCC, UNEP). By regularly participating in climate conferences and engaging in lobbying activities – to get the UNFCCC to officially adopt the JT concept – members of the trade union constituency became recognized and respected members of the international climate community.

Explaining growing non-union interest towards JT

As was previously highlighted, until recently, non-union actors involved in the international climate space rarely took up the JT concept. At the UNFCCC level, it was essentially and almost exclusively a trade union demand. The Paris climate conference clearly signals its wider adoption by non-union actors both inside and on the margins of the negotiation halls. In the run-up to and during the Paris COP, a wide array of non-state actors representing very different constituencies, interests and approaches called for its inclusion in the final agreement, placing it on a par with more established ideas and concepts – particularly in the field of human rights (indigenous rights, women's rights) and traditional knowledge. These included representatives from the business community, such as We Mean Business or the B Team,[13] as well as respected and influential think tanks, such as E3G or WRI. Practically all of the key environmental and development NGOs and networks – from the more moderate to the more radical – publicly endorsed the 'just transition' concept. This marks a significant shift when compared to earlier climate conferences and in particular COP15 in Copenhagen, where only a handful of non-union groups explicitly referred to the JT concept. We can identify four main reasons behind JT's growing popularity outside the union movement.

The first relates to the movement's growing presence within the international negotiation space and recognition as a legitimate and credible partner in the debate. This is the consequence of sustained efforts – in particular by Sustainlabour and the ITUC, with backing from the ILO and UNEP – to mainstream environmental – and in particular climate – concerns within the union community and to build up union capacities in international environmental processes. JT, as we have shown, was a core component of this mainstreaming effort. As a credible and trusted partner, the ITUC was able to build up and consolidate a network of delegates, UNFCCC, NGO and intergovernmental representatives who were prepared to vouch for the JT concept's inclusion in the negotiations process. The ITUC also took part in a variety of informal and formal coalitions.

Secondly, the Paris Agreement's reference to JT further legitimises the concept and encourages others to use it. This is compounded by the fact that JT is consistent with the agreement's voluntary and bottom-up approach to change, and the wider narrative on the combined economic, social and environmental benefits of climate action, especially in the energy field. E3G, for example, highlights that, given the influence of the coal industry, a JT framework is essential if we are to secure the support of workers and their communities for a green industrial transformation towards a low-carbon economy.[14]

Thirdly, the current economic and political climate – marked by high unemployment, rising inequalities and rampant popular support for right wing, and often climate sceptic, populists – emphasizes the need to better integrate labour and social justice concerns into the climate agenda. Brexit in the United Kingdom and Donald Trump's presidential victory in the United States are stark reminders of the widening gap between policy-makers and the most vulnerable sections of society. In the climate field, this will require policies that contribute to solve the social implications of climate change and climate action. As WRI writes in a statement published shortly after the electoral victory of Donald Trump, "one of the clearest messages from the [US presidential] campaign is that America must address inequality within our society and reshape our economy so that all people can thrive."[15] By introducing social justice concerns into the climate change debate, JT constitutes a useful tool towards this end. As Newell and Mulvaney explain, the global energy landscape's rapid and profound restructuration will require just transition policies that secure energy access to those who don't have it, and justice for those who work in or whose livelihoods depend on the fossil fuel economy – and who, as we saw in the United States, actively supported a climate-sceptic candidate to the White House.[16]

Challenges

The aforementioned factors have undoubtedly contributed to raise JT's – and through it the international trade union movement's – profile among non-union stakeholders active in the climate arena. Most groups simply mention the concept without seeking to shape its meaning or re-appropriate it for the pursuit of their agendas. In other words, while a growing number of actors are referring to the JT concept, only a handful of groups dedicate human and financial resources to thoroughly reflect on its meanings and implications and to translate it into concrete actions. While in doing so they undoubtedly contribute further root social concerns into the climate agenda, they also pose new challenges for the ITUC. As we saw, JT acted as a formidable tool to mobilize unions and workers on climate change. From the moment that JT goes from being a trade union concept to an 'unbranded' one, the risk is of undermining JT's awareness-raising role within the trade union movement.

Among the handful of groups that have actively re-appropriated and acted upon the concept, some abstain from referring to organised labour altogether. At the international level, a noteworthy example is the Just Transition Co-Learning and Strategy Development Collaborative. Launched in late 2015 by members of the EDGE Funders Alliance, the Collaborative brings together funders, civil society organisations and networks

> to inform and deepen understanding of just transition narratives and practice, encouraging funders to learn from, engage with and provide funding to community-based groups, movement-support organizations and policy advocacy institutes building and promoting systemic alternatives at local, national and international levels.[17]

The Collaborative builds on pre-existing campaigns, and most notably the US-based Climate Justice Alliance's Our Power Campaign (OPC), which pushes for community-centred JT initiatives such as local agriculture, zero waste, public transportation, efficient, affordable and durable housing and clean community energy, among others.[18] As with OPC, the Collaborative adopts a 'systems-change' approach that prioritizes and empowers frontline communities. By communities, they mean "indigenous peoples, people of color, new immigrants and working class peoples." In addition to promoting a bottom-up approach to JT and targeting a broader set of actors than just workers, their approach to JT sidelines labour organisations, and particularly national unions. This has a lot to do with trade unions' perceived rigidity and conservatism that are ill adapted to contemporary problems. As David Bollier writes in his report 'A Just Transition and Progressive Philanthropy,' the main drivers of the JT will be "flexible players in open, fluid environments – as players in dynamic, collaborative *movements*."[19]

What the Just Transition Collaborative example tells us is that the JT concept's development can produce new understandings of what the concept should stand for; new understandings that while broadly keeping with its original 'social' emphasis marginalize unions instead of placing them at the centre of the equation. By depriving unions of the JT concept, initiatives such as these risk demobilizing them and undermining years of efforts to engage them in the international climate and environmental agenda. While the Just Transition Collaborative is a relatively isolated case, and given JT's growing popularity, there is a genuine danger of seeing other actors take up the concept in a similar fashion. The question becomes of seeing whether the ITUC and its allies are able to preserve the JT concept's trade unionist credentials. The ITUC's decision to launch a Just Transition Centre in 2016 marks a clear effort to do so.[20] Whether or not this will be sufficient, however, remains to be seen.

Notes

1 UN Framework Convention on Climate Change, *The Paris Agreement* (2016), available online at http://unfccc.int/paris_agreement/items/9485.php.
2 P. Newell and D. Mulvaney, "The Political Economy of the 'Just Transition'", 179(2) *The Geographical Journal* (2013) 132; R. Felli and D. Stevis, "Global Labour Unions and Just Transition to a Green Economy", 15(1) *International Environmental Agreements: Politics, Law and Economics* (2015) 29, at 32.
3 E. Morena, 'Les reconfigurations environnementales du syndicalisme: construction de positions et stratégies globales', in J. Foyer (ed.), *Regards croisés sur Rio+20: La modernisation écologique à l'épreuve* (2015); A. Rosemberg, "Climate Change and Labour: The Need for a 'Just Transition'", 2(2) *Journal of Labour Research* (2010) 125.
4 Important milestones in its international development include the first Trade Union Assembly on Labour and Environment in Nairobi (2006), the recognition of trade unions as a distinct major group at the UNFCCC (2008), the recognition of just transition in the "shared vision" document in the run-up to the Copenhagen COP (successive COP decisions systematically included references to the JT), the passing of a historic resolution on the need to "[combat] climate change through sustainable development and just transition" at the ITUC's second World Congress in Vancouver, and its inclusion in the COP16 declaration (Cancun, 2010).
5 D. R. Fisher and A. M. Galli, 'Civil Society', in K. Backstrand and E. Lovbrand (eds.), *Research Handbook on Climate Governance* (2015) 97.
6 A. Rosemberg, 'Developing Global Environmental Union Policies Through the ITUC', in N. Räthzel and D. Uzzell (eds.), *Trade Unions in the Green Economy* (2012) 19.
7 J. Foyer, *Regards croisés sur Rio+20: La modernisation écologique à l'épreuve* (2015).
8 S. Barca, "Greening the Job: Trade Unions, Climate Change and the Political Ecology of Labour," in R. L. Bryant (ed.), *The International Handbook of Political Ecology* (2015) 387.
9 L. M. Murillo, "Making the Environment a Trade Union Issue," in N. Räthzel and D. Uzzell (eds.), *Trade Unions in the Green Economy: Working for the Environment* (2012).
10 Morena, *supra* note 3. Representatives from the ILO and UNEP as well as sympathetic national governments (especially Spain) were also involved.
11 S. R. King, *United Nations Environment Programme, Terminal Evaluation of the UNEP-Sustainlabour Project on Strengthening Trade Union Participation in International Environmental Processes* (2012), available online at https://wedocs.unep.org/rest/bitstreams/840/retrieve, at 1.

12 This includes a course on "Green Jobs for a Just Transition to Low-Carbon and Climate Resilient Development", see International Training Centre of the International Labour Organisation, Green jobs for a just transition to low-carbon and climate resilient development, available online at www.ilo.org/wcmsp5/groups/public/---ed_emp/---gjp/documents/event/wcms_459900.pdf.
13 International Trade Union Confederation, Historic partnership of business, NGOs, faith groups and trade unions sign commitment to just transition dialogue for a zero carbon future, 30 November 2015, available online at www.ituc-csi.org/historic-partnership-of-business?lang=en.
14 S. Schulz and J. Schwartzkopff, Making the Just Transition Happen, *E3G* (12 January 2015), available online at www.e3g.org/library/making-the-just-transition-happen.
15 World Resource Institute, *STATEMENT: WRI Statement on Donald Trump as the Next President of the United States* (9 November 2016), available online at www.wri.org/news/2016/11/statement-wri-statement-donald-trump-next-president-united-states.
16 Newell and Mulvaney, *supra* note 2.
17 EDGE Funders Alliance, *Transitioning to the Next Economy: A Just Transition Co-Learning and Strategy Collaborative*, available online at http://www.edgefunders.org/wp-content/uploads/2015/10/Just-Transition-Collaborative-Summary2.pdf
18 Grassroots Global Justice Alliance, *Climate Justice Alliance: Our Power Campaign*, available online at http://ggjalliance.org/ourpowercampaign.
19 D. Bollier, *News and Perspectives on the Commons, Progressive Philanthropy Needs to Spur System Change* (5 February 2016), available online at bollier.org/blog/progressive-philanthropy-needs-spur-system-change.
20 Its purpose is "to deliver and build the social dialogue for a just transition" by "[bringing] together and [supporting] unions, businesses, companies, communities and investors in social dialogue . . . for a fast and fair transition to zero carbon and zero poverty." International Trade Union Confederation, Just Transition Centre, available online at www.ituc-csi.org/just-transition-centre?lang=en.

Part V
Regional case studies

22

'There Is No Time Left'

Climate change, environmental threats, and human rights in Turkana County, Kenya

Katharina Rall and Felix Horne

Introduction

Over the past century, the average annual temperature on earth has increased, the oceans have warmed, snow and ice caps have diminished, and sea levels have risen.[1] Climate change is being felt in countries throughout the world, from low-lying countries, such as Bangladesh and the Maldives, to temperate countries in the northern hemisphere, to countries in Africa's arid and semi-arid Sahel.[2]

Scientists believe Africa will be one of the continents most vulnerable to climate change, with average temperatures expected to increase throughout this century, and with drier sub-tropical regions warming more than the moister tropics.[3] In its latest report, issued in November 2014, the Intergovernmental Panel on Climate Change (IPCC) states that the continued global emission of greenhouse gases will increase the likelihood of severe, pervasive, and irreversible impacts for people and ecosystems in Africa, including but not limited to its effects on water, food, and health.[4]

Human rights bodies, scientific experts, governments, and civil society have recognized that worldwide, climate change is having, and will continue to have, a devastating impact on the ability of people to enjoy their basic human rights and the capacity of governments to fulfill their obligations to realize those rights.[5] The Paris Agreement explicitly acknowledges "that climate change is a common concern of humankind" and that parties should "respect, promote and consider" human rights, including the rights of indigenous peoples, as well as "migrants, children, persons with disabilities and people in vulnerable situations."

The earth's surface temperature is projected to rise further, heat waves are likely to occur more often and last longer, extreme precipitation events will become more intense and frequent, the ocean will continue to warm and acidify, and the global mean sea level will continue to rise.[6] These changes, and the costs associated with them, threaten the ability of governments, particularly in low- and middle-income countries, to progressively realize the full range of human rights.[7] Indigenous populations, marginalized groups, women, and people with disabilities – populations that are already vulnerable to human rights abuses – will face the biggest challenges adapting to a changing climate.[8]

Countries with tropical or subtropical climates (such as those in Africa) are projected to experience the effects of climate change most intensely, and low-income countries are least able to prevent and prepare for the impact of climate change.[9] According to the IPCC, climate change is expected to lead to disproportionate increases in ill health in developing countries with low income.[10]

African regional institutions have long highlighted the vulnerability of African states to the effects of climate change. Starting in 2007, the African Union has passed a number of resolutions expressing concern about the impact of climate change and the ability of countries to respond to its consequences.[11] In 2009, ministers from more than 30 African countries agreed that "while Africa has contributed the least to the increasing concentration of greenhouse gases in the atmosphere, it is the most vulnerable continent to the impacts of climate change and has the least capacity to adapt."[12] The African Commission on Human and Peoples' Rights also passed a resolution in 2009, urging the inclusion of human rights safeguards into climate change laws and treaties and committing the Working Group on Extractive Industries, Environment and Human Rights Violations in Africa to carry out a study on the impact of climate change on human rights in Africa.[13]

Case study: impacts of climate change in Turkana County, Kenya

This chapter is based on research in Turkana County in northwestern Kenya.[14] Along with Marsabit and Samburu Counties, Turkana County is home to Lake Turkana, the largest desert lake in the world. Turkana County borders South Sudan, Ethiopia, and Uganda and is globally renowned as the cradle of humankind: in Turkana County and the Omo Valley in southern Ethiopia, archeologists have found the oldest ancestors to modern humans, dating back more than one million years ago. However, today, Turkana County is home to a rapidly growing population that is among the poorest in Kenya.[15]

The population is predominately indigenous Turkana people and pastoral, relying on livestock herding. Some Turkana fish in the waters of Lake Turkana, while others reside in the county's towns. Their traditional reliance on natural resources for food and livelihood, the historic marginalization of the region, and the lack of infrastructure make them especially vulnerable to any changes in the environment.[16] Turkana County has had a long history of chronic malnutrition and some of the poorest health indicators in Kenya. According to government records, the Turkana county population has grown dramatically in the last two decades, increasing in the past few years from an estimated 855,393 people in 2009 to 1,256,152 people in 2015.[17]

Turkana County has long experienced periods of cyclical drought. However, increasing temperatures and shifting precipitation patterns, combined with population growth and threats to Lake Turkana from hydroelectric and irrigation projects in Ethiopia, present significant, long-term challenges for the Turkana County and Kenyan national governments.

Over the past several decades, Kenyan government data show a clear trend in increasing average temperatures in the country as a whole and in Turkana County.[18] While global mean temperatures are estimated to have increased by 0.8°C (1.5°F) in the past century, in Turkana County minimum and maximum air temperatures have increased by between 2°C and 3°C (3.5°F and 5.5°F) between 1967 and 2012.[19] Rainfall patterns have also changed: the long rainy season has become shorter and drier and the short rainy season has become longer and wetter, while overall annual rainfall remains at low levels.[20] These patterns appear consistent with scientific evidence suggesting a correlation between increasing overall temperatures on the one hand, and droughts and more extreme rainfall on the other.[21]

Industrial and agricultural development across Turkana's northern border with Ethiopia also poses threats that could affect the realization of rights of the Turkana people. Over the past several years, Ethiopia has embarked on a massive plan for dams, water-intensive irrigated cotton and sugar plantations, and other infrastructure in Ethiopia's Omo River Basin, which provides 90 percent of the water in Lake Turkana. These developments are predicted to dramatically reduce the water supply of Lake Turkana: the planned irrigation projects alone could reduce by up to 50 percent the Omo River's total flow and the lake could recede into two small pools. Following the development of dams and plantations in Ethiopia's lower Omo Valley, Lake Turkana's water levels have already dropped by approximately 1.5 meters since January 2015, and further reduction is likely without urgent efforts to mitigate the impact of Ethiopia's actions.[22]

Reduced water levels in Lake Turkana will have a devastating impact on the environment and people of Turkana County. Dramatic reductions in freshwater input from the Omo River into Lake Turkana will increase levels of salinity in the lake and raise water temperatures, decimating fish breeding areas and mature fish populations. Higher air temperatures will increase rates of evaporation, further increasing salinity while reducing biological productivity.

Impact of climate change on human rights of indigenous people in Turkana

Human Rights Watch conducted research in Turkana County between April 2014 and February 2015, interviewing 40 people, including indigenous pastoralists, fishers, health clinic staff, students, teachers, local civil society activists, and police officers. In addition, Human Rights Watch reviewed international, Kenyan, and Turkana County laws, policies, and development plans, including the Turkana County Development Plan and the Kenya National Climate Change Action Plan, and met with Turkana County and Kenyan national government officials.

Our research finds that climate change, in combination with existing political, environmental, and economic development challenges in Turkana, has had an impact on the Turkana indigenous people. The pastoralists and fishers in Turkana County interviewed by Human Rights Watch painted a picture of a population struggling to survive in an inhospitable climate, ill equipped to adapt to increasing changes in climate and livelihoods and receiving little support from government or civil society. They stated that changes in climate are already having a significant impact on their welfare in multiple ways. In combination with rapid population growth and industrial development, climate change is exacerbating the already significant challenges the Turkana face in securing sufficient water, food, and health.

Right to water

Increased temperatures and unpredictable rainy seasons have placed increased pressure on water resources, resulting in less dry season grazing land, diminished livestock herds, and increased competition over grazing lands. Pastoralists told Human Rights Watch that prolonged and more frequent droughts have exacerbated already difficult access to potable water, making every day a struggle for survival. One elder living near Lake Turkana described his reliance on the lake in times of drought. He said,

> In the past, I came to the lake because my animals had all died. I came telling myself I can get something to feed my stomach here. What will happen when I arrive and the lake has been closed [disappeared]? How will I survive when my animals have died and the lake

> has disappeared? How will I survive when the drought sweeps me away and sends me to my grave?[23]

Traditionally, in times of drought, many pastoralist communities dig in dry riverbeds for water. However, communities now report longer and more severe droughts. As a result, they must dig deeper or walk further. Women and girls often walk extremely long distances to dig for water in dry riverbeds. Human Rights Watch visited a girls' school where the girls must walk several kilometers every day to reach a dry riverbed where they dig for water and then transport 25-liter jerry cans back to the school.

Right to food and livelihood

Livestock herding and fishing, Turkana's two main livelihoods, are extremely susceptible to changes in environment – shifts in water availability, temperature, and other environmental variables can have a devastating impact on livelihoods.[24] While people in Turkana County have long experienced cyclical drought and famine, a number of people told Human Rights Watch that more intense and frequent droughts and a rainy season that was shorter than in recent years has reduced grazing lands and limited water sources, necessary for dry season livestock herding. They pointed out that this has a variety of consequences for herders and can have self-reinforcing negative effects.

Parents expressed hopelessness while describing the challenges in accessing enough food to feed themselves and their children. They said that the death of their livestock put an extra burden on households that traditionally rely in part on meat and milk for sustenance. Community members routinely cited hunger and malnutrition as among the most severe challenges they face. A male elder from one of the areas hardest hit by the drought said,

> There is no time left due to hunger in Turkana: for children, women and everybody. Now, if you decide to help, you may find no one. There isn't much time left for us.[25]

Right to health

Turkana has had a long history of chronic malnutrition and some of the poorest health indicators in Kenya, with minimal health investment and infrastructure, staff, and services for its largely mobile pastoralist population, all of which is exacerbated by a growing population.[26] Climate change is now creating added challenges for the health of the Turkana people.

Pastoralists told Human Rights Watch about a wide range of illness that they and their children suffered, including stomachaches and diarrhea, malaria, malnutrition, and trachoma. They said that these illnesses were worse with the most recent droughts, and that when community members are sick they often have to walk long distances to reach a medical clinic. A pastoralist, living near Lake Turkana, said,

> There is malaria in this area. There is no hospital here that can take in emergencies. Even when the kids fall sick, there is no help unless taken to Kolokol [nearest health facility, 40 km away]. Sometimes the patient dies on the way.[27]

Government obligations and response

Climate change is undermining the ability of governments to protect and fulfill people's basic rights. This impact is felt most by those parts of the population that are already most vulnerable due to factors such as gender, age, poverty, minority status, and disability. Under international, and regional, human rights law, the principle of equality and non-discrimination obliges all states to take specific measures to identify and address such vulnerabilities.[28]

Economic, social, and cultural rights, such as the right to food, water, health, and livelihood, are subject to progressive realization based on available resources. This is in recognition of the fact that states require sufficient resources, and time, to meet their obligations to respect, protect, and fulfill these rights. However, the Committee on Economic, Social and Cultural Rights has stressed that the steps required to meet this goal be "deliberate, concrete and targeted as clearly as possible towards meeting the obligations" in the International Covenant on Economic, Social and Cultural Rights (ICESCR) and that there is an obligation to "move as expeditiously and effectively as possible towards that goal."[29]

A human rights approach to climate change adaptation provides guiding principles for addressing the current impacts of climate change in line with a country's human rights obligations. Box 22.1 illustrates elements of rights-based adaptation planning and implementation.[30]

Box 22.1 Rights-based adaptation planning

1 Equality and non-discrimination

The principle of non-discrimination and equality is central to the international human rights framework. It requires that no group or individual should be excluded from adaptation planning and that priority in allocating limited public resources should be given to those who do not have access or who face discrimination in the enjoyment of their basic human rights.

2 Progressive realization

States need to take deliberate steps to ensure progressive realization of economic, social, and cultural rights. They should devote the necessary planning, and financial and institutional resources to climate change adaptation to avoid retrogression from existing human rights standards. During periods of growth, states should plan for the long-term realization of the human rights so as to build resilience for times of acute crisis.

3 Active, free, and meaningful participation and access to information

Looking at climate change adaptation planning from a human rights perspective mandates that individuals and communities should have access to information and participate in decision-making. Poor people and members of marginalized groups are frequently excluded from decision-making regarding adaptation planning. Community participation in the planning and design of adaptation programs is important to ensure that plans are relevant and appropriate, and thus ultimately sustainable.

4 Accountability

A central feature of a human rights–based approach to adaptation planning is its focus on accountability, which underlines the obligations of the state, as duty bearer, to respect and fulfill human rights. In practice, accountability requires the development of laws, policies, institutions, administrative procedures, and mechanisms of redress to promote and protect human rights.

Integrating these principles into adaptation planning entails steps to

- Identify especially vulnerable individuals and marginalized social groups through disaggregated data according to prohibited grounds of discrimination (human rights risk assessment)
- Ensure that the process of identifying disadvantaged individuals and groups is inclusive and participatory
- Fully include vulnerable groups and individuals in all levels of adaptation planning, as well as implementation processes
- Understand and address their unique needs through targeted and differentiated interventions
- Ensure that adaptation activities do not inadvertently worsen their vulnerability
- Redress power imbalances and other underlying structural causes of differential vulnerability within and between households

- Identify and address barriers and reasons for lack of access to participation planning
- Set specific targets for disadvantaged groups
- Ensure that the adaptation plan foresees monitoring mechanisms measuring the reduction of inequalities
- Monitor the increase or decrease in inequalities

The impact that climate change is having in Turkana County on access for the population to water, food, health, and livelihood creates specific challenges for the government in meeting its obligations to protect and fulfill the human rights of the Turkana population, and in particular to progressively realize economic and social rights in the region. To the extent that the Kenyan and Turkana County governments are failing to address the disproportionate burdens that climate change imposes on specific vulnerable populations, including women, children, individuals with disabilities, and indigenous populations, the authorities may not be meeting their obligations to ensure equality under the law and non-discrimination in the enjoyment of those rights. As the Office of the High Commissioner of Human Rights pointed out, "irrespective of the additional strain climate change-related events may place on available resources, States remain under an obligation to ensure the widest possible enjoyment of economic, social and cultural rights under any given circumstances."[31]

While Kenya's national and county governments have acknowledged the impact that climate change has had and will have on people's lives, the Turkana County and the Kenyan national government have been struggling to address the human rights consequences of climate change and other environmental developments. Existing health, development, and climate change policies largely fail to address the disproportionate impact that these changes are likely to have on Kenya's most marginalized populations, including Indigenous Peoples.

The Kenyan national government has made several efforts towards developing a national policy addressing the effects of climate change, including a National Climate Change Response Strategy (2010) and a National Climate Change Action Plan 2013–2017.[32] The 2010 National Climate Change Response Strategy (NCCRS) was the first Kenyan national policy document to fully acknowledge climate change.[33] While the Strategy did mention the "urgent needs of vulnerable socioeconomic groups," it did not propose any concrete steps to identify and address those vulnerabilities or suggest specific measures of the impact of climate change on marginalized groups and individuals.[34]

The National Climate Change Action Plan 2013–2017 (NCCAP) identifies certain sectors of Kenya's economy that are particularly vulnerable to climate change impacts, including pastoralist livestock, water, and health, and acknowledges that every Kenyan has the right to a clean and healthy environment.[35] The NCCAP also names several groups and individuals that are particularly vulnerable to climate change (including urban poor, women, children, and pastoralists).[36] It also acknowledges that States Parties to the UNFCCC are required to prepare and implement a National Adaptation Plan (NAP), and said that such a plan would be finalized during 2013.[37] However, as far as Human Rights Watch is aware, the National Adaptation Plan has not yet been published.

The Kenyan Climate Change Law was finally passed in May 2016. If rigorously carried out, it could improve coordination and governance of national and local policies related to climate change and ensure that the rights of Indigenous Peoples are respected. The law mandates the participation of a representative from a "marginalized community" who has "experience in matters relating to indigenous knowledge" as a member of the new National Climate Change Council.[38] Yet, the post has not been filled.

At the local level, the Turkana County Integrated Development Plan 2013–2018 aims to "ensure the County's preparedness and capacity to respond" to the effects of climate change including "high infant mortality. . . , increased resource based conflicts, [and] increased morbidity."[39] In interviews with Human Rights Watch, Turkana County officials acknowledged the impact of climate change on the county and the need to design adaptation policies to realize the rights of pastoralists.[40] At the county level, the 2013 devolution agreement that led to the decentralization of many government functions has created many new responsibilities for the county government, which gives it an expanded role in addressing the human rights effects of climate change and other environmental threats. Yet, the county government has done little to integrate climate change adaptation strategies for vulnerable populations.[41]

The government of Kenya is not alone in facing climate change and other environmental threats. Kenya should recognize that these changes will likely impede its ability to realize the human rights of many people. As a regional leader, Kenya should assess this impact, identify which individuals and communities are most vulnerable, and take steps to reduce this vulnerability and ensure that human rights standards are integrated into adaptation plans. Throughout this process, the government should ensure meaningful participation and access to relevant information for affected groups and refrain from investments and implementing actions that could undermine people's rights.

Notes

1 The Intergovernmental Panel on Climate Change, *Climate Change 2014: Synthesis Report, Contribution of Working Groups I, II and III to the Fifth Assessment Report of the Intergovernmental Panel on Climate Change* (2014), available online at www.ipcc.ch/pdf/assessment-report/ar5/syr/SYR_AR5_FINAL_full.pdf (last visited 2 February 2017) 40. The Report found that "[w]arming of the climate system is unequivocal, and since the 1950s, many of the observed changes are unprecedented over decades to millennia." Further, the "atmosphere and ocean have warmed, the amounts of snow and ice have diminished, and sea level has risen."

2 Germanwatch Global Climate Risk Index, *Who Suffers Most From Extreme Weather Events? Weather-Related Loss Events in 2013 and 1994 to 2013* (2015), available at http://germanwatch.org/en/download/10333.pdf (last visited 25 February 2015). See also Union of Concerned Scientists, *Climate Hot Map* (2013), available at www.climatehotmap.org.

3 The Intergovernmental Panel on Climate Change, *Climate Change 2007: Impacts, Adaptation, and Vulnerability, Contribution of Working Group II to the Fourth Assessment Report of the Intergovernmental Panel on Climate Change* (2007), available at www.ipcc.ch/pdf/assessment-report/ar4/wg2/ar4_wg2_full_report.pdf, at 435. Already in 2007, the IPCC stated, "Africa is one of the most vulnerable continents to climate change and climate variability, a situation aggravated by the interaction of 'multiple stresses', occurring at various levels, and low adaptive capacity (high confidence)."

4 The Intergovernmental Panel on Climate Change, *Climate Change 2014: Impacts, Adaptation, and Vulnerability, Part B: Regional Aspects. Contribution of Working Group II to the Fifth Assessment Report of the Intergovernmental Panel on Climate Change* (2014), available online at www.ipcc.ch/pdf/assessment-report/ar5/wg2/WGIIAR5-PartB_FINAL.pdf, at 1199–1265.

5 UN Special Procedures of the Human Rights Council, *The Effects of Climate Change on the Full Enjoyment of Human Rights* (30 April 2015), available at www.thecvf.org/wp-content/uploads/2015/05/humanrightsSRHRE.pdf; OHCHR, *Report of the Office of the United Nations High Commissioner for Human Rights on the Relationship Between Climate Change and Human Rights*, UN Doc. A/HRC/10/61, 15 January 2009, available at https://documents-dds-ny.un.org/doc/UNDOC/GEN/G09/103/44/PDF/G0910344.pdf?OpenElement (last visited 22 June 2015); OHCHR, *Zeid Urges Climate Change Ambition as Paris Deal Enters Into Force* (3 November 2016), available at www.ohchr.org/EN/NewsEvents/Pages/DisplayNews.aspx?NewsID=20822&LangID=E#sthash.K7rgSAVa.dpuf; UN Special Procedures of the Human Rights Council, *The Effects of Climate Change on the Full Enjoyment of Human*

Rights (30 April 2015), available at www.thecvf.org/wp-content/uploads/2015/05/humanrightsS-RHRE.pdf; HRC Res. 32/L/34 (2016); HRC Res. 29/L/21 (2015); HRC Res. 26/L.33 (2014); HRC Res. 18/22 (2011); HRC Res. 10/4 (2009); HRC Res. 7/23 (2008).

6 IPCC, *Climate Change 2014*, *supra* note 1.

7 OHCHR, *Report of the Office*, *supra* note 5, at 40.

8 *Ibid.*, at 40.

9 *Ibid.*

10 IPCC, *Climate Change 2014: Impacts, Adaptation, and Vulnerability, Part A: Global and Sectoral Aspects, Contribution of Working Group II to the Fifth Assessment Report of the Intergovernmental Panel on Climate Change* (2014), available online at www.ipcc.ch/pdf/assessment-report/ar5/wg2/WGIIAR5-PartA_FINAL.pdf, at 19.

11 AU Dec. 134 (VII), January 2007; AU Dec. 236 (XII), January 2009; AU Dec. 257 (XII), January 2009; AU Dec. 342 (XVI), January 2011; AU Dec. 448 (XIX), January 2012.

12 UNEP, *Nairobi Declaration on the African Process for Combating Climate Change* (May 2009), available online at www.unep.org/roa/Amcen/Amcen_Events/3rd_ss/Docs/nairobi-Decration-2009.pdf.

13 African Commission on Human and Peoples' Rights, *Resolution 153 on Climate Change and Human rights and the Need to Study Its Impact in Africa*, UN Doc. ACHPR/Res.153(XLVI)09 (November 2009), available online at http://old.achpr.org/english/resolutions/resolution153_en.htm (last visited 1 February 2017); Working Group on Extractive Industries, *Environment and Human Rights, 271: Resolution on Climate Change in Africa* (2014), available online at www.achpr.org/sessions/55th/resolutions/271/.

14 Human Rights Watch, *There Is No Time left: Climate Change, Environmental Threats, and Human Rights in Turkana County, Kenya* (2012), available online at www.hrw.org/report/2015/10/15/there-no-time-left/climate-change-environmental-threats-and-human-rights-turkana.

15 Z. Mwangi, *Keyna National Bureau of Statistics, Highlights of the Socio-Economic Atlas of Kenya* (10 November 2014), available online at www.knbs.or.ke/index.php?option=com_phocadownload&view=category&download=673:highlights-of-the-socio-economic-atlas-of-kenya&id=116:the-socio-economic-atlas-of-kenya&Itemid=599 (last visited 30 November 2014); Kenya National Bureau of Statistics and Society for International Development, *Exploring Kenya's Inequality, Turkana County* (2013), available online at http://inequalities.sidint.net/kenya/wp-content/uploads/sites/2/2013/09/Turkana.pdf, at 12.

16 D. K. Kalinaki, *The East African, Wean Turkana Off Aid, Get the People Working* (26 February 2014), available online at www.theeastafrican.co.ke/news/Wean-Turkana-off-aid-get-the-people-working/-/2558/2222678/-/wtodh5z/-/index.html.

17 Turkana County Government, *First County Integrated Development Plan* (2013/14–2017/18), at xxxiii. On file at Human Rights Watch.

18 Government of Kenya, *National Climate Change Response Strategy* (April 2010), available online at www.environment.go.ke/wp-content/documents/complete%20nccrs%20executive%20brief.pdf, at 28. For the north and northeast, the government data show an increase of the maximum temperature by 0.7–1.8°C and the minimum temperature by 0.1–0.7°C.

19 S. Avery, *African Studies Centre, Lake Turkana and Lower Omo: Hydrological Impacts of Major Damn & Irrigation Developments* (October 2012), available online at www.africanstudies.ox.ac.uk/sites/sias/files/documents/Volume%20I%20Report.pdf 1, at 123.

20 S. Avery, *African Studies Centre, What Future for Lake Turkana? The Impact of Hydropower and Irrigation Development on the World's Largest Desert Lake* (2012), available online at www.africanstudies.ox.ac.uk/what-future-lake-turkana.

21 K. E. Trenberth, 'Changes in Precipitation With Climate Change', 47 *Climate Research* (2010) 123–138.

22 Human Rights Watch, *Ethiopia: Dams, Plantations a Threat to Kenyans Lake Turkana Water Levels Down, Further Drop Expected*, available online at www.hrw.org/news/2017/02/14/ethiopia-dams-plantations-threat-kenyans.

23 Human Rights Watch interview with O.P. Kalokal, September 2014.

24 Avery, *What Future*, *supra* note 20.

25 Human Rights Watch interview with O.T., location withheld, September 2014.

26 Kenya National Bureau of Statistics (KNBS) and Society for International Development (SID), *Exploring Kenya's Inequality: Pulling Apart or Pooling Together?* (2013), available online at http://inequalities.sidint.net/kenya/wp-content/uploads/sites/2/2013/09/Turkana.pdf (last visited 2 February 2017). See also The Health Project Policy, *Turkana County: Health at a Glance* (May 2015), available online at www.healthpolicyproject.com/pubs/291/Turkana%20County-FINAL.pdf; J. M. Brainard, *Health and Development in a Rural Kenyan Community* (1991).

27 Human Rights Watch interview with R.O., Eliye Springs, September 2014.

28 All of the major human rights treaties contain legal obligations to ensure non-discrimination and, where it exists, end discrimination. For example, Article 2 (2) of the International Covenant on Economic, Social and Cultural Rights (ICESCR) specifies that "The States Parties to the present Covenant undertake to guarantee that the rights enunciated in the present Covenant will be exercised without discrimination of any kind as to race, colour, sex, language, religion, political or other opinion, national or social origin, property, birth or other status." The International Covenant on Civil and Political Rights (ICCPR) includes an almost identical guarantee. For regional human rights law, see African [Banjul] Charter on Human and Peoples' Rights, adopted 27 June 1981, entered into force 21 October 1986, OAU Doc. CAB/LEG/67/3 rev.5, 21 I.L.M. 58 (1982). Art. 2 stipulates: "Every individual shall be entitled to the enjoyment of the rights and freedoms recognised and guaranteed in the present Charter without distinction of any kind such as race, ethnic group, colour, sex, language, religion, political or any other opinion, national and social origin, fortune, birth or any status." The Protocol to the African Charter on Human and Peoples' Rights on the Rights of Women in Africa provide for special protection of women from discrimination: see Protocol to the African Charter on Human and Peoples' Rights on The Rights of Women in Africa (the Maputo Protocol), adopted by the 2nd Ordinary Session of the Assembly of the Union, Maputo, 13 September 2000, entered into force 25 November 2005, CAB/Leg/66.6.

29 CESCR, General Comment No. 3, *The Nature of States Parties' Obligations*, UN Doc. E/1991/23 annex III (1991).

30 This summary of a human rights–based approach to climate change adaptation is adapted from the following basic human rights principles: Equality and non-discrimination; right to participation and information; sustainability and non-retrogression; accountability; and the rule of law. For a similar framework applying these principles to the realization of the right to water and sanitation, see UN Special Rapporteur on the human right to safe drinking water and sanitation, Realising the human rights to water and sanitation: A handbook by the UN Special Rapporteur Catarina de Albuquerque (2014), available at www.ohchr.org/Documents/Issues/Water/Handbook/Book4_Services.pdf. For a human rights impact assessment prior to adopting and implementing policies affecting the right to health, see Gillian MacNaughton and Paul Hunt, Impact Assessments, Poverty and Human Rights: A Case Study Using the Right the Highest Attainable Standard of Health, Health and Human Rights Working Paper Series No. 6 (2006), available at www.who.int/hhr/Series_6_Impact%20Assessments_Hunt_MacNaughton1.pdf (last visited 2 February 2015). For a human rights–based approach to development cooperation programming, see OHCHR, Frequently asked questions on a human rights–based approach to development cooperation (2006), available at www.ohchr.org/Documents/Publications/FAQen.pdf.

31 OHCHR, Report of the Office, *supra* note 5, at 40. The OHCHR further argues that "a State party in which any significant number of individuals is deprived of essential foodstuffs, of essential primary health care, of basic shelter and housing, or of the most basic forms of education would be failing to meet its minimum core obligations and, *prima facie*, be in violation of the Covenant" (citing CESCR general comment No. 3, para. 10).

32 Government of Kenya, *National Climate Change Response Strategy* (2010), available online at www.environment.go.ke, at 5. The 2010 National Climate Change Response Strategy (NCCRS) was the first Kenyan national policy document to fully acknowledge climate change. It defined its objective as "enhanc[ing] understanding of the global climate change regime," and "assess[ing] the evidence and impacts of climate change in Kenya."

33 Government of Kenya, *supra* note 32, at 5.

34 *Ibid.*, at 12.

35 Government of Kenya, *National Climate Change Action Plan 2013–2017* (2013), available online at http://kccap.info/index.php?option=com_phocadownload&view=category&download=320:national-climate-change-action-plan&id=35:executive-reports&Itemid=72, at 25.

36 *Ibid.*, at 27, 41, 50.

37 *Ibid.*, at 9, 45.

38 *The Climate Change Act* (2016), at para. 7(2)(h).

39 Turkana County Government, *First County Integrated Development Plan* (2013/14–2017/18) 25. On file at Human Rights Watch.

40 Human Rights Watch interview with Peter Ekai Lokoel, Deputy Governor, Lodwar, September 2014 and August 2016; Human Rights Watch interview with Beatrice A. Moe, County Executive for the Ministry of Water Service, Irrigation and Agriculture, Lodwar, September 2014.
41 At the time of Human Rights Watch's visit in Turkana in August 2016, no tangible progress had been made with regard to the new Turkana County Integrated Development plan. Human Rights Watch interview with Peter Ekai Lokoel, Deputy Governor, Lodwar, August 2016.

23

Human rights and climate change

Focusing on South Asia

Vositha Wijenayake

Introduction

With the intensity of climate change impacts felt increasingly across the globe, the existing vulnerabilities among populations are more exposed, and aggravated. South Asia holds over one-fifth of the world's population. High rates of population growth, and natural resource degradation, with continuing high rates of poverty and food insecurity have made South Asia one of the most vulnerable regions to the impacts of climate change.[1] Impacts of climate change will be felt across multiple sectors and have adverse effects on livelihoods, as well as the standard of life of communities in the region. Temperature rise will negatively impact crop yields in tropical parts of South Asia where these crops are already being grown close to their temperature tolerance threshold. While direct impacts are associated with rise in temperatures, indirect impacts due to water availability and changing soil moisture status and pest and disease incidence are likely to be felt.[2] These impacts will in turn influence the rights of those living in the region, highlighting the need for protection of these rights.

Food security, access to natural resources, health, ownership of land, and gender equality will be impacted by climate change increasing already existing vulnerabilities. Such impacts create risks for the protection of many rights, such as the right to life, the right to food, the right to access to natural resources, and right to a livelihood.

Impacts of climate change on human rights in the region

Climate change creates many adverse impacts on the South Asian region, including abnormal monsoon patterns. More frequent and intense storms have aggravated natural disasters and climate change impacts in recent years. Bearing the brunt of these are more than 600 million poor, who depend on climate-sensitive sectors including agriculture, forestry, and traditional fishing for much of their day-to-day needs. With changes in the global climate system likely to continue into the next century, geography, high population density, and immense poverty will continue to make South Asia especially vulnerable.[3]

The impacts of climate change felt across the South Asian region threatens many human rights that need to be protected. However, for the need of brevity, this chapter will focus primarily on rights that are violated due to food insecurity, climate-induced migration, as well as threat to life.

Living in South Asia

The right to life is one of the key rights that are impacted due to climate change. The right to life is explicitly protected under the Universal Declaration of Human Rights[4] (UDHR), the International Covenant on Civil and Political Rights (ICCPR), and the Convention on the Rights of the Child (CRC).[5] According to the Fourth Assessment Report of the Intergovernmental Panel on Climate Change (IPCC AR4), climate change impacts will increase the death count in the region due to disease and injury from heatwaves, floods, storms, fires, and droughts. Climate change will also exacerbate weather-related disasters, which already have devastating effects on people and their enjoyment of the right to life, particularly in the developing world. An estimated 262 million people were affected by climate disasters annually from 2000 to 2004, of whom over 98% lived in developing countries.[6]

Equally, climate change will affect the right to life through an increase in hunger and malnutrition and related disorders impacting child growth and development, and cardio-respiratory morbidity and mortality related to ground-level ozone.[7]

Impacts on the right to life in South Asia can be seen through the number of deaths that have been suffered by countries due to extreme and slow-onset events resulting from climate change and through health impacts of people in the region. The cities of South Asian countries are vulnerable to water- and vector-borne infectious diseases that are climate sensitive. Diseases like yellow fever, cholera, dengue, diarrhea, and malaria are generally sensitive to climate change. The modeling results suggest that the mortality rate for the region caused by dengue, malaria, and diarrhea would increase over time as a consequence of climate change. Morbidity and deaths from such diseases could increase in the future under all scenarios.[8]

Between 1990 and 2008, more than 750 million people in South Asia were affected by at least one type of natural disaster, resulting in almost 230,000 deaths. The Himalayan region of the Indian subcontinent has become even more vulnerable to natural disasters spawned by melting glaciers, which form high-altitude lakes that can suddenly breach and cause catastrophic flooding downstream. The frequency of such events, called glacial lake outburst floods (GLOF), has increased in recent decades. At the other end of extreme weather events is a predicted increase in the duration and intensity of droughts, particularly in the arid and semi-arid areas of Bangladesh and India.[9]

According to the 2010 World Development Report, diseases linked to climate, namely malnutrition, diarrheal diseases, and vector-borne illnesses (especially malaria) already represent a huge health burden in some regions, particularly Africa and South Asia.[10] It is further added that the increased burden felt from climate change will be most consequential for the poor. The estimated additional 150,000 deaths a year attributable to climate change in recent decades may be just the tip of the iceberg.[11]

Feeding South Asia

Agriculture is one of the key sectors vulnerable to climate change in South Asia. Changes in precipitation patterns (timing and amount) increase the likelihood of short-run crop failures and long-run production declines, posing a serious threat to food security. Tropical and subtropical regions of Bangladesh, Bhutan, India, and Sri Lanka are projected to be vulnerable to increasing temperature and CO_2 level, with a decline in rice yield of as much as 23% by 2080.[12]

Furthermore, in 2007, IPCC AR4 indicated that cereal crop production in South Asian countries could decline by 4–10% by 2100.[13] Research conducted by International Alert[14] provides that India has already seen a fall in wheat and rice production due to temperature increases, and the agriculture-dependent communities are being severely impacted.

It is believed that, as a consequence of climate change, the potential for food production is projected initially to increase at middle to high latitudes with an increase in global average temperature in the range of 1–3°C. However, at lower latitudes, crop productivity is projected to decrease, increasing the risk of hunger and food insecurity in the poorer regions of the world. Poor people living in developing countries are particularly vulnerable given their disproportionate dependency on climate-sensitive resources for their food and livelihoods.[15]

Food insecurity and associated food price fluctuations already represent a systemic source of risk that is expected to increase with climate change.[16] According to the 2010 World Development Report, food prices are expected to be higher and more volatile in the long run. Poor people, who spend up to 80% of their money on food, probably will be hit hardest by the higher food prices.[17] And this will have grave impacts on the countries of South Asia.

The World Bank has estimated that a 2°C increase in average global temperature would put "between 100 million and 400 million more people at risk of hunger and could result in over 3 million additional deaths from malnutrition each year". Considering the influence of rapid population growth and urbanization, the risk of hunger is projected to remain very high in several developing countries.[18] As many as 160,000 people will die annually in India by 2050 due to food insecurity, followed closely by Bangladesh.[19]

The right to food is explicitly mentioned under the International Covenant on Economic, Social and Cultural Rights (ICESCR), the CRC, and the Convention on the Rights of Persons with Disabilities (CRPD) and implied in general provisions on an adequate standard of living of the Convention on the Elimination of All Forms of Discrimination against Women (CEDAW) and the International Convention on the Elimination of All Forms of Racial Discrimination (CERD).[20] Given the threats to food security in the region that will lead to violations of the right to food, it remains important to understand how the region will address climate change impacts of food security and protect the rights of its people.

To address the needs of food security, South Asia's regional initiative for cooperation, the SAARC, has provided policy guidance on how to address situations of difficulty. The SAARC is a regional intergovernmental organization and geopolitical union of nations in South Asia of which the members are Afghanistan, Bangladesh, Bhutan, India, Nepal, the Maldives, Pakistan, and Sri Lanka. It was established in 1985 with the objective of promoting development of economic and regional integration as per the Charter of SAARC.

The Declaration of the 18th SAARC Summit of 2014, titled Kathmandu Declaration,[21] aims to address the issue of food security in the region. It provides under the section on agriculture and food security that

> The Heads of State or Government agreed to increase investment, promote research and development, facilitate technical cooperation and apply innovative, appropriate and reliable technologies in the agriculture sector for enhancing productivity to ensure food and nutritional security in the region. They also underscored the importance of promoting sustainable agriculture.

The Declaration further expands access to food grains at time of emergency, which would help marginalized farmers and vulnerable groups access food in times of difficulty. It provides, "the Leaders directed to eliminate the threshold criteria from the SAARC Food Bank Agreement so as to enable the Member States to avail food grains, during both emergency and normal time food difficulty."

South Asia on the move

The greatest single impact of climate change might be on human migration. The report estimated that by 2050, 150 million people could be displaced by climate change–related phenomena, such as desertification, increasing water scarcity, and floods and storms. In some parts of South Asia, extreme weather events have in the past led to significant population displacement, and changes in the incidence of extreme events will amplify the challenges and risks of such displacement.[22]

Sea level rise is the most obvious climate-related impact in coastal areas. The Maldives made a submission to the UN Human Rights Council[23] in 2008, stating that its right to life was being threatened due to the effects of climate change. According to the submission, under a scenario in which the sea level rose by 0.49 meters, 15% of Malé would be inundated by 2025, and half of the island flooded by 2100.

The Fifth Assessment Report (AR5) of the IPCC provides with very high confidence that, due to sea level rise projected throughout the 21st century and beyond, coastal systems and low-lying areas will increasingly experience adverse impacts such as submergence, coastal flooding, and coastal erosion. It also adds that climate change will have significant impacts on forms of migration that compromise human security.[24]

Forced migration impacts many rights of these populations, among which are liberty, security, and the right to an adequate standard of living. Article 3 of the UDHR guarantees the right to life, liberty, and security of a person, while Article 11 of the ICESCR guarantees the right of everyone to an adequate standard of living for himself and his family, including adequate food, clothing, and housing, and to the continuous improvement of living conditions. In South Asia, many countries are affected by migration, with Bangladesh being one of the most impacted. The country faces issues related to migration due to climate change impacts on a daily basis, and losses and damages related to such migration as well.

The Kathmandu Declaration of the SAARC specifically mentions the need to focus on climate change impacts on the region when it

> directed the relevant bodies/mechanisms for effective implementation of SAARC Agreement on Rapid Response to Natural Disasters, SAARC Convention on Cooperation on Environment and Thimphu Statement on Climate Change, including taking into account the existential threats posed by climate change to some SAARC Member States.[25]

However, given that many migrate in search of work due to impacts of climate change felt in some countries, this could also be considered a policy initiative that could be relevant to addressing climate change. The Declaration states that the States of the region agreed to collaborate and cooperate on safe, orderly, and responsible management of labor migration from South Asia to ensure safety, security, and well-being of their migrant workers in the destination countries outside the region.

While there exists regional policies that have been developed to address the issue of displacement and migration by the SAARC, their implementation is yet to materialize. In order to uphold the rights of the displaced, it will be important to see the implementation of these policies and plans of action. With the region's political conflicts affecting the migration policies of countries, if implemented these actions would help many impacted by climate change and forced to migrate to address their vulnerabilities and protect their rights.

South Asia's vulnerable

Climate change impacts are felt more severely by some parts of the communities than others, and women and children fall under this category. UNDP's Human Development Report 2007/2008 states that women's historic disadvantages – their limited access to resources, restricted rights, and a muted voice in shaping decisions – make them highly vulnerable to climate change. The nature of that vulnerability varies widely, cautioning against generalization. But climate change is likely to magnify existing patterns of gender disadvantage.[26]

It is established that women, particularly elderly women and girls, are affected more severely and are more at risk during all phases of weather-related disasters: risk preparedness, warning communication and response, social and economic impacts, recovery, and reconstruction.[27] The death rate of women is markedly higher than that of men during natural disasters. This is particularly the case in disaster-affected societies in which the socio-economic status of women is low.[28] Rural women are particularly affected by the effects on agriculture and deteriorating living conditions in rural areas. According to UNDP, Indian women born during a drought or flood in the 1970s were 19% less likely to ever attend primary school, when compared with women of the same age who were not affected by natural disasters.[29]

Further, children will be amongst the worst affected by climate change in South Asia, and the impacts of climate change on children are equally significant as for women. According to UNICEF, South Asia will account for one-quarter of the world's children, with 614 million children under the age of 18.[30] Like women, children have a higher mortality rate as a result of weather-related disasters.[31] In addition to the right to life of the children being impacted by climate change, other rights are also violated, threatening their right to education.

Displacement due to climate change, and temporary housing due to impacts of climate change, strike severe blows on the lives of women and children. In some Bangladeshi focus groups, both boys and girls aged 12 to 17 reported sexual and physical abuse by relatives following disasters. Girls and boys have provided information on how these situations were affecting them, and how they felt unable to share these experiences with family or friends. The village surveys in Sri Lanka found that 15–20% of parental caregivers had gone abroad for work, because of the lack of opportunities at home as a result of poor harvests, climate-related disaster losses, and overwhelming poverty.[32]

CEDAW[33] guarantees the basic human rights and fundamental freedoms, the right to education and the rights to employment and health of women, while the CRC obliges States to take action to ensure the realization of all rights in the Convention for all children in their jurisdiction, including measures to safeguard children's right to life, survival, and development through, *inter alia*, addressing problems of environmental pollution and degradation. However, implementation, and the protection of these rights, have a long way to go in many countries.

Another group in the region highly affected by climate change are farmers. They experience crop loss and damage due to climate change, and there are also high levels of suicide reported in countries due to crop failure from impacts such as droughts.

As a response to losses and damages that farmers suffer due to slow-onset events and extreme weather events, Sri Lanka has been implementing a crop insurance scheme since 1958. The current form in implementation, which is the amended version of 1958,[34] introduced in 1973, provides farmers compensation for damages to their crops.

The Government of Sri Lanka enacted the Agricultural Insurance Law No.27 of 1973,[35] which replaced the then-existing Crop Insurance Scheme that had been run on a pilot basis since 1958. The scheme is implemented in three phases, which includes insurance of the paddy crop in the entire island, insurance of livestock, insurance of selected subsidiary food crops, and insurance of non-traditional food drops. The objective of the insurance was to indemnify farmers against loss, so as to stabilize farm income thereby promoting agricultural production, and in turn ensuring that their rights are protected. The financing of the scheme is set as a partially subsidized venture, with administrative costs being borne by the state. The premium per acre is fixed based on average yields and damage rates in different localities. The premiums should be paid once every season (*Yala* and *Maha* are the two seasons of paddy cultivation).

The most recent agriculture policy of Sri Lanka, the National Agricultural Policy, also focuses on ways to lessen the burdens of farmers. Examples such as agricultural credit could contribute to helping farmers address the impacts of climate change, as it provides the possibility for them to finance their agricultural activities even in a situation where the previous crops have been impacted by the adverse effects of climate change. The new policy aims to strengthen rural credit institutions connected with farmers' investments, savings, and risk management, introduce simple procedures in providing loan facilities for agricultural activities and agro-based industries, ensure availability of credit for farmers at concessionary interest rates, and establish a mandatory share for agricultural credit in the state banks' overall lending for the benefit of the farming communities. The policy further highlights the need for introducing appropriate agricultural insurance schemes to protect farmers from the risks associated with natural calamities.

However, while the insurance schemes have been present for over half a decade to address impacts on agricultural activities such as droughts and floods, the coordination and distribution of compensation when the scheme is implemented has presented difficulties for farmers. This points to the indispensable need for the government to better coordinate the implementation of existing solutions which focus on vulnerable groups.

Conclusion

South Asia is one of the most vulnerable regions in the world. With its diversity and its large population, it is important to address the impacts of climate change with urgency if the rights of the communities most vulnerable to climate change are to be protected. While the impacts are visible, and countries already suffer the brunt of adverse climatic impacts, a blind eye is turned to these impacts in some instances of policy implementation. It will be vital to focus on the impacts of climate change on the human rights of populations and address them through climate policies developed by countries after ratification of the Paris Agreement. While vulnerabilities exist, the way ahead would be to make available opportunities for protecting the rights of individuals in a much more effective manner, seeing climate actions from a human rights perspective, when policies related to Nationally Determined Contributions and the commitments under the Paris Agreement are implemented.

Notes

1 R. Lal *et al.* (eds.), *Climate Change and Food Security in South Asia* (2011).
2 *Ibid.*
3 M. Ahmed and S. Suphachalasai, *ADB, Assessing the Costs of Climate Change and Adaptation in South Asia* (June 2014), available online at www.adb.org/sites/default/files/publication/42811/assessing-costs-climate-change-and-adaptation-south-asia.pdf.
4 Universal Declaration of Human Rights, 10 December 1948.
5 Article 6 International Covenant on Civil and Political Rights (ICCPR).
6 United Nations Development Programme (UNDP), Human Development Report 2007/2008, Fighting Climate Change: Human Solidarity in a Divided World, at 8, cited in OHCHR, 'Understanding Human Rights and Climate Change' (2015), Submissions of the Office of the United Nations High Commissioner for Human Rights to the 21st Conference of Parties to the United Nations

Framework Convention on Climate Change, available online at www.ohchr.org/Documents/Issues/ClimateChange/COP21.pdf.

7 IPCC, *AR4 Working Group II (WGII) Report*, available online at www.ipcc.ch/publications_and_data/ar4/wg2/en/contents.html, at 393.

8 Ahmed and Suphachalasai, *supra* note 3.

9 *Ibid.*

10 World Bank, *World Development Report* (2010) 95, heading 'Keep People Healthy'.

11 *Ibid.*

12 Ahmed and Suphachalasai, *supra* note 3.

13 *Ibid.*

14 Smith and Vivekananda, *International Alert, a Climate of Conflict: The Links Between Climate Change, Peace and War* (November 2007), available online at www.international-alert.org/sites/default/files/publications/A_climate_of_conflict.pdf.

15 IPCC, *AR4 Synthesis Report*, at 48, cited in OHCHR, 'Understanding Human Rights and Climate Change', *supra* note 6.

16 World Bank, *supra* note 10, at 107, heading 'Provide Safety Nets for the Poor and Vulnerable'.

17 World Bank, *supra* note 10, at 168, heading 'Rising energy, water and agricultural prices could spur innovation and investment in increasing productivity'.

18 International Council on Human Rights Policy (ICHRP), *Climate Change and Human Rights: A Rough Guide* (2008), available online at www.ichrp.org/files/reports/45/136_report.pdf.

19 D. D'Monte, *The Third Pole, Climate Change Will Claim 160,000 Lives a Year in India by 2050* (8 August 2016), available online at www.thethirdpole.net/2016/08/08/climate-change-will-claim-160000-lives-a-year-in-india-by-2050/.

20 Art. 11 International Covenant on Economic, Social and Cultural Rights (ICESCR); Art. 24, para. 2 (c) Convention on the Rights of the Child (UNCRC); Art. 25(f) and Art. 28, para. 1 Convention on the Rights of Persons with Disabilities (CRPD); Art. 14, para. 2(h) Convention on the Elimination of All Forms of Discrimination against Women (CEDAW); Art. 5(e) International Convention on the Elimination of All Forms of Racial Discrimination (ICERD).

21 SAARC, *Kathmandu Declaration* (2014), available online at www.saarc-sec.org/press-releases/18th-saarc-summit-declaration/121/.

22 IPCC, *Fifth Assessment Report*, Chapter 12, Executive Summary (2014), available online at www.ipcc.ch/report/ar5/wg1/, at 758.

23 Maldives, Human Rights Council Resolution 7/23, Submission of the Maldives to the Office of the UN High Commissioner for Human Rights, 25 September 2008.

24 IPCC, *Fifth Assessment Report*, Chapter 12, *supra* note 22, at 758.

25 South Asian Association for Regional Cooperation, 18th Summit, Kathmandu Declaration 2014 (November 2014) available online at http://mea.gov.in/Uploads/PublicationDocs/24375_EIGHTEENTH_SUMMIT_DECLARATION.pdf

26 UNDP, *supra* note 6.

27 OHCHR, *Understanding Human Rights and Climate Change*, *supra* note 6.

28 Neumayer and Plümper, 'The Gendered Nature of Natural Disasters: The Impact of Catastrophic Events on the Gender Gap in Life Expectancy, 1981–2002', 97(3) *Annals of the American Association of Geographers* (2007) 551.

29 UNDP, *supra* note 6.

30 UNICEF, *State of the World's Children Report 2010* (Statistical Annex) (January 2011), available online at www.unicef.org/about/annualreport/files/Communication_AR_2010.pdf.

31 OHCHR, *Understanding Human Rights and Climate Change*, *supra* note 6.

32 Harris and Hawrylyshyn, *Overseas Development Institute, Climate Extremes and Child Rights in South Asia: A Neglected Priority* (October 2012), available online at www.odi.org/publications/6838-climate-extremes-child-rights-south-asia-neglected-priority.

33 CEDAW.

34 UNCTAD, *General Framework for Agricultural Insurance in Sri Lanka*, UN Doc. UNCTAD/INS/27/Add.2 (October 1979).

35 *Ibid.*

24

Climate change and the European Court of Human Rights

Future potentials

Heta Heiskanen

Introduction

The European Court of Human Rights (ECtHR) has established case law on environmental matters[1] and extraterritorial human rights obligations[2] that provides guiding principles for climate change litigation. This chapter analyses the suitability of the ECtHR for climate change and human rights litigation from three perspectives. The first perspective seeks inspiration from the *Urgenda* case to assess what theoretical opportunities are available to establish climate change liability before the ECtHR for a single state towards its citizens (territorial liability).[3] The second perspective analyses how a single state could be held liable to individuals residing outside the territoriality of the responsible state (extraterritorial liability). The third perspective concerns the possibilities of establishing the shared responsibility of several states.[4]

The basis for the responsibility of a single state in a climate change case: territorial liability or extraterritorial liability

Territorial liability

The interpretation of the European Convention on Human Rights is not done in isolation of international development. According to the ECtHR's own case law, it is well established that it "can and must" take into account the development of international law, and it does so especially in new areas of protection.[5]

In addition, due to the subsidiary role[6] of the ECtHR, it draws inspiration and guidance from domestic rulings.[7] One possible guiding model for climate change rulings is provided in the Dutch landmark case, *Urgenda*. Despite of the significance of the *Urgenda* ruling, it should be noted that it has been appealed by the Dutch government,[8] so the reconsideration may change the conclusions.

The *Urgenda* ruling provides guiding principles for climate change litigation on how to assess whether a state has been aware of climate change and the human rights obligations related to it.[9] In addition, it acts as a guide on how to establish the responsibility of the state towards its own citizens, even though there has not yet been physical damage resulting from the failure of precautionary measures. Besides, the *Urgenda* case established the ruling that the individual responsibility of the Netherlands is not diminished even though climate change results from the actions of multiple actors.[10] The assessment of the awareness of the state and the formula of establishing the duty to undertake preventive measures in the *Urgenda* case is in line with the current green jurisprudence of the ECtHR. The assessment of the awareness of the state is closely connected to the doctrine of consensus,[11] whereas the precautionary measures have a strong connection to the doctrine of positive obligations.

Assessment of the state's awareness of climate change

The jurisprudence of the ECtHR has established a requirement that the state should take preventive measures to protect rights if the authorities are aware or should have been aware of real and immediate threats to life. The burden of proof on establishing this can be on the side of the applicant or the state.[12]

The *Urgenda* case provides guidance on how to build consensus argumentation[13] on the awareness of the state about climate change. The first set of evidence in the *Urgenda* case was to prove that there is a scientific basis for the urgent demand to take actions to diminish the level of greenhouse gas emissions. These materials included scientific reports from both international and domestic institutions. The second set of evidence included materials on the relevant obligations related to the international climate change legal and policy framework as well as European climate change policy. The Hague District Court found that the involvement of the Netherlands in the UN and EU climate change agreements and policy measures proved that the state should have been aware of the risks of climate change since 1997, and without any doubt since 2007. All of these documents showed that the Netherlands was involved in international policy-making that undoubtedly required the state's knowledge of the risks of climate change.

The assessment conducted in the *Urgenda* case is similar to the *Brincat and Others v. Malta* case, which was a case presented to the ECtHR on asbestos. The ECtHR used a consensus assessment in order to determine whether Malta knew about the health risks related to asbestos. The ECtHR assessed both domestic and international scientific knowledge of asbestos, as well as Malta's membership of the International Labour Organization (ILO).[14] The similarities in the assessment of the *Urgenda* and *Brincat* cases could lower the threshold for the ECtHR to conduct a similar assessment in the context of climate change and to conclude that there is a scientific and international consensus on the existence of climate change and its impacts on human rights.

Positive obligations as criteria for the assessment of the adequacy of the adopted measures

The approach in the *Urgenda* case focused on the question of whether the state had fulfilled its duties on prevention of climate change. The ruling of the *Urgenda* case concluded that there is no necessity to have a substantive violation if procedural rights are not respected. In the ECtHR, a similar approach is connected to the positive obligations of the state.

The current doctrine of positive obligations requires states "to take all appropriate steps to safeguard life for the purpose of Article 2", including "a legislative and administrative framework", and they must "govern licensing, setting up, operation, security and supervision of the activity" and "make it compulsory for all those concerned to take practical measures to ensure the effective protection".[15] Measures under the positive obligation doctrine have included the right of access to environmental information;[16] the establishment of safety zones;[17] the implementation of effective risk assessment; the establishment of a coherent supervisory system, including an emergency warning system; the establishment of specific mutual agreements for cooperation between authorities crossing the borders of Member States; the control of private parties' practical steps to ensure that river channels are clean[18] and the safety or transfer of buildings.[19]

In the context of climate change, positive obligations can include a variety of duties. For example, current global and regional climate change agreements can be interpreted to constitute such mutual agreements to cooperate and control the private parties to which the positive obligations refer.[20] In addition, implementing effective risk assessment and access to environmental information could require the state to prepare adequate studies on the relationship of its policies and its impacts on climate change and human rights. In specific circumstances, where there is a risk of climate change causing a rise in sea levels, the state should inform its citizens and establish a monitoring and warning system.

Extraterritorial liability of a single state in the context of climate change

Boyle has stated that if one state can be identified as causing extraterritorial environmental harm, according to international environmental law, "costs can be redirected back to the Polluting State in full, emphasizing the responsibility of states to control sources of environmental harm".[21]

However, the ECtHR has been cautious in extending its jurisprudence of extraterritoriality outside of the limited context of military operations, extraditions and expulsions, so there is no current jurisprudence concerning environmental matters.[22] Nevertheless, the current doctrine provides general guiding principles for climate change litigants.

It should be noted that the extraterritorial approach of the ECtHR differs from its own logic established in non-extraterritorial cases. Thus, a successful climate change claim would have to comply with the current extraterritoriality doctrine. The extraterritoriality doctrine requires the following elements: there are "exceptional circumstances" resulting from "acts of (–) authorities", the acts may take place inside or outside national boundaries,[23] the acts have negative effects outside the territory of the responsible state[24] and the state should have effective control over the person or an area.[25]

In formulating the environmental extraterritoriality claim, it is important to be aware that the ECtHR has accepted the application of the doctrine only in respect to acts that are conducted by state actors.[26] Thus, the criteria might not be fulfilled in cases where there is only a failure of the state to control private parties.

Establishing shared liability?

The basis in differentiated fault doctrine

The jurisprudence of the ECtHR has established shared responsibility in specific circumstances. For example, in the human trafficking case of *Rantseva v. Cyprus and Russia*, the ECtHR considered the differentiated faults of the states.[27] Oxana Rantseva was trafficked for the purpose of sexual exploitation from Russia to Cyprus, where she died. The ECtHR found that Cyprus as the state of destination failed to protect Rantseva from trafficking and to sufficiently investigate her death. In addition, the ECtHR found that Russia had an obligation as the state of origin to sufficiently investigate how the trafficking of Rantseva took place from its borders.

It should be noted that the Court made references to other international treaties and also allowed third-party interventions (INTERIGHTS). Both of these facts had an impact on the judgment. For example, Hodson states that through third-party interventions, NGOs "fulfil a role of assisting the Court in new areas of law where the impact is particularly broad. They provide comparative analysis and practical information that the parties may be unable to marshal and the Court would otherwise be unable to acquire." Hodson has argued that while NGOs are not the "life blood" of the Court, they still have a "meaningful and largely overlooked impact".[28]

In addition, in *El-Masri* v. *The Former Yugoslav Republic of Macedonia*, Macedonia was responsible under Article 3 of the ECHR for failure to take adequate preventive measures, even though the CIA had a dominant role in committing the acts of torture.[29] The ECtHR emphasized the responsibility of Macedonia to "carry out an effective investigation" and extended its responsibility to even cover the ill-treatment of the applicant after his transfer to the United States.

In theory, similar logic could be applicable in the context of climate change, where the fault could be divided between two or more states. State authorities X could have an obligation to ensure that activities are not transferred to such country Y where the legislation and administrative measures do not provide adequate safeguards of absolute rights. In parallel, country Y, where the heavy pollution would be transferred, should take action to ensure that the rights are not violated.

The basis in joint enterprise

Another doctrine of shared responsibility concerns joint ventures. In *Hess v. the United Kingdom*,[30] the Four Allied Powers were committed inherently to joint conduct, illustrated by the decision-making body and actual control over the person. However, while the Commission acknowledged the *de facto* existence of shared activity in *Hess*, it was not at the time ready to establish a division of "joint authority" between the states involved.

In *Hussein v Albania and twenty other States*, the ECtHR continued its cautious approach by emphasizing the dominant role of the US in the arresting process, giving the presence of the European coalition a secondary role.[31] The threshold used in the case requires active and direct involvement and a common act of joint enterprise instead of sole participation in a joint enterprise.[32] A strict reading of the case would imply that joint action and intent are not present in the context of climate change because the phenomenon has developed over the years without proper joint control.

Concluding remarks

The ECtHR does not yet have jurisprudence on climate change and human rights. However, current jurisprudence on positive obligations, extraterritoriality and shared responsibility provides guiding principles for climate change litigation. These guiding principles provide a model for litigants on how to establish whether the state is aware of climate change, what precautionary measures under human rights law states are obliged to take in order to prevent violations, how to argue that the state has a responsibility towards citizens of another state and under which conditions the responsibility could be shared.

So long as there are no climate change judgments from the ECtHR, these principles are only guiding. The actual development of climate change–related human rights case law under the European Convention on Human Rights requires strategic litigation, such as that done in the *Urgenda* case in the Netherlands. Currently, there are several other pending cases in Europe, such as the *Swiss senior* case, the Swedish *Magnolia* case, the *Klimaatzaak* case in Belgium and the *People v. Arctic Oil* in Norway.[33] Similar to the *Urgenda* case, these cases seek to secure the recognition of the link between human rights obligations and climate action. In relation to the new areas of protection, the ECtHR often waits for domestic and international developments in order to be consistent before it stretches its interpretation to new fields. Thus, the importance of the rulings resulting from current strategic litigation cannot be overemphasized.

Notes

1 See, for example, O. Pedersen, 'The Ties That Bind: The Environment, the European Convention on Human Rights and the Rule of Law' 16 *European Public Law* (2010) 571.
2 See, for example, Heiskanen and Viljanen, 'Reforming the Strasbourg Doctrine on Extraterritorial Jurisdiction in the Context of Environmental Protection', 11 *European Law Reporter* (2014) 285.
3 The Hague District Court ruled that the Netherlands has a duty to take more action to reduce greenhouse gas emissions in its territory. The Court ruled that the Netherlands should ensure that domestic emissions are at least 25% lower than 1990 levels by 2020. The Dutch Government has appealed the decision. Urgenda C/09/456689, 13–1396, 24 May 2015, Rechtbank Den Haag, The District Court of the Hague, see in particular paras 3.2.m 4.45, 4.46, 4.49, 4.52, 4.74.
4 See, for example, Boyle, 'Making the Polluter Pay? Alternatives to State Responsibility in the Allocation of Transboundary Environmental Costs', in F. Francioni (ed.), *International Responsibility for Environmental Harm* (1991) 378; Francioni, 'Exporting Environmental Hazard Multi-National Enterprises: Can the State of Origin Be Held Responsible?', in F. Francioni (ed.), *International Responsibility for Environmental Harm* (1991) 279.
5 ECtHR, *Al-Adsani v. United Kingdom*, Appl. no 35763/97, Judgment of 21 November 2001, para 55; ECtHR, *Demir and Baykara v. Turkey*, Appl. no 34503/97, Judgment of 12 November 2008, paras 147–151. All ECtHR decisions are available online at http://hudoc.echr.coe.int/.
6 For the subsidiarity role of the ECtHR, see Helgesen, 'What Are the Limits to the Evolutive Interpretation of the European Convention on Human Rights', 31 *Human Rights Law Journal* (2011) 275, at 279.
7 J. Polakiewicz, 'The Status of the Convention in National Law', in R. Blackburn and J. Polakiewicz (eds.), *Fundamental Rights in Europe, The ECHR and Its Member States, 1950–2000* (2001) 49.
8 Urgenda, available online at http://us1.campaign-archive2.com/?u=91ffff7bfd16e26db7bee63af&id=c5967d141c&e=46588a629e.
9 Pedersen, *supra* note 1.
10 Urgenda C/09/456689, *supra* note 3, at 4.97.
11 Consensus may refer to the shared consensus of the majority of the States Parties in relation to certain values and moral principles, in comparison to the domestic legislation and policy of the States Parties, on the assessment of the existence of international treaties or on 'scientific reports by universities and government agencies, expert opinions, and experiential testimony'. Dzehtsiarou, *European Consensus: A Way of Reasoning?* University College Dublin Law Research Paper No 11/2009 (28 May 2009); West and Schultz, 'Learning for Resilience in the European Court of Human Rights: Adjudication as an Adaptive Governance Practice', 20 *Ecology and Society* (2015) 31; see also ECtHR, *Brincat and Others v. Malta*, Appl. nos. 60908/11, 62110/11, 62129/11, 62312/11 and 62338/11; Judgment of 24 July 2014, paras 105–107; ECtHR, *Fagerskjöld v. Sweden*, Appl. no. 37664/04, Decision of 26 February 2008; ECtHR, *Vilnes and Others v. Norway*, Appl. nos. 52806/09 and 22703/10, Judgment of 5 December 2013, para 244.
12 See ECtHR, *Aksoy v. Turkey*, Appl. no 21987/93, Judgment of 18 December 1996, para 61; ECtHR, *Abdulaziz, Cabales and Balkandali v. the United Kingdom*, Appl. no 9214/80, 9473/81and 9474/81, Judgment of 28 May 1985, para 78; ECtHR, *Creangă v. Romania*, Appl. no. 29226/03, 23 February 2012, para 88.
13 Urgenda C/09/456689, *supra* note 3.
14 ECtHR, *Brincat and Others v. Malta*, Appl. nos. 60908/11, 62110/11, 62129/11, 62312/11 and 62338/11, Judgment of 24 July 2014, paras 9, 37–40, 105, 106.
15 *Ibid.*, para 101.
16 See ECtHR, *Guerra v. Italy*, Appl. no 14967/89, Judgment of 19 February 1998, paras 57–60; ECtHR, *Brincat and Others v. Malta*, Appl. nos. 60908/11, 62110/11, 62129/11, 62312/11 and 62338/11, Judgment of 24 July 2014, para 114; ECtHR, *Grikovskaya v. Ukraine*, Appl. no. 38182/03, Judgment of 21 July 2011, paras 67, 69.
17 ECtHR, *Kolyadenko and Others v. Russia*, Appl. nos. 17423/05, 20534/05, 20678/05, 23263/05, 24283/05 and 35673/05, Judgment of 28 February 2012, para 173.

18 See Mowbray, *The Development of Positive Obligations Under the European Convention on Human Rights by the European Court of Human Rights* (Oxford: 2004) 2–3; Domelly, 'Positive Obligations and Privatization', 61 *Northern Ireland Legal Quarterly* 3(2) 2012, pp. 215, 216, 218.
19 ECtHR, *Kolyadenko and Others v. Russia*, Appl. nos. 17423/05, 20534/05, 20678/05, 23263/05, 24283/05 and 35673/05, Judgment of 28 January 2012, paras 168–172, 185.
20 See, for example, A. Kiss and D. Shelton, *International Environmental Law* (1991) 131.
21 Boyle, *supra* note 4, at 379.
22 See M. Milanovic, 'The Spatial Dimension: Treaties and Territory', in C. J. Tams, A. Tzanakopoulos, A. Zimmermann and A. E. Richford (eds.), *Research Handbook of the Law of Treaties* (2014) 203–209; S. Karagiannis, 'The Territorial Application of Treaties', in D. B. Hollis (ed.), *Oxford Guide to Treaties* (2012) 322–323.
23 The question of whether the act is committed inside or outside the state border is not necessary as the responsibility may be established currently under both circumstances. See ECtHR, *Soering v. the United Kingdom*, Appl. no 14038/88, Judgment of 7 June 1989; ECtHR, *Vilvarajah and Others v. the United Kingdom*, Appl. no. 13163/87, 13164/87, 13165/87, 13447/87 and 13448/87, Judgment of 30 October 1991; ECtHR, *Bankovic v. Belgium*, Appl. no 5220/99, Judgment of 12 December 2001; ECtHR, *Illich Sanchez Ramirez v. France*, App. no. 28780/95, Judgment of 24 June 1996; ECtHR, *Hirsi Jamaa Others v. Italy*, Appl. no. 27765/09, Judgment of 23 February 2012.
24 See ECtHR, *Loizidou v. Turkey*, Appl. no 15318/89, Judgment of 23 March 1995, paras 134–136; ECtHR, *Al-Skeini and Others v. the United Kingdom*, Appl. no. 55721/07, Judgment of 7 July 2011. For literature, see, for example, M. Milanovic, *Extraterritorial Application of Human Rights Treaties, Law, Principles and Policy* (Oxford: Oxford University Press, 2011) 126–127.
25 ECtHR, *Loizidou v. Turkey*, Appl. no 15318/89, Judgment of 23 March 1995, para 62.
26 Vennemann, 'Application of International Human Rights Conventions to Transboundary State Acts', in R. M. Bratspies and R. A. Mills (eds.), *Transboundary Harm in International Law* (2006) 297; ECtHR, *Bankovic v. Belgium*, Appl. no 5220/99, Judgment of 12 December 2001.
27 ECtHR, *Rantsev v. Cyprus and Russia*, Appl. no 25965/05, Judgment of 7 January 2010.
28 L. Hodson, *NGOs and the Struggle for Human Rights in Europe* (2011) 152–153, at 371.
29 ECtHR, *El-Masri v. The Former Yugoslav Republic of Macedonia*, Appl. no 39630/09, Judgment of 13 December 2012, paras 206, 211.
30 ECtHR, *Hess v. United Kingdom*, Appl. no 6231/73, Judgment of 28 May 1975.
31 ECtHR, *Hussein v. Albania and twenty other States*, Appl. no 23276/04, Judgment of 14 March 2006.
32 *Ibid.*
33 Klimaseniorinnen, *Swiss Senior Case* (2016), available online at http://klimaseniorinnen.ch/wp-content/uploads/2016/10/Gesuch-um-Erlass-Verfuegung_Sperrfrist.pdf; J. Palmblad and P. Björnstrand, *Magnolia Summons Application* (2016), available online at https://drive.google.com/file/d/0BwNst9QrJa18Y2x6X1hMYmJmSEk/view; A. Neslen, *Norway Faces Climate Lawsuit Over Arctic Oil Exploration Plans*, available online at www.theguardian.com/environment/2016/oct/18/norway-faces-climate-lawsuit-over-oil-exploration-plans; Klimaatzaak, *De Rechtszaak*, available online at www.klimaatzaak.eu/nl/de-rechtszaak/#klimaatzaak.

25

Are Europeans equal with regard to the health impact of climate change?

Isabell Büschel

The correlation between climate change and human health

In line with Simon Caney's affirmation that any account of climate change (CC) impacts that ignores its implications for people's enjoyment of human rights is fundamentally incomplete and inadequate,[1] we try in this chapter to outline the link between CC and human health[2] and to demonstrate that CC has a human rights impact, especially with regard to the rights to the protection of health and to equal treatment of persons in Europe.

Climate-borne threats for the population of Daredevil's health: true concerns in a ficticious case

December 2020. As a sad record this year, the Kingdom of Daredevil, Member State of the European Union (EU), registers 55,000 premature deaths among its citizens due to bad air quality in association with extreme heat waves. It is established that repeated air pollution peaks associated with ground-level ozone lead to respiratory deficiencies and premature deaths.[3] In the case of the Kingdom of Daredevil, the number of premature deaths has almost doubled since 2016. Researchers argue that there is a correlation with CC, as on the one hand heat peaks have been increasing by 5.5 degrees Celsius over the past half century and private investment into sustainable housing (for example, in the form of adequate isolation against heat and cold) could not be unblocked since the 2008 crisis severely hit the country's economy and households. On the other hand, road traffic has been steadily increasing and the government still favours the use of diesel vehicles over alternatively fuelled cars and trucks. On top of that, unemployment affects all levels of society, especially young professionals whose qualifications do not match the needs for skilled workforce in both natural and social sciences in order to adequately address the sanitary, scientific and legal challenges of CC. Because of wrong or short-term-view political choices particularly in the fields of environment, health, energy and transport the Kingdom of Daredevil failed to build resilience in the face of pathologies linked to extreme air pollution and heat levels.

Demonstrated risks for human health provoked by climate change

Admittedly, not all climate-related changes are negative for human health. Milder temperatures will lead to less cold-related fatalities and a more comfortable indoor environment during winter. Also, the productivity of outdoor workers is expected to increase because of milder winters, and more precipitation will promote agriculture and food production.[4]

However, it is undisputable that CC largely constitutes a threat to human health in Europe.[5] This threat may be caused by direct effects on human health, e.g. changes in the incidence of allergic diseases[6] or diseases transmitted by insects (mosquitoes and ticks), and by indirect effects, such as changes in water and air quality, or the impacts from extreme weather conditions.[7] Since 1998, floods have caused some 700 deaths within the EU, the displacement of approximately half a million people and at least €25 billion in insured economic losses.[8] Cold temperatures are expected to lead to an increase in suicidality. Combined with coronary thrombosis (possibly leading to stroke) and respiratory disorders, they also contribute to other types of excess winter deaths, together with other factors such as influenza, social class and per capita gross national product.[9] There are serious concerns and evidence that CC could amplify existing mental disorders and especially addictions and suicide rates.[10] On top of that, wrong energy policy choices and unaffordable electricity tariffs may cause fuel poverty, which in turn may negatively affect individuals' health and well-being.

Climate change as a cause for health inequalities

Under the abovementioned circumstances, inequalities are likely to be caused by the effects of CC on populations' health across Europe, having as consequence violations of the rights to the protection of human health and to equal treatment provided for by Articles 35 and 20–26 of the Charter of Fundamental Rights of the European Union (CFREU),[11] as well as national constitutions. Such violations may be caused by either inequity in the exposure to health risk or uneven coverage of health care costs related to the effects of CC.

Inequitable exposure to health risks

Differences across European Member States due to geography, demography and levels of sustainable development[12] may exacerbate the uneven distribution of health impacts caused by CC among populations and regions. As a consequence, vulnerable population subgroups are likely to be hit more severely in some Member States of the EU than in others: children, elderly, people with chronic disease and socially/economically disadvantaged individuals.[13] For example, mortality increases during heat waves from 7.6% to 33.6%.[14] The EuroHEAT project demonstrates that mortality is even higher when ozone levels are high, especially among the elderly (75–84 years).[15] Winter mortality is higher in countries with a warmer winter climate due to housing standards that are not thermally efficient to retain heat. An increase in premature deaths during winter is observed in Portugal (28%), Ireland and Spain (each 21%), the UK (18%), Greece (18%) and Italy (16%), whereas the populations of Finland, Germany and the Netherlands are far less exposed to this health risk due to housing standards that are prepared for cold climate.[16]

Besides the question of ethical justifiability of this challenge to equal treatment, there is a legal issue with regard to effectiveness in respect of the right to protection of human health and equal treatment. According to Article 35, sentence 2 of the CFREU, "a high level of human health protection shall be ensured in the definition and implementation of all the Union's policies and activities". As Member States are legally bound by the respect of the principles of primacy of EU law and of loyal cooperation, this high level of human health protection must be reflected in national policies and laws. If it is established that "a high level" of human health protection does not necessarily mean the highest attainable level,[17] another question is whether it is tolerable and legal that this level differs across Europe's regions.[18] As long as there is no common EU legislation on patients' rights other than the directive on cross-border healthcare, there is no legal guarantee for equality in access to healthcare. Consequently, differing premature death rates for identical risks related to CC, such as extreme temperatures or excess air pollution levels, are currently causing differences in treatment of EU citizens according to the place where they live and work.

Coverage of country-specific climate change–related health expenses

In the face of CC, a phenomenon that is universal and affects without exception the various geographic zones of the EU, the health expenses that it causes or contributes to are being covered according to national health insurance schemes.[19] In other words, coverage for treatment of allergies or cardiovascular disorders provoked by exposure to pollen, nitrogen dioxide and particulate matter and/or by extreme weather events differs from one Member State to another. Even though the sufferings from CC are similar across countries, access to treatment and cost coverage are different due to the principle of subsidiarity[20] that prevails in the definition and implementation of public health policies within the EU according to Article 168, paragraph 1, sentence 2 of the Treaty on the Functioning of the European Union.[21] Due to the restricted competences attributed to the EU in this area, the Commission's role has traditionally been limited to supporting the EU Member States' efforts to protect and improve the health of their citizens and to ensure the accessibility, effectiveness and resilience of their health systems. However, the EU can be considered empowered to legislate to tackle climate-related health issues based on Articles 191 to 193 of the Treaty on the functioning of the EU. According to these provisions, EU environmental policy should contribute to pursuing the objectives of preserving, protecting and improving the quality of the environment; protecting human health; prudently and rationally utilising natural resources; and promoting measures at the international level to deal with regional or worldwide environmental problems, in particular combating climate change. There is an urgent need for EU legislation to be adopted on these grounds given that CC-related migration to and within EU territory will increasingly stress Member States' health care systems.[22]

How has the EU so far been addressing the challenge of health inequalities caused by CC?

EU adaptation policies as a means to combat health inequalities

The following measures are highlighted among thosewhich the EU has adopted to address and prevent health inequalities in the framework of adaptation policies to CC:

a Decision no. 1386/2013/EU of the European Parliament and of the Council of 20 November 2013 on a General Union Environment Action Programme to 2020 'Living well, within the limits of our planet'.[23] This 7th Environment Action Programme sets at its paragraph 2.1(c) as one of the priority objectives to "safeguard the Union's citizens from environment-related pressures and risks to health and well-being". In the Annex, the Union commits to "transforming itself into an inclusive green economy that secures growth and development, safeguards human health and well-being, . . . reduces inequalities" (point 10) and to "update targets in line with the latest science and seek more actively to ensure synergies with other policy objectives in areas such as climate change, mobility and transport, biodiversity" (point 47). While the European Commission recognizes that enhancing climate resilience can have important benefits for public health, it stresses at the same time the need for adequate management of the synergies and potential trade-offs between climate-related and other environmental objectives. Taking the example of air quality, it alerts about the risk that switching to certain lower carbon emission fuels such as biofuels in response to climate-related considerations could lead to substantial increases in particulate matter and dangerous emissions. This is because currently most biofuels are produced from land-based crops, which causes concern over increased consumption of biofuels requiring agricultural expansion at a global scale, in turn leading to additional carbon emissions (this effect is called Indirect Land Use Change, or ILUC). In this respect, the EU's policy of using biodiesel for transport is bad practice, as it is set to increase Europe's overall transport emissions by almost 4% instead of cutting CO_2 emissions, which is equivalent to putting around 12 million additional cars on Europe's roads in 2020.[24] In London, medical doctors recently called for a ban on diesel vehicles to stop this cause of premature deaths.[25]

b In the Communication "A Budget for Europe 2020", the European Commission commits to mainstreaming CC into overall Union spending programmes and to direct at least 20% of the Union budget to climate-related objectives.[26] According to Directorate-General "Climate Action" of the European Commission, at least 20% of the EU budget for 2014–2020 – as much as €180 billion – should be spent on climate change–related action.[27] To achieve this increase, mitigation and adaptation actions are to be integrated into all major EU spending programmes, in particular cohesion policy, regional development, energy, transport, research and innovation and the Common Agricultural Policy.[28] Furthermore, it is foreseen that the EU's development policy also contributes to achieving the 20% overall commitment, with an estimated €1.7bn in 2014–2015 and €14bn over the years 2014–2020 for climate spending in developing countries.[29]

c Regulation no. 282/2014 on the establishment of a Third Programme for the Union's action in the field of health (2014–2020) is meant to, "in particular in the context of the economic crisis, contribute to addressing health inequalities . . . through actions under the different objectives and by encouraging and facilitating the exchange of good practices" (Recital 10).[30] The sharing of information and lessons learned across countries and sectors concerns, for example, behavioural strategies such as clothing, drink, food; scheduling daily work; seasonal migration; food safety and water quality. Furthermore, health education and training is to be promoted with respect, for example, to urban/spatial planning, building design, natural cooling systems etc.

d In the *Roadmap to a Resource Efficient Europe,*[31] the European Commission invites Member States' governments to phase out environmentally harmful subsidies (EHS), with tax reductions or exemptions being one example (point 3.4). Whereas this Roadmap is not of a binding legal nature itself, it serves as a strategy document intended to achieve the EU's climate commitments laid down in binding legal acts over which the European Commission enjoys power of enforcement.[32] As one step in the direction of phasing out EHS can be considered the 2016 report by the German Environment Agency revealing a paradox in national government policy, namely showing that subsidies granted by the German government are worth €57 billion work against climate policy.[33]

Outlooking remarks

As of today, the health impact of CC is insufficiently addressed by research.[34] Only seven research projects have been funded by the EU between 2004 and 2010 about public health knowledge on extreme weather events, such as heat waves and cold spells and their environmental consequences (e.g. floods, wildfires, air pollution).[35] Can Europe afford, on moral, economic and legal grounds, to not grant sufficient funding today to addresstomorrow's consequences of CC on human health?

Notes

1 S. Caney, 'Climate Change, Human Rights, and Moral Thresholds', in S. M. Gardiner *et al.* (eds.), *Climate Ethics: Essential Readings* (2010) 163, at 173; Report to the European Parliament's Directorate-General for External Policies of the Union, *Human Rights and Climate Change: EU Policy Options* (2012), available online at www.europarl.europa.eu/committees/fr/studiesdownload.html?languageDocument=EN&file=76255; C. Cournil and A-S. Tabau (eds.), *Changements climatiques et droits de l'homme, Larcier* (2013).

2 WHO Regional Office for Europe, *Protecting Health in Europe from Climate Change*, World Health Organization, 2008, available online at www.euro.who.int/__data/assets/pdf_file/0016/74401/E91865.pdf.

3 European Environment Agency, *Air Quality in Europe – 2016 Report* (2016), available online at www.eea.europa.eu/publications/air-quality-in-europe-2016, at 58.

4 European Commission, *Commission Staff Working Document Accompanying Document to the White Paper Adapting to Climate Change: Towards a European Framework for Action. Human, Animal and Plant Health Impacts of Climate Change*, SEC(2009) 416, 1.4.2009, available online at http://ec.europa.eu/health/ph_threats/climate/docs/com_2009-147_en.pdf, at 4.

5 D. Paci, *Human Health Impacts of Climate Change in Europe*, Report for the PESETA II Project (2014), available online at http://publications.jrc.ec.europa.eu/repository/bitstream/JRC86970/lfna26494enn.pdf (last visited 24 November 2016); European Commission, *Commission Staff Working Document on Health Security in the European Union and Internationally*, 23.11.2009, SEC(2009) 1622 final, available online at http://ec.europa.eu/health/preparedness_response/docs/commission_staff_healthsecurity_en.pdf, at 4. On the universal level, WHO has conducted extensive research and reporting on the health impacts of CC including reports on the effects of CC on selected causes of death; gender, CC and health; CC and health systems strengthening; climate mitigation health co-benefits; social dimensions of CC; air pollution; and nutrition.

6 Climate and associated land-use change will affect the range of allergenic species and the timing and length of the pollen season, and plant productivity and pollen production may be increased by elevated CO_2 levels, which is expected to lead to at least the doubling of sensitization to ragweed across Europe from 33 million currently to 77 million people by 2041–2060, I. R. Lake *et al.*, 'Climate Change and Future Pollen Allergy in Europe', *Environmental Health Perspectives* (2016) 1, at 5.

7 European Commission, *Public Health: Climate Change: Policy*, available online at http://ec.europa.eu/health/climate_change/policy_en.

8 European Commission, *Public Health: Climate Change: Flooding in Europe: Health Risks*, available online at http://ec.europa.eu/health/climate_change/extreme_weather/flooding_en.

9 European Commission, *Public Health: Climate Change: Cold Spell*, available online at http://ec.europa.eu/health/climate_change/extreme_weather/cold_weather/index_en.htm.

10 A. C. Willox *et al.*, 'Examining Relationships Between Climate Change and Mental Health in the Circumpolar North', 15 *Regional Environmental Change* (2014) 169–182; F. Burque and A. C. Willox, 'Climate Change: The Next Challenge for Public Mental Health?', 26 *International Review of Psychiatry* (2014) 415–422, 415; K. N. Fountoulakis *et al.*, 'Relationship of Suicide Rates With Climate and Economic Variables in Europe During 2000–2012', 15 *Annals of General Psychiatry* (2016) 1, at 4.

11 The CFREU is the catalogue of fundamental rights of the EU. It was solemnly proclaimed by the heads of state at the Nice European Council on 7 December 2000. Since the entry into force of the Treaty of Lisbon on 1 December 2009, it became part of EU primary law, meaning that it has the same binding legal effect on EU institutions and national governments as the EU treaties themselves. Its content is consistent with the European Convention of Human Rights adopted by the Council of Europe in 1950.

12 Bluszcz, 'Classification of the European Union Member States According to the Relative Level of Sustainable Development', 50 *Quality & Quantity* (2016) 2591, at 2603.

13 CC is likely to alter health inequalities and to affect in an uneven manner especially 'children, those working outdoors, the elderly, women and people with a pre-existing illness': European Commission, *supra* note 4, at 7.

14 European Commission, *Public Health: Climate Change: Public Health Responses to Heat Waves*, available online at http://ec.europa.eu/health/climate_change/extreme_weather/heatwaves/index_en.htm#fragment1.

15 *Ibid.*

16 European Commission, *supra* note 9.

17 Court of Justice of the EU, Dow AgroSciences e.a. / Commission, T-475/07, Rec. p. II-5937, Decision of 9 September 2011, point 149: "the Community institutions are bound by their obligation, under the first subparagraph of Article 152(1) EC, to ensure a high level of human health protection. That high level does not necessarily, in order to be compatible with that provision, have to be the highest that is technically possible (Case C-284/95 Safety Hi-Tech [1998] ECR I-4301, paragraph 49)"; I. Büschel, 'Les rapports entre santé et libertés économiques fondamentales dans la jurisprudence de la Cour et du Tribunal de Première Instance des Communautés Européennes' (8 October 2009) (Thèse pour le doctorat en droit public, Aix-Marseille University).

18 European Union Agency for Fundamental Rights, *Inequalities and Multiple Discrimination in Access to and Quality of Healthcare* (2013), available online at https://fra.europa.eu/sites/default/files/inequalities-discrimination-healthcare_en.pdf.

19 T. K. Hervey and J.V. McHale, *European Union Health Law: Themes and Implications* (Cambridge: Cambridge University Press, 2015) 47.

20 According to the principle of subsidiarity, enshrined in Article 5 of the Treaty on EU, decisions are taken as closely as possible to the citizen and the EU does not take action (except in the areas that fall within its exclusive competence), unless it is more effective than action taken at the national, regional or local level.

21 "Union action, which shall complement national policies, shall be directed towards improving public health, preventing physical and mental illness and diseases, and obviating sources of danger to physical and mental health."

22 Migration related to CC to the extent that it contributes to tensions over scarce resources, land loss and border disputes, conflicts over energy sources, tensions between those whose emissions caused CC and those who will suffer the consequences of CC, political radicalisation in weak or failing States: *Paper from the High Representative and the European Commission to the European Council, Climate Change and International Security* (14 March 2008), S113/08, section II, available online at www.consilium.europa.eu/uedocs/cms_data/docs/pressdata/en/reports/99387.pdf.

23 European Parliament and Council, Decision no. 1386/2013/EU on a *General Union Environment Action Programme to 2020 'Living Well, Within the Limits of Our Planet'* (20 November 2013), available online at http://eur-lex.europa.eu/legal-content/EN/TXT/HTML/?uri=CELEX:32013D1386&qid=1483814084474&from=EN (last visited 7 January 2017).

24 Transport & Environment, 'Globiom: The Basis for Biofuel Policy Post-2020' (2016), available online at www.transportenvironment.org/sites/te/files/publications/2016_04_TE_Globiom_paper_FINAL_0.pdf.

25 BBC News, *Doctors Call for Ban on Diesel Engines in London*, available online at www.bbc.com/news/uk-england-london-38274792.

26 European Commission, *Communication to the European Parliament, the Council, the European Economic and Social Committee and the Committee of the Regions. A Budget for Europe 2020. Part I*, COM(2011) 500 final, 29.6.2011, available online at http://eur-lex.europa.eu/resource.html?uri=cellar:d0e5c248-4e35-450f-8e30-3472afbc7a7e.0011.02/DOC_3&format=PDF; European Commission, *Communication to the European Parliament, the Council, the European Economic and Social Committee and the Committee of the Regions. A Budget for Europe 2020. Part II: Policy Fiches*, COM(2011) 500 final, 29.6.2011, available online at http://ec.europa.eu/budget/library/biblio/documents/fin_fwk1420/MFF_COM-2011-500_Part_II_en.pdf.

27 European Commission, *Public Health: Climate Action: Supporting Climate Action Through the EU Budget*, available online at https://ec.europa.eu/clima/policies/budget_en.

28 *Ibid.*

29 *Ibid.*

30 European Parliament and Council, *Regulation (EU) No 282/2014 of the on the Establishment of a Third Programme for the Union's Action in the Field of Health (2014–2020) and Repealing Decision No 1350/2007/EC* (11 March 2014), available online at http://eur-lex.europa.eu/legal-content/EN/TXT/HTML/?uri=CELEX:32014R0282&qid=1483815912360&from=EN.

31 European Commission, *Communication to the European Parliament, the Council, the European Economic and Social Committee and the Committee of the Regions, Roadmap to a Resource Efficient Europe*, COM(2011) 571 final, 20.9.2011, http://eur-lex.europa.eu/legal-content/EN/TXT/?uri=CELEX:52011DC0571.

32 Such as the "2020 Climate & Energy Package": European Commission, *Public Health: Climate Action: 2020 Climate & Energy Package*, available online at http://ec.europa.eu/clima/policies/strategies/2020_en#tab-0-0.

33 L. Köder and A. Burger (eds.), *Umweltschädliche Subventionen in Deutschland* (2016), Umweltbundesamt, available online at www.umweltbundesamt.de/sites/default/files/medien/2546/publikationen/umweltschaedliche_subventionen_2016_.pdf.

34 K. L. Ebi, Jan C. S. and J. Rocklöv, 'Current Medical Research Funding and Frameworks Are Insufficient to Address the Health Risks of Global Environmental Change', 15 *Environmental Health* (2016) 108.

35 A list of these EU-funded projects is available online at http://ec.europa.eu/health/climate_change/extreme_weather/flooding/index_en.htm.

26

Integrating a human rights-based approach to address climate change impacts in Latin America

Case studies from Bolivia and Peru

Andrea Rodriguez and María José Veramendi Villa

Introduction

Latin America is one of the regions that has been most affected by climate change. People in situations of vulnerability, such as indigenous communities, have been most affected. As a result of climate change impacts, their human rights have been and/or are at risk of being violated.

Based on a limited analysis performed to date, we have examined two specific case studies – one in Peru and one in Bolivia – to determine the climate change impacts on two communities, the existing legislation on climate change and human rights, and the extent to which effective measures were taken to mitigate such impacts.

Through analysis of climate- and human rights-related papers and materials, this chapter provides an analysis that demonstrates the negative impacts climate change has on the fulfillment of human rights. It also evaluates State interventions and legal frameworks to address the issue.

Climate change impacts in Bolivia

Bolivia is one of the countries most vulnerable to the effects of climate change in Latin America. Glacier retreat in the Andean Region, severe drought, and increased and heavier rainfall are among the impacts affecting the country and its people.[1] Climate change is exacerbating existing development-related issues that threaten the livelihood of the Bolivian population, particularly the most vulnerable. For example, glacier retreat in the Andes is currently affecting valuable water sources for main cities, including Bolivia's capital, La Paz.[2]

The recent disappearance of the second largest lake in Bolivia is one of the most alarming examples of the destructive power of climate change.

The case of Lake Poopó

Lake Poopó is situated in the department of Oruro, Bolivia. It is the second largest lake in the country (2,337 square kilometers). It is the home to more than 200 species of birds, fish, and plants.[3] In December 2015, Lake Poopó dried up completely.

The Bolivian government claims that this unfortunate situation can be attributed to weather changes caused by El Niño, mining activities in the area, and climate change caused primarily by developed nations.[4] The latter is supported by arguments made by the scientific community, in which it has been estimated that the lake has warmed, on average, 0.23 degrees Celsius each decade since 1985.[5] The warming of the lake has led to the disappearance of many species and has forced local communities to relocate. This is the case with the Uru-Murato people.

The human rights dimension

The Uru-Murato people are the oldest indigenous ethnic group living in the area. They are known as "men from the lake," given their close relationship with and dependence on the natural resources of Lake Poopó.

Since the lake's disappearance, this ethnic group has been forced to leave their communities and find work in nearby mines or salt flats. It is estimated that only 636 Uru-Murato remain in the area and nearby villages.[6]

The lake's warming and subsequent disappearance has had serious implications on the fulfillment of indigenous rights, particularly those related to "the protection of their cultural identity, religious beliefs, spiritualities, practices and customs, and their own world view,"[7] as recognized under the Bolivian Constitution. The Uru-Murato people can no longer fish or trade. Fishing was the primary activity and source of food and income for this ethnic group.[8]

Rights to water and food security have also been undermined. Unable to fish, these communities are forced to change their diets and adapt to other sources of food, which are scarce in the Bolivian flatlands.

The Uru-Muratos' rights to self-determination and territoriality, as proclaimed in the Bolivian Constitution, have also been affected.[9] They are forced to migrate and seek alternate ways of living that do not necessarily correspond with their traditional lifestyle. The Uru-Muratos are a clear example of a community displaced by climate change.

Climate change and human rights policies in Bolivia

Bolivia adopted a new constitution in 2009 into which a wide range of rights providing constitutional protection for the environment, Mother Earth, and Indigenous Peoples were incorporated.[10] Chapter II states the fundamental rights of Bolivians, including the right to water and food, the right to an adequate habitat and home, and the right to universal and equitable access to basic services. Chapter IV describes the particular rights of rural native Indigenous Peoples, including the right to a healthy environment.

In 2010, Bolivia passed the Rights of Mother Earth Law; and in 2012, the Mother Earth Law and the Integral Development to Live Well Law.[11] These legal instruments declare both Mother Earth and her life systems as titleholders of rights. These laws incorporate climate change perspectives into general environmental and socio-economic legislative frameworks.

In 2014, a disaster risk management law also was approved,[12] incorporating climate change estimates into national, regional, and local risk management strategies. A national fund was created to finance prevention projects.

Bolivia ratified the United Nations Framework Convention on Climate Change (UNFCCC) in October 1994. In a recent effort to contribute to the success of the Paris Agreement, in 2015 Bolivia submitted its Intended Nationally Determined Contribution, which states that the government will make fair and ambitious efforts to address the impacts of climate change.[13] However, these efforts will only be ambitious if means of implementation, including financial support, are made available through mechanisms of international cooperation.[14]

Bolivia is also a signatory of the major human rights treaties of the United Nations and the Inter-American System on Human Rights, including the American Convention on Human Rights in the area of Economic, Social and Cultural Rights, the "Protocol of San Salvador" and its Additional Protocol. This instrument recognizes the right to live in a healthy environment and to have access to basic public services. It also states the duties of State Parties to "promote the protection, preservation, and improvement of the environment."[15]

Furthermore, the Bolivian Constitution states that any international treaty ratified by Bolivia prevail over national law. The right to water is recognized as a human right, and the State assumes responsibility to manage, regulate, protect, and plan the adequate and sustainable use of water resources.[16]

As a response to this critical issue, the Bolivian government adopted a General Framework for the Management of Lake Poopó.[17] This legal instrument will provide a plan to distribute water to different sectors, including people living in nearby areas. It is difficult to estimate whether this plan will be sufficient to address the problem, particularly the human rights impacts. The framework attempts to provide equal redress to people and other sectors affected by the depletion and contamination of the water, including the mining sector. However, it does not give priority to the Uru-Murato people.

Concrete measures to compensate and reduce the impacts on the Uru-Muratos have not been made beyond the adoption and implementation of the legal instrument, through which the government hopes to remediate the lake's problem. No formal human rights complaint has been made to strengthen support for local affected communities.

The case of Uru-Murato people is a clear example of how communities in vulnerable areas are being affected despite the existence of national and international regulatory frameworks that secure and protect the rights of Indigenous Peoples. The fundamental rights of the Uru-Muratos have been affected and will continue to be affected as the planet continues to warm. Whether these issues can be directly attributed to climate change or to poor management of the lake's resources, the government of Bolivia will require assistance to ensure its most vulnerable people can adapt to the impacts of climate change without having their human rights undermined.

Unfortunately, government measures to address this issue have so far proved to be insufficient. Given the particular fragility and vulnerability of the region, it is vital for the international community to strengthen their climate change commitments in an effort to avoid situations such as this one.

Climate change impacts in Peru

Peru is another country highly vulnerable to the impacts of climate change. In 2004, the Tyndall Centre for Climate Change research released a report indicating that Peru is one of the top 10 most vulnerable countries in the world, and the third most vulnerable in Latin America, to the impacts of climate change.[18] According to the Grantham Research Institute on Climate Change and Environment at the London School of Economics, "Peru reportedly experiences more natural disasters than any other country in Latin America. Since 1970, it has experienced over 100 droughts, floods, mudslides, frosts, earthquakes and volcanic eruptions resulting in tens of thousands of deaths."[19] These hydrometeorological phenomena have increased more than sixfold between 1997 and 2006.[20] Among these, the so-called El Niño phenomenon is happening more often and with more intensity.

Peru has rich ecological and climatic diversity, featuring 27 of the world's 32 climates.[21] The *Cordillera Blanca* (White Mountain Range) is home to 71 percent of the world's tropical glaciers.[22] Twenty-two percent of the surface area of Peru's glaciers has disappeared in the past 30 years.[23] As indicated by the World Bank,

> [t]ropical glaciers play a key role in regulating water in the Andean region. During droughts or the dry season, the [glaciers] provide abundant water for human consumption, agriculture and hydroelectric energy. One result of accelerated glacier retreat is the formation of new lakes in the highlands, which increase the risk of floods and landslides, in addition to hindering the glaciers' ability to regulate water, thereby increasing water shortages.[24]

One such example is the case of Lake Palcacocha.

The case of Lake Palcacocha

The Palcaraju and Pucaranra glaciers lay above Lake Palcacocha. In December 1941, a large piece of Palcaraju fell into the lake, causing a wave that flooded the city of Huaraz, killing more than 5,000 people.[25] Tropical glaciers are melting rapidly due to climate change, and it's possible that what happened that December might happen again. A 2013 report from the National Institute of Civil Defense indicates that between January 2011 and October 2012, 11 Supreme Decrees were issued declaring a state of emergency for Lake Palcacocha due to high water levels and the subsequent risk posed to residents of Huaraz.

A 2013 analysis on glacial hazards in Huaraz simulated the worst-case scenario and a smaller melting event. Both resulted in extensive flooding of the city. The study found that "[b]ecause of the inundation depth and the velocity of the flow, most of the area of the city that experiences flooding will have a very high hazard level, putting both lives and property at risk."[26]

Saul Luciano Lliuya is a farmer and tour guide in the Cordillera Blanca who, in March 2015, filed a complaint letter with German energy utility company RWE demanding compensation for the impact its activities have on climate change and glaciers. Lliuya alleged that RWE's contribution to climate change is causing glacial melting and therefore putting Lake Palcacocha and the city of Huaraz, his home, at high risk of flooding. Lliuya's claim goes hand-in-hand with RWE's definition of the company as Europe's "biggest single emitter of CO_2."[27]

In May 2015, RWE issued an official reply to the letter asserting that the claims lack a legal basis and that the company is not responsible for any damages. On November 24, 2015, Lliuya filed a lawsuit against RWE at the Regional Court in Essen, Germany, where the company is headquartered. Lliuya claims that RWE is partially responsible for glacial melting in the Andes and for placing his home, located at the foot of the mountains, at risk. In his claim, he requests financial contributions from RWE, with a payment proportional to the company's contribution to climate change, as well as safety measures for Lake Palcacocha.[28]

In the course of the lawsuit, RWE has continued to deny responsibility, arguing that "individual emissions cannot be tied to glacial melting in the Andes." RWE further claims that "local temperatures have even decreased, meaning there would be no warming that could be tied to increased glacial melting [and] ... that flooding no longer poses a risk for the Andean city of Huaraz."[29]

On December 15, 2016, the Essen District Court dismissed Lliuya's claims because, according to the judge, "legal causality" was not demonstrated.[30] The lawyer representing Lliuya indicated that "legal causality does exist" and that they will "most likely appeal."[31]

The human rights dimension

Huaraz, with a population over 100,000, is the capital city of the department of Ancash, located in the Cordillera Blanca. Huaraz is located on the flood path of Lake Palcacocha. It has been and continues to be at risk of being affected by a disaster such as the one that occurred in 1941.

A possible overflow of Lake Palcacocha and the consequent flooding of the city of Huaraz will cause severe human rights impacts on its population.

The rights of people will be affected if glacier melting continues, including the right to life – given the human toll that a potential flood could cause without any early warning systems in place – and the right to water – since the city's main source of drinking water is the Paria River, which flows from Lake Palcacocha.

A flood could cause mass displacement and, in turn, affect the ways of life of the traditional and indigenous populations living in the impacted area. The lack of comprehensive adaptation plans, on both the national and local levels, would have a direct impact on the survival of the city of Huaraz and its residents.

Climate change and human rights policies in Peru

The Peruvian Constitution of 1993 establishes that every person has the right to a balanced and appropriate environment for the development of life.[32] Peru is also a signatory of the major human rights treaties of the United Nations and the Inter-American System on Human Rights. Article 11 of the Additional Protocol to the American Convention on Human Rights in the area of Economic, Social and Cultural Rights, known as the "Protocol of San Salvador," establishes that: "1. Everyone shall have the right to live in a healthy environment and to have access to basic public services. 2. The States Parties shall promote the protection, preservation, and improvement of the environment."[33]

Peru has existing obligations to respect and guarantee human rights, as well as to adopt legislation or other measures to give effect to the rights enshrined in those treaties.

On the climate change front, Peru ratified the UNFCCC in 1992 and created the National Commission on Climate Change in 1993. In 2002, it ratified the Kyoto Protocol and in 2003 adopted Decree No. 086–2003 PCM – National Strategy on Climate Change.[34] It is worth noting that the Strategy does not refer to the human rights impacts of climate change. It also faced a number of challenges for its implementation. In 2015, the Strategy was updated and approved via Decree No. 011–2015 of the Ministry of Environment.[35] The new strategy identifies glacial retreat and reduced access to the associated water resources as climate risks.

In 2011, the National Disaster and Risk Management System (SINAGRED)[36] was established. The Decree by which the System was created states that any policies, strategies, or plans related to climate change are part of the National Disaster and Risk Management Policy.

Peru submitted its INDC in 2015, which INDC recognizes that "seven basins studied in the "Cordillera Blanca" (mountain range) have exceeded a critical transition point in their retreat."[37] Despite acknowledging the high vulnerability of the country to the impacts of climate change, nowhere in the INDC is there a reference to the impact on human rights and/or any hint of using a human rights–based approach when addressing the impacts of climate change and/or the measures to adapt to it.

Finally, Peru ratified the Paris Agreement[38] on July 21, 2016, the preamble of which provides guidance for the respect, promotion, and consideration of existing human rights obligations in its implementation. This is of particular importance when designing adaptation plans for the population that will be adversely affected by receding glaciers and potential floods caused by Lake Palcacocha.

Conclusions and recommendations

Climate change impacts cannot be attributed to a single or specific event. From a liability standpoint, it is difficult to assess who should be responsible for the negative impacts of climate change and the required remediation of human rights violations.

While there are indications that Peru and Bolivia have taken measures to ensure legal and institutional frameworks are in place, effective implementation is required to address climate-related impacts and guarantee the protection of human rights, particularly for the most vulnerable populations.

The narrative presented indicates that enhanced efforts are needed by the international community for more ambitious climate change efforts. This does not translate into disregarding the responsibility of developing countries over their own obligations to protect human rights from climate change impacts. Efforts to mainstream a human rights dimension in climate change policy are needed, as are mitigation measures to reduce the risk of climate change in the fulfilment of human rights.

Effective and ambitious climate measures are particularly important for the survival of vulnerable indigenous groups. Many of these groups are reaching a tipping point and, without the support and commitment of the international community to effectively tackle climate change, disappearance will be the only possible result.

Notes

1 Oxfam, *Bolivia: Climate Change, Poverty and Adaptation* (2009), available online at www.oxfam.org/sites/www.oxfam.org/files/file_attachments/bolivia-climate-change-adaptation-0911_4.pdf.
2 Hoffmann Dirk, *Press Release: El retroceso de los glaciares bolivianos pone en peligro a comunidades* (2016), available online at www.egu.eu/news/293/el-retroceso-de-los-glaciares-bolivianos-pone-en-peligro-a-comunidades/.
3 BBC World, *¿Cómo se secó el Poopó, el segundo lago más grande de Bolivia?* (2016), available online at www.bbc.com/mundo/noticias/2015/12/151223_ciencia_bolivia_lago_poopo_desaparicion_sequia_wbm.
4 *Ibid.*
5 C. Nicholas, Climate Change Claims a Lake and an Identity, *New York Times* (2016), available online at www.nytimes.com/interactive/2016/07/07/world/americas/bolivia-climate-change-lake-poopo.html.
6 *Ibid.*
7 Art 30 (II 2), *Chapter IV: Rights of the Nations and Rural Native Indigenous Peoples, Constitution of Plurinational State of Bolivia* (2009), available online at www.constituteproject.org/constitution/Bolivia_2009.pdf.
8 S. Gerardo, *Amazonia y sus Etnias 2010*, at 154.
9 Art 30 (I), *supra* note 7.
10 *Ibid.*
11 Law N 071, Law of the Rights of Mother Earth (2010) and Law No 300, the Mother Earth Law and Integral Development to Live Well (2012).
12 Law 601, Bolivian Disaster Risk Management Law (2014).
13 UN Framework Convention on Climate Change, *Intended Nationally Determined Contribution of the Plurinational State of Bolivia* (2015), available online at http://www4.unfccc.int/submissions/INDC/Published%20Documents/Bolivia/1/INDC-Bolivia-english.pdf.
14 *Ibid.*
15 A-52: Additional Protocol to the American Convention on Human Rights in the area of Economic, Social and Cultural Rights "Protocol of San Salvador", 17 November 1988 (General Assembly,

Organization of American States), Treaty Series, N° 69, (entered into force 16 November 1999) (ratified by Bolivia on October 2006).

16 See Constitution of Plurinational State of Bolivia, *supra* note 7.

17 Estado Plurinominal de Bolivia, *Plan director de la cuenca del Lago Poopó plantea la recuperación del lago y el manejo integral de la cuenca* (2015), available online at www.mmaya.gob.bo/index.php/noticias/notas,1502.html.

18 W. N. Adger *et al.*, *New Indicators of Vulnerability and Adaptive Capacity* (2004), Tyndall Project IT1.11. Technical Report 7, United Kingdom: Tyndall Centre, available online at www.tyndall.ac.uk/sites/default/files/it1_11.pdf.

19 Granthan Research Institute on Climate Change and the Environment at the London School of Economics and Political Science, *Climate Change Legislation in Peru*, available online at www.lse.ac.uk/GranthamInstitute/wp-content/uploads/2015/05/PERU.pdf, at 6.

20 Servindi, *Perú es el tercer país más vulnerable del mundo al cambio climático*, available online at www.servindi.org/actualidad/99300.

21 Ministerio del Ambiente, *Portalde Cambio Climàtico: En el Perú*, available online at http://cambioclimatico.minam.gob.pe/cambio-climatico/sobre-cambio-climatico/que-impactos-tiene/en-el-peru/.

22 The World Bank, *Peru Prepares to Address Andean Glacier Retreat* (25 March 25 2013), available online at www.worldbank.org/en/news/feature/2013/03/25/peru-prepares-to-face-the-retreat-of-andean-glaciers.

23 *Ibid.*

24 *Ibid.*

25 M. A. Somos-Valenzuela *et al.*, *Inundation Modeling of a Potential Glacial Lake Outburst Flood in Huaraz, Peru*, 14-01 CRWR Online Reports, The University of Texas at Austin (March 2014) 7, available online at http://hdl.handle.net/2152/27738.

26 *Ibid.*

27 RWE, *Climate Protection*, available online at www.rwe.com/web/cms/en/1904186/rwe/responsibility/environment/climate-protection/.

28 Germanwatch, *Peruvian Farmer Sues German Utility RWE Over Dangers Related to Glacial Melting* (24 November 2015), available online at https://germanwatch.org/en/11302.

29 Germanwatch, *The Case of Huaraz: RWE Denies Responsibility for Climate Damages in the Andes – Court Hearing This Autumn* (3 June 2016), available at online https://germanwatch.org/en/12291.

30 Germanwatch, *Regional Court Dismisses Climate Lawsuit Against RWE – Claimant Likely to Appeal*, available online at https://germanwatch.org/en/13234.

31 *Ibid.*

32 Art 2, para 22, *Constitución Política del Perú de 1993*, available online at http://www4.congreso.gob.pe/ntley/Imagenes/Constitu/Cons1993.pdf.

33 "Protocol of San Salvador", *supra* note 15 (ratified by Peru on 17 May 1995) (instrument of ratification deposited by Peru on 4 June 4 1995).

34 Congreso de la República, *DECRETO SUPREMO No 086–2003-PCM, Aprueban la Estrategia Nacional sobre Cambio Climático* (2013), available online at http://www2.congreso.gob.pe/sicr/cendocbib/con4_uibd.nsf/63F2FF2A354CBBB305257C9E005A7203/$FILE/086-2003-pcm.pdf.

35 Ministerio de Ambiente del Peru, *Estrategia Nacional ante el Cambio Climático* (2015), available online at www.minam.gob.pe/wp-content/uploads/2015/09/ENCC-FINAL-250915-web.pdf.

36 Instituto National de la Defensa Civil (INDECI), *DECRETO SUPREMO No 048-2011-PCM que crea el Sistema Nacional de Gestión del Riesgo de Desastres (SINAGERD)* (2011), available online at www.indeci.gob.pe/objetos/secciones/MQ==/Mw==/lista/MzEx/MzE0/20111013154908 1.pdf.

37 UN Framework Convention on Climate Change, *Intended Nationally Determined Contribution (INDC) from the Republic of Peru*, available online at http://www4.unfccc.int/ndcregistry/PublishedDocuments/Peru%20First/iNDC%20Per%C3%BA%20english.pdf.

38 El Peruano, *DECRETO SUPREMO N° 058–2016-RE, Ratifican el Acuerdo de París*, available online at http://busquedas.elperuano.com.pe/normaslegales/ratifican-el-acuerdo-de-paris-decreto-supremo-n-058-2016-re-1407753-12/.

27

Connecting human rights and short-lived climate pollutants

The Arctic angle

*Sabaa A. Khan**

Introduction

The reduction of greenhouse gas emissions (GHGs) has long been the exclusive focal point of climate change activism and regulatory efforts. While carbon dioxide (CO_2) emissions are without a doubt the most important cause of rising temperatures, anthropogenic emissions of other gaseous and aerosol substances also have an immense impact on the rate of global warming. Advances in knowledge on the prominent climate warming and human health effects of short-lived climate pollutants (SLCPs), such as black carbon (BC), have pushed this type of emissions into a forefront global regulatory concern.

SLCP emissions have both direct and indirect impacts on the enjoyment of human rights. As major contributors to indoor and outdoor air pollution, SLCP emissions present toxicological effects to humans and the environment, thus directly threatening the human rights to life and to health. From the climate perspective, SLCP emissions cause significant temperature increases close to where they are emitted, and thus contribute to local and regional ecosystem changes such as rising water levels, volatile temperature fluctuations and changes in the lifecycles of flora and fauna. In turn, these kind of ecosystem disruptions can have severe consequences on human livelihoods. Specifically, they may threaten the human rights to life, to culture and to property, by curtailing access to traditional food sources[1] and by forcing the displacement of entire communities. In this respect, SLCP emissions indirectly aggravate human rights risks associated to climate change.

The inaction of states to regulate emissions of SLCPs can thus be seen as very intimately impacting a broad range of human rights. As explained by the International Court of Justice in its *Gabcikovo-Nagymaros* decision, environmental protection is a precondition to the enjoyment of *all* human rights:

> the protection of the environment is . . . a vital part of contemporary human rights doctrine, for it is a sine qua non for numerous human rights such as the right to health and the right to life itself. It is scarcely necessary to elaborate on this, as damage to the environment can impair and undermine all the human rights spoken of in the Universal Declaration and other human rights instruments.[2]

This chapter looks at the connections between SLCP emissions and human rights in the context of Arctic climate change. It discusses the effects of SLCP emissions on human health and global warming, highlighting the acute impact of BC emissions on the Arctic environment and on the human rights of Arctic communities. It further examines the potential of the Arctic Council's new soft law efforts to mitigate SLCPs to contribute to the protection of human rights in the region.

The science and law of short-lived climate pollutants

For well over a decade, Arctic Indigenous Peoples and scientific researchers have been signaling to us the fragility of the Arctic region to global warming and highlighting the actions necessary to address the latter's pervasive destructive effects on human health and the ecosystem.[3] While the issue of reducing GHG emissions has dominated climate change discourses, contemporary scientific knowledge emphasizes that alongside GHGs, microscopic particulate matter known as aerosols are also significant contributors to global warming. An emerging body of research points to the urgency of curtailing emissions of short-lived climate pollutants[4] (SLCPs), a subset of greenhouse gases and particulates (black carbon, methane, tropospheric ozone and hydrofluorocarbons) that make a substantial contribution to global warming and can be pinpointed to both anthropogenic activities and natural sources.[5]

BC, commonly known as soot, is a light-absorbing ultrafine particle emitted from the incomplete burning of fossil fuels, biofuels and biomass, and has been identified as "the second most important individual climate-warming agent after carbon dioxide."[6] Released directly into the air from anthropogenic activities such as shipping traffic, agricultural and forest burning, and wood or coal combustion, BC is also closely associated with respiratory and cardiovascular human health risks and cancer.[7] BC particles not only threaten human health by intensifying air pollution, they warm the atmosphere by darkening ice and snow surfaces and thereby reduce the earth's capacity to radiate sunlight back to space.[8]

SLCP governance approaches

The view that SLCP mitigation measures should be considered an integral part of the broader climate governance agenda is becoming more widespread, as evidenced by the inclusion of SLCP actions in the Nationally Determined Contributions (NDCs) submitted by a number of countries to the Paris Agreement adopted under the United Nations Framework Convention on Climate Change (UNFCCC).[9] Still, there does not appear to be momentum at the international institutional level to extend the UNFCCC framework to SLCP emissions further than what states might decide to include in their NDCs. Moreover, aerosols are not explicitly covered under the scope of the UNFCCC regime. The UNFCCC defines emissions restrictively, as "the release of greenhouses gases and/or their precursors,"[10] thereby excluding aerosols.[11]

Global cooperative efforts to reduce different types of SLCP emissions have taken place within the context of international and regional legal regimes as well as through transnational, multistakeholder voluntary initiatives, such as the Climate and Clean Air Coalition, the Global Alliance for Clean Cookstoves and the Global Methane Initiative. In terms of international institutions and regimes, methane and hydrofluorocarbons (HFCs) do fall under the regulatory scope of the UNFCCC. Most recently, measures to phase down HFCs were adopted by 197 countries under the Montreal Protocol on Substances that Deplete the Ozone Layer.[12]

With regards to BC, the International Maritime Organization is looking into the regulation of BC emissions from international shipping operations in the Arctic region, yet no specific control measures have been proposed to date.[13] The Gothenburg Protocol adopted under the Convention on Long-Range Transboundary Air Pollution[14] (CLRTAP) is the only international legal instrument to adopt emission reduction targets specifically for black carbon. Under its 2012 amendment, the Gothenburg Protocol obliges contracting states to prioritize the reduction of black carbon emissions in their regulatory efforts to control particulate matter.[15]

In general, this growing landscape of SLCP mitigation measures follows the prevalent trend in international environmental governance of organizational and regime fragmentation.[16] The overall mitigation landscape shows states and non-state actors selectively addressing SLCPs within many institutional platforms, albeit with little coherence, vague institutional interplay and questionable enforceability concerning black carbon measures in particular.[17]

In the meantime, the urgency of regulating black carbon emissions for climate, atmospheric and human health protection is become increasingly evident. The World Health Organization estimates that global emissions of black carbon contribute to 4.3 million deaths annually from household air pollution and 3.7 million deaths from ambient air pollution.[18] In relation to global warming, anthropogenic emissions of black carbon have the most powerful climate forcing effect, following emissions of carbon dioxide.[19] Since the atmospheric lifetime of black carbon and other SLCPs lasts from only a few days to months, reducing their emissions generates immediate, localized benefits in terms of protecting human health and slowing the rate of long-term warming.[20]

Black carbon impacts in the Arctic

BC pollution is widely acknowledged to have a particularly detrimental effect on vulnerable ecosystems such as the Arctic, where melting and warming are accelerated by BC emission sources from both within and outside the Arctic region.[21] From a biodiversity perspective, accelerated Arctic warming is linked to changes in circumpolar vegetation and an earlier snowmelt, which is expected to have a profound negative impact on plant and animal biodiversity, reindeer husbandry and tourism, implying significant transformations to the livelihoods of Indigenous Peoples.[22]

Arctic emissions and human rights

The absence of effective international and even national regulatory frameworks to address anthropogenic emissions of SLCPs poses an immediate human rights threat to Arctic inhabitants. The negative outcomes of climate change are especially impactful on Arctic Indigenous Peoples, due to their unique cultural, social and economic reliance on the Arctic environment that is crucial to sustaining their livelihoods and indigenous ways of life.[23] Building upon earlier Inuit human rights activism on CO_2 emissions, Arctic Indigenous Peoples have initiated a second wave of legal activism to target emissions of BC.

Arctic Indigenous Peoples have played a leading role in bringing the issue of climate change to the attention of human rights bodies. In 2005, a petition was filed to the Inter-American Commission on Human Rights (IACHR) on behalf of all the Inuit of the Arctic regions of Canada and the US. Relying closely on the findings of the 2004 Arctic Impact Assessment Report, the petition alleged that the United States' failure to regulate CO_2 emissions essentially resulted in profound climatic changes to the Arctic ecosystem, leading to the violation of fundamental human rights of the Inuit protected under the American Declaration of the Rights and Duties of Man and other international instruments, including "rights to the benefits of culture, to property, to the preservation of health, life, physical integrity, security, and a means of subsistence, and to residence, movement, and inviolability of the home."[24] Rather than constituting an action for monetary compensation or redress, the petition was filed in the hope of engaging the United States in meaningful dialogue on introducing legal measures to address climate change.[25]

Although the petition was deemed inadmissible, the Inuit initiative did convince the IACHR to establish a hearing aimed at strengthening the regional human rights body's understanding of the linkage between climate change and human rights, ultimately paving the way for global discussions on climate change as a human rights issue.[26] Seven years later, in 2013, the Arctic Athabaskan Council (AAC; a Permanent Participant of the Arctic Council that represents approximately 45,000 Indigenous Peoples spread across 76 communities in Alaska, Yukon and the Northwest Territories) filed a petition to the IACHR against the government of Canada, for undermining the human rights of Athabaskan peoples by failing to implement effective regulatory measures on black carbon emissions. The petition, which has yet to be considered admissible, targets Canada based on contemporary scientific findings confirming that BC emissions from within or close to the Arctic have a stronger regional climate impact than emissions from distant sources.[27]

The request for relief is modeled after the earlier Inuit Circumpolar Council (ICC) petition. The AAC petition attempts to establish a strong causal link between BC emissions from Canada and the Arctic climate changes that affect Athabaskan peoples' enjoyment of human rights. In this sense, the AAC petition builds upon the main weakness of the ICC petition. While the ICC pursued the US based on the latter's status as the world's highest CO_2 emitting nation, the petition itself acknowledged that "the actual correlation between cumulative emissions and temperature increase is subject to some uncertainty."[28] Even though there has been no explicit explanation from the IACHR for the failure of the 2005 petition, the lack of clear scientific evidence linking US emissions of CO_2 to the Arctic climatic changes adversely impacting Inuit rights is likely to have been an important decisional factor.[29]

Despite the strong scientific evidence and indigenous testimony and knowledge that underlie the AAC petition, it is possible that persisting scientific uncertainty regarding the precise role of global and regional emissions of BC in Arctic climate change, along with limitations in our current understanding of the complex physical and chemical processes affecting atmospheric levels of SLCPs,[30] present obstacles to the admissibility of the petition.

Moreover, since the petition was filed Canada has undertaken efforts to address BC emissions through the Arctic Council. In 2015, the member states of the Arctic Council adopted a voluntary agreement to collectively address emissions of BC and methane. Although the Arctic Council's *Framework for Action on Enhanced Black Carbon and Methane Emission Reductions*[31] does not create legally binding obligations upon the eight Arctic Council states to reduce BC or methane emissions, it embodies climate mitigation goals under a common vision and provides Arctic Council observer states an opportunity to implement the agreement as well.

While the Arctic Council is neither an international nor a regional organization, it is a unique high-level political platform that has led in recent years to the signing of two international treaties and thus can be said to have an evolving, legal dimension. With six Indigenous Peoples' organizations engaged in the Arctic Council as Permanent Participants (including the Arctic Athabaskan Council) and an evolving, increasingly internationalized roster of observer states and organizations, the Council provides a truly distinct and inclusive space for international cooperation on environmental issues. There are no other fora at the international level that centralize the participation of Arctic Indigenous Peoples to the same degree. In the adoption of the Framework, Arctic indigenous representatives were central participants, alongside Canada and other Arctic Council members, in the development and adoption of a soft law instrument through which Arctic states have committed themselves to take concrete measures on BC and methane. In this way, the Council has provided Arctic Indigenous Peoples a valuable pathway to spurring regional and international regulatory action on black carbon mitigation, a problem that lies at the nexus of climate change and human rights and remains inadequately addressed within existing multilateral environmental agreements. It is still important to recall that despite the adoption of the Framework, the petition has not lost its relevance. Its admissibility to the IACHR would constitute a clear affirmation of the human rights dimension of SLCP emissions and, more broadly, of global climate change. While the preamble of the Framework recognizes that Arctic warming is "leading to fundamental changes to the environment and human living conditions in both the Arctic and around the world" and also notes the "climate, health and economic benefits" of reducing black carbon and methane, the agreement does not expressly make a connection to human rights. In this regard, the future admissibility of the petition certainly remains of immense value to Arctic communities.

The Arctic council framework for action on black carbon and methane

The Framework expresses Arctic states' commitment "to take enhanced, ambitious, national and collective action" on the reduction of BC and methane emissions, as well as their intention to "adopt an ambitious, aspirational and quantitative collective goal on black carbon" within a two-year time frame. It institutes a national reporting system that requires Arctic states to develop and submit national action plans and mitigation strategies, as well as inventories on black carbon and methane emissions to the Arctic Council Secretariat. Furthermore, the Framework sets out an iterative two-year review process by an Expert Group that is meant to assess progress towards the common vision on a periodic basis and provide relevant recommendations for the continuous enhancement of mitigation actions.

A promising aspect of the Framework is that is recognizes the regional environmental and human health impacts of non-Arctic black carbon and methane emissions, and thus calls upon Arctic observer states to actively implement the agreement. To date, eight observer states (including India, one of the largest BC-emitting nations), as well as the EU, have submitted reports in response to the Framework. If the geographical scope of the Framework can be extended to high-emitting countries such as India and China, the effectiveness of this soft law mitigation approach would be significantly strengthened.[32] In fact, it is estimated that 40% of Arctic BC comes from sources located in East and Southeast Asia.[33] The transboundary pathways of BC air pollution and warming pose an acute challenge from a regulatory perspective, highlighting the truly universal dimension of the problem of Arctic climate change and the need for global mitigation measures.[34]

In addition to engaging with Arctic Council observer states, the Framework calls upon the private sector, civil society, other governments as well as financial institutions to implement the framework to the extent possible. As such, it is an inclusive and dynamic transnational agreement that cuts across all scales of governance, from the international to the local. Indeed, the Framework does not deliver concrete regional or international legal measures to curb black carbon emissions, but this is not an impossible outcome given the ongoing legal evolution of the Arctic Council and its ability to serve as a negotiating forum for international treaties. In light of the great difficulty at arriving at substantive collective commitments in global environmental negotiations, the Arctic Council Framework represents what is perhaps a more realistic approach to global problems, such as black carbon, by avoiding the traditional spatial and substantive limitations of international law. While international legal approaches often reflect a "general failure to see global problems across the multiple scales upon which they are experienced and ordered,"[35] the transnational approach of the Framework highlights the significance of local and regional, social and environmental considerations, and builds outwards to engage stakeholders on a global level.

Conclusion

Our scientific knowledge of the Arctic climate effects of black carbon emissions is still in early evolution. At the same, there is no doubt that black carbon emissions need to be reduced globally in order to protect the human rights of Arctic Indigenous Peoples. The transboundary nature of SLCP pollution means that even if Canada were to implement ambitious regulatory measures to control black carbon emissions, the human rights of Arctic Indigenous Peoples would remain at risk. Collective action is imperative, and in this regard, the adoption of the Arctic Council Framework should be considered a progressive contribution towards global SLCP mitigation. In particular, the fact that both petitioners to the IACHR (the ICC and the AAC) are Permanent Participants of the Arctic Council, and thereby fully engaged in the development and evolution of the Framework, reveals the unique and still relatively unexplored institutional potential of the Arctic Council to deliver inclusive and collective actions towards abating the impact of climate change on human rights. The Arctic Council was established to enhance regional cooperation on "common Arctic issues, in particular issues of sustainable development and environmental protection of the Arctic."[36] This mandate intimately entwines with the human rights of Arctic communities. Adoption of the Framework is a concrete example of how cooperation on sustainable development and environmental protection through the Council inherently holds an empowering potential concerning the human rights of Arctic citizens who have been historically marginalized from lawmaking processes.

Notes

* This research has been conducted with funding from the Academy of Finland (Decision #285389).

1 For example by making traditional hunting or fishing grounds inaccessible or more dangerous to access. See M. Nuttall *et al.*, 'Hunting, Herding, Fishing and Gathering: Indigenous Peoples and Renewable Resource Use in the Arctic', in C. Symon (lead ed.), *Arctic Climate Impact Assessment* (2005) 649.

2 Gabcikovo-Nagymaros Project (Hungary/Slovakia), Judgment, 25 September 1997, ICJ Reports (1997) 7, at paras 91–2.

3 One of the first comprehensive studies on the Arctic impacts of climate change was the 2004 Arctic Climate Impact Assessment, undertaken upon the request of the ministers of the Arctic Council. The multidisciplinary assessment which included both natural and social science perspectives drew upon the work of over 250 scientists and 6 representative groupings of Arctic indigenous peoples. See Arctic Monitoring and Assessment Programme, *Impacts of a Warming Arctic: Arctic Climate Impact Assessment* (2004), available online at www.amap.no/documents/doc/impacts-of-a-warming-arctic-2004/786.

4 Also known as short-lived climate forcers.

5 Intergovernmental Panel on Climate Change, Working Group I (IPCC/WG1), *The Physical Science Basis* (2013), available online at www.climatechange2013.org/images/report/WG1AR5_ALL_FINAL.pdf ; United Nations Environment Program/World Meteorological Organization, *Integrated Assessment of Black Carbon and Tropospheric Ozone* (2011), available online at www.ccacoalition.org/fr/file/638/download?token=Cmpbe41v; AMAP / P. K. Quinn *et al.*, *The Impact of Short-Lived Climate Pollutants on Arctic Climate* (2008); D. Zaelke and N. Borgford-Parnell, 'The Importance of Phasing Down Hydrofluorocarbons and Other Short-Lived Climate Pollutants', 5(2) *Journal of Environmental Studies and Sciences* (2015) 169; A Hu *et al.*, 'Mitigation of Short-Lived Climate Pollutants Slows Sea-Level Rise', 3(8) *Nature Climate Change* (2013) 730; D. Shindell *et al.*, 'Simultaneously Mitigating Near-Term Climate Change and Improving Human Health and Food Security', 335 *Science* (2012) 183; J. K. Shoemaker *et al.*, 'What Role for Short-Lived Climate Pollutants in Mitigation Policy', 342(6164)

Science (2013) 1323; United States Environmental Protection Agency, *Report to Congress on Black Carbon* (March 2012), available online at https://www3.epa.gov/airquality/blackcarbon/2012report/fullreport.pdf; Economic Commission for Europe, *Executive Body for the Convention on Long-Range Transboundary Air Pollution, Report by the Co-Chairs of the Ad-hoc Expert Group on Black Carbon*, ECE/EB.AIR/2010/7 (30 September 2010), available online at www.unece.org/fileadmin//DAM/env/documents/2010/eb/eb/ece.eb.air.2010.7.e.pdf.

6 T. C. Bond *et al.*,'Bounding the Role of Black Carbon in the Climate System: A Scientific Assessment', 118(11) *Journal of Geophysical Research: Atmospheres* (2013) 5380, at 5490.

7 European Environment Agency, *Status of Black Carbon Monitoring in Ambient Air in Europe*, EEA Technical Report No.18 (2013), available online at www.eea.europa.eu/publications/status-of-black-carbon-monitoring; World Health Organization, *Health Effects of Black Carbon* (2012), available online at www.euro.who.int/__data/assets/pdf_file/0004/162535/e96541.pdf.

8 V. Ramanathan and G. Carmichael, 'Global and Regional Climate Changes Due to Black Carbon', 1 *Nature Geoscience* (2008) 221.

9 These include Mexico, Chile, Nigeria, Cote d'Ivoire, Cameroon and Mauritius.

10 Art1(4), United Nations Framework Convention on Climate Change 1992, 1771 UNTS 107.

11 For a discussion of the prospects of including aerosols under the UNFCCC regime, see S.A. Khan, 'The Global Commons Through a Regional Lens: The Arctic Council on Short-Lived Climate Pollutants', 6(1) *Transnational Environmental Law* (2017) 131.

12 See 28th Meeting of the Parties to the Montreal Protocol on Substances that Deplete the Ozone Layer, Decision XXVIII/1: Further Amendment of the Montreal Protocol and Decision XXVIII/2: Decision related to the amendment phasing down hydrofluorocarbons.

13 Thus far, the Sub-Committee on Pollution Prevention and Response has adopted a definition of black carbon and is still in the process of refining the measurement protocols it has developed. See IMO, *Sub-Committee on Pollution Prevention and Response, 2nd Session, 19–23 January, 2015 Meeting Summary* (23 January 2015), available online at www.imo.org/en/MediaCentre/MeetingSummaries/PPR/Pages/PPR-2.aspx; *IMO Sub-Committee on Pollution Prevention and Response, 3rd Session 15–19 February, 2016, Meeting Summary* (19 February 2016), available online at www.imo.org/en/MediaCentre/MeetingSummaries/PPR/Pages/PPR-3rd-Session.aspx.

14 Convention on Long-Range Transboundary Air Pollution, Geneva, 13 November 1979, (entered into force 16 March 1983). The CLRTAP was adopted in 1979 under the United Nations Economic Commission for Europe and to date has been ratified by 51 UNECE members.

15 ECE, Executive Body for the Convention on Long-range Transboundary Air Pollution, 1999 Protocol to Abate Acidification, Eutrophication and Ground-level Ozone to the Convention on Long-range Transboundary Air Pollution, as amended on 4 May 2012, UN doc. ECE/EB.AIR/114, 6 May 2013.

16 On the fragmentation of international environmental law see G. Loibl, 'International Environmental Regulations: Is a Comprehensive Body of Law Emerging or Is Fragmentation Going to Stay?', in G. Hafner and I. Buffard (eds.), *International Law Between Universalism and Fragmentation: Festschrift in Honour of Gerhard Hafner* (2008) 783; H. van Asselt, *The Fragmentation of Global Climate Governance: Consequences and Management of Regime Interactions* (2014); T. Gehring, 'Treaty-Making and Treaty Evolution', in D. Bodansky, J. Brunnée and E. Hey (eds.), *Oxford Handbook of International Environmental Law* (2007) 467. For a comprehensive account of the international legal landscape for SLCPs, see Y. Yamineva and K. Kulovesi, 'Keeping the Arctic White: The Current Legal Landscape for Reducing Short-Lived Climate Pollutants in the Arctic Region and Opportunities for Future Development' (unpublished draft, filed with author).

17 Khan, *supra* note 11.

18 World Health Organization, *Reducing Global Health Risks Through Mitigation of Short-Term Climate Pollutants* (2015), available online at www.who.int/phe/publications/climate-reducing-health-risks/en/.

19 Bond *et al.*, *supra* note 6.

20 Arctic Marine Assessment Program, *Summary for Policy-Makers: Arctic Climate Issues 2015 – Short-Lived Climate Pollutants* (2015), available online at www.amap.no/documents/doc/summary-for-policy-makers-arctic-climate-issues-2015/1196.

21 Arctic Council, *An Assessment of Emissions and Mitigation Options for Black Carbon: Technical Report of the Task Force on Short-Lived Climate Forcers* (2011); Sand *et al.*, 'Arctic Surface Temperature Change to Emissions of Black Carbon Within Arctic or Midlatitudes', 118 *Journal of Geophysical Research: Atmospheres* (2013) 7788.

22 AMAP / Quinn *et al.*, *supra* note 5.

23 See M. Nuttall *et al.*, *supra* note 1.
24 ICC, *Petition to the Inter American Commission on Human Rights Seeking Relief From Violations Resulting From Global Warming Caused by Acts and Omissions of the United States* (7 December 2005) 5, available online at www.inuitcircumpolar.com/uploads/3/0/5/4/30542564/finalpetitionicc.pdf.
25 ICC Chair and Petitioner Sheila Watt-Cloutier stated the petition's intent "to encourage and to inform". See Watt-Cloutier cited in H. M. Osofksy, 'The Inuit Petition as a Bridge?', 31(2) *American Indian Law Review* (2006/2007) 675, at 687.
26 *Ibid.*, see also E. A. Kronk Warner and R. S. Abate, 'International and Domestic Law Dimensions of Climate Justice for Arctic Indigenous Peoples', 43 *Revue générale de droit* (2013) 113.
27 AAC, *Petition to The Inter-American Commission on Human Rights Seeking Relief from Violations of the Rights of Arctic Athabaskan Peoples Resulting From Rapid Arctic Warming and Melting Caused by Emissions of Black Carbon by Canada* (2013) 15, available online at http://earthjustice.org/sites/default/files/AAC_PETITION_13-04-23a.pdf.
28 ICC, *supra* note 24, at 69.
29 V. de la Rosa Jaimes, 'The Arctic Athabaskan Petition: Where Accelerated Arctic Warming Meets Human Rights', 45(2) *California Western International Law Journal* (2015) 213, at 259.
30 See AMAP, *supra* note 20.
31 Arctic Council, *Framework for Action on Enhanced Black Carbon and Methane Emission Reductions, Annex 4, Iqaluit, 2015 SAO Report to Ministers* (2015) (Framework), available online at https://oaarchive.arctic-council.org/handle/11374/610.
32 K. Kupiainen *et al.*, 'Black Carbon and Other Short-Lived Climate Pollutants in the Arctic – Consequences of Current Regulatory Frameworks to Emissions and Impacts', study presented at the 17th IUAPPA World Clean Air Congress (29 August 2016), Busan (Korea).
33 AMAP *supra* note 20, at 9.
34 *Ibid.*, as noted by the AMAP, "emissions from Arctic States are responsible for about a third of black carbon's warming effects in the Arctic, largely from direct effects of black carbon in the region, including enhanced melting of ice and snow. The other two-thirds come primarily from black carbon emissions outside Arctic countries that raise global temperature and thus affect the Arctic indirectly."
35 Khan, *supra* note 11, at 19.
36 Global Affairs Canada, *Declaration on the Establishment of the Arctic Council, Ottawa, Canada* (19 September 1996), available online at www.international.gc.ca/arctic-arctique/ottdec-decott.aspx?lang=eng.

28

Climate change and human rights in the Commonwealth Caribbean

Case studies of The Bahamas and Trinidad and Tobago

Lisa Benjamin and Rueanna Haynes

Introduction

Caribbean small island states are some of the most vulnerable in the world to climate change, and the human rights of residents of these countries have arguably already been impacted. Most Commonwealth Caribbean countries have written constitutions and institutional structures which protect fundamental rights and freedoms, though not in regard to human rights violations due to climate change. Recent events in both The Bahamas and Trinidad and Tobago have impacted the rights of residents in these countries and exemplify the need for these states to institute policies and protections specifically in the context of human rights and climate change. This chapter is separated into four main sections. The first sets out the particular vulnerabilities of Caribbean states to climate change, highlighting the issue of loss and damage. The second provides a brief overview of constitutional protection of human rights in Commonwealth Caribbean constitutions. The third and fourth sections provide examples of weather events in The Bahamas, and Trinidad & Tobago, which illustrate some of the ongoing impacts from climate change that these states are currently enduring.

Vulnerabilities to climate change and loss and damage

Caribbean countries are diverse in their historical background, systems of governance, language, economies and topographies. They geographically span from The Bahamas in the north to Suriname in the south.[1] Many Commonwealth Caribbean countries are also small island developing states, or SIDS, and some, such as The Bahamas and Trinidad and Tobago, are also archipelagic nations. Despite their diversity, these countries share a number of commonalities, including environmental and socio-economic vulnerabilities. These include:

- low-lying areas vulnerable to sea level rise and storm surges;
- geographic positions strongly affected by tropical storms and cyclones;

- high temperatures;
- scarce land resources;
- considerable dependence on scarce or depleted fresh groundwater resources;
- small natural resource bases;
- concentrations of population and infrastructure along coastal areas;
- dependence on a narrow range of export products;
- heavy dependence on imports;
- susceptibility to international trade and commodity price fluctuations;
- small domestic markets and limited ability to develop economies of scale;
- limited opportunities for economic diversification;
- high transport and communication costs (particularly acute in archipelagic nations like The Bahamas and Trinidad and Tobago);
- limited public budgets and dependence on foreign capital to finance development; and
- weak institutional structures and limited human capacity.[2]

These countries are also particularly vulnerable to the impacts of climate change,[3] and their socio-economic vulnerabilities often constrain their ability to adapt to the negative impacts of climate change. These vulnerabilities are likely to increase as global temperature increases trend towards exceeding the 1.5°C threshold[4] which is so important to SIDS. In 2010,[5] Parties to the UNFCCC agreed to review the adequacy of the 2°C temperature goal in relation to a 1.5°C goal beginning from 2013 to 2015 and periodically thereafter. The outcome of this two-year process resulted in the report of the Structured Expert Dialogue (SED) on the 2013–2015 review. After an in-depth examination of the scientific information available from the Fifth Assessment Report of the IPCC and other recognised sources of scientific data on climate change, the report of the SED co-facilitators determined that the existing 0.85°C temperature increase since 1880 had already resulted in impacts that exceeded the adaptation capacity of many people and ecosystems.[6] The report also concluded that the 'guardrails' of 2°C were not safe,[7] and for small islands, the difference in projected risks between 1.5°C and 2°C of warming was significant.[8] SIDS have already experienced negative impacts from climate change, some of which exceed the ability of these small states to adapt to or recover from them. As a result, SIDS have been advocating for some time for a mechanism to address loss and damage due to climate change through the Alliance of Small Island States, or AOSIS.[9]

While there is no universally accepted definition of loss and damage, it is often referred to as 'negative effects of climate variability and climate change that people have not been able to cope with or adapt to'.[10] These impacts are often understood to include both extreme events such as tropical cyclones, floods and storm surges, as well as slow-onset events such as sea level rise, ocean acidification and increased land, air and ocean temperatures. The term loss and damage includes the concept of loss, which refers to permanent and irreversible loss, such as the loss of a freshwater resource or loss of a coral reef. It also includes the concept of damage, which includes damage that can be repaired or restored, but where such activities are constrained due to financial, technical or other limitations.[11] Impacts due to climate change[12] often exceed the financial, human and ecological capacities of vulnerable states and put excessive strain on public budgets. These impacts also put increasing pressure on, and remove financing from, other developmental objectives of these states such as health care, poverty reduction and, ironically, climate change adaptation efforts, as shorter-term and more affordable options for addressing climate change damage are prioritised over long-term adaptation solutions which tend to be more costly. According to Lyster, after disasters developing countries are often faced with depleted tax bases, as well as declining foreign reserves and credit ratings, which makes borrowing more difficult and expensive.[13] These countries therefore often 'borrow domestically' by diverting post-disaster capital from other domestic programmes,[14] which can include developmental priorities such as environmental protection, health, education and poverty reduction, thereby decreasing their climate resiliency. The government response to the damage caused by the flooding in Trinidad and Tobago discussed below is an example of this type of phenomenon. The estimated costs of reconstruction after both Hurricanes Joaquin and Matthew in The Bahamas have put considerable strain on public budgets. Negative impacts from climate change therefore both hamper existing developmental goals through the damage they cause and hinder these states from investing in efforts to improve their resiliency to withstand future impacts.

As Roberts and Pelling note, loss and damage involve 'intolerable risks', which are risks to core social objectives such as health, welfare, security and sustainability.[15] Impacts from climate change are already threatening the human rights of residents of these countries. These include the right to life, which is threatened by injury and death from heatwaves, floods, storm surge and hurricanes, but also other rights such as the right to food, water, health, shelter and access to other key resources.[16] The U.N. General Assembly noted that in the context of small island states, climate change–related sea level rise and extreme events are threatening the habitability and, in the longer term, territorial existence of these states, impacting their citizens' rights to self-determination.[17] The projected increase in tropical cyclones, exacerbated by sea level rise, will pose direct threats to the lives and wellbeing of coastal inhabitants in these countries, and impacts may include increased exposure to hazardous weather conditions, flooding, submergence of settlements, coastal erosion as well as loss of freshwater resources.[18] Increased water temperatures and ocean acidification may lead to declines in marine biodiversity and ecosystem production[19] and ultimately threats to communities and states dependant on coastal resources for their economic wellbeing. It is precisely these types of impacts that have been observed in The Bahamas and Trinidad and Tobago, and which serve as case studies to highlight the connection between human rights and climate change impacts. Both of these states are Commonwealth Caribbean states, and both have written constitutions which seek to preserve and protect human rights.

Constitutional protection of human rights in the Commonwealth Caribbean and the question of environmental human rights

Commonwealth Caribbean countries became independent through the mechanism of written constitutions. These constitutive documents were often crafted with the oversight of the British Imperial legislatures in London as part of the decolonization process,[20] and are therefore considered non-autochthonous.[21] Many of these constitutions, therefore, follow the same model[22] and include a fundamental rights and freedoms chapter. These often include provisions to protect human rights, such as the right to life, freedom from slavery or indentured labour and inhuman treatment, privacy, freedom of expression and protection of private property.[23] Despite their vulnerabilities and high dependence on biological diversity, it is curious that most of these countries do not have justiciable environmental rights provisions in their constitutions. This may be due, in part, to the fact that many of these constitutions were written in the 1960s and 1970s, before the environmental movement had taken hold, and that, at present, constitutional reform exercises are politically sensitive. Since these original constitutions were adopted, only two countries (Guyana and Jamaica) have enacted entirely new constitutions under their own, sovereign legislatures.[24] Articles 25 and 26 of the Constitution of the Republic of Guyana are exceptional, as they provide for a duty of citizens to participate in protecting the environment, and an obligation of the State to make rational use of its resources.[25] The recently amended Jamaican constitution includes a substantive right to enjoy a healthy and productive environment free from the threat of injury or damage from environmental abuse and degradation of ecological heritage.[26] The Constitution of The Bahamas does not contain environmental rights provisions.[27]

For Trinidad and Tobago, whereas the preamble of the Constitution speaks to the pursuit of social justice and the equitable distribution of material resources, it is silent on the question of specific environmental rights, including in relation to the chapter on fundamental human rights and freedoms.[28] Nevertheless, the Environmental Management Act of 2000 references government commitment to achieving economic growth in accordance with sound environmental practices and includes the principle of the preservation of the environment for present and future generations as one of the objectives of the legislation.[29] In Trinidad and Tobago, the issue of the inclusion of a human right to a clean and healthy environment arose in discussions on constitutional reform in 1987. Commissioners appointed to oversee discussions on reform requested the development of a proposal on the inclusion of an environmental element in the constitution of Trinidad and Tobago. It should be noted that at the time, the Environmental Management Act was under discussion but had not yet been enacted. The proposal called for the inclusion in the Constitution of 'a class right of protection by the State of the Natural Environment, and terrain essential to life and posterity'.[30] This class right would then be enforceable both against the State for failure to protect and against a citizen or group of citizens for damaging State lands. The State would also derive from this right the authority to declare an 'Environmental Emergency'.

The basis of such a provision was to be the duty to protect the environment grounded in the United Nations Declaration of Human Rights, which speaks to exercising rights and freedoms in keeping with the 'general welfare in a democratic society'.[31] The provision was also to be in keeping with the need to protect the environment for present and future generations. At the time, discussions on constitutional reform were not successfully concluded. Discussions on constitutional reform in Trinidad and Tobago to date[32] have not addressed the issue of an environmental element for the Trinidad and Tobago constitution.[33] It should be noted that the Environmental Management Act of Trinidad and Tobago 2000[34] makes provision for 'emergency situations' and 'emergency response activities'.

Moreover, the National Environmental Policy of 2006 references the fundamental human rights chapter of the Trinidad and Tobago constitution and sets out that 'basic environmental, health and development principles are interdependent and in harmony with the Constitution of the Nation.'[35] Of course, these disparate references to the linkage between human rights and environmental principles cannot be viewed as justiciable and have not been sufficient to form the basis for the development of the inclusion of environmental rights in the constitution of Trinidad and Tobago.

Throughout the region, most other environmental provisions either are included in draft versions of amendments to constitutions[36] or are preambular, non-justiciable provisions.[37] None of these environmental provisions refers specifically to climate change, although a few do refer to inter-generational equity. The hesitation regarding the inclusion of a justiciable environmental constitutional provision may be due to the concern by some governments of their limited capacity to realize and provide for such rights.[38] This timidity directly reflects the vulnerable circumstances and capacity constraints of these states. The observed and projected impacts from climate change largely exceed the capacity of these states to manage. In the advent of justiciable environmental human rights, the traditionally vertical relationship between human rights and the State may require these vulnerable states to be directly responsible to residents for impacts from climate change, for which they bear little responsibility. Whatever the cause, the connection between human rights and climate change has not been vociferously argued domestically within these states, nor enshrined in their constitutions. Despite the lack of legal and administrative provisions regarding human rights and climate change impacts, and specifically loss and damage, in these vulnerable states, residents there are already experiencing climate-related impacts which are negatively impacting their human rights. The impacts of Hurricane Joaquin in The Bahamas, and a flooding event in the southeast coast of Trinidad and Tobago, due to sea level rise and extreme rainfall, are provided to illustrate the types of impacts on human rights that these events have already caused within these states.

An extreme event in The Bahamas: Hurricane Joaquin

In early October 2015, Hurricane Joaquin hit the South and Southeastern islands of The Bahamas as a Category 4 storm, and 'devastated' five islands: Long Island, Acklins, Crooked Island, Rum Cay and San Salvador.[39] It was the strongest October hurricane to hit The Bahamas since 1866, and the strongest Atlantic hurricane with non-tropical origins ever seen in the satellite area.[40] The strength of the storm was attributed to its movement over very warm waters. Sea surface temperatures in the area were almost 1.1°C higher than normal, and were the warmest ever recorded during the September period.[41] The storm intensified quickly, almost overnight. This meant that residents had very little time to prepare for such a strong storm, which then moved very slowly, hovering over these islands for two days. These factors contributed to the large scale of damage experienced.[42] The damage was due primarily to storm surges and high winds of up to 130 mph.[43] The storm surge was estimated at 12–15 feet in Rum Cay, Crooked Island and Acklins, and water marks were noted as high as 15 or 20 feet on some islands.[44]

The damage was 'catastrophic' in some areas.[45] Thirty-four lives were lost,[46] and hundreds of people had to be evacuated. Electrical outages were experienced in most islands where entire settlements were 'completely obliterated', and their fishing fleets washed ashore and were destroyed.[47] Approximately 85% of one settlement in Crooked Island was entirely destroyed.[48] In Long Island, over two-thirds of the island remained inundated under 4–6 feet of water almost a week after the storm had passed.[49] Almost 70% of Crooked Island was flooded with at least 5 feet of water.[50] Critical public infrastructure such as airports, health clinics, schools, docks, roads, as well as electrical and telecommunications networks were negatively impacted. The airports in Rum Cay and San Salvador were destroyed, and coastal roads were washed away, making emergency and reconstruction efforts more difficult.

These impacts affected the rights of residents in a number of ways. Many residents lost their homes, leading to evacuation and migration hundreds of mile away to the capital. Entire settlements were destroyed and may never be rebuilt, leading to a loss of community and a way of life. More than a year later, many of these communities have not been rebuilt. Excessive flooding contaminated freshwater supplies in Crooked Island with fecal matter, affecting residents' right to water. A number of these island communities depend on subsistence fishing and farming. These activities also act as economic generators of local economies. Many families have small herds of pigs, sheep and goats which they rely on to supplement their diets. The destruction of entire fishing fleets wiped out the fishing industry in Long Island. Excess flooding contaminated land traditionally used for agriculture near the coast, and many small ruminants were killed,[51] impacting these residents' rights to food and self-determination. It is estimated that the economic impact in the country exceeded $100 million.

Almost a year later, in October 2016, Hurricane Matthew devastated the central and northern islands in The Bahamas. While the islands affected by Hurricane Joaquin were largely spared, the cumulative damage of these two hurricanes was tremendous. The official reports regarding the damage caused by Hurricane Matthew are not yet publicly available, but it is estimated that Matthew alone caused approximately $500 million in damage.[52] The estimated damage of the two hurricanes combined is currently between approximately $800 million to $1 billion, although these figures could be exceeded by including uninsured losses.[53]

Trinidad and Tobago: the flooding of the Manzanilla roadway

Over the weekend of 14–16 November 2014, torrential rainfall in southern Trinidad caused flooding of the Nariva Swamp in the southeast of the island. This flooding, described as the worst experienced in the last 45 years, swept away 3 km of roadway, damaged nearby agricultural lands and severely damaged coastal homes. In some areas where the seas and floodwaters merged over the roadway the water was reported to be as much as 5 feet high,[54] leaving a number of residents stranded in their homes and creating difficulties for emergency services. The Office of Disaster Preparedness and Management (ODPM) insistently assured the population that Trinidad and Tobago was not under tropical storm threat, watch or warning. The service described the event as a 'weather condition'.[55] Nevertheless, the area was inaccessible for more than a week following the event and, as a result of continued intermittent rainfall and tidal activity, the flooding did not subside for over five days. Given the location of the roadway and increasing sea level rise over the years, the area was prone to seasonal flooding during the rainy season in the months of July to December. Similar damage had occurred in the 1970s in this area, but residents and officials[56] agreed that this was the worst that they had experienced in their lifetimes. Further sea level rise along the coast of the affected area is expected with future warming and increasing climate change.[57] This will make future events of this nature increasingly dangerous and damaging for residents if sustainable measures are not put in place to address the vulnerability of the area.

Government efforts to repair the damage focused on the roadway, which was an important access route connecting north east and south east areas of the island and served as a service road for large vehicles used for transportation in economic activities related to oil and gas exploration. Closure of the road for a temporary period resulted in inconvenience to thousands of residents who were forced to use a much longer alternative route, and it also caused the roadways of the alternative route to be vulnerable to damage by the re-routing of large transportation vehicles for oil- and gas-related activities.

Initially, government representatives outlined plans to replace the roadway first with a short-term temporary road that would cost around US$5.8 million,[58] swiftly followed by a longer-term project for modernising the infrastructure in the area. Plans for longer-term, more expensive solutions, however, quickly devolved into a cheaper, short-term plan that was described as 'part of an attempt to strike a balance – the need to provide a quick fix with aspirations of investing in a more thoughtful long-term solution for the road'.[59] This, in spite of resident concerns[60] that government intervention should focus on the longer-term protection of the coastal road and the communities in its vicinity. This flooding event severely damaged homes, curtailed access to the area because of the road closure and resulted in the loss of private property, as well as the flooding of agricultural lands. Consequently, residents' rights in relation to adequate standards of living, adequate food and housing were affected. A justiciable duty to protect the environment, in line with what had been suggested in the proposal to the constitutional commission of 1987, could have strengthened the case made by citizens for more sustainable government action in response to the damage caused by the flooding and the need to address the overall vulnerability of the area. In this regard, it should be noted that the damage to private property and agricultural lands because of coastal zone erosion has been identified as a major problem in Trinidad and Tobago in the context of climate change impacts.[61] The 2014 Integrated Coastal Zone Management Policy lists as one of its objectives the '[c]onduct [of] coastal vulnerability and risk assessments and incorporat[ion of] appropriate preventative and adaptive measures into all planning and management policies and decision-making processes to account for projected changes in climate, particularly increases in sea level'.[62] Given the uncertainty of data available for determining the effects of projected changes in climate on weather events, determining the precise nature of coastal vulnerability at the local level is a challenging exercise at best. This reality, especially for small island developing states, is one of the key constraints that impede adaptation planning and implementation.[63]

It is estimated that 60% of small-scale economic activities significant for the support of human lives are located in the coastal zone, and approximately 50% of the country's national transportation arteries – coastal roads, bridges and ports – are also in this highly vulnerable area.[64] As average global temperatures increase, the human rights component of the negative impacts of climate change will undoubtedly become an increasingly pressing issue in Trinidad and Tobago, as in other island states. The effects of this event are mild in comparison to the damage wreaked by Hurricane Joaquin in the Bahamas. Nevertheless, it serves as a poignant example of the scale of the population's vulnerability in these areas to extreme weather events, which are likely to occur on a more frequent and devastating basis on a warmer planet.[65]

Conclusion

Impacts from climate change are already causing incidents of loss and damage in SIDS. Extreme weather events are quickly becoming the 'new normal' in these states, leading to significant human rights impacts of residents there. The connection between human rights and climate change, however, has yet to be deeply enshrined in constitutions within these countries. Given the expense both financially and politically of constitutional referendums, it is unlikely that many more Commonwealth Caribbean SIDS will make attempts to include justiciable environmental provisions in their constitutions, leaving impacted residents with fewer legal options for redress. Many residents rely heavily on government assistance for reconstruction and rebuilding of their homes and livelihoods, but public budgets in these states are already overstretched. Governments are struggling to rebuild and repair public infrastructure such as roads, hospitals and airports, which are negatively impacted by climate events.[66] The costs of these impacts will also make climate adaptation efforts less feasible in these countries.

Given the lack of redress within constitutions, domestic policies and legislation may be the primary mechanism to address the connection between human rights and climate change. The accumulation of elevation data, coastal and hazard mapping, the establishment of hazardous zones, vulnerability analysis, and loss and damage and relocation policies could all provide these states with tools to begin to address the human rights impacts from climate change. Given the dualist nature of Commonwealth states, domestic legislation to implement the Paris Agreement could also include mechanisms to address the issue. Existing policy and legislative mechanisms providing procedural human rights, such as public participation in environmental policy- and decision-making, could also provide avenues for redress.

Climate adaptation, data accumulation and implementing loss and damage policies are expensive endeavours, and support from international sources for these activities is currently limited. Recent decisions[67] of the Green Climate Fund to aim for a 50/50 allocation of funding between adaptation and mitigation over time, with priority for SIDS, LDCs and African states, provides some glimmer of hope on the horizon. Nevertheless, in the absence of a clear roadmap for future sources of funds and an urgent scaling up of the finance that has been pledged, the options for financing urgently needed for adaptation actions in vulnerable countries remain limited. Additionally, member states of the UNFCCC and the ad hoc working group on the Paris Agreement are yet to find consensus on the need to link loss and damage to climate finance.

Despite these constraints, it is likely that impacts from climate change and incidents of loss and damage will only become more frequent in SIDS. As a result, mechanisms to address human rights impacts from climate change are likely to become a new developmental priority for SIDS.

Notes

1 UNEP, *Climate Change in the Caribbean and the Challenge of Adaptation* 6 (2008), available online at www.pnuma.org/deat1/pdf/Climate_Change_in_the_Caribbean_Final_LOW20oct.pdf.
2 *Ibid.* See also L. Benjamin and M. Stevenson, 'A Greener Future for Caribbean Constitutions? The Bahamas as a Case Study', XXI (2) *Widener Law Review (WLR)* (2015) 217, at 220–221.
3 The Fifth Assessment Report of the IPCC sets out with very high confidence that 'coastal systems and low-lying areas will increasingly experience submergence, flooding and erosion throughout the 21st century and beyond, due to sea level rise. And moreover that some low-lying developing countries and small island states are expected to face very high impacts that could have associated damage and adaptation costs of several percentage points of gross domestic product.' See IPCC, Climate Change 2014: Synthesis Report, Contribution of Working Groups I, II and III to the Fifth Assessment Report of the Intergovernmental Panel on Climate Change [Core Writing Team, R.K. Pachauri and L.A. Meyer (eds.)].
4 R. McKie, The Guardian, *World Will Pass Crucial 2C Global Warming Limit, Experts Warn* (2016), available online at www.theguardian.com/environment/2015/oct/10/climate-2c-global-warming-target-fail (last visited 30 September 2016).
5 UNFCCC, *Decision 1/CP.16*, available online at https://unfccc.int/resource/docs/2010/cop16/eng/07a01.pdf (last visited 27 January 2017), at paras 4, 138 and 139.
6 UNFCC, *Report on the structured expert dialogue on the 2013–2014 Review* (4 May 2015), at paras 37, 39. See also L. Benjamin and A. Thomas, '1.5 to Stay Alive? AOSIS and the Long Term Temperature Goal in the Paris Agreement', 7 *IUCN Academy of Environmental Law eJournal (IUCNAEL)* (2016) 122, at 123.
7 *Ibid.*, at para 40.
8 Benjamin and Thomas, *supra* note 6, at 124.
9 As far back as 1991, during the Intergovernmental Negotiation Committee's work on establishing the UNFCCC, AOSIS submitted a proposal for a mechanism to deal with loss and damage which included an insurance-based component, as well as rehabilitation, compensation and risk-management components. See also UNFCCC, *Adoption of the Paris Agreement*, UN Doc. FCCC/CP/2015/L.9 (12 December 2015), Article 8.
10 A. Durand and S. Huq, A Simple Guide to the Warsaw International Mechanism on Loss and Damage, *ICCAD*, available at www.icccad.net/wp-content/uploads/2015/09/A-simple-guide-to-the-Warsaw-International-Mechanism.pdf (last visited 30 September 2016); M. Burkett, 'Reading Between the Red Lines: Loss and Damage and the Paris Outcome' 6 *Climate Law (CL)* (2016) 118, at 199.
11 L. Rajamani, 'Addressing Loss and Damage From Climate Change Impacts', L(30) *Economic & Political Weekly (EPW)* (2015) 17; S. Surminski and A. Lopez, 'Concept of Loss & Damage of Climate Change - a New Challenge for Climate Decision-Making? A Climate Science Perspective', 7(3) *Climate and Development* (2015) 267, at 269.
12 The IPCC Fifth Assessment Report lists 'rising ocean temperature, ocean acidification, and loss of Arctic sea ice' as well as sea level rise and coastal flooding including storm surges as key hazards for SIDS, with the risk of resulting in 'loss of coral cover, Arctic species, and associated ecosystems with reduction of biodiversity and potential losses of important ecosystem services, death, injury, and disruption to livelihoods, food supplies, and drinking water', among others. See IPCC, Climate Change 2014: Impacts, Adaptation, and Vulnerability, at Summaries, Frequently Asked Questions, Cross-Chapter Boxes, and Table TS.3, p. 59.
13 R. Lyster, 'A Fossil Fuel Funded Climate Disaster Response Under the Warsaw International Mechanism for Loss and Damage Associated With Climate Change', 4(1) *Transnational Environmental Law (TEL)* (2015) 125, at 140–141.
14 *Ibid.*
15 E. Roberts and M. Pelling, 'Climate Change-Related Loss & Damage: Translating the Global Policy Agenda for National Policy Processes', *Climate and Development* (2016) 1, at 4, http://dx.doi.org/10.1080/17565529.2016.1184608.
16 GA Res. 10/61, 15 January 2009; UNEP and Columbia Law School, *Climate Change and Human Rights* (2015), available online at www.unep.org/newscentre/Default.aspx?DocumentID=26856&ArticleID=35630&l=en.
17 GA Res. 10/61, 15 January 2009.

18 UNEP and Columbia Law School, *supra* note 16, at 4.
19 *Ibid.*
20 W. Dale, 'The Making and Remaking of Commonwealth Constitutions', 42 *International and Comparative Law Quarterly (ICLQ)* (1993) 67.
21 Benjamin and Stevenson, *supra* note 2, at 235.
22 Robinson argues that together these written constitutions form a type of 'familial relationship' between Commonwealth Caribbean states. See T. Robinson, 'A Caribbean Common Law', 49(2) *Race & Class (Race Cl)* (2007) 118, at 122.
23 For example, see Chapter III, Protection of Fundamental Rights and Freedoms of the Individual, The Constitution of the Commonwealth of The Bahamas, available at www.oas.org/juridico/mla/en/bhs/en_bhs-int-text-const.pdf.
24 D. O'Brien and S. Wheatle, 'Post-Independence Constitutional Reform in the Commonwealth Caribbean and a New Charter of Fundamental Rights and Freedoms for Jamaica', 4 *Public Law* (2012) 683.
25 Constitution of the Co-operative Republic of Guyana, Act 1980, February 20, 1980, although Anderson has expressed doubt over whether these provisions actually confer enforceable rights. See W. Anderson, 'Environmental Protection of Nariva Swamp and Guiana Island: The Need for the Greening of Caribbean Constitutions', 10 *Caribbean Law Review (CLR)* (2000) 225, at 231.
26 Article 13(1)(l), Constitution of Jamaica.
27 For more information on the process of constitutional reform including an environmental rights provision, see Benjamin and Stevenson, *supra* note 2.
28 Ch. 1.01, Constitution of the Republic of Trinidad and Tobago, Preamble.
29 Ch. 35:05, *The Environmental Management Act of Trinidad and Tobago*, available online at http://rgd.legalaffairs.gov.tt/Laws2/Alphabetical_List/lawspdfs/35.05.pdf.
30 M. McShine, Documents Incorporating an Environmental Element in the Constitution of Trinidad and Tobago, 1989 High Court Library Trinidad and Tobago (updated 1998).
31 GA Res. 217A Paris, 10 December 1948, Article 29, at para 2.
32 In 2006, a Draft Constitution was laid before the House of Representatives but no further action was taken. In 2009, 'A Working Document on Constitutional Reform for Public Consultation' was drafted, but no further action was taken on this document. Both documents are available at www.ttparliament.org/publications.php?mid=32.
33 This is perhaps unsurprising as the initial discussion of an environmental element for the Trinidad and Tobago constitution was linked to a specific incident which resulted in public outcry and subsequent litigation all the way to the Privy Council (Lopinot Limestone Cases – High Court 2720 of 1980; Court of Appeal No. 89 of 1982 and Privy Council Law Reports 1988, Appeal Cases at page 45) concerning State refusal to grant a quarry license for activities to take place on State Lands, citing inconsistency with the then State Conservation Policy which was, at the time, unknown to the contractor. The Privy Council allowed the State's Appeal and remitted the plaintiff's application for planning permission for due consideration according to law.
34 *Supra*, note 29, at sections 2 and 25.
35 Republic of Trinidad and Tobago, *National Environmental Policy* (2006), available at www.ema.co.tt/new/images/policies/national-environmental-policy2006.pdf.
36 After considerable delay, provisions in bills to amend Grenada's constitution to protect the environment were rejected by referendum on 26 November 2016. See http://grenadaconstitutionreform.com/. See also CARICOM Today, *'No' Dominates in Grenada's Constitutional Referendum Reform* (25 November 2016), available online at http://today.caricom.org/2016/11/25/no-vote-dominates-in-grenadas-constitutional-reform-referendum/.
37 It is important to note that most Commonwealth Caribbean countries have ratified the Charter of the Organization of American States as well as the American Declaration on the Rights and Duties of Man and the American Convention on Human Rights, but not the Additional Protocol to the American Convention on Human Rights in the Area of Economic, Social and Cultural Rights (the Protocol of San Salvador, 1998).
38 For example, see Constitutional Commission, Commonwealth of The Bahamas, *Report of the Constitutional Commission into a Review of The Bahamas Constitution* (2013), available online at www.bahamas.gov.bs/wps/wcm/connect/7c2fe440-cb66-4327-9bf3-432131510cc4/Constitution+Commission+Report+2013_8JULY2013.pdf?MOD=AJPERES, at 23. See also Benjamin and Stevenson, *supra* note 2, at 231.
39 R. Berg, 'Hurricane Joaquin', *National Hurricane Center Tropical Cyclone Report* (12 January 2016), 1.
40 *Ibid.*, at 1.
41 *Ibid.*, at 2.

42 Simmons and King, 'Impact of Hurricane Joaquin in The Bahamas', *Department of Meteorology* (2015) 2 (on file with author).
43 *Ibid.*, at 2.
44 Berg, *supra* note 39, at 5; Simmons and King, *supra* note 42, at 11.
45 Simmons and King, *supra* note 42, at 11.
46 These are primarily attributed to the sinking of the cargo ship *The Faro* during the storm.
47 Simmons and King, *supra* note 42, at 11.
48 *Ibid.*, at 12.
49 Berg, *supra* note 39, at 8.
50 Berg, *supra* note 39, at 7.
51 J. Simmons and A. King, *supra* note 42, at 14.
52 See R. Jones Jr., Government to Borrow $150 Million, Storm Damage Pegged at $500 Million, *The Nassau Guardian* (20 October 2016), available online at www.thenassauguardian.com/news/68626-govt-to-borrow-150m.
53 S. Komolafe, Hurricane Matthew: Time to Rethink Financing, *The Nassau Guardian* (18 October 2016), available at www.thenassauguardian.com/opinion/op-ed/68576-hurricane-matthew-time-to-rethink-financing- (last visited 29 January 2017); N. Hartnell, Insurers: Matthew Loss 'Could Exceed' $400 Million, *The Tribune* (12 October 2016), available online at www.tribune242.com/news/2016/oct/12/insurers-matthew-loss-could-exceed-400m/.
54 R. Banwarie, Floods Eat Away Road, Newsday (17 November 2014), available online at archives.newsday.co.tt/news/0,203158.html.
55 ODPM, *Public Advisory – Monday 17th November, 2014–10:00am* (17 November 2014), available online at http://odpm.gov.tt/node/821.
56 One senior official of the Ministry of Works and Transport noted that it was the worst damage he had ever witnessed to a road in his 34 years as an engineer with the Highways Division. He also noted that the damage was not as a result of poor work on the road but as a result of the unpredicted amount of rainfall. See Staff Editor, Catastrophic Damage to Manzanilla-Mayaro Road in T&T After Heavy Rain, *Starbroek News* (18 November 2014), available online at www.stabroeknews.com/2014/news/regional/11/18/catastrophic-damage-manzanilla-mayaro-rd-tt-heavy-rain/.
57 In the Integrated Coastal Zone Management (ICZM) Policy Framework of Trinidad and Tobago (2014), the policy identifies the Manzanilla area as highly vulnerable to any further sea level rise. See ICZM Steering Committee, *Ministry of Environment and Water Resources, Integrated Coastal Zone Management Draft Policy Framework* (April 2014), available online at www.ima.gov.tt/home/images/docs/Ingrated_Coastal_Zone_Mment_Policy_Framework_Minister_April_2014.pdf, at 25.
58 C. Matroo, *Trinidad and Tobago Newsday, a East Still Under Flood Alert* (20 November 2014), available online at www.newsday.co.tt/news/0,203158.html.
59 M. Powers, *Trinidad and Tobago Guardian, Shouting Match Between MP, Residents Over Manzan Road Repair* (14 December 2014), available online at www.guardian.co.tt/news/2014-12-14/shouting-match-between-mp-residents-over-manzan-road-repair.
60 *Ibid.*
61 The Second National Communication of Trinidad and Tobago to the UNFCCC notes, 'Coastal erosion is a major problem in Trinidad. It undermines the stability of structures including roads, vacation and residential homes and defence works. It removes coastal agricultural land, destroys recreational areas (beaches) and causes loss of habitat.' See Government of the Republic of Trinidad and Tobago, *Second National Communication of the Government of the Republic of Trinidad and Tobago to the United Nations Framework Convention on Climate Change* (April 2013), available online at www.eldis.org/go/home&id=71907&type=Document#.WM2dYYSFif4 (last visited 30 October 2016).
62 See Government of the Republic of Trinidad and Tobago, ICZM, *supra* note 57.
63 IPCC AR5, *supra* note 3, at 19.
64 *Ibid.*, at 8.
65 IPCC AR5 notes that 'Extreme precipitation events over most of the mid-latitude land masses and over wet tropical regions will very likely become more intense and more frequent'. See *ibid.*, at 11.
66 See Jones, *supra* note 52.
67 See Green Climate Fund, *6th Board Meeting of the Green Climate Fund, Policies and procedures for the initial allocation of Fund resources 19–21*, Decision B.06/05 (February 2014), available online at www.green-climate.fund/documents/20182/24940/GCF_B.06_05_-_Policies_and_Procedures_for_the_Initial_Allocation_of_Fund_Resources.pdf/c58faf81-d780-4918-94d4-407a49af501e.

Part VI
Future directions

29

Mobilizing human rights to combat climate change through litigation

Abby Rubinson Vollmer

Introduction

Human rights claims are being used increasingly as a means to hold governments and corporations accountable for harm to the environment, including climate change.

Historically, human rights bodies were often asked to hear claims of violations of civil and political rights. Yet over the past few decades, United Nations and other human rights bodies have made clear that all human rights are universal and indivisible.[1] Thus, under the eyes of the law, economic, social, and cultural rights are just as important as civil and political rights. Human rights bodies now regularly hear cases about economic, social, and cultural rights, as well as solidarity rights, which include environmental rights. Although some human rights bodies have been reluctant to consider environmental harm from climate change within the scope of their purview, in recent years UN and regional human rights bodies have increasingly recognized the implications that environmental harm, including from climate change, can have on human rights.

Nonetheless, climate change has posed a challenge for some human rights bodies.[2] For instance, the Inter-American Commission on Human Rights (IACHR) has received but not admitted two petitions on climate change. However, in December 2015, coinciding with the United Nations Framework Convention on Climate Change negotiations of the Paris Agreement, the IACHR issued a statement expressing concern about the effects of climate change on human rights and acknowledging that it has received hundreds of cases demonstrating that "climate change is a reality that is affecting the enjoyment of human rights in the region."[3] Also in 2015, some national courts began to consider and decide climate cases involving human rights claims. These developments set the stage for additional national courts, as well as human rights bodies like the IACHR, to hear cases premised on climate-related harms that implicate human rights in the future. This chapter provides some practical considerations in bringing such cases, reviews existing human rights–based climate change cases, and reflects on the future of human rights litigation to prevent or redress climate change–related harms.

Conceptualizing climate change harms as human rights claims: relevant facts and law

Constructing a human rights–based climate change case entails casting climate harms as the relevant facts and conceptualizing human rights as the relevant law. Impacts of climate change that threaten or violate human rights include sea level rise, droughts, heat waves, desertification, fires, the spread of tropical and vector-borne diseases, and more frequent and extreme floods, storms, and other natural disasters.[4] These harms can provide the factual basis for a human rights claim where the harm (i) infringes on a human right; (ii) is sufficiently severe; and (iii) the causal link to climate change is sufficiently clear. The standards for severity of harm and causation vary by jurisdiction, but these elements are often relevant considerations for courts.

International human rights treaties, such as the major United Nations human rights treaties, provide a solid set of rights that can form the basis for a case, where a country has either ratified the relevant treaty or enshrined its rights in its national law.[5] Those rights include the rights to life, health, food, water, housing, property, culture, freedom of residence and movement, and nationality, as well as procedural rights, such as the rights of access to information and justice, and the right to participation. The rights of Indigenous Peoples, women, and children also enjoy protection under international human rights law and can suffer from climate change impacts.

To understand how a climate change–related harm might infringe on a human right, consider the example of temperature rise, along with repeated intense droughts, in the Turkana region of Kenya.[6] These impacts implicate the rights to water, food, livelihood, health, and security where they interfere with people's ability to realize or enjoy those rights.[7] The Maldives's submission to the Office of the High Commissioner for Human Rights provides a useful table showing which human rights (in UN human rights treaties) certain climate change effects might implicate.[8]

Causation has proven challenging in many lawsuits, given that climate change can combine with other factors in producing its effects. However, some courts have begun to recognize contributory action as a basis for liability. Sectors that clearly generate climate change impacts include fossil fuels, transportation, agriculture, and forestry. In addition, scientific data on climate change, including reports by the Intergovernmental Panel on Climate Change (IPCC), continue to become more specific, decisive, and robust. A relatively recent scientific study links 63% of historic, industrial carbon dioxide and methane emissions to 90 state, state-owned, and investor-owned entities, the so-called Carbon Majors.[9] Climate attribution studies are also becoming more prevalent.[10]

Building the lawsuit: essential elements

Litigation related to climate change and human rights can take many forms and proceed in many fora.[11] International law provides a basis for cases in international human rights tribunals and treaty bodies, while many countries' national constitutions provide human rights protections that enable domestic courts to hear claims of climate change effects' harm to human rights.

To build a potential case (human rights–based or otherwise), questions of who, what, and where are essential to answer. The "who" element takes on two dimensions: who can sue and who can be sued. International human rights law generally provides individuals, and in some instances, groups, with "standing" to complain of violations to their human rights. Several human rights treaties establish that States subject themselves to complaints if they are alleged to have violated provisions of treaties they have ratified. Human rights law generally does not provide jurisdiction over private actors, but States' duty to protect human rights entails ensuring that third parties, including private actors, do not interfere with enjoyment of human rights. In addition, States' domestic legal systems provide jurisdiction over private actors on constitutional or human rights claims.[12]

Regarding the "what," as explained above, human rights law need not expressly recognize climate change to provide a basis for a claim. Rather, what matters is whether an applicable human rights treaty or law protects the human rights at issue. "Applicable" is important given that treaties only bind States that have ratified them (Parties), and most treaties and laws have jurisdictional limitations. Beyond such procedural limitations, the law must also apply to the substance of the harm. Thus, for instance, if climate change resulted in the spread of a vector-borne disease, thereby infringing on the right to health, then legal action is possible under an international body like the Committee on Economic, Social, and Cultural Rights (CESCR), which hears communications related to rights enshrined in the International Covenant on Economic, Social, and Cultural Rights (ICESCR). To be admissible to the CESCR, the State that allegedly violated the rights would need to have ratified both the ICESCR and its Optional Protocol, which provides jurisdiction to the CESCR to hear communications against States that ratify it. Litigation would also be possible under the domestic laws of a country that has enshrined the right to health in its national constitution or other laws, but the success of such a claim would depend on that country's willingness to find the harm cognizable as a violation of the right to health.

With the relevant law identified, the next step (still part of the "what") is to determine whether the harm at issue rises to the level of a violation of that law. The text of the law itself, as well as past decisions of the body charged with hearing claims under that law, will be instructive here. For instance, keeping with the right to health example, the text of ICESCR and decisions of the CESCR would be relevant in submitting a communication to the CESCR. The text of domestic law and decisions under that law (if considered under that legal system) would be relevant in bringing a case at the national level. In some instances, particularly under domestic law, this inquiry can stray into other areas of the law. In Chile, for example, whose constitution protects the right to a healthy environment, one way to establish a violation of this right is to show that the activity in question exceeded allowable pollution limits under its environmental quality standards.[13]

The question of "where" largely follows from the determinations of who and what. Identifying the proper parties and relevant law will often answer the question of where to bring the case, because those elements will establish which fora have jurisdiction to hear the claim. For example, if a fisherwoman in India sought to claim that the Indian government's failure to adequately respond to climate change resulted in fish kills that infringed on her right to a means of subsistence, and found that Indian law protects the right to livelihood and clean environment under its constitutional right to life, then she would need to bring her claim in a forum that has jurisdiction to hear those claims, e.g., India's Supreme Court or National Green Tribunal.[14]

Elements such as the actors, existence of relevant law, severity of the harm, and forum are all key determinants of whether a lawsuit will be admissible and successful on the merits.

Surveying the field: current cases

Notably, every country has ratified at least one of the major United Nations human rights treaties, signifying that they uphold those rights in their national legal systems,[15] and at least 117 countries have established a national human rights commission.[16] As explained in the "Conceptualizing Climate Change Harms as Human Rights Claims" section above, these bodies can provide a means of recourse for people who suffer climate change–related harm, where that harm rises to the level of violating their human rights in a way cognizable by those bodies.

Of the growing number of climate change–related lawsuits, several have asserted human rights claims. Of those cases that have asserted human rights claims, only a few have succeeded in achieving a favorable ruling on those claims. While this outcome may sound discouraging to those seeking to bring human rights claims related to climate change, it may well reflect the relatively new nature and understanding of both climate change and its implications on human rights. As domestic courts and human rights bodies develop their understanding and recognition of these implications, they will be better equipped to find human rights violations based on climate change–related harm where they exist. Similarly, as victims of climate change–related harm develop their understanding of that harm's effects on their rights, they will be better equipped to avail themselves of courts and human rights bodies to hear their claims.

Climate change cases at both the domestic and international levels have included human rights claims. The discussion below describes these cases in order from greatest to least recognition of human rights claims, as one would write a legal brief, by focusing on the most relevant cases.

First, touching very briefly on these cases, at the domestic level, one case in Pakistan considered human rights as part of constitutional and fundamental rights claims. One case in the Netherlands raised human rights obligations under the European Convention on Human Rights, among other claims. Largely following the Netherlands' case, one case in Belgium and another in Switzerland also included claims based on obligations under the European Convention. A Kiribati national asserted human rights arguments in seeking asylum in New Zealand, based on climate change harms to Kiribati. In the Philippines, Greenpeace Southeast Asia and other groups filed a petition before the Philippines' Commission on Human Rights against 47 investor-owned major carbon producers regarding the human rights implications of climate change and ocean acidification in the Philippines and those companies' responsibilities for them.[17] Two domestic cases, one in Norway and one in Sweden, also involve human rights claims relating to fossil fuels. The Inter-American Commission on Human Rights has received two petitions on behalf of Arctic Indigenous Peoples regarding human rights violations caused by climate change effects in the North American Arctic. Each of these cases is discussed below, with a view to replicability in other cases.

Ashgar Leghari v. Federation of Pakistan (Sept. 2015)

In *Ashgar Leghari v. Federation of Pakistan*, a subsistence farmer claimed the government violated his constitutional rights by failing to act on climate change. The farmer sued the government of Pakistan for violations of his fundamental and constitutional rights – including the rights to life, healthy environment, human dignity, property, and information – due to its inaction on climate change. Asserting that "climate change is a serious threat to water, food and energy security of Pakistan which offends the fundamental right to life under . . . the Constitution," along with other rights, Mr. Leghari challenged the government's failure to make progress on its National Climate Change Policy, 2012 and the Framework for Implementation of Climate Change Policy (2014–2030).[18]

On September 4, 2015, the Lahore High Court agreed with Mr. Leghari, noting that "[c]limate [c]hange is a defining challenge of our time and has led to dramatic alterations in our planet's climate system," with "heavy floods and droughts [in Pakistan] raising serious concerns regarding water and food security." The court explained that in legal and constitutional terms, this was a "clarion call for the protection of fundamental rights of the citizens of Pakistan, in particular, the vulnerable and weak segments of the society who are unable to approach this Court."[19] The court thus found the State's "delay and lethargy" in implementing the Framework "offend[ed] the fundamental rights of the citizens which need to be safeguarded." Consequently, the court ordered the government of Pakistan to appoint a climate change focal person in each of its relevant ministries, have each ministry develop a list of adaptation action points, and establish a national Climate Change Commission.[20]

Calling for a move from environmental justice to climate change justice, the court declared that the

> [r]ight to life [which includes the right to a healthy environment], right to human dignity, right to property and right to information under . . . the Constitution read with the constitutional values of political, economic and social justice provide the necessary judicial toolkit to address and monitor the Government's response to climate change.[21]

This case demonstrates the ability of a judge to order and obtain swift, concrete action on climate change:

> Within one month of having heard Ashgar Leghari's case, Judge Shah had summoned all of the country's main officials before him; appointed named climate change focal points for each government department; addressed individual capacity needs of departments; and appointed a named Climate Change Commission to ensure implementation [of Pakistan's climate change framework].[22]

Although this case's success was likely unusual in terms of the speed and concreteness of the results it obtained, it is theoretically replicable where a nation has (i) a constitution enshrining relevant rights (such as the rights to life, human dignity, property, and information); (ii) an unfulfilled mandate to act on climate change; and (iii) a willing judiciary.

Perhaps testing this theory, in April 2016, a seven-year-old sued the Pakistani government, alleging violations of her constitutional and human rights due to the government's actions and inaction related to carbon pollution from coal, transportation, and industrial activity.[23]

Urgenda Foundation v. the State of the Netherlands (June 2015)

In *Urgenda Foundation v. the State of the Netherlands*, an organization representing 886 individuals sued the government for inaction on climate change. Beyond asserting the Netherlands' failure to take adequate action on climate change, Urgenda claimed that the resulting emissions levels infringed on certain human rights, among other claims. The human rights claims pertained to the right to life and the right to health and respect for private and family life, both under the European Convention on Human Rights (ECHR), and the rights of future generations and the duty of care regarding the livability of the country and protection and improvement of the living environment, both under Dutch law.[24] The court also referenced environmental law principles derived from rulings of the European Court on Human Rights (European Court).[25]

The court held that Urgenda could not rely directly on the ECHR because those rights apply only to natural persons and Urgenda was a legal, not natural, person.[26] Notwithstanding, it considered the ECHR and the European Court's interpretation of those rights relevant to its interpretation of Dutch civil law.[27] The court reasoned it could rely on the jurisprudence of the European Court because "the State has the obligation to protect its citizens from it by taking appropriate and effective measures" when, as in that case, "there is a high risk of dangerous climate change with severe and life-threatening consequences for man and the environment."[28] Ultimately, the court ordered the Dutch government to limit GHG emissions to 25% below 1990 levels by 2020. The Dutch government appealed the ruling in October 2015.[29]

This case illustrates the applicability of human rights law to harms resulting from climate change – in this case, a State's failure to adequately address climate change. While the court did not find violations of human rights law, the fact that human rights law informed its analysis shows that other courts may also find it relevant in their consideration of climate change–related harms. Moreover, the reasons the Dutch court did not directly consider the human rights claim would not necessarily arise in other cases. For instance, Urgenda's inability as a legal person to assert human rights claims would be irrelevant if a natural person were to file the lawsuit. In addition, where the harm more closely affects the rights at issue (as opposed to there, where policy inaction was the basis for the rights violations), a court might be more inclined to find violations.

VZW Klimaatzaak v. Kingdom of Belgium, et al. (pending)

The Belgian case *VZW Klimaatzaak v. Kingdom of Belgium* draws significantly on the Urgenda case in the Netherlands in asserting the government's failure to take adequate climate action, to the detriment of human rights. In December 2014, the organization Klimaatzaak wrote a letter to the Belgian government demanding that it take all necessary measures to guarantee a reduction in domestic greenhouse gas emissions to 40% below 1990 levels by 2020. Not satisfied with the response, in April 2015, Klimaatzaak sued the Belgian government, including for violations of the right to life under the ECHR, and the right to health and respect for private and family life, under the ECHR and the Belgian constitution.[30] The plaintiffs also alleged inconsistencies with the precautionary and prevention principles, as well as negligence. The case is currently pending.

Verein KlimaSeniorinnen Schweiz and others v. Federal Council of the Swiss Confederation and others (pending)

Also drawing heavily on the *Urgenda* case, the Swiss case *Verein KlimaSeniorinnen Schweiz and others v. Federal Council of the Swiss Confederation and others*, filed in October 2016, argues that the government's national mitigation target does not meet the standards set forth by the scientific community and therefore constitutes a violation of its duties. In that case, 459 elderly women sued the Swiss government, alleging that its climate policy violates their constitutional and human rights due to effects of heat waves and other climate change impacts.[31] In particular, the case highlights that failure to take adequate climate action threatens the right to life under the Swiss constitution and the ECHR, and the right to private and family life under the ECHR.[32] The plaintiffs, whose organization name translates to Senior Women for Climate Protection, base their legal application primarily around the recognition of the health implications associated with climate change – in particular for elderly people. While principles of environmental law feature among the legal grounds in the complaint, references to human rights are central to the request for legal remedies. As in the Urgenda complaint, the plaintiffs seek an order that the government strengthen its mitigation policy to meet the lower range of the IPCC's suggested targets (25% emissions reductions by 2020). The application has yet to be considered by a court.

Ioane Teitiota v the chief executive of the ministry of business, innovation and employment (July 2015)

In *Ioane Teitiota v The Chief Executive of the Ministry of Business, Innovation and Employment*, a Kiribati national applied for asylum in New Zealand based on harm that climate change was causing to Kiribati.[33] Specifically, in his 2012 complaint, he claimed "his homeland, Kiribati, [wa]s facing steadily rising sea water levels as a result of climate change," giving rise to a fear "that, over time, the rising sea water levels and the associated environmental degradation will force the inhabitants of Kiribati to leave their islands." Rejecting claims based on the 1951 United Nations Refugee Convention, the International Covenant on Civil and Political Rights (ICCPR), and New Zealand law, the High Court and Court of Appeals both found that Mr. Teitiota failed to qualify for refugee status based on the impacts of climate change on Kiribati. Affirming, in July 2015 the Supreme Court found Mr. Teitiota would not face "serious harm" if returned to Kiribati and found no evidence that Kiribati was failing to "take steps to protect its citizens from the effects of environmental degradation to the extent it c[ould]." Finding the ICCPR inapplicable on the facts of that case and a lack of jurisdiction to review the appellate court's rulings on New Zealand law, the Supreme Court dismissed the case. However, it noted that its dismissal "did not mean that environmental degradation resulting from climate change or other natural disasters could never create a pathway into the Refugee Convention or protected person jurisdiction."

Thus, this case leaves the door open for possible international law claims by people who are forced from their country, or fear return, due to environmental degradation from climate change, at least in New Zealand. Indeed, in a case in which citizens of Tuvalu faced deportation from New Zealand, *In re: AD (Tuvalu)*, the Immigration and Protection Tribunal cited the Kiribati case as "expressly acknowledg[ing]" that climate change impacts may affect enjoyment of human rights and noted the "wide accept[ance]" of this reality.[34] While the tribunal did not expressly rule on the basis of human rights, it found that the facts taken together constituted "exceptional circumstances" that warranted granting asylum based in part on climate-based harms the applicants claimed they would suffer if sent to Tuvalu.[35]

Greenpeace Southeast Asia et al. v. Carbon Majors – Commission on Human Rights of the Philippines (pending)

In contrast to the cases above, which involve claims brought against governments, several individuals and organizations filed a groundbreaking petition against the major carbon producers. In September 2015, Greenpeace Southeast Asia and other groups filed a complaint before the Commission on Human Rights of the Philippines against 47 investor-owned corporations that are among the largest historic emitters of greenhouse gases. The complaint asks the Commission to investigate the human rights implications of climate change and ocean acidification in the Philippines and those companies' responsibility for resulting human rights violations or threats. Furthermore, the complaint requests that the Commission (i) ask the companies for plans on how they will eliminate, remedy, and prevent human rights violations resulting from climate change in the future; (ii) monitor communities most vulnerable to climate change in the Philippines; and (iii) recommend that policymakers and lawmakers develop corporate reporting standards on human rights issues related to the environment.

In December 2015, the Commission on Human Rights announced that it would launch an investigation as requested by the petitioners. In July 2016, the Commission ordered the companies to respond to the petition. Twenty-one companies complied, yet only one company (Rio Tinto) acknowledged the fact-finding nature of the investigation. (All of the other companies that responded challenged the Commission's ability to exercise jurisdiction and to conduct an investigation on these matters.) In December 2016, the Commission confirmed that it would move ahead with a national public inquiry and hold public hearings.[36] In February 2017, the petitioners submitted a consolidated reply to the corporate responses. The Commission expects to hold hearings in 2017.

Greenpeace Nordic Association and Natur og Ungdom (Nature & Youth) v. The Government of Norway represented by the Ministry of Petroleum and Energy (pending)

In October 2016, Greenpeace Nordic Association and Natur og Ungdom, with an alliance including Indigenous Peoples, youth groups, famed scientist James Hansen, sued the Norwegian government for opening up the Barents Sea to fossil fuel exploration.[37] In their petition, the plaintiffs assert that the government's decision to allow oil drilling there constitutes a violation of the government's obligations with regards to environmental protection and human rights.[38] In particular, the petition relies on article 112 of the Norwegian constitution, which provides a right of present and future generations to "an environment that is conducive to health and to a natural environment whose productivity and diversity are maintained."The petition emphasizes that this provision must be interpreted in light of relevant international human rights instruments, in particular articles 2 and 8 of the ECHR, enshrining the rights to life and private and family life, and article 12 of the ICESCR, enshrining the right to the enjoyment of the highest attainable standard of physical and mental health. The petition contends that the production licenses Norway awarded would result in increased emissions incompatible with Norway's commitment under the Paris Agreement. Trial is set to take place in November 2017.[39]

PUSH Sverige, Fältbiologerna and others v. The Government of Sweden (pending)

The responsibility of a national government to prevent the extraction of fossil fuels as a matter of human rights is also at the core of a petition filed in September 2016 against the Swedish government.[40] In this case, known as Magnolia, youth plaintiffs denounced the decision of the state-owned enterprise Vattenfall to sell one of its lignite coal mines in Germany to a private operator. The plaintiffs argue that such a sale would result in the future exploitation of the mine and that it was the government's duty, based on its international climate commitments and its human rights obligations, to prevent such a deal and to ensure that the extraction of lignite coal is discontinued or takes place under strict standards. Referring to recent scientific literature, the summons emphasize that the continued exploitation of lignite coal is not compatible with the objective of keeping temperatures well below 2°C as agreed in the Paris climate agreement. The plaintiffs assert that, by allowing the commercial deal to take place, the government failed to meet its human rights obligations under the Swedish constitution, the ECHR, and the European Social Charter. They also argue that the Swedish government failed to guarantee the respect of the UN Guiding Principles on business and human rights by the state-owned operator.

Petitions to the Inter-American Commission on Human Rights

At the international level, the Inter-American Commission on Human Rights (Inter-American Commission), an organ of the Organization of American States, has received two petitions related to climate change harms. In both cases, indigenous people in the Arctic filed the petitions: *Inuit Circumpolar Conference v. United States* in 2005 for acts and omissions causing global warming and *Arctic Athabaskan Council v. Canada* in 2013 for inadequate regulation of black carbon emissions. Both cases alleged violations of the rights to culture, property, health, and means of subsistence.[41]

The Inter-American Commission declined to rule on the 2005 petition, though it did provide a hearing for the petitioners, giving the case acclaim as the one that "put a human face on the effects of climate change."[42] The Inter-American Commission has not yet responded to the 2013 petition. However, in a statement issued in December 2015, the Commission expressed concern about the effects of climate change on human rights and acknowledged that it "has received hundreds of cases related to conflicts over land and water and threats to food sovereignty which evidence that climate change is a reality that is affecting the enjoyment of human rights in the region."[43] This development indicates that the Commission better understands the relevance of climate change to the cases it considers, and it may signal a greater willingness to hear cases alleging human rights claims based on climate change harms.

Assessing the future: lessons learned and challenges to confront

Given the clear impacts of climate change on the enjoyment of human rights, and States' corresponding human rights obligations, courts can be an appropriate place to seek compliance with those obligations. The climate change–related cases discussed in the previous section illustrate this potential. However, some hurdles remain on the path to justice.

First, human rights cases based on climate change–related harm are rare so far, and some human rights bodies have been reluctant to recognize such cases as within their purview. However, this is changing. The Inter-American Commission's 2015 statement on climate change and human rights demonstrates its evolving understanding of this link and its applicability to its cases. In addition, countries including India, Kenya, Chile, and Nigeria have found that environmental harm violated human rights, and they may be well positioned to consider climate change–related harm as the basis for future claims. Indeed, a study by the Environmental Law Alliance Worldwide (ELAW) identified Brazil, Colombia, Ecuador, India, Kenya, Mexico, Nigeria, and Pakistan as countries whose legal systems would be well suited to hear climate change damages cases based on constitutional rights and involving private actors.[44]

Furthermore, in countries whose courts lack a strong recognition of human rights, victims of human rights violations from climate change–related harms may be able to seek relief based on constitutional rights that reach those violations without directly bringing human rights claims as such.[45] For instance, in *Juliana v. US*, a U.S. district judge held that plaintiffs had "adequately alleged infringement of a fundamental right" under the U.S. constitution by claiming that "governmental action is affirmatively and substantially damaging the climate system in a way that will cause human deaths, shorten human lifespans, result in widespread damage to property, threaten human food sources, and dramatically alter the planet's ecosystem."[46] In reaching this conclusion, the judge noted she had "no doubt that the right to a climate system capable of sustaining human life is fundamental to a free and ordered society [and] quite literally the foundation 'of society, without which there would be neither civilization nor progress.'"[47] (The Panjey petition to India's National Green Tribunal recently cited this language.)

Second, proving causation remains a considerable challenge. This is particularly true for cases aiming to tie developed countries' emissions to climate-related harms in developing countries. However, some courts and human rights bodies have not let this get in the way, such as the *Urgenda v. Netherlands* case and the Philippines' investigation against the Carbon Majors. Moreover, the increase in scientific evidence tracing quantified emissions to particular actors, such as the Carbon Majors report, suggests that plaintiffs are inching closer to obtaining sufficient proof to link climate change impacts to states or businesses. With that element of the causal chain established, plaintiffs should be able to show that the climate change impacts, if sufficiently severe, infringe on human rights.

Third, climate-related legislation requirements are often not strong or concrete enough to provide the basis for a viable lawsuit. For instance, a law might require a government to "take measures" or "develop a plan" without specifying minimum standards or otherwise enforceable content for those measures or plan, or to "consider" doing something. However, over time and in accordance with Paris Agreement commitments, legislation – at least in some countries – will likely contain more concrete requirements that could actionable if not met, including by the resulting effects on human rights. Furthermore, cases like *Leghari v. Pakistan* and *Urgenda v. Netherlands* provide examples of courts taking action for a State's failure to make sufficient progress or take adequate action toward achieving an obligation.

Finally, some architectural considerations include that communities confronting climate-based harm do not necessarily organize themselves or identify in that way, and that, from a climate justice perspective, it may be unfair to bring cases against developing countries that have contributed relatively little to climate change–related harms.

Conclusion

As the UN Office of the High Commissioner for Human Rights (OHCHR) has explained, States clearly have human rights obligations related to climate change: "Because of the impacts of climate change on human rights, States must effectively address climate change in order to honour their commitment to respect, protect and fulfil human rights for all."[48] The OHCHR further acknowledged that, because "climate change mitigation and adaptation measures can have human rights impacts[,] all climate change-related actions must also respect, protect, promote and fulfil human rights standards."[49]

Businesses have human rights responsibilities in this context, too: "It is not only States that must be held accountable for their contributions to climate change but also businesses which have the responsibility to respect human rights and do no harm in the course of their activities."[50]

Thus, when States or businesses fail to meet their obligations or responsibilities, courts and human rights bodies can provide a means for recourse. Courts and human rights bodies will likely increasingly serve this function in the years to come, as they develop their recognition of climate change's implications on human rights, and as communities and individuals enhance their understanding of climate change–related harms they face in human rights terms.

Notes

1 See, e.g., United Nations General Assembly, *Vienna Declaration and Programme of Action* (12 July 1993), available online at www.ohchr.org/EN/ProfessionalInterest/Pages/Vienna.aspx, at para. I.5; United Nations, *Human Rights Are Universal, Indivisible, Interdependent, Secretary-General Stresses in Video Message to International Journalists' Round Table* (8 December 1997), available online at www.un.org/press/en/1997/19971208.HR4344.html.

2 See J. Gordon, 'Inter-American Commission to Hold Hearing After Rejecting Inuit Climate Change Petition', 7 Sustainable Development Law & Policy (2007), available online at http://digitalcommons.wcl. american.edu/cgi/viewcontent.cgi?article=1239&context=sdlp/

3 Organization of American States, *IACHR Expresses Concern Regarding Effects of Climate Change on Human Rights* (2 December 2015), available online at www.oas.org/en/iachr/media_center/PReleases/2015/140.asp.

4 Intergovernmental Panel on Climate Change, *Fifth Assessment Report (AR5)* (2014), available online at www.ipcc.ch/report/ar5/.

5 See, e.g., International Covenant on Economic, Social and Cultural Rights (ICESCR), 16 December 1966, 6 I.L.M. 360, 365, 993 U.N.T.S. 3; International Covenant on Civil and Political Rights, 16 December 1966, 6 I.L.M. 368, 999 U.N.T.S. 171; Convention on the Elimination of All Forms of Racial Discrimination, 21 December 1965, 660 U.N.T.S. 195; Convention on the Elimination of All Forms of Discrimination Against Women, 18 December 1979, 19 I.L.M. 33; Convention on the Rights

of the Child, 20 November 1989, 28 I.L.M. 1448, 1577 U.N.T.S. 3; United Nations General Assembly, Convention on the Rights of Persons with Disabilities, 24 January 2007, A/RES/61/106.

6 Human Rights Watch, *There Is No Time Left: Climate Change, Environmental Threats, and Human Rights in Turkana County, Kenya* (15 October 2015), available online at www.hrw.org/report/2015/10/15/there-no-time-left/climate-change-environmental-threats-and-human-rights-turkana, at 52–53.

7 *Ibid.*

8 Maldives Submission to the Office of the High Commissioner for Human Rights Under Human Rights Council Resolution 7/23, *Human Rights and Climate Change* (25 September 2008), available online at www.ohchr.org/Documents/Issues/ClimateChange/Submissions/Maldives_Submission.pdf.

9 See R. Heede, 'Tracing Anthropogenic Carbon Dioxide and Methane Emissions to Fossil Fuel and Cement Producers, 1854–2010', 122 *Climate Change* (2014) 229. ("The analysis presented here focuses attention on the commercial and state-owned entities responsible for producing the fossil fuels and cement that are the primary sources of anthropogenic greenhouse gases that are driving and will continue to drive climate change. The results show that nearly two-thirds of historic carbon dioxide and methane emissions can be attributed to 90 entities.")

10 See, e.g., J.T. Abatzoglou and A. P. Williams, 'Impact of Anthropogenic Climate Change on Wildfire across Western US Forests', 113(42) *Proceedings of the National Academy of Sciences* (10 October 2016) 11770, available online at www.pnas.org/content/113/42/11770.abstract (last visited 30 October 2016). Finding climate change nearly doubled forest fire area (contributing to an additional 4.2 million hectares) in the western continental United States from 1984 to 2015.

11 I confine this chapter to litigation based on human rights claims. Litigation related to climate change using administrative, civil tort, or criminal claims is also possible and indeed occurring. My focus here is solely on using human rights law to combat climate change.

12 See, e.g., *Shri Bodhisattwa Gautam v. Miss Subhra Chakraborty* [1996] SCC (1) 490, Supreme Court of India, available online at http://indiankanoon.org/doc/642436/ ("Fundamental Rights can be enforced even against private bodies and individuals."); See also Environmental Law Alliance Worldwide (ELAW), 'Holding Corporations Accountable for Damaging the Climate' (2014), available online at www.elaw.org/system/files/elaw.climate.litigation.report.pdf, at 6 ("[C]ourts in Brazil, Colombia, Ecuador, India, Kenya, and Mexico will hold or are likely to hold private entities liable for violations of fundamental rights").

13 Chile, Law N°19,300 General Bases of the Environment, published in the Official Gazette (9 March 1994), available online at www.eisourcebook.org/cms/February%202016/Chile%20General%20Environment%20Law%201994.pdf.

14 See *Olga Tellis and Others v. Bombay Municipal Council*, AIR 1986 SC 180, Supreme Court of India, available online at https://indiankanoon.org/doc/709776/, at para 2.1 ("The sweep of the right to life conferred by Article 21 [of the Indian constitution] is wide and far reaching . . . An . . . important facet of that right is the right to livelihood, because, no person can live without the means of living, that is, the means of livelihood."); *Shantistar Builders v. Narayan Khimalal Totame*, AIR 1990 SC 630, Supreme Court of India, available online at https://indiankanoon.org/doc/1813295/ (last visited 30 October 2016) ("Basic needs of man have traditionally been accepted to be three – food, clothing, and shelter. The right to life is guaranteed in any civilized society. That would take within its sweep the right to food, the right to clothing, the right to decent environment and a reasonable accommodation to live in."); *Bhopal Gas Peedith Mahila Sanghathan v. Union of India* [2012] 8 SCC 326, Supreme Court of India, available online at http://indiankanoon.org/doc/178436640/ (instructing that environmental cases be filed before the National Green Tribunal).

15 See Conventions, *supra* note 5.

16 See Global Alliance of National Human Rights Institutions, *Chart of the Status of National Institutions, accreditation status as of 5 August 2016*, available online at www.ohchr.org/Documents/Countries/NHRI/ChartStatusNHRIs.pdf.

17 Greenpeace Southeast Asia *et al.*, *Petition Requesting for Investigation of the Responsibility of the Carbon Majors for Human Rights Violations or Threats of Violations Resulting from the Impacts of Climate Change*, filed 22 September 2015, amended 9 May 2016, available online at www.greenpeace.org/seasia/ph/PageFiles/735232/Climate_Change_and_Human_Rights_Petition.pd.

18 *Ashgar Leghari v. Federation of Pakistan*, Lahore High Court (4 September 2015), available online at https://elaw.org/system/files/pk.leghari.090415_0.pdf.

19 *Ibid.*, at para. 6.

20 *Ashgar Leghari v. Federation of Pakistan*, *supra* note 20, at para. 8.

21 *Ibid.*, at para. 7. The court further noted that these fundamental rights and principles "include within their ambit and commitment, the international environmental principles of sustainable development, precautionary principle, environmental impact assessment, inter and intra-generational equity and public trust doctrine."
22 Ma. Mehra, *Pakistan Ordered to Enforce Climate Law by Lahore Court* (20 September 2015), available online at www.climatechangenews.com/2015/09/20/pakistan-ordered-to-enforce-climate-law-by-lahore-court/.
23 *Rabab Ali v. Federation of Pakistan & Another*, Supreme Court of Pakistan (April 2016), available online at www.elaw.org/system/files/Pakistan%20Climate%20Case-FINAL.pdf.
24 See Arts 2, 8, European Convention on Human Rights (ECHR); Book 5, Section 37 and Book 6, Section 162 of the Dutch Civil Code; Art. 21, Constitution of the Netherlands,
25 *Urgenda Foundation v. the State of the Netherlands*, HAZA C/09/00456689, The Netherlands, District Court of the Hague [2015], available online at http://uitspraken.rechtspraak.nl/inziendocument?id=ECLI:NL:RBDHA:2015:7196, at paras. 4.47–4.50 (citing Council of Europe 2012, Manual on Human Rights and the Environment).
26 *Ibid.*, at para 4.45.
27 *Ibid.*, at paras 4.46, 4.52.
28 *Ibid.*, at para. 4.74.
29 See Urgenda/State Letter of Judgment dated June 24, 2015 (in Dutch), available online at www.rijksoverheid.nl/binaries/rijksoverheid/documenten/kamerstukken/2015/09/01/kabinetsreactie-vonnis-urgenda-staat-d-d-24-juni-jl/kabinetsreactie-vonnis-urgenda-staat-d-d-24-juni-jl.pdf; see also M. Darby, *Dutch Government to Appeal Court Ruling on Climate Change*, 1 September 2015, www.climatechangenews.com/2015/09/01/dutch-government-to-appeal-court-ruling-on-climate-change/.
30 Klimaatzaak, *Le Proces [The Lawsuit]*, available online at http://klimaatzaak.eu/fr/le-proces/.
31 Swiss Info, *Grandmothers Sue Switzerland in Climate Complaint* (25 October 2016), available online at www.swissinfo.ch/directdemocracy/global-warming_grandmothers-sue-switzerland-in-climate-complaint/42544428.
32 *Verein KlimaSeniorinnen Schweiz and others v. Federal Council of the Swiss Confederation*, Federal Department of the Environment, Transport, Energy and Communications, Federal Office for the Environment and Swiss Federal Office for Energy, filed 25 October 2016, available online at http://klimaseniorinnen.ch/einreichung-der-klage/.
33 *Teitiota v Ministry of Business Innovation and Employment*, NZSC 107 (20 July 2015), available online at www.nzlii.org/nz/cases/NZSC/2015/107.html.
34 *In re: AD (Tuvalu)* [2014] NZIPT 501370–371 (14 June 2014), available online at https://forms.justice.govt.nz/search/IPT/Documents/Deportation/pdf/rem_20140604_501370.pdf, at para. 28 (citing AF (Kiribati) [2013] NZIPT 800413), at para. 63.
35 *Ibid.*, para. 30.
36 Greenpeace Philippines, *Petitioners' Consolidated Reply to the Respondent Carbon Majors in the National Public Inquiry Being Conducted by Commission on Human Rights of the Philippines* (14 February 2017), available online at www.greenpeace.org/seasia/ph/press/releases/Petitioners-consolidated-reply-to-the-respondent-Carbon-Majors-in-the-National-Public-Inquiry-being-conducted-by-Commission-on-Human-Rights-of-the-Philippines.
37 A. Nelsen, *Norway Faces Climate Lawsuit Over Arctic Oil Exploration Plans* (18 October 2016), available online at www.theguardian.com/environment/2016/oct/18/norway-faces-climate-lawsuit-over-oil-exploration-plans.
38 *Greenpeace Nordic Association and Natur og Ungdom (Nature & Youth) v. The Government of Norway* represented by the Ministry of Petroleum and Energy, filed 18 October 2016, available online at www.greenpeace.org/norway/Global/norway/Arktis/Dokumenter/2016/legal_writ_english_final_20161018.pdf.
39 Greenpeace Nordic Association and Natur og Ungdom (Nature & Youth), *The People vs. Arctic Oil: a climate court case against the Norwegian government for opening new oil fields in the Arctic* (2 June 2017, available online at http://www.greenpeace.org/norway/Global/norway/Hav/2017_ClimateLawsuit_Media%20Briefings/Media%20Briefing%20Lawsuit.pdf.
40 PUSH Sverige, *Fältbiologerna and others v. The Government of Sweden*, filed 15 September 2016, available online at www.xn—magnoliamlet—1cb.se/info/the—magnolia—lawsuit—application—in—english/ [Magnolia Case].
41 For further discussion of these cases see S. Jodoin *et al.*, 'Look Before You Jump: Assessing the Potential Influence of the Human Rights Bandwagon on Domestic Climate Policy' (this volume).

42 See, e.g., J. Geiger, 'Nation Builder: Environment – Sheila Watt-Cloutier', *The Globe and Mail* (1 January 2010), available online at www.theglobeandmail.com/news/national/sheila-watt-cloutier/article1317996/; S. Atapattu, *Human Rights Approaches to Climate Change: Challenges and Opportunities* (2015); V. Morin, 'Fighting Climate Change From Inside the World's Air Conditioner', *Huffington Post* (17 June 2016), available online at www.huffingtonpost.com/entry/arctic-climate-change_us_57643076e4b015db1bc92c5f (last visited 30 October 2016); See also A. C. Hu, 'More Than a Mascot: Indigenous People and Climate Politics', *Harvard International Review* (2 June 2015), available online at http://hir.harvard.edu/more-than-a-mascot-indigenous-people-and-climate-politics/; B. Duff-Brown, 'Inuit Aim to Put a Human Face on the Effects of Climate Change', *Boston Globe* (1 March 2007), available online at http://archive.boston.com/news/world/articles/2007/03/01/inuit_aim_to_put_a_human_face_on_the_effects_of_climate_change/.

43 Organization of American States, *supra* note 3.

44 See ELAW, *supra* note 12.

45 See, e.g., *Juliana v. United States*, No. 6:15-cv-1517, ELR (D. Or. November 10, 2016).

46 *Ibid.*

47 *Ibid.*, quoting *Obergefell v. Hodges*, 135 S. Ct. 2584, 2598 (2015) (citations omitted).

48 United Nations Office of the High Commissioner for Human Rights, *Human Rights and Climate Change*, available online at www.ohchr.org/EN/Issues/HRAndClimateChange/Pages/HRClimateChangeIndex.aspx.

49 *Ibid.*

50 *Ibid.*

30

Human rights and land-based carbon mitigation

Kate Dooley

Introduction

The 2015 Paris Agreement on climate change saw the international community agree to ambitious climate goals, with commitments to keeping temperature rise 'well below' 2°C above pre-industrial levels, and to 'pursue efforts' for 1.5°C.[1] It is now well recognized that a temperature rise of 2°C cannot be considered 'safe', with even a 1.5°C temperature rise exceeding thresholds of dangerous climate change.[2] This means that greenhouse gas emissions must effectively be reduced to zero. The long-term goal towards achieving this has been articulated in the Paris Agreement as 'balancing emissions from sources with removals from sinks',[3] which suggests a greater role for land, forests and other ecosystems in removing carbon from the atmosphere.

Increased competition for natural resources over the past few decades has contributed to an increase in environmental resource-related conflicts. The risk that climate mitigation activities will significantly increase demand for land in the future raises fundamental concerns over the ability of existing institutions to govern multiple and competing demands for land. As a recent UN report has noted, 'as the global demand for natural resources grows, the environment is becoming a new frontline for human rights and our common future'.[4]

This chapter examines the threat that land-based climate mitigation actions could pose to human rights and the potential for human rights legal frameworks to guard against these risks. On the whole, I argue that strengthening land rights, and the rights of those who protect land and the environment, directly contributes to combating climate change, making a rights-based approach to land use a crucial element of implementing the Paris Agreement.[5]

The role of land in achieving human rights and sustainable development objectives

Achieving the aim of the Paris Agreement to balance sinks and sources is dependent on land use, as are a number of the UN's 2030 Sustainable Development Goals (SDGs), such as those related to ending poverty (SDG 1), food security and sustainable agriculture (SDG 2), and sustainable land use (SDG 15).[6] Sustainable use of land, including for small-holder and subsistence agriculture, as well as halting and reversing the global loss of forests, will be essential for limiting warming to 1.5°C. The questions that are raised (and not answered) by the Paris Agreement are, what will be the *scale* of reliance on land-based mitigation activities, *how* they will be implemented, and what are the likely social and environmental *impacts*? The cross-cutting principles in the Paris Agreement, including human rights and the rights of Indigenous Peoples, the right to health, and the rights to development and food security, are particularly relevant in addressing these questions.[7] Hence, I argue that the ability of the international community to achieve climate and development objectives relies on the protection of human rights that both depend on, and relate to, the environment.

The goal of 'balancing emissions and removals' in the Paris Agreement effectively means that the same volume of anthropogenic emissions must be drawn back out of the atmosphere through terrestrial sinks, as is released through human activity. This is also referred to as 'net-zero emissions'.[8] In this sense, the Paris Agreement represents an increase in the reliance on land-based mitigation compared to the Kyoto Protocol (KP), where land-based mitigation was limited to specific activities (afforestation, reforestation and deforestation since 1990).[9] The inclusion of land-based carbon sinks in the KP was a last-minute political compromise and was widely seen as weakening the level of ambition.[10] By contrast, in the Paris Agreement land-based sinks 'have moved from being a politically contested add-on to occupying center stage in the long-term mitigation goal'.[11] While this shift recognizes the crucial role that land will play in limiting warming to 1.5°C, 'the push to lock up potentially hundreds of billions of tons of carbon in land – most of it poor-world land locking up rich-world emissions – is set to emerge as a major issue in climate politics'.[12]

Land-use demand in climate mitigation pathways for 2°C and 1.5°C

To understand the scale of land-use mitigation expected to meet the goals of the Paris Agreement, we first need to understand the carbon budget, which is the maximum amount of CO_2 that can be released to the atmosphere for a given expected temperature rise. From 2015, there was a very small carbon budget left – approximately 553 $GtCO_2$ for 2°C, and only 217 $GtCO_2$ for 1.5°C, according to the 2016 UNEP Gap Report.[13] Balancing sources and sinks, some would think, allows us to extend the carbon budget by removing carbon from the atmosphere via photosynthesis, and storing it in land and trees. This idea has been used to argue that we can exceed our carbon budget, and then remove the extra emissions from the atmosphere.[14] This is known as 'negative emissions'. Indeed, almost all scenarios for 2°C in the Fifth IPCC Assessment Report, and all scenarios for 1.5°C, include large amounts of negative emissions.[15] In fact, many mitigation scenarios suggest emitting up to three times our remaining carbon budget, and then removing two-thirds of this,[16] raising questions of scale and feasibility.

While negative emissions could be achieved in many different ways, including mechanical and chemical processes, currently all modelled mitigation scenarios which include carbon removal rely on land-based options: industrial forest plantations that increase the forest carbon stock; or bioenergy, where the emissions from burning the biomass are captured and geologically sequestered, known as BECCS.[17] Relying on land-based mitigation to generate the scale of carbon removal typically considered in long-term mitigation assessments would require large areas of productive land – representing a doubling of all human harvest of biomass if BECCS were used at the scale proposed in many climate models.[18] Estimates of the land requirements for such proposals vary, although 500 million hectares is commonly cited – equivalent to one-third of the world's currently cultivated land area.[19] Given that competition for productive land is already a global concern,[20] proposed land use on this scale for future sequestration of emissions from current fossil fuels use raises profound questions of equity and impacts on human rights.

Human rights and the environment

The international human rights legal framework provides a relevant body of rights related to human rights and the environment. The Universal Declaration of Human Rights (UDHR) along with the International Covenant on Civil and Political Rights (ICCPR) and the International Covenant on Economic, Social and Cultural Rights (ICESCR) enshrine rights to exercising fundamental freedoms, such as the rights to expression, privacy, association and peaceful assembly, as well as a collective dimension of rights, such as the rights to work, food, shelter and health. The draft declaration on Human Rights and the Environment notes the duty of states to respect and ensure the right to a secure, healthy and ecologically sound environment, necessary for the enjoyment of a vast range of human rights.[21] The specific rights of Indigenous Peoples and tribal communities are recognized in the UN Declaration on the Rights of Indigenous Peoples (UNDRIP) and the International Labour Organization Convention No. 169, which include customary ownership and usage rights to collectively held lands and territories, and the right to free, prior and informed consent over any actions that could impact on these access and usage rights.[22]

According to the UN Special Rapporteur on Human Rights and the Environment, the duties of states relating to the environment drawn from international agreements and human rights bodies are threefold. These include the procedural obligations of states to assess the environmental impacts on human rights, to facilitate participation in environmental decision-making and to provide access to remediation; substantive obligations of states to adopt legal and institutional frameworks that protect against environmental harm; and non-discrimination relating to protection of vulnerable groups, including women, children and Indigenous Peoples.[23] In addition, the 1998 Declaration on Human Rights Defenders (DHRD) recognizes the legitimacy of the defense of environmental rights, stating that 'everyone has the right, individually and in association with others, to promote and to strive for the protection and realization of human rights and fundamental freedoms at the national and international levels'.[24]

The scale of expected reliance on land use for climate mitigation has the potential to adversely affect these rights, particularly with respect to food security, the rights and livelihoods of Indigenous Peoples and local communities, and biodiversity protection. The IPCC has already noted that a large-scale increase in land use from mitigation activities may conflict with these objectives,[25] and has identified bioenergy as an emergent global risk to food security and ecosystems due to indirect land-use change.[26]

Globally, 65% of the world's land area is under customary or traditional ownership, although the area legally recognized by governments is a small fraction of this, making those with customary land rights vulnerable to dispossession.[27] Increased pressure for land puts indigenous communities, who often lack clear legal recognition of their rights over the resources located on their lands, at the front lines of land grabbing and resource exploitation, exacerbating poverty, food insecurity and increasing the likelihood of violent conflict.

The threats to land rights, and the people and communities who strive to defend them, are growing. In 2015, Global Witness reported a 59% increase in murders of those defending environmental and land rights compared to the previous year, with approximately three murders each week.[28] Almost half of these were defenders of environmental, land and Indigenous Peoples' rights, yet these figures undoubtedly underestimate the true extent of threats and risks facing environmental and human rights defenders.[29] Further reports in 2016 indicated a 'sharp rise in the number of killings and the overall increase in human rights atrocities against poor rural communities embroiled in land conflicts',[30] with at least 200 land and environmental defenders reported murdered in 2016,[31] and this rate continuing into 2017.[32] Coinciding with this rise in violence, the Special Rapporteur on the Situation of Human Rights Defenders delivered a major report in 2016 that raised the alarm about the increasing and intensifying violence against environmental and human rights defenders, stating 'the scale of killings indicates a truly global crisis', and calling on states to 'address the disturbing trend of increasing violence, intimidation, harassment and demonization of the brave individuals and groups who strive to defend and promote environmental and land rights'.[33]

Climate mitigation, human rights and land

Forest protection to mitigate climate change has so far been advanced under the UNFCCC through REDD+, which has adopted a safeguards approach to avoiding negative impacts from REDD+ policies and interventions. Adopting safeguards is an unusually prescriptive step for the UNFCCC, as safeguards are generally related to 'conditionalities' imposed by financial and development aid entities to prevent and mitigate any harm that may result from funded activities.[34] Reporting on how safeguards are being 'addressed and respected' in the implementation of REDD+ activities is a pre-condition for results-based payments,[35] meaning that, while not formally legally binding on UNFCCC parties, the REDD+ safeguards could be interpreted as a legal requirement for those who seek international payments to carry out REDD+ activities.[36]

The application to specific funded activities makes the REDD+ safeguards narrower in scope than international human rights obligations. Yet the REDD+ safeguards specify that actions should be 'consistent with' the objectives of 'relevant international conventions and agreements', which Savaresi interprets as an implicit reference to human rights obligations in international treaties that UNFCCC parties have ratified or may ratify in the future.[37] Jodoin notes that while the REDD+ safeguards conceive of land tenure issues and participatory rights to be central to the development of national REDD+ strategies and also require respect for the knowledge and rights of Indigenous Peoples and local communities in the context of REDD+, they adopt weaker language than international human rights obligations.[38] Opinions on the value of the REDD+ decisions for protecting the rights of Indigenous Peoples and local communities remains divided.

The expected reliance on land use to meet the goals of the Paris Agreement is set to contribute to increasing global demand for land and, in a context of increased violence and threats to rural communities, will exacerbate these conflicts if mitigating action is not taken. While the REDD+ safeguards do not offer adequate protections against the increasing pressures faced by environmental defenders, the development of these safeguards has raised the awareness of environmental and human rights protections in international law,[39] thereby strengthening the normative context of human rights obligations and 'reflecting the clear advancement of international legal norms relating to the status and rights of Indigenous Peoples and the participation of local communities'.[40] However, achieving the ambitious climate goals of the Paris Agreement will require both an institutional approach that further strengthens human rights protections, particularly in relation to collectively held lands, and a policy response that minimizes reliance on land-based mitigation options that have the potential to exacerbate competition for land and conflict.

On the institutional side, securing Indigenous Peoples' land and resource rights is central to protecting forests, and it is well known that halting tropical forest loss and restoring degraded forests is key to avoiding dangerous climate change.[41] The lack of legal recognition of land rights has been identified as one of the root causes of abuses suffered by environmental and human rights defenders, in particular for indigenous communities.[42] Rather than a safeguards approach, which applies only to specific policies and programs receiving international finance, an approach is needed that addresses the substantive obligations of states to adopt legal and institutional frameworks that recognize and uphold the rights of Indigenous Peoples and local communities, including to collectively held lands and resources, participation in decision-making, and the right to free, prior and informed consent. Research has shown that community-owned and managed forests, incorporating localized knowledge and decentralized decision-making, result in multiple benefits: securing livelihoods, conserving biodiversity and reducing conflict, while also storing carbon.[43] Collectively held lands, whether legally recognized or customarily claimed, hold around one-quarter of the carbon stored in the world's tropical forests,[44] yet countries have made slow progress on land tenure reform despite years of work on REDD+ implementation.[45]

In terms of climate policy response options to reduce the reliance on land-based mitigation, research suggests that allowing degraded forests to regenerate, restoring natural ecosystems and community managed reforestation, has the potential to draw down enough atmospheric carbon to prevent the need for more riskier negative emissions options that would compete with other land uses, such as BECCS.[46] However, carbon removal from forest and ecosystem restoration can only be achieved at the lower end of the expected scale of cumulative negative emissions this century, particularly for a 1.5°C pathway. This means that carbon sequestration in land and forests cannot be used to offset ongoing emissions from energy and industry, and that radical emission reductions will be needed to keep society within the given carbon budgets for 2°C and 1.5°C.[47] Minimizing the scale of reliance on negative emissions, while strengthening policy responses to conserve and protect forests, such as REDD+, can help to avert some of the serious risks of social and environmental impacts from mitigation activities that rely on the large-scale use of land.

Conclusion

Urgent solutions are needed to combat the growing threats to land rights and those that defend them. The adoption of a conducive institutional and legal framework that recognizes land rights and participatory rights of forest communities is the first element for guaranteeing a safe and enabling environment for environmental defenders. International human rights law gives clarity to the obligations of states to protect environmental and human rights defenders from violations committed by state and non-state actors.[48] Jodoin, Hansen and Hong point out that much of the literature overlooks the many ways in which a human rights framework may be most relevant to the climate change regime: in its ability to ensure that laws, policies, programs and projects adopted to mitigate or adapt to climate change respect, protect and fulfill human rights.[49] Given the potential for climate mitigation actions to increase competition for land, and thereby exacerbate land-related conflict, a critical component of climate action must include reviewing domestic legislation to ensure adequate protections for the individuals, communities and Indigenous Peoples who act as environmental and human rights defenders. This will include enacting laws where gaps are identified, and reviewing and repealing laws that facilitate the exploitation of natural resources and threaten the rights of those affected.[50]

Of key concern in the global regulatory framework for climate change is overstressing certain notions, such as 'sovereignty' and 'country-driven', that may legitimize states' inadequate formulation of the domestic regulatory framework for the protection of Indigenous Peoples' land rights.[51] A key role for the international legal framework is to enable domestic law to be brought into line with international law and for states to be held responsible for international standards and obligations. In a context where the rights of Indigenous Peoples are particularly threatened by the impacts of climate change and of ill-designed mitigation projects, the Paris Agreement does not offer a strong enough framework to protect human rights in mitigation actions undertaken. A human rights–based approach is required to guide domestic legislation and the work of environmental protection agencies and other relevant institutions.[52]

Protecting the rights of landholders and those who defend them is crucial to maintaining natural ecosystems, which in turn are crucial for achieving the climate goals in the Paris Agreement. In the words of the Special Rapporteur on the Situation of Human Rights Defenders, 'the fulfilment of the international community's commitment to the protection of the environment is premised on the empowerment of environmental and human rights defenders.'[53] Strengthening land rights and the rights of those who defend land and environment will be key in the fight against climate change.

Notes

1 Paris Agreement to the United Nations Framework Convention on Climate Change, Art 2.
2 United Nations Framework Convention on Climate Change (UNFCCC), *Report of the Structured Expert Dialogue on the 2013–2015 Review* (2015).
3 Paris Agreement, *supra* note 1, Art 4.
4 United Nations General Assembly (UNGA), *Report of the Special Rapporteur on the Situation of Human Rights Defenders*, UN Doc. A/71/2821 (2016) 8. [The Situation of Human Rights Defenders]
5 Centre for International Environmental Law and Environmental Investigation Agency, *A Rights-Based Approach to Land-Use in a Future Climate Agreement: Policy and Implementation Framework* (2015), available online at www.ciel.org/wp-content/uploads/2015/06/LandUse_RBA_20May2015.pdf.
6 *United Nations Sustainable Development Goals*, available online at www.un.org/sustainabledevelopment/sustainable-development-goals/.
7 Paris Agreement, *supra* note 1, at preamble.
8 J. Rogelj *et al.*, 'Zero Emission Targets as Long-Term Global Goals for Climate Protection', 10 *Environmental Research Letters* (2015) 1.
9 Article 3.3 of the KP limits consideration of greenhouse gas emissions by sources and removals by sinks to afforestation, reforestation and deforestation since 1990. See: Kyoto Protocol to the United Nations Framework Convention on Climate Change 1997, 2303 UNTS 148 (1998).
10 K. Dooley and A. Gupta, 'Governing by Expertise: The Contested Politics of (Accounting for) Land-Based Mitigation in a New Climate Agreement', 17(4) *International Environmental Agreements: Politics, Law and Economics* (2017) 483.
11 *Ibid.*, at 14.
12 F. Pearce, *Going Negative* (2016) 4, available online at www.fern.org/sites/fern.org/files/Going%20negative%20version%202.pdf.
13 United Nations Environment Programme (UNEP), *The Emissions Gap Report* (2016).
14 Rogelj *et al.*, *supra* note 8.
15 K. Anderson and G. Peters, 'The Trouble With Negative Emissions', 354 *Science* (2016) 182.
16 Hare *et al.*, *Australia: Emissions Set to Soar by 2020*, Climate Action Tracker Policy Brief 1 (2014).
17 P. Williamson, 'Emissions Reduction: Scrutinize CO_2 Removal Methods', 530 *Nature* (2016) 153.
18 H. Haberl *et al.*, 'Bioenergy: How Much Can We Expect for 2050?', 8 *Environmental Research Letters* (2013) 031004S.
19 P. Smith *et al.*, 'Biophysical and Economic Limits to Negative CO_2 Emissions', 6 *Nature Publishing Group* (2015) 42.
20 S. Nilsson, 'Availability of Cultivable Land to Meet Expected Demand in Food, Fibre and Fuel', in *The Global Need for Food, Fibre and Fuel Land Use Perspectives on Constraints and Opportunities in Meeting Future Demand* 151 (2012) 37; T. Searchinger and R. Heimlich, 'Avoiding Bioenergy Competition for Food Crops and Land', Instalment 9 of *Creating a Sustainable Food Future* (2015) 1.
21 The Draft Declaration of Principles on Human Rights and the Environment, E/CN.4/Sub.2/1994/9, Annex I (1994), para 22, as noted in: UN Commission on Human Rights, *Human Rights and the Environment* (1994).

22 UNGA, *United Nations Declaration on the Rights of Indigenous Peoples*, A/RES/61/295 (2007), Arts 26, 27 and 32; International Labour Organization (ILO), *Indigenous and Tribal Peoples Convention C169* (1989), Arts 14–16.
23 UNGA, *Report of the Independent Expert on the Issue of Human Rights Obligations Relating to the Enjoyment of a Safe, Clean, Healthy and Sustainable Environment*, John H. Knox, A/HRC/28/61 (2013).
24 UNGA, *United Nations Declaration on the Right and Responsibility of Individuals, Groups and Organs of Society to Promote and Protect Universally Recognized Human Rights and Fundamental Freedoms*, A/RES/53/144 (1998), Art 1.
25 P. Smith *et al.*, 'Agriculture, Forestry and Other Land Use (AFOLU)', in O. Edenhofer *et al.* (eds.), *Climate Change 2014: Mitigation of Climate Change. Contribution of Working Group III to the Fifth Assessment Report of the Intergovernmental Panel on Climate Change* (2014) 811.
26 M. Oppenheimer *et al.*, 'Emergent Risks and Key Vulnerabilities', in O. Edenhofer *et al.* (eds.), *Climate Change 2014: Mitigation of Climate Change. Contribution of Working Group III to the Fifth Assessment Report of the Intergovernmental Panel on Climate Change* (2014) 1039.
27 Rights and Resources Initiative, *Who Owns the World's Land?* (2015), available online at www.rightsandresources.org/wp-content/uploads/GlobalBaseline_web.pdf.
28 Global Witness, *On Dangerous Ground* (2015), available online at www.globalwitness.org/fr/reports/dangerous-ground/
29 The Situation of Human Rights Defenders, *supra* note 4, at 6.
30 R. Chandran, *Deaths of Land Rights Defenders Treble in a Year as Violence Surges, Says Report* (12 December 2016), available online at www.reuters.com/article/us-asia-landrights-killings-idUSKBN14113K?il=0.
31 Global Witness, *Defenders of the Earth* (2017), available online at www.globalwitness.org/en/campaigns/environmental-activists/defenders-earth/
32 The Guardian, *98 Environmental Defenders Have Been Killed so far in 2017*, available online at www.theguardian.com/environment/ng-interactive/2017/jul/13/the-defenders-tracker.
33 The Situation of Human Rights Defenders, *supra* note 4, at 3.
34 A. Savaresi, 'The Legal Status and Role of Safeguards', in C. Voigt (ed.), *Research Handbook on REDD+ and International Law* (2016) 126.
35 UNFCCC, Work programme on results-based finance to progress the full implementation of the activities referred to in decision 1/CP.16, paragraph 70, Decision 9/CP.19, at para 4.
36 Savaresi, *supra* note 34, at 133.
37 *Ibid.*, at 137.
38 S. Jodoin, 'The Human Rights of Indigenous Peoples and Forest-Dependent Communities in the Complex Legal Framework for REDD+', in C. Voigt (ed.), *Research Handbook on REDD+ and International Law* (2016) 157.
39 D. D. Pugley, 'Rights, Justice, and REDD+: Lessons From Climate Advocacy and Early Implementation in the Amazon Basin' (this volume).
40 Jodoin, *supra* note 38, at 172.
41 F. Seymour and J. Busch, *Why Forests? Why Now? The Science, Economics, and Politics of Tropical Forests and Climate Change* (2016).
42 The Situation of Human Rights Defenders, *supra* note 4, at 36.
43 A. Chhatre and A. Agrawal, 'Trade-Offs and Synergies Between Carbon Storage and Livelihood Benefits From Forest Commons', 106 *Proceedings of the National Academy of Sciences* (2009) 17667; H. Ding *et al.*, *Climate Benefits, Tenure Costs: The Economic Case for Securing Indigenous Land Rights in the Amazone* (2016), available online at www.wri.org/publication/climate-benefits-tenure-costs; C. Stevens *et al.*, *Securing Rights, Combating Climate Change: How Strengthening Community Forest Rights Mitigates Climate Change* (2014), available online at www.wri.org/sites/default/files/securingrights-full-report-english.pdf.
44 Rights and Resources Initiative, *Toward a Global Baseline of Carbon Storage in Collective Lands* (2016), available online at http://rightsandresources.org/wp-content/uploads/2016/10/Toward-a-Global-Baseline-of-Carbon-Storage-in-Collective-Lands-November-2016-RRI-WHRC-WRI-report.pdf.
45 Pugley, *supra* note 39.
46 K. Dooley and S. Kartha, Land-based negative emissions: risks for climate mitigation and impacts on sustainable development, *International Environmental Agreements: Politics, Law and Economics*, in press, (2017).
47 K. Anderson and A. Bows, 'A New Paradigm for Climate Change', 2 *Nature Climate Change* (2012) 639.
48 *Ibid.*, at 6.

49 S. Jodoin, K. Hansen and C. Hong, 'Displacement Due to Responses to Climate Change: The Role of a Rights-Based Approach', in F. Crépeau and B. Mayer (eds.), *Research Handbook on Climate Change, Migration and the Law* (2017) 205–237.
50 The Situation of Human Rights Defenders, *supra* note 4, at 6.
51 A. Jegede, 'Protecting Indigenous Peoples' Land Rights in Global Climate Governance' (this volume).
52 A Rights-Based Approach to Land-Use, *supra* note 5.
53 The Situation of Human Rights Defenders, *supra* note 4, at 16.

31

Climate change

Human rights and private remedies

Nathalie Chalifour, Heather Mcleod-Kilmurray, and Lynda M. Collins

Introduction

As a matter of global environmental governance, climate change may be viewed as the quintessential public law problem. It is an ecospheric threat involving every nation in the world and requiring political cooperation at the highest level to achieve a viable and just solution.[1] Yet climate change also engages interests that have historically been viewed as matters of private law: the loss of private property, harm to individual economic interests and personal injury. Like private law, international human rights law also focuses on the protection of individual rights and interests. It makes some sense, then, that a volume on human rights approaches to climate change would include a chapter on private law remedies.

Before the advent of international human rights law, before the creation of international or domestic environmental law, the private law systems of tort, delict and their cultural counterparts throughout the world were the individual's only means of legal redress for wrongful injury.[2] This chapter will argue that private law remedies remain highly relevant in the Anthropocene era[3] and have real potential in the area of climate change. We will focus specifically on the common law field of torts, but many of our conclusions can be generalized to any system of private remedies.

Strange bedfellows? Environmental rights in private law

Private law (be it tort, delict or some other variant) is defined as that branch of the law concerning relationships between private parties. Unlike the explicitly rights-conscious field of human rights law, private law has multiple facets and functions, some of which would seem unconnected to the realm of rights. For claimants, tort law is most often seen as a 'compensation machine', i.e. a useful vehicle for repairing losses. For defendants, tort liability is a cost of doing business and a determinant of commercial and industrial choices. Yet whatever its pragmatic functions may be, tort law has always been a rights-based system, and a growing body of scholarship identifies rights as an ongoing and significant preoccupation of tort law today.[4] As one scholar has put it, "tort law . . . is the institution that determines what our legal rights and obligations are to one another, independent of voluntary agreements or assumed legal roles and, as such, affects our expectations of ourselves and of others."[5]

The tort of negligence protects people from harm caused by unreasonably risky behavior, while nuisance protects a basic entitlement to environmental quality in the places where people live.[6] While debates remain regarding the efficacy of environmental tort law,[7] there is no doubt that tort doctrine, in particular the law of nuisance, protects a basic set of environmental entitlements (e.g. the right to be free from noxious fumes, water pollution and unreasonable noise).[8] Indeed, "nuisance theory and case law is the common law backbone of modern environmental law."[9] For the most part, tort law is not explicit in its treatment of environmental rights, but exceptions can be found. In the early nuisance case of *Radenhurst v. Coate*, for example, the Court of Chancery of Upper Canada held that "[i]t is a plain common law right to have the free use of the air in its natural unpolluted state."[10] Substitute 'climate' for 'air' and this statement reads like a manifesto for climate rights. Conceptually, then, private remedies are consistent with a human rights approach to climate change. On the pragmatic level, private law approaches offer a promising complementary tool for enhancing the efficacy of global climate governance and redressing individual injuries wrought by climate change.

Scholars have identified at least two unique advantages of tort law as a mechanism for environmental protection, both of which apply in the climate context. First, compared to statutory environmental law, civil suits more frequently produce a financially significant result and/or effective injunctive relief.[11] The astronomical damage awards that could theoretically be made in climate change suits have an unprecedented potential to force corporate emitters or producers to internalize the climate costs of their activities.[12] Even a few successful climate suits could theoretically transform the economic equation, making low-carbon strategies a fiscal necessity. Given that corporations operate under the pressure of powerful legal and financial incentives to maximize profit, this ability to align fiscal and climate interests is profoundly important. Injunctive relief may be even more significant, requiring corporations to cease or limit carbon emissions in order to respect the private law rights of others. Second, tort law does not depend for its enforcement on the political priorities of any given government. This independence from political pressures is particularly important in the climate change context. Climate regulation requires a level of economic and industrial reorganization that may be daunting to governments which operate under the imperative of reelection. Where courts are willing and able to enforce them, private law rights may provide a crucial safety net where regulatory efforts are inadequate.

Challenges to private law remedies: lessons from past climate suits

Several climate tort suits have already been attempted. Although the *Urgenda* case is the only one which has been successful to date, the others have played important roles in raising awareness of climate change and alerting governments and emitters that people are prepared to go to court to protect their climate rights and interests.

Most of the climate tort litigation has taken place in the United States.[13] Plaintiffs encountered barriers that prevented the cases from reaching a full trial on the merits. While several of these barriers may be particular to the US system (such as rules of displacement), obstacles such as strict rules of standing, the political question doctrine (called 'justiciability' in Canada) and problems of extraterritoriality may be found in many legal systems and will need to be overcome to reach the remedies stage of any climate tort action. In fact, several of these barriers were eventually overcome in appellate courts in some of these suits, and the cases were therefore successes in that respect.[14] Ultimately, while none of the private law cases in the United States to date has been successful, none has gone to full trial, so the potential of private remedies for climate rights infringements remains to be fully tested in the US (and remains to be explored in Canada).

Plaintiffs argued that climate change was a public nuisance in all four of the key US climate tort cases of *Connecticut v American Electric Power*,[15] *California v General Motors Corp.*,[16] and *Comer v Murphy Oil USA Inc.*[17] and *Native Village of Kivalina v ExxonMobil.*[18] The court in *General Motors* found that there was no such thing as a federal common law claim for global warming nuisance. The Court of Appeal in *American Electric* held that the plaintiffs had "properly alleged public nuisance" under federal common law when they claimed that "the emissions constitute continuing conduct that may produce a permanent or long lasting effect, and . . . Defendants know or have reason to know that their emissions have a significant effect upon a public right,"[19] although the Supreme Court held that the Clean Air Act had displaced federal common law (leaving open the question of whether a public nuisance claim against the emitters could be brought under state common law). It is arguable, therefore, that without the procedural and other technical hurdles encountered, a private law remedy might have been obtained for this climate tort.

In *Comer*, actual property damage had already occurred, as a result of Hurricane Katrina, so the plaintiffs included claims in private nuisance and trespass. They also raised negligence, alleging that the defendants "have a duty to conduct their business so as to avoid unreasonably endangering the environment, public health, public and private property and the citizens of Mississippi; breached this duty by emitting substantial quantities of GHGs, and . . . these emissions caused"[20] public and private property damage. The appeal court held that the plaintiffs had standing and that the private nuisance, trespass and negligence claims were justiciable questions, sending the matter back for trial.[21] Again, therefore, the viability of these tort claims remains to be finally determined.

Finally, *Juliana v US*[22] is a climate lawsuit being brought on behalf of a group of youth between the ages of 8 and 19 by Our Children's Trust, an association of young environmental activists, and Dr. James Hansen, acting as a guardian for future generations. The claims are based on the youths' rights to life, liberty and property, but also on the common law public trust doctrine. The plaintiffs argue that the US government, having known for more than 50 years about the dangers of emissions to the climate, has failed to protect essential public trust resources such as the atmosphere and oceans. Judge Aiken of the federal district court in Oregon denied the defendants' motion to dismiss, so this may be the first case in which a common law claim is heard on the merits in the United States.[23]

If climate tort claims made it to the merits stage, what remedies could a court usefully provide to protect or compensate for human rights harms? Plaintiffs will have to emphasize that courts are not being asked to set climate policy in such cases, but rather to compensate for existing harms to persons and property by applying the usual tests of reasonableness and foreseeability to questions such as standards of care in negligence or unreasonable interference in nuisance. As the appeal court in *Comer* stated, "common law tort rules provide long-established standards for adjudicating the nuisance, trespass and negligence claims at issue, the policy determinations underlying those common law tort rules present no need for nonjudicial policy determinations to adjudicate this case."[24] Providing compensatory and perhaps even punitive damages against climate polluters would create significant deterrence by rebalancing the costs and benefits of climate harms.

Tort claims can lead also to injunctions, which as a private law remedy can significantly and directly change the behavior of emitters or producers. For example, the successful ENGO plaintiffs in *Urgenda* did not seek damages but rather made their negligence claim to get a court to order the government of the Netherlands to create a climate policy that would allow it to meet its duty of care to the public to prevent dangerous climate change. The court was not dissuaded by arguments based on economic cost or political questions. Interestingly, the case was argued based on the combination of negligence law and human rights and environmental law.[25]

This negligence remedy will have direct impact on climate law and emissions behavior in the Netherlands, and is a precedent for other countries to emulate.[26]

Some have suggested that since climate harms are collective, the remedies should also be group based. Farris has suggested a no-fault Climate Compensation Fund,[27] while Kysar proposes judicial creativity in dealing with group harms such as climate change and "lost cultures, language and territorial homelands".[28]

Some climate suits are also being brought expressly on human rights grounds, rather than tort law.[29] However, if these are also unsuccessful, it may fall back to tort law to try to recover appropriate remedies for human rights harms resulting from climate change.

The way forward: developing private law remedies for climate change

The recent landmark *Urgenda* ruling[30] demonstrates that tort law can be used to hold governments responsible to take meaningful action to reduce GHG emissions. Indeed, climate-related litigation is growing rapidly as citizens aim to compel their governments to respond, and those whose lives and property are impacted by climate change seek redress from the courts. While these cases will allow courts to test some of the principles of existing jurisprudence and civil codes as applied to climate-related facts, and some cases will likely succeed, the reality is that many will fail for the reasons outlined above. Governments may therefore wish to enact legislation to overcome some of the limitations of existing law. In this section, we discuss one potential reform which could pave the way for those suffering from climate-related harm to seek a remedy through private law, namely the development of climate damages legislation.

Although there are some important differences, there is a compelling analogy to be made between legislation to facilitate recovery from the tobacco industry and potential climate damages legislation. Large producers of tobacco avoided liability for the human health harms caused by tobacco for many years, initially by not sharing evidence about tobacco's links to cancer and seeding doubt about the causal links through public relations and communications campaigns. Eventually, when evidence of the links between cancer and smoking could no longer be denied, claimants began to sue. The tobacco industry defended such claims on numerous fronts, including suggesting claimants' cancer was caused by other factors and that smokers assumed the risk of smoking. Given the burdens imposed by tobacco on public health, US states filed lawsuits against the major tobacco manufacturers, resulting in a major settlement between 46 states and four of the largest tobacco companies in 1998.[31]

Because Canada has publicly funded health care, there is a direct link between the costs to the health-care system and smoking-related cancers. In order to facilitate recovery of those costs from the tobacco industry, all provinces and territories (with the exception of the Yukon) have developed damage recovery statutes.[32] These laws enable the government to recover damages from tobacco producers on behalf of the health care system. The statutes also facilitate the awarding of damages to individual claimants by modifying evidentiary and causation standards for both government and individual plaintiffs.[33] The B.C. law (which was the first) was challenged by the industry, but the Supreme Court of Canada confirmed the constitutionality of the law (provinces are entitled to enact legislation to recover damages occurring within their boundaries) and the provisions aimed at modifying evidentiary requirements.[34]

Governments could follow the tobacco example and enact climate damages legislation to address some of the challenges involved in seeking private law remedies for climate-related harm. For instance, such legislation could allow aggregate actions, permitting recovery for injury to a class of persons such as those people living in low-lying coastal regions or in cities where climate-related flooding has caused extensive damage. This could address concerns about plaintiff indeterminacy. Climate damages legislation could also reverse the onus of proof related to causation, requiring courts to presume that a given extreme weather event is caused by climate change, rather than compelling the plaintiff to establish this causal link. Climate damages legislation could establish rules relating to market share liability, so that responsibility can be apportioned appropriately among major GHG emitters or producers over a given period of time. Such legislation could also render certain types of evidence, such as geographical, epidemiological and sociological, admissible. Gage and Byers offer some additional examples of the ways in which climate damages legislation could alleviate the burden on claimants, including clarifying standing rules, creating new or adapting existing causes of action, recognizing or creating legal rights and/or duties to form the basis of liability, and defining defendants.[35]

One of the advantages of private law remedies is that, unlike statutory fines or penalties, they compensate a victim directly.[36] In addition, damage awards have the potential to be very large, especially if they are part of a class action, which can send a powerful signal to actors causing the harm.[37] However, if there are significant barriers to seeking private law remedies because of the nature of the problem and the harms caused, as with tobacco and climate change, it is critical to address these. This is especially important from a social justice perspective, given that the victims of climate change are often from vulnerable groups already experiencing discrimination.

Parties to the Paris Agreement formally recognized that in spite of mitigation and adaptation efforts, climate change will harm people and property around the world. Whether due to sudden events like floods and hurricanes, or slow-onset harms such as sea level rise and desertification, this 'loss and damage' will especially harm those developing countries which have limited capacity to deal with the effects.[38] As the international community works to develop mechanisms for addressing this 'loss and damage' at a global level, governments should ensure that individuals experiencing loss and damage domestically can seek compensation through private law.

Conclusion

Private law has an important role to play in protecting human rights from the far-reaching ill effects of climate change. While private law and global harms may at first appear to be strange bedfellows, the unique advantages of tort (and its analogues) make it a potentially powerful tool for those affected by climate change. The emerging jurisprudence of climate tort law suggests that these claims are viable in many jurisdictions and may play an important complementary role in our collective efforts to respect, protect and fulfill the human right to a safe climate.

Notes

1 See D. Stabinsky, 'Climate Justice and Human Rights' (this volume).
2 See L. M. Collins and H. McLeod-Kilmurray, 'Toxic Torts in Historical Perspective', in L. M. Collins and H. McLeod-Kilmurray (eds.), *The Canadian Law of Toxic Torts* (2014) 39.
3 See e.g., R. E. Kim and K. Bosselmann, 'International Environmental Law in the Anthropocene: Towards a Purposive System of Multilateral Environmental Agreements', 2(2) *Transnational Environmental Law* (2013) 285.
4 D. Nolan and A. Robertson (eds.), *Rights and Private Law* (2012); R. Stevens, *Torts and Rights* (2007); D. Friedman and D. Barak-Erez (eds.), *Human Rights in Private Law* (2003); J. Thomas, 'Which Interests Should Tort Protect?' 61(1) *Buffalo Law Review* 1 (2013).
5 Thomas, *Ibid.*, at 3.
6 See A. M. Linden and B. Feldthusen, *Canadian Tort Law*, 9th ed (2011) 570; G. Pun and M. I. Hall, *The Law of Nuisance in Canada* (2010) 3, at 18–19; B. Bilson, *The Canadian Law of Nuisance* (1991) 4, at 142, 156.

7 See e.g., J. B. Ruhl, 'Making Nuisance Ecological', 58 *Case Western Reserve Law Review* (2008) 753; D. S. Wilgus, 'The Nature of Nuisance: Judicial Environmental Ethics and Landowner Stewardship in the Age of Ecology', 33 *McGeorge Law Review* (2001) 99; L. A. Halper, *Untangling the Nuisance Knot* (1998); J. McLaren, 'The Common Law Nuisance Actions and the Environmental Battle: Well-Tempered Swords or Broken Reeds?', 10(3) *Osgoode Hall Law Journal* (1972) 505; See also A. D. K. Abelkop, 'Tort Law as an Environmental Policy Instrument', 92(2) *Oregon Law Review* (2014) 381; K. N. Hylton, 'When Should We Prefer Tort Law to Environmental Regulation?', 41 *Washburn Law Journal* (2002) 515.

8 See Linden and Feldthusen, *supra* note 6, at 570.

9 W. H. Rodgers, *Environmental Law*, 2nd ed. (1994) 2. See also H. M. Green, 'Common Law, Property Rights and the Environment: A Comparative Analysis of Historical Developments in the United States and England and a Model for the Future', 30 *Cornell International Law Journal* 541 (1997).

10 *Radenhurst v. Coate* (1857), CarswellOnt 8, 6 Gr. 139, at para 7.

11 Collins and McLeod-Kilmurray, *supra* note 2, c. 1.

12 See *ibid.*

13 The Sabin Centre for Climate Change Law at Columbia Law School maintains a very useful compilation of climate litigation in all areas, including tort law: for US litigation, see U.S. Litigation Database, *Sabin Center for Climate Change Law*, available online at http://columbiaclimatelaw.com/resources/u-s-litigation-database/; for non-US litigation see Non-U.S. Climate Change Litigation, *Sabin Center for Climate Change Law*, available online at http://wordpress2.ei.columbia.edu/climate-change-litigation/non-us-climate-change-litigation/.

14 In *Connecticut v. American Electric Power*, 406 F Supp 2d 265 (SD NY 2005), reversed 582 F 3d 309 (2d Cir, 2009), reversed 131 S Ct 2527 (US, 2011), the Court of Appeals of the Second Circuit did find that the plaintiffs had standing and that there was no political question problem, and this was upheld by the Supreme Court. The Court of Appeal of the Fifth Circuit made similar findings in *Comer v. Murphy Oil USA Inc*, 839 F Supp 2d 849 (SD Miss, 2012).

15 *Connecticut v. American Electric Power*, *supra* note 14. Connecticut and seven other US states, as well as New York City and several NGOs, tried to sue five power companies, seeking the remedy of an injunction, for their continuing emissions that contributed to the public nuisance of global warming.

16 *California v. General Motors Corp* (2007), WL 2726871 (ND Cal, 2007). The state of California tried to sue the major US car manufacturers in public nuisance, seeking damages and declaratory relief, for their contributions to GHG emissions and resulting climate change.

17 *Comer v. Murphy Oil USA Inc.*, 839 F Supp 2d 849 (SD Miss, 2012). Residents and property owners along the Mississippi Gulf Coast tried to bring a class action against energy, fossil fuels and the chemical producers for their emissions which contribute to climate change and thus to the strength of Hurricane Katrina. They claimed in both public and private nuisance, but also in negligence, unjust enrichment, civil conspiracy, fraudulent misrepresentation and concealment and trespass, and they sought compensatory and punitive damages.

18 *Native Village of Kivalina v. ExxonMobil*, 663 F Supp 2d 863 (ND Cal, 2009), affirmed 696 F 3d 849 (9th Cir, 2012), leave to appeal refused USSC May 2013. The village tried to sue energy, oil and utility defendants in federal public nuisance for damages to pay for the costs of relocating their entire village due to the elimination of protective sea ice by climate change, caused in part by the emissions of the defendants.

19 *Connecticut v. American Electric Power*, Connecticut 2d Cir, *supra* note 14, at 352–353.

20 *Comer v. Murphy Oil USA Inc*, *supra* note 14, at 861.

21 Eight of sixteen judges recused themselves due to ties with the fossil fuel industry, so there was no judicial quorum to hold the en banc review which the Second Circuit granted of its decision. Since the Second Circuit had vacated its decision when it granted the en banc review, the decision of the district court was the controlling law, and further appeal to the Supreme Court was refused due to a finding of res judicata – see Case No. 12–60291 (5th Cir, May 14, 2013).

22 *Juliana v. US*, No. 15–1517, 2016 WL 6661146 (D. Or. Nov. 10, 2016)

23 The trial is expected to take place in 2017.

24 *Comer v. Murphy Oil USA Inc*, *supra* note 14, at 875.

25 R. Cox (lawyer for Urgenda), *A Climate Change Litigation Precedent: Urgenda Foundation v. The State of the Netherlands*, CIGI Papers No. 79 (November 2015), available online at www.cigionline.org/sites/default/files/cigi_paper_79.pdf.

26 See, e.g., Klimaatzaak case in *Belgium and Zoe & Stella Foster v. Washington Department of Ecology* (Superior Court of the State of Washington for King County, no. 14–2–25295–1 SEA), both cited in Cox, *supra* note 25, at 14.

27 M. Farris, 'Compensating Climate Change Victims: The Climate Compensation Fund as an Alternative to Tort Litigation', 2 *Sea Grant Law and Policy* (2009–10) 49, at 60.
28 D.A. Kysar, 'What Climate Change Can Do About Tort Law', 41 *Environmental Law* (2011) 1, at 67.
29 See Petition to the Commission on Human Rights of the Philippines Requesting for Investigation of the Responsibility of the Carbon Majors for Human Rights Violations or Threats of Violations Resulting from the Impacts of Climate Change, *Greenpeace* (2015), available online at www.greenpeace.org/seasia/ph/PageFiles/105904/Climate-Change-and-Human-Rights-Complaint.pdf; See also K. Dooley, 'Human Rights and Land-Based Carbon Mitigation' (this volume).
30 *The Urgenda Foundation v. State of the Netherlands* (Ministry of Infrastructure and the Environment), C/09/456689/HA ZA 13–1396 (24 June 2015).
31 See National Association of Attorneys General, *Master Settlement Agreement* (1998), available online at http://web.archive.org/web/20080625084126/www.naag.org/backpages/naag/tobacco/msa/msa-pdf/1109185724_1032468605_cigmsa.pdf.
32 See, for e.g., Tobacco Damages and Health Care Recovery Act, [SBC 2000] c. 30.
33 Collins and McLeod-Kilmurray, *supra* note 2, at 223.
34 *British Columbia v. Imperial Tobacco Canada Ltd.*, (2005) 2 SCR 473.
35 A. Gage and M. Byers, *Payback Time? What the Internationalization of Climate Litigation Could Mean for Canadian Oil and Gas Companies* (2014), available online at http://wcel.org/sites/default/files/publications/Payback%20Time.pdf, at 36.
36 Collins and McLeod-Kilmurray, *supra* note 2, at 177.
37 *Ibid.*
38 See Art. 8, *United Nations Framework Convention on Climate Change, Adoption of the Paris Agreement*, U.N. Doc. FCCC/CP/2015/L.9/Rev. 1 (12 December 2015).

32

Towards responsible renewable energy

Assessing 50 wind and hydropower companies' human rights policies in the context of rising allegations of abuse

Eniko Horvath and Kasumi Maeda

Introduction

The Paris Climate Agreement and the Sustainable Development Goals are driving commitments to climate action and universal access to energy globally. Renewable energy will be central to achieving these commitments, and with falling technology costs, investments in the sector are already on the rise. A welcome 198 GBP billion was invested worldwide in renewable energy in 2015, marking also the largest annual increase of clean energy implementation.[1] Renewable energy is expected to become the largest source in global electricity production by 2030.[2]

But this increase in investment is not without problems: there is also a rapid rise in allegations of human rights abuses linked to renewable energy projects. In the past 10 years, the Business & Human Rights Resource Centre approached companies in the wind and hydropower sectors 115 times to seek responses to human rights allegations by NGOs and local communities. Renewable energy projects, including hydropower and wind farms,[3] are associated with serious human rights abuses including in Central and South America, East Africa and Southeast Asia. Ninety-four of these allegations took place after 2010, and over 50% of the allegations were related to operations in Central and South America. Though hydropower projects face the most allegations, we found that the number of allegations against wind power projects is increasing. Local communities are faced with some of the most damaging impacts, including dispossession of their lands, undermined livelihoods, threats and intimidation, killings and displacement, among other abuses.

Many of the allegations against wind and hydropower companies are caused by lack of adequate human rights due diligence and impact assessment procedures in place, as established by the UN Guiding Principles on Business and Human Rights (UNGPs). Key to understanding the potential impact of renewable projects is rigorous consultation with affected communities and respecting Indigenous Peoples' right to free, prior and informed consent (FPIC). Failure to

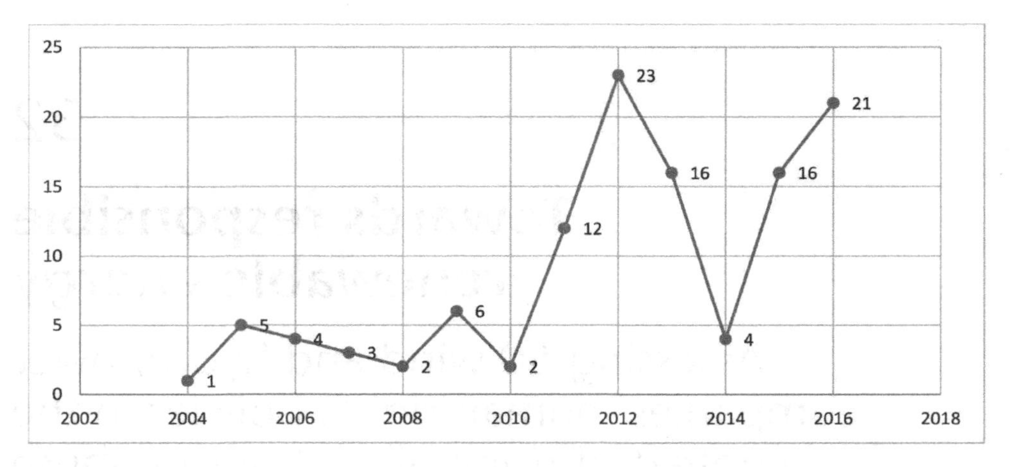

Figure 32.1 Total number of approaches to wind and hydropower companies with human rights allegations

consult adequately and address potential abuse is driving rising levels of community resistance to renewable projects in many countries. In turn, this will cause project delays, as well as financial, legal and reputational penalties for companies.

The sector has the opportunity to avoid these costs, but to do so it must act now to radically strengthen its human rights due diligence prior to investment. The success of the industry and our transition to a low-carbon economy depend upon it. In October 2016, the Business & Human Rights Resource Centre conducted an outreach to 50 renewable energy companies with a set of 10 questions on their approach to human rights, focusing on community engagement.[4] We received responses from 20 companies and conducted desk-based research on all 50. We contacted a broad range of companies that work in wind and hydropower, selecting them based on a mix of company renewable energy generating capacity (in MW) and regions of operation. Previous outreach was conducted in May 2016 with the same questionnaire to 35 companies.

The analysis below highlights two key points:

- The wind and hydropower sectors are not well developed in terms of human rights due diligence, despite substantial risks in their operations and their reputation as the "clean" actors of the broader energy sector.
- There are a small number of leading companies, including Statkraft and Engie, who have credible human rights policies and practices. And while the leaders still have more to do, the rest of the sector should learn quickly from them to avoid future risks.

Below we focus our analysis on four key aspects of human rights due diligence, which offer immediate opportunities for the sector to improve performance and mitigate risks: designing policy and practice in line with international standards; introducing rigorous consultation with affected communities and respect land rights, especially of Indigenous Peoples; enforcing and monitoring the implementation of policy and procedures in all installations; and ensuring any government-sanctioned initiatives that label renewable energy projects as clean enforce rigorous human rights safeguards.

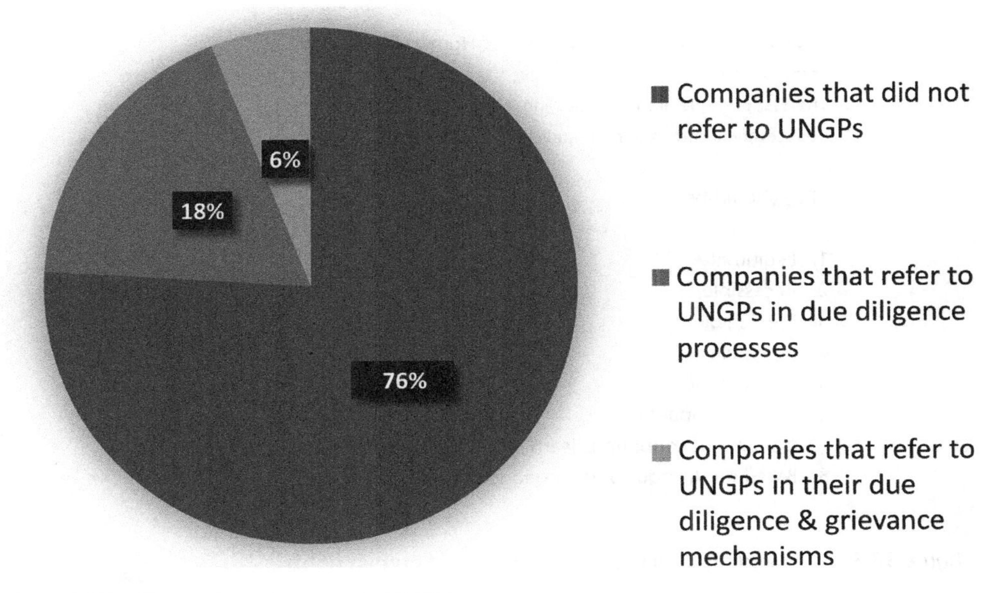

Figure 32.2 Company references to UNGPs

Reference to international standards

Thirty-three of the 50 companies we contacted had publicly available human rights policies. The degree of commitments in these policies ranged widely, from Statkraft who has shown an explicit commitment to the UNGPs, the Organisation for Economic Co-operation and Development (OECD) Guidelines, and International Finance Corporation (IFC) Performance Standards, to Renovalia Energy who failed to show the existence of any human rights policies.

Only 50% of the companies we researched referred to any international standards. A small group of 12 companies referred to the UNGPs specifically, including Acciona, EDP Renewables, Eletrobras, Engie, EPM, E.ON, Isagen, Statkraft, Statoil, Suncor, Vattenfall and Vestas.

The UNGPs emphasise that companies have a responsibility to avoid causing or contributing to adverse human rights impacts and to take steps to prevent, mitigate and remedy impacts. Of the 12 companies that referred to the UNGPs, most of them did so in reference to their due diligence processes or human rights policies. Only three companies (Isagen, Statoil and Vattenfall) went further by referring to the UNGP obligation to "access to remedy" and having grievance mechanisms available to communities and workers (see Figure 32.2 for breakdown of company references to UNGPs).

Ensuring effective grievance mechanisms are in place is an essential part of protecting the rights of Indigenous Peoples and local communities. These mechanisms also serve as an early alert mechanism for companies around issues that could turn into costly conflicts.

The UNGPs outline a set of effectiveness criteria for companies' operational grievance mechanisms – yet only two companies referred to these. Vattenfall stated that the company is currently aligning its policies, including a grievance mechanism, to the UNGPs. Isagen stated that its grievance mechanism was developed in line with the UNGPs. All renewable energy companies have an interest in examining their grievance mechanisms in light of the UNGPs' effectiveness criteria.

UN Guiding Principles on Business & Human Rights

The 31st principle of the UN Guiding Principles set out eight "effectiveness criteria" for non-judicial grievance mechanisms.

They should be:

1. Legitimate
2. Accessible
3. Predictable
4. Equitable
5. Transparent
6. Rights-compatible
7. A source of continuous learning
8. Based on dialogue and engagement

Figure 32.3 UN Guiding Principles on business and human rights

Weak commitment to community consultations

Our outreach to companies revealed a weak commitment to community consultations in the renewable energy sector.

Although 34 companies we reached out to have some commitment to consult with local communities, their policies and practices varied significantly. Community consultations ranged from a social and environmental impact assessment at the start of the project, to continuous consultation via meetings from the development phase, construction, and during the operation phase of the project.

Companies such as Isagen establish relationships with communities at the early stages of the project, prior to construction, in order to identify who would be impacted. Isagen provided details on their processes, showed a commitment to respect the social and cultural lives of community members, and indicated that it takes a rights-based approach to identifying who would be affected.

Acciona, a utility company based in Spain, carries out consultations with local communities throughout the project cycle. Ontario Power Generation consults communities and indigenous people in various ways, including holding public information meetings, giving council presentations, making information available on their website, performing in-person visits, sharing reports and maintaining an ongoing relationship with those affected communities throughout the project cycle.

Others such as Enel Green Power have a commitment to consultation as per national legislation but do not have a structured process in place. However, 16 companies do not have any form of publicly available consultation commitments in place. The majority of these are in the wind power sector.

A structured and rigorous consultation process can act as a basis for preventing future project delays and costs due to conflicts with local communities, and at its best create a strong and mutually beneficial relationship between communities and the company.

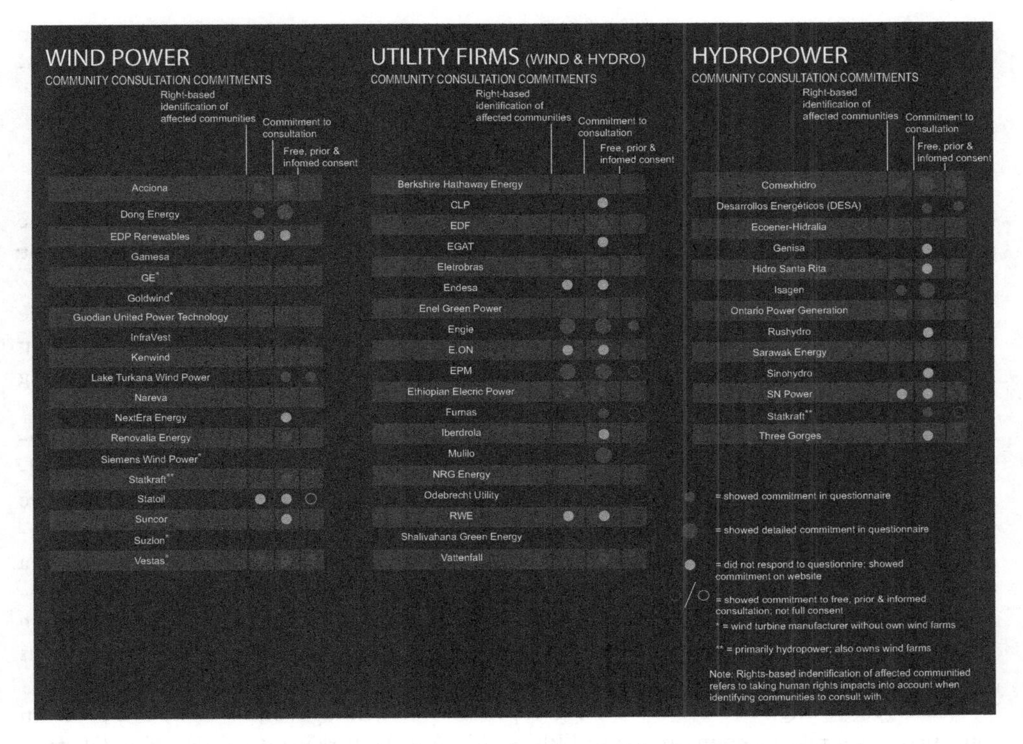

Figure 32.4 Summary of the outreach to wind, utility and hydropower companies

Inadequate consultations can delay project development, and prevent construction and operations from running smoothly. In Myanmar, the Mong Ton Dam has been linked to concerns around inadequate consultations, and the local Shan villagers claim the dam would flood and destroy 100 communities.[5] The villagers have held protests against the Snowy Mountains Engineering Corporation (SMEC), which was commissioned to conduct an environmental and social impact assessment. Villagers allege that SMEC is helping push the project ahead without proper consultations. In January 2015, 16 Shan villagers submitted a statement in Bangkok warning that the project could fuel conflict in an already unstable region. The letter also claims that SMEC surveyors gave villagers "gifts," which the villagers saw as bribes, to get them to sign documents they did not understand. SMEC denies these allegations and holds that it continues to reach out to local civil society organizations without success.

On the other hand, renewable energy development can provide great economic opportunities for locals if they are given more of a say in the project. In contrast to other wind farms in Oaxaca, Mexico, the Ixtepec wind project is a community-owned indigenous wind power project.[6] The community of Ixtepec reached out to Grupo Yansa to build a wind farm in the area. Yansa proposed a site where agricultural impact would be minimal, and conducted an environmental impact assessment. Yansa seeks to directly involve communities in the construction and operation of the wind farms, giving them control of the renewable energy sources. The energy will be sold to the national grid at a guaranteed fixed price, providing community members with an income and an opportunity for economic and social growth. Fifty per cent of the earnings will go to community members to compensate land owners and pay for community programmes. This approach demonstrates that if renewable energy companies keep human rights considerations at their core, it is possible to transition to clean energy whilst keeping abuses at bay.

Policies need to be reflected on the ground

A small group of five companies out of the 50 we reached out to expressed a commitment to the internationally agreed ILO Convention 169 standard to obtain free, prior and informed consent: Comexhidro, Desarrollos Energéticos (DESA), Engie, Lake Turkana Wind Power and Vestas.

The lack of commitment is especially concerning as many of the allegations about renewable energy projects are related to inadequate consultation and the failure of companies to obtain FPIC.[7] Though FPIC is a primary right for indigenous people, civil society experts including Oxfam recommend applying it to all affected communities.[8]

FPIC is underpinned by the UN Declaration on Indigenous Peoples Rights and ILO Convention 169. This right entails providing early access to information about projects planned for development in Indigenous Peoples' lands, and obtaining consent from them at their free will before construction begins. Oxfam America's Community Consent Index shows that the extractives sector has taken strides in committing to FPIC, illustrating the feasibility of this commitment at a company level.[9] Although the extractives sector is far from being free of conflicts with local communities, renewable energy companies have an opportunity to learn from extractives' experience in putting FPIC commitments in place. There is a strong need for clear language in commitments and in plans to implement FPIC in all sectors affecting local communities.

However, writing FPIC into human rights policies is not enough. Three of the companies in our survey that made a commitment to FPIC – DESA, Lake Turkana Wind Power and Vestas – have allegations against them that claim the companies did not adequately consult and obtain consent from local communities. The distinction between consultation and consent is key here: meaningful FPIC goes beyond discussions with the local population and requires the community to agree to and give consent for the development of the project. This may require the company to amend their original project development plans and provide adequate compensation to the people. Once the policies are in place at a corporate level, continuous engagement with affected communities and open channels of communication are essential for responsible renewable energy and mitigation of human rights abuses.

DESA, a hydropower company based in Honduras, claims it had obtained FPIC for a "run-of-the-river" hydropower project called Agua Zarca in Honduras, but also admits there is a group within the affected community that opposes the project. FPIC should not be based on what the majority say, but apply to all community members, and companies must continue to constructively engage with those locals who oppose the project and obtain their consent before beginning construction.

A grave reminder of the consequences of a flawed FPIC process was the killing of Berta Cáceres, indigenous community leader and human rights defender who was protesting the Agua Zarca hydropower project, earlier this year. COPINH, the organisation she worked with, continues to protest the project and faces threats and assassination attempts.[10] DESA stated in their response to our questionnaire that all security forces directly or indirectly hired by DESA abide by human rights policies and that FPIC was secured for the project. There is an apparent disconnect between the company's policies and what is happening on the ground.

Lake Turkana Wind Power is poised to build the largest wind farm in Africa, with a planned capacity of 310 MW covering 162 km^2. The project has been linked to allegations of increased alcoholism, prostitution and violence due to an influx of people into the area, as well as claims that the company has not respected Indigenous Peoples' rights, land rights, and free, prior and informed consent.[11] Lake Turkana Wind Power previously held that the pastoralist groups affected by the project are not indigenous and therefore FPIC was not necessary. More recently, the company has stated that they respect Indigenous Peoples' rights and would consult with affected communities regardless of whether they are indigenous.[12] The project continues to be stalled as of February 2017 and is subject to ongoing legal disputes.

However, other examples illustrate that it is possible to keep human rights at the core of the transition to clean energy. Statkraft proved its ability to be open to change when the Jijnjevaerie Saami village submitted a complaint to the Swedish and Norwegian National Contact Points (NCPs) to the OECD Guidelines concerning the company's planned wind power project in the area where the Saami village engages in reindeer herding.[13] While the NCPs found that Statkraft complied with the OECD Guidelines, they also provided recommendations for how the company could work in a manner that promotes Indigenous Peoples' rights. Following the OECD complaint, Statkraft reports that it was able to reach an agreement with Saami villagers.

In another instance, the Cerro de Oro hydropower project in Mexico was cancelled and other sites were pursued due to concerns from indigenous communities. The project was co-owned by Comexhidro and the Latin Power III Fund and financed by the Overseas Private Investment Corp (OPIC).[14] Indigenous communities submitted a complaint to OPIC's Office of Accountability over concerns about impacts on their safety, access to water and fishing areas. After dialogues with OPIC and local and regional government officials, the project was suspended. As of August 2014, the company continues to respect the communities' decision to reject the project. Comexhidro is also developing the Puebla hydropower plant in Mexico, where they consulted with local communities prior to obtaining permits. Comexhidro has developed an ongoing relationship with the local community and created a local office with both foreign and local workers to handle complaints and mitigate human rights impacts.

Engie also exemplified its commitment to securing FPIC when the company began to develop an offshore wind farm in France, off the islands of Yeu and Normoutier. Although the local fishermen originally opposed the project, after 2.5 years of consultations and negotiations, Engie was able to receive consent from the community.

UN clean development mechanism not a guarantee against human rights abuses

Governments and international funders also have a responsibility to ensure renewable energy projects they fund do not cause or contribute to human rights abuses. The UNGPs clearly establish the state duty to protect human rights potentially affected due to private sector activities. A number of international instruments also specify governments' duties regarding respecting Indigenous Peoples' rights, including ILO Convention 169.

Any government-funded or -sanctioned initiative that label renewable energy projects as clean or responsible also need to ensure that human rights are protected on the ground. The UN Clean Development Mechanism (CDM) is a global environmental investment scheme that funds emission-reducing projects.[15] However, many projects backed by this scheme have come under criticism for their human rights impacts, and the CDM does not have a process to ensure project developers respect human rights. Carbon Market Watch follows the development of CDM-registered projects and finds that many of them are linked to abuse of local communities' rights and environmental damage.[16]

Of the companies we reached out to, 31 have CDM-registered projects. Eleven of these have human rights allegations associated with them, including Oaxaca wind farms, Foum El Oued wind farm and Santa Rita hydropower project. Many of these allegations are tied to the abuse of the right to land, right to FPIC and Indigenous Peoples' rights.

Any future iteration of CDM must adopt rigorous human rights safeguards. John Knox, UN Special Rapporteur on human rights and the environment, has also emphasised the importance of human rights safeguards in the new climate financing mechanism, in a letter to UNFCCC.[17] His letter emphasises that simply because a project is considered "clean" does not mean that it is good for the people it affects.

Progress since May 2016

Over a five-month period, we have found that a handful of companies are moving in the right direction. When we first reached out to companies in May, we contacted 35 firms involved in renewable energy project development with the same questionnaire. Notably, some companies that did not respond to our previous outreach, such as Ontario Power Generation, EDF and Electrobras, disclosed information the second time about how they approach community engagement. Isagen, Lake Turkana Wind Power and Vestas updated their answers in the second outreach as well. Specific changes since the last outreach include:

- Isagen is now reviewing their human rights policies with an independent third party to ensure that it is aligned with the UNGPs.
- Lake Turkana Wind Power previously stated that FPIC was only necessary for indigenous people. However, they now recognise the need to engage in free, prior and informed consultations with *all* affected communities.
- Vestas has disclosed additional information about their social due diligence process to "identify, prevent and mitigate risk and impacts of project affected communities."

This improvement in transparency is a crucial first step to ensure the protection of human rights. We encourage more companies in the renewable energy sector to disclose information on their human rights policies and practices in recognition of their responsibility to provide clean energy that respects human rights.

Conclusion

Our results reveal the alarming lack of transparency and implementation of human rights responsibilities in the renewable energy sector. Without commitments in place as a first step and subsequent implementation on the ground, communities have no way of ensuring their rights are protected and of expressing their grievances about projects being built on their land.

Although 34 out of 50 companies have some commitment to consult with local communities, these commitments vary significantly and the majority are weak or non-existent. Only 5 out of 50 refer to respecting Indigenous Peoples' internationally recognised right to FPIC. Moreover, there is a disconnect between companies' policies and practice at the project site level. Three out of the five companies that have FPIC commitments in place have faced challenges about this commitment on the ground. Failing to undertake these consultations in a rigorous manner can cause project delays as well as financial, legal and reputational costs to companies.

Governments and international funding mechanisms also have a significant role to play, as their stamp of approval does not guarantee that projects will be free from abuse. Thirty-one companies we reached out to had projects registered with the UN's Clean Development Mechanism funding renewable energy projects; however, 11 of these faced human rights allegations including the abuse of the right to FPIC and right to land and violence against communities.

Based on the findings of this study, several key recommendations are put forward to ensure that the transition to renewable energy is fast but also fair. First, renewable energy companies must adopt and disclose human rights policies and due diligence procedures in line with the UNGPs. This should include a commitment to rigorous community consultation procedures, including obtaining FPIC from communities affected by renewable energy projects. Companies should also introduce grievance mechanisms in line with the UNGPs' effectiveness criteria, designed with communities and workers. One of the ways companies can lift human rights standards across the industry would a sector-wide initiative on human rights.

Next, the UN Framework Convention on Climate Change (UNFCCC) and future Conference of Parties (COP) should adopt rigorous human rights safeguards for projects financed by the Clean Development Mechanism, Sustainable Development Mechanism and Green Climate Fund (or other funding mechanisms for renewable energy projects). They must also ensure that criteria adopted by Global Climate Action Agenda high-level champions include a requirement to act according to the UNGPs.[18]

Last, governments implementing the Paris Climate Agreement and Sustainable Development Goals must adopt and enforce human rights safeguards in national energy policies facilitating renewable energy projects. Reporting on how climate actions are taking human rights impacts into consideration in Intended Nationally Determined Contributions (INDCs) will further ensure that standards are met. Further, governments are encouraged to ratify the ILO Convention 169 on Indigenous Peoples and ensure respect for the right to FPIC.

Notes

1 A. Neslen, Renewable Energy Smashes Global Records in 2015, Report Shows, *The Guardian* (1 June 2016), available online at www.theguardian.com/environment/2016/jun/01/renewable-energy-smashes-global-records-in-2015-report-shows.

2 Coal and Gas to Stay Cheap, But Renewables Still Win Race on Costs, *Bloomberg New Energy Finance* (13 June 2016), available online at https://about.bnef.com/press-releases/coal-and-gas-to-stay-cheap-but-renewables-still-win-race-on-costs/.

3 Recognising concerns by human rights organizations about the negative impacts of large dams on climate change and human rights on the one hand and the continued classification of these projects as renewable or clean energy by international funders on the other, hydropower companies have been included in this research as far as they claim their projects are green, clean or renewable.

4 Renewable Energy & Human Rights: Outreach to Companies, *Business & Human Rights Resource Centre* (2016), available online at https://business-humanrights.org/renewable-energy-human-rights.

5 Myanmar: Mong Ton Dam Impacts Shan Communities by Salween River, *Business & Human Rights Resource Centre* (2016), available online at https://business-humanrights.org/en/myanmar-mong-ton-dam-impacts-shan-communities-by-salween-river.

6 Mexico: Ixtepec Wind Project Impacts Communities in Oaxaca, *Business & Human Rights Resource Centre* (2016), available online at https://business-humanrights.org/en/mexico-ixtepec-wind-project-impacts-communities-in-oaxaca.

7 Case Studies: Renewable Energy & Human Rights, *Business & Human Rights Resource Centre* (2016), available online at https://business-humanrights.org/en/case-studies-renewable-energy.

8 Oxfam's Community Consent Index 2015 provides more information on FPIC. See below.

9 Community Consent Index 2015, *Oxfam* (23 July 2015), available online at www.oxfam.org/sites/www.oxfam.org/files/file_attachments/bp207-community-consent-index-230715-en.pdf.

10 N. Lakhani, *Honduran Activists Survive Attacks Months After Berta Cáceres Murder* (11 October 2016), available online at www.theguardian.com/world/2016/oct/11/honduran-activists-survive-assassination-berta-caceres.

11 Kenya: Report on renewable energy projects' impacts on indigenous communities; Lake Turkana Wind Power Responds, *Business & Human Rights Resource Centre* (2016), available online at https://business-humanrights.org/en/kenya-report-on-renewable-energy-projects-impacts-on-indigenous-communities-lake-turkana-wind-power-responds.
12 Lake Turkana Wind Power: Renewable Energy & Human Rights, *Business & Human Rights Resource Centre*, available online at https://business-humanrights.org/en/lake-turkana-wind-power-renewable-energy-human-rights.
13 Saami villagers submit complaint to Swedish & Norwegian NCPs Over Land Rights Impacts of Statkraft Wind Farms, *Business & Human Rights Resource Centre* (2016), available online at https://business-humanrights.org/en/saami-villagers-submit-complaint-to-swedish-norwegian-ncps-over-land-rights-impacts-of-statkraft-wind-farm.
14 Mexico: Cerro de Oro Dam Impacts Indigenous Communities Near Santo Domingo River, *Business & Human Rights Resource Centre* (2016), available online at https://business-humanrights.org/en/mexico-cerro-de-oro-dam-impacts-indigenous-communities-near-santo-domingo-river.
15 Clean Development Mechanism (CDM), *United Nations Framework Convention on Climate Change*, available online at http://unfccc.int/kyoto_protocol/mechanisms/clean_development_mechanism/items/2718.php.
16 Harmful CDM Projects, *Carbon Market Watch*, available online at http://carbonmarketwatch.org/category/project-campaigns/.
17 Knox, *Letter From the Special Rapporteur on Human Rights and the Environment* (4 May 2016), available online at http://srenvironment.org/wp-content/uploads/2016/06/Letter-to-SBSTA-UNFCCC-final.pdf.
18 Submission on Road Map for Global Climate Action, *Business & Human Rights Resource Centre and Center for International Environmental Law* (2016), available online at http://unfccc.int/files/parties_observers/submissions_from_observers/application/pdf/670.pdf.

33

Intersectionalities, human rights, and climate change

Emerging linkages in the practice of the UN human rights monitoring system

Joanna Bourke Martignoni

The reflexive relationship between climate change and various forms of inequality is increasingly being highlighted by the international system for the promotion and protection of human rights. Human rights monitoring mechanisms, ranging from the United Nations Human Rights Council and its Special Procedures to the Treaty Monitoring Bodies and the Office of the High Commissioner for Human Rights, have noted that discrimination and inequalities constitute both causes and consequences of climate change–related injustices.[1]

Intersectionality is a concept that has been used by social theorists to critique the manner in which laws and policies developed from a single identity perspective ignore, and thereby reinforce and perpetuate, the experiences of oppression faced by people who exist at the crossroads of multiple characteristics.[2] Intersectional approaches acknowledge that discrimination and inequality are dynamic and interactive constructions rather than fixed in time and space.[3] Beyond their deployment as observational aids designed to render visible complex power relations, intersectional methods also serve a transformative purpose in that they seek to redress the structural factors and institutions that nurture and sustain inequalities.[4]

The utility of intersectionality as an analytical methodology and as a tool for inclusive, innovative policy-making has gained wide acceptance from scholars and activists working from the standpoint of critical race and feminist studies.[5] The adoption of intersectional approaches to non-discrimination by international human rights monitoring mechanisms is, however, still in its infancy.[6] The application of an intersectional paradigm to human rights analyses of the impacts of climate change is even less developed, although there has been some work done on intersectionalities and climate change from a socio-ecological perspective.[7]

This chapter seeks to outline the ways in which the UN human rights monitoring mechanisms are beginning to address intersectional forms of inequality and to map some pathways for future work by the international human rights system on the issue of intersectionalities within the context of climate change.

The first section briefly introduces theories of intersectionality and the manner in which these have been applied within the context of climate change policies. The role of the UN human rights treaty monitoring bodies and their increasing engagement with the idea of intersectional discrimination is then addressed in the second section. The work of the Committee on the Elimination of Discrimination Against Women (CEDAW Committee) and of the Committee on the Rights of the Child (CRC Committee) is highlighted, as these are the institutions that have, thus far, produced the most significant interpretive guidance on the issue of intersectional discrimination and on the relationship between human rights and climate change. The contribution of the Human Rights Council and its Special Procedures to the development of intersectional approaches to climate change is outlined in the third section of the chapter. The conclusions examine the potential of intersectionality as an analytical tool and as a form of policy and legislative practice that could be developed to advance discussions within the UN human rights monitoring system on the linkages between climate change, sustainable development and human rights.

Intersectionality, equality and climate change

> *Intersectionality is a form of resistant knowledge developed to unsettle conventional mindsets, challenge oppressive power, think through the full architecture of structural inequalities and asymmetrical life opportunities, and seek a more just world.*[8]

Social theorists have demonstrated that gender, ethnicity, disability, age, socio-economic status and other identity characteristics are reflections of social hierarchies that shape political and economic power and promote or limit an individual's enjoyment of rights and capabilities.[9] These accounts offer a critical counterpoint to traditional approaches to anti-discrimination law and social policy-making, which tend to reduce inequality to a singular attribute and, in the process, fail to identify and confront the complex power relations that constitute and reinforce oppression.[10]

From a socio-ecological perspective, a number of authors have drawn attention to the need for policies and legislation to better reflect and value the mutually dependent relationship between humans and nature.[11] These theorists, many working within an eco-feminist paradigm, have sought to highlight the ways in which economic, political and social institutions have been built to serve the interests of powerful elites in exploiting the environment.[12] To this end, a number of authors have used intersectional forms of analysis to draw out the systems of power grounded in anthropocentrism, racism, classism, colonialism and sexism that underlie most human-nature interactions. These studies interrogate dominant modes of production, consumption and natural resource use and the ingrained assumptions about human behaviour and the characteristics of a 'good life' upon which these are based.[13]

Recent advocacy and analytical work on intersectionality and climate change has stressed the need to incorporate the experiences of indigenous and rural women, children and youth, and other historically marginalized groups within the development and monitoring of climate change mitigation and adaptation policies and programmes.[14] These interventions seek to emphasize the interconnectedness of demands for social equality and environmental sustainability, and they incorporate the environment as a stakeholder that must be given 'standing' within decision-making structures. The expansion of intersectional methods to encompass the interests of non-human species constitutes a significant theoretical and practical development.[15]

As with the work being done in socio-environmental studies, the application of intersectional methods within discussions on human rights and climate change will require the adoption of a 'radically cross disciplinary stance'.[16] It remains to be seen whether the international human rights system is sufficiently permeable to respond to the demands and the challenges to existing institutional structures that this kind of interdisciplinarity presents.

UN human rights treaty monitoring bodies and intersectional discrimination

The adoption of the core international human rights treaties and the subsequent practice these instruments have inspired has broadened the reach of international anti-discrimination law. Developments in equality law have occurred not only in terms of the definition of the acts or omissions that may constitute discrimination, but also in relation to the grounds upon which discrimination must be prohibited.

The oldest of the multilateral human rights treaties, the Convention on the Elimination of Racial Discrimination (CERD), prohibits distinctions based on 'race, colour, descent, or national or ethnic origin'.[17] The two International Covenants, the International Covenant on Civil and Political Rights (ICCPR) and the International Covenant on Economic, Social and Cultural Rights (ICESCR), added an open-ended list to the categories of prohibited discrimination that includes, 'sex, language, religion, political or other opinion, national or social origin, property, birth or other status'.[18]

Treaty provisions on non-discrimination in international and regional instruments now routinely refer to 'other status', in order to leave the potential categories of prohibited grounds open, thus allowing the mechanisms a certain degree of latitude in their responses to newly identified forms of inequality and oppression. The category of 'other status' has been read as including a number of different attributes, such as age,[19] disability,[20] migrant or refugee status,[21] place of residence,[22] health situation,[23] status of deprivation of liberty,[24] sexual orientation,[25] physical appearance[26] and poverty.[27]

In recent years, a number of the human rights treaty monitoring bodies, including the CERD Committee,[28] the Human Rights Committee (HRC)[29] and the CESCR,[30] have begun to mention forms of multiple and intersectional discrimination within their work. The incorporation of intersectional approaches to the interpretation of equality guarantees in international human rights law is occurring, albeit in a piecemeal, ad hoc manner that varies depending on the mechanism concerned.

Committee on the Elimination of Discrimination Against Women

The CEDAW Committee has gradually begun to move from a 'double discrimination' paradigm towards a more fully fledged approach to intersectionality that recognizes the existence of 'intersecting and compounded forms of disadvantage'.[31] Several of the Committee's decisions in individual cases submitted under the Optional Protocol to the Convention on the Elimination of All Forms of Discrimination Against Women have recommended transformative remedies to address situations of intersectional discrimination against women.[32]

In its more recent interpretive General Recommendations, the Committee emphasizes the obligations of States Parties to take positive measures to ensure that policies and legislation respond to the particular circumstances of women-headed households, widows, religious and ethnic minority women, African-descendent and indigenous women, women with disabilities, elderly women and other groups.[33] In its 2016 General Recommendation no. 34 on the rights of rural women, the Committee notes, 'Women working in rural areas, including peasant, pastoralists, migrants, fisherfolk and landless, also suffer disproportionately from intersecting forms of discrimination.'[34] The CEDAW Committee's General Recommendation no. 37 on disaster risk reduction in a changing climate, due to be adopted in 2018, also addresses the need for coherent policy and legislative frameworks on climate change mitigation and adaptation that reflect the experiences of diverse groups of women, including those in rural and urban settings, indigenous women, women with disabilities, girls and older women.[35]

While the Committee has pointed out the need for States parties to address the 'macroeconomic roots' of gender inequality and climate change, it has not yet developed specific policy recommendations concerning the ways in which these phenomena should be tackled.[36] In the absence of far-reaching proposals for institutional reform, the treaty body tends to return to well-worn techniques for redressing inequalities within existing power structures and frameworks, such as the adoption of participatory policy-making processes, temporary special measures including quotas, interventions to confront harmful stereotypes and discriminatory social and cultural norms and practices, legislative reform and the collection of disaggregated data.[37] These recommendations are certainly important and constitute the building blocks for the application of an intersectional paradigm within the Committee's work; however, they do not contribute to the radical transformation of power relations that intersectionality theorists have frequently proposed.[38]

Committee on the Rights of the Child

Like the CEDAW Committee, the Committee on the Rights of the Child has also begun to widen the scope of its exploration of children's rights to address the experiences of particular groups of children, including indigenous children, migrant and asylum-seeking children, girl children, children with disabilities and adolescents.[39]

The CRC Committee has recently started drawing attention to the impact of climate change on children in its Concluding Observations and Recommendations on the periodic reports of States Parties to the Convention on the Rights of the Child.[40] In its assessment of the report of Fiji in 2014, the Committee noted that 'children face more acute risks from disasters and are more vulnerable to climate change than adults' and highlighted the situation of children and families living in coastal and low-lying areas, particularly those located in the outer islands, who are experiencing loss of land and freshwater resources as well as reduced opportunities for agriculture and subsistence living.[41]

In 2016, the Committee held a Day of General Discussion on children's rights and the environment.The event included an examination of the impact of environmental harm on the rights of children in 'vulnerable situations' and it also addressed the need to ensure that 'children of different ages, gender and social backgrounds . . . participate in decisions and actions to prevent, respond and adapt to environmental harm'.[42]

Intersectionality in the human rights council

The Human Rights Council has consistently referred in its resolutions on climate change to the need to take steps to respect, protect and fulfill the human rights of 'marginalized groups' and people in 'vulnerable situations'.[43] High-level discussion panels on climate change within the Council have also examined intersectionalities in connection with the right to health and children's rights.[44]

A number of the Special Procedures of the Human Rights Council have engaged in intersectional forms of analysis. In relation to climate change and intersectionalities, the Special Rapporteur on Human Rights and the Environment has compiled a report that summarizes the work of various human rights mechanisms on the topic.[45] Of particular note in this regard are the reports of the former and current Special Rapporteurs on the Right to Food – both of which have drawn attention to the specific impacts of climate change on the rights to food for rural women, small-holder farmers, people living in poverty and indigenous communities.[46]

The Special Rapporteur on Water and Sanitation has also highlighted the fact that intersectional inequalities are exacerbated by climate change, particularly for people living in poverty.[47]

Recommendations related to the human rights impact of climate change and the obligations of governments to take effective measures, in particular to protect 'marginalized and vulnerable groups' in connection with the accessibility of the rights to food, water and sanitation, have been made within the framework of the Universal Periodic Review process.[48]

The Human Rights Council and the treaty bodies have seemingly embraced the concept of intersectional inequality as part of their discourse in connection with climate change. There is little indication, however, that the human rights mechanisms have begun to engage in the substantially more difficult process of moving beyond the rhetorical acknowledgement of intersectionality as an observational aid to render visible the experiences of particular groups in order to use this knowledge to challenge and reform existing power structures.

Intersectionality, climate change and international human rights mechanisms: future directions

There is ongoing debate as to the utility of intersectionality as a practical tool for activists, policy makers, legislators and the judiciary. Objections to the use of intersectional methodologies to inform policies and law-making have mainly centered on the concern that splitting identity-based groups into smaller sub-divisions (e.g. ethnic minority women or older persons with disabilities) will dilute the political power of group-based equality demands.[49] There is also a degree of skepticism as to the ability of institutions and policy makers to successfully capture the complexities of an intersectional analysis within policies and laws. Others have argued that the simple addition of further sub-groups to existing anti-discrimination frameworks is not sufficient to transform the structures of power and domination that create and reinforce inequalities.[50]

Despite this reticence concerning the application of intersectional models in certain circles, there are signs that international human rights mechanisms are beginning to consider intersectionalities within a number of contexts.[51] The development of intersectional approaches to discrimination by the various components of the international human rights monitoring system has the potential to change the way in which these mechanisms observe, measure, prevent and remedy different forms of inequality.[52]

The international human rights system is increasingly recognizing the need to engage with the environmental and climate change communities. It is now widely accepted that the impacts of climate change are not equally shared and that particular segments of the population – generally those people already facing violations of basic human rights – will be most affected. The application of an intersectional approach to human rights and climate change has the potential to promote substantive equality through responsive and inclusive planning, policies, legislation and remedies. Intersectional methodologies highlight positive obligations for States and other duty bearers to identify marginalized and particularly affected population groups and to ensure their participation in the conceptualization, implementation and monitoring of climate policies and legislation.

Beyond the imperative of inclusivity – something that intersectional and rights-based approaches share – an intersectional lens may be used to observe and question the normative assumptions that underlie social, economic and political institutions and forms of behavior. When deployed in this way to unsettle dominant narratives about socio-economic development and resource use, an intersectional analysis could be used to inspire radical shifts in thinking about the relationship between human rights and climate change.

Notes

1 HRC, Res. 32/33, 18 July 2016; HRC, Res. 29/15, 22 July 2015; HRC, Res. 26/27, 15 July 2014; HRC, Res. 18/22, 17 October 2011; HRC, Res. 10/4, 25 March 2009; HRC, Res. 7/23, 28 March 2008; Joint Statement by the UN Special Procedures on the occasion of World Environment Day, 5 June 2015; Committee on the Elimination of Discrimination against Women (CEDAW), General Recommendation no. 34 on the rights of rural women (2016); Convention on the rights of the Child (CRC), Day of General Discussion on children's rights and the environment, 23 September 2016; HRC, Report of the Independent Expert on human rights and the environment A/HRC/25/53, March 2014; UN Office of the High Commissioner for Human Rights, webpage on human rights and climate change.

2 K. Crenshaw, 'Mapping the Margins: Intersectionality, Identity Politics, and Violence Against Women of Color', 43 *Stanford Law Review* (1991) 1241; R. Kapur, 'The Tragedy of Victimization Rhetoric: Resurrecting the "Native" Subject in International Post-Colonial Feminist Legal Politics', 15 *Harvard Human Rights Journal* (2002) 1; N. Yuval-Davis, 'Intersectionality and Feminist Politics', 13(3) *European Journal of Women's Studies* (2006) 193.

3 S. Cho, K. W. Crenshaw and L. McCall, 'Toward a Field of Intersectionality Studies: Theory, Applications, and Praxis', 38(4) *Signs* (2013) 785.

4 V. M. May, *Pursuing Intersectionality: Unsettling Dominant Imaginaries* (2015) 21; B. Smith, 'Intersectional Discrimination and Substantive Equality: A Comparative and Theoretical Perspective', 16 *The Equal Rights Review* (2016) 73.

5 A-M. Hancock, 'When Multiplication Doesn't Equal Quick Addition: Examining Intersectionality as a Research Paradigm', 5(1) *Perspectives on Politics* (2007) 63; K. Davis, 'Intersectionality as Buzzword: A Sociology of Science Perspective on What Makes a Feminist Theory Successful', 9(1) *Feminist Theory* (2008) 67; O. Hankivsky, *Intersectionality* 101 (2014).

6 I. Truscan and J. Bourke-Martignoni, 'International Human Rights Law and Intersectional Discrimination', 16 *The Equal Rights Review* (2016) 103; J. E. Bond, 'International Intersectionality: A Theoretical and Pragmatic Exploration of Women's Human Rights Violations', 52 *Emory Law Journal* (2003) 71.

7 K. Vinyeta, K. P. Whyte and K. Lynn, *Climate Change Through an Intersectional Lens: Gendered Vulnerability and Resilience in Indigenous Communities in The United States* (2015); A. Kaijser and A. Kronsell, 'Climate Change through the Lens of Intersectionality', 23(3) *Environmental Politics* (2014) 417; E. Cudworth and S. Hobden, 'Beyond Environmental Security: Complex Systems, Multiple Inequalities and Environmental Risks', 20 *Environmental Politics* (2011) 42.

8 V. M. May, *supra* note 4, at xi.

9 G. Quinn, 'Reflections on the Value of Intersectionality to the Development of Non-Discrimination Law', 16 *The Equal Rights Review* (2016) 63.

10 Cho, Crenshaw and McCall, *supra* note 3.

11 C. Merchant, The Death of Nature. Women, Ecology and the Scientific Revolution (1980); Cudworth and Hobden, *supra* note 7.

12 A. Salleh, *Ecofeminism as Politics. Nature, Marx and the Postmodern* (1997); R. Twine, 'Intersectional Disgust? Animals and (Eco)feminism', 20(3) *Feminism and Psychology* (2010) 397.

13 Kaijser and Kronsell, *supra* note 7, at 429.

14 See, e.g., K. Vinyeta, K. P. Whyte and K. Lynn, *supra* note 7; E. Walsh, 'Why We Need Intersectionality to Understand Climate Change', *Intercontinental Cry* (8 June 2016); A. Bohr, 'Climate Change's Intersectionality', *Whitman Wire* (11 February 2016).

15 Kaijser and Kronsell, *supra* note 7.

16 N. Lykke, *Feminist Studies: A Guide to Intersectional Theory, Methodology and Writing* (2010) 22.

17 Art 1(1), UN General Assembly, International Convention on the Elimination of All Forms of Racial Discrimination, 21 December 1965, United Nations, Treaty Series, vol. 660, 195.

18 Art 2(1), UN General Assembly, International Covenant on Civil and Political Rights (ICCPR), 16 December 1966, United Nations, Treaty Series, vol. 999, 171; Art 2(2), UN General Assembly, International Covenant on Economic, Social and Cultural Rights (ICESCR), 16 December 1966, United Nations, Treaty Series, vol. 993, 3.

19 Committee against Torture (CAT), General Comment No. 2, Implementation of Article 2 by States Parties, UN Doc. CAT/C/GC/2, 24 January 2008, at paras 21–22; CAT, General Comment No. 3, Implementation of Article 14 by States Parties, UN Doc. CAT/C/GC/3, 13 December 2012, at paras 8, 32; Committee on the Elimination of Discrimination against Women (CEDAW Committee), General Recommendation No. 24, Women and Health, 1999, para 24; Committee on Economic, Social

and Cultural Rights (CESCR), General Comment No. 6, The economic, social and cultural rights of older persons, 7 October 1996, at para 12; CESCR, General Comment No. 20, Non-discrimination in economic, social and cultural rights, UN Doc. E/C.12/GC/20, 2 July 2009, at para 29.

20 Human Rights Committee (HRC), General Comment No. 2, Reporting Guidelines, 28 July 1981, at para 13; CEDAW Committee, General Recommendation No. 24, Women and Health, 1999, at para 25.

21 Committee on the Elimination of Racial Discrimination (CERD Committee), *General Recommendation No. 31, the Prevention of Racial Discrimination in the Administration and Functioning of the Criminal Justice System*, UN Doc. CRPD/C/GC/31 (20 August 2004).

22 CESCR, General Comment No. 20, *Non-Discrimination in Economic, Social and Cultural Rights*, UN Doc. E/C.12/GC/20 (2 July 2009), at para 34.

23 *Ibid.*, at para 33.

24 *Ibid.*, at para 27.

25 CAT, General Comment No. 3, *Implementation of Article 14 by States Parties*, UN Doc. CAT/C/GC/3 (13 December 2012), at paras 8, 32.

26 CERD Committee, *Concluding Observations: The Dominican Republic*, UN Doc. CERD/C/DOM/CO/13–14 (19 April 2013), at para 16.

27 CEDAW Committee, *Concluding Observations: Peru*, UN Doc. CEDAW/C/PER/CO/6 (2 February 2007), at para 36.

28 CERD Committee, General Recommendation No. 25, *Gender Related Dimensions of Racial Discrimination* (20 March 2000).

29 HRC, General Comment No. 18, *Non-Discrimination* (10 November 1989). While the Committee does not expressly refer to intersectional or multiple forms of discrimination, it insists upon substantive (de facto) equality as the standard that it uses to assess state compliance with obligations under Articles 2 and 26 of the ICCPR. The Committee therefore argues that differential treatment is justified in order to redress inequalities and in paragraph 8, it cites the example of the provisions in Article 6(5) of the ICCPR that prohibit the death sentence being carried out on pregnant women.

30 CESCR, General Comment No. 20, *Non-Discrimination in Economic, Social and Cultural Rights*, UN Doc. E/C.12/GC/20 (2 July 2009), para 17, '[s]ome individuals or groups of individuals face discrimination on more than one of the prohibited grounds, for example women belonging to an ethnic or religious minority. Such cumulative discrimination has a unique and specific impact on individuals and merits particular consideration and remedying.' The CESCR does not go further to suggest which particular measures States must take to consider or remedy 'cumulative' or multiple discrimination.

31 Bourke Martignoni and Truscan, *supra* note 6.

32 *Ibid.*

33 *Ibid.*

34 CEDAW, General Recommendation no. 34 (2016), UN Doc. CEDAW/C/GC/34, at para. 14.

35 CEDAW, Draft General Recommendation no. 37 (2017) on disaster risk reduction in a changing climate.

36 CEDAW, *supra* note 34, para. 10.

37 *Ibid.*

38 Quinn, *supra* note 9.

39 Committee on the Rights of the Child (CRC Committee), General Comment No. 11, *Indigenous Children and Their Rights Under the Convention*, UN Doc. CRC/C/GC/11 (12 February 2009); CRC Committee, *General Comment No. 6, Treatment of Unaccompanied and Separated Children Outside Their Country of Origin*, UN Doc. CRC/C/GC/2005/6, 1 September 2005; CRC Committee, *General Comment No. 20 on the Implementation of the Rights of the Child During Adolescence*, UN Doc. CRC/C/GC/20, 6 December 2016.

40 CRC Committee, *Concluding Observations on Grenada*, CRC/C/GRD/CO/2, 22 June 2010, at paras 51–52; CRC Committee, *Concluding Observations on Mauritius*, CRC/C/MUS/CO/3–5, 27 February 2015, at paras 57–58; CRC Committee, *Concluding Observations on Kenya*, CRC/C/KEN/CO/3–5, 21 March 2016, at para. 56.

41 CRC Committee, *Concluding Observations on Fiji*, CRC/C/FJI/CO/2–4 (CRC, 2014), at paras 55–56.

42 CRC Committee, Day of General Discussion 'Children's Rights and the Environment' (23 September 2016).

43 HRC, *supra* note 1.

44 HRC, *Panel Discussion on Climate Change and the Rights of the Child*, 1 March 2017; HRC, *Panel Discussion on Climate Change and the Right to Health*, 3 March 2016.

45 HRC, *Report of the Independent Expert on Human Rights and the Environment*, A/HRC/25/53, 30 December 2013.
46 HRC, *Report of the Special Rapporteur on the Right to Food*, A/HRC/22/50, 24 December 2012; UNGA, *Interim Report of the Special Rapporteur on the Right to Food*, A/70/287, 5 August 2015; HRC, *Report of the Special Rapporteur on the Right to Food*, A/HRC/31/51, 29 December 2015.
47 HRC, *Report of the Special Rapporteur on the Human Right to Safe Drinking Water and Sanitation*, A/HRC/18/33, 2 August 2011.
48 See, e.g., Recommendations by the Philippines to Kiribati, A/HRC/29/5/Add.1 (UPR, 2015) and by Fiji to the Marshall Islands, A/HRC/30/13/Add.1 (UPR, 2015).
49 D. Petrova, 'Intersectionality', 16 *The Equal Rights Review* (2016) 5.
50 *Ibid.*
51 Bourke Martignoni and Truscan, *supra* note 6, at 103.
52 Petrova, *supra* note 49.

34

Climate change, human rights, and divestment

Basil Ugochukwu

Introduction

The questions that this chapter raises are relatively intuitive. They deserve answers nonetheless because doing so helps to accomplish two goals. First is to relate carbon intensity to climate change and therefore to its human rights impacts. Second is to show how divesting from carbon-intensive businesses could reduce carbon emissions and therefore improve human rights conditions.

The chapter is structured as follows: after this introduction, the second section examines the relationship between carbon emissions and climate change. For the purposes of the chapter, carbon is defined to include all the known greenhouse gases that drive climate change. The discussion does not, however, include black carbon. Thereafter, the third section looks at some of the strategies that have been utilized both internationally and in the domestic context to deal with the challenge of carbon emissions. The fourth section examines the extent that divestment from carbon-intensive businesses could be considered a human rights strategy. The last section will contain some reflections by way of a conclusion.

Carbon emissions and climate change

There are at least three different ways that corporations could be exposed to the climate discourse for good or for ill. In the first place, there are businesses that exist based solely on their involvement in activities that are the known drivers of climate change, that is, those whose business models cannot survive a day without generating greenhouse gases. This includes all the Carbon Majors dealing in coal, oil and gas who by extracting fossil fuels and selling them to be burnt are held responsible for roughly two-thirds of the entire global carbon pollution, which is also the biggest cause of climate change.[1] Research done in 2014 actually traced the lion's share of cumulative global CO_2 and methane emissions since the industrial revolution began to the largest multinational and state-owned producers of crude oil, natural gas, coal and cement.[2] Combating emissions is therefore strategic and significant for combating global warming and protecting the earth's integrity.

Scientists and corporate managers have warned that over the coming years, the assets of fossil fuel companies would lose substantial fractions of their value because most of the coal, oil and gas reserves that they own will have to be left in the ground (or "stranded")[3] if the world were to meet the goal of keeping temperatures under 2°C compared to pre-industrial levels.[4] This lost value would obviously form a significant fraction of the $2.5 trillion that climate change could wipe away from total global assets, with dire consequences for the global economy.[5] The results from the likelihood of these events happening are already trickling in and, as expected, have been startling. For example, in 2016 alone, major Australian corporations reported massive write-downs on fossil fuel–related assets and loans to the tune of $13.8 billion, covering such projects as offshore petroleum, natural gas and coal.[6]

Nonetheless, the world faces a huge dilemma in dealing with this challenge. This is demonstrated in two recent reports indicating two conflicting paths that are being pursued concurrently, but which only tend to result in diverging outcomes. An article by four Oxford University scholars published in 2016 warned that to have a chance of avoiding 2°C in global temperature, no new fossil fuel plants could be built after 2017.[7] Yet only a few short months before the report, a different research done on the same issue showed that 1,500 coal-fired power plants were either under construction or being planned around the world.[8] This latter report also stated on the flipside that air pollution from coal currently causes an estimated 800,000 premature deaths annually, and planned coal plants would increase such deaths by 130,000 people per year.[9]

Secondly, the very existence and operation of certain businesses may not necessarily depend on their involvement in extensive greenhouse gases emitting activities; yet those businesses could be concerned about the impact of such carbon-intensive activities on the climate and therefore feel an ethical obligation to ameliorate that impact any which way they could. Corporations in this category (including banks, shipping and air transport companies, and all other businesses whose energy use contributes to GHG emissions) could do this by reducing energy and water use and supporting partners whose business practices make the most sense, environmentally speaking.[10] They could execute this goal through exercising an abundance of caution in the kinds of goods they purchase and how they use them or by generating their own energy from renewable sources and managing their waste in the most environmentally responsible manner possible. Other strategies could include changing corporate behaviour in other environmentally conscious ways, developing new products for new markets, participating in GHG emissions trading and offsetting schemes and carrying out other activities such as public education on climate change.[11]

Thirdly, there are corporations unlike the two categories described above whose exposure to climate change owes to their management of investment funds whose owners they owe a fiduciary duty to invest in a risk-free and ethically responsible manner.[12] Such businesses are not built strictly on carbon-intensive activities for their survival but could be exposed to them nevertheless through the securities investment portfolios that they manage. An example of such entity is the California State Teachers' Retirement System (CALSTRS) which, while it had investments in carbon emissions-intensive corporations, in July 2016 committed up to $2.5 billion to low-carbon strategies in US, non-US developed and emerging equity markets.[13] While I will return to an analysis of the motivation for this decision and its underlying calculation in latter parts of this chapter, it is sufficient for the moment to state that CALSTRS said it was motivated to take this decision in part because of its support for environmental, social and governance (ESG) principles.[14]

In this last category could be included the insurance industry, which as the governor of the Bank of England, Mark Carney, notes has been dealing with the reality of climate change when other sectors have only been debating the theory.[15] Similarly involved are entities like pension funds or university endowments,[16] for example, which increasingly must account for ESG risks in their investment decision-making. Those calculations could be financially material both positively (when high ESG performance correlates to improved long-term financial performance)[17] and negatively (when ignoring them can result in lower returns or losses which could be direct, as in "stranded assets," or indirect, as in the exposure of investee companies to regulatory investigations, litigation, regulatory changes and reputation/brand damage).[18] While the emissions discussed in this chapter could result from activities in any other of the four different business models identified above, the chapter is more specifically focused on those in the first category, that is, the more emission-intensive ones.

The chapter will now proceed to exemplify how carbon emissions could be framed as a human rights question. In doing this, an example is drawn from a complaint that was filed in 2015 by Greenpeace Southeast Asia and Philippines Rural Construction Movement, together with 20 individuals before the Philippine Human Rights Commission. In it, the petitioners claimed that climate change interferes with the enjoyment of fundamental rights and demanded the accountability of those contributing to the problem.[19] Respondents to the petition included 50 companies identified as investor-owned Carbon Majors. There were fossil fuel companies and cement as well as energy producers. Among them were Chevron, ExxonMobil (USA), BP (UK), Royal Dutch Shell (the Netherlands), Conoco Philips (USA), Suncor and Encana (Canada) and several others alleged to be "responsible for the majority of global CO_2 and methane emissions in the earth's atmosphere."[20] Based on their stated contribution to global warming and climate change, petitioners requested that the Carbon Majors be held accountable for violating or threatening to violate the human rights of Filipinos in various ways. Among the rights petitioners claimed had been infringed are the rights to life, the highest attainable standard of physical and mental health, food, water, sanitation, adequate housing and self-determination.

Long before more recent cases and at a time when climate change had not engaged global attention as it does today, similar connection had indeed been drawn in other contexts between gas flaring (a form of carbon emission) and environmental degradation. One of the earliest cases in this regard was launched before the African Commission on Human and Peoples' Rights by a non-governmental organization on behalf of the minority Ogoni peoples against the government of Nigeria. While neither the petition nor the ruling of the Commission made any explicit reference to carbon emissions in relation to climate change, some authors have discussed the case within the rubric of climate change litigation.[21] The communication alleged that the military government ruling Nigeria at the time was directly involved in oil production in Ogoniland through its oil company, the Nigerian National Petroleum Corporation (NNPC), causing environmental degradation, air contamination through gas flaring and health problems to the Ogonis. The NNPC was majority shareholder in a government consortium that also included the Anglo-Dutch giant fossil fuel company Shell.

The communication claimed that the oil consortium exploited oil reserves in Ogoniland without regard for the health or environment of the surrounding communities, discharging toxic wastes into the environment and local waterways and thereby breached applicable international environmental standards. It was further alleged that

> The consortium . . . neglected and/or failed to maintain its facilities [thereby] causing numerous avoidable [oil] spills in the proximity of villages. The resulting contamination of water, soil and *air* . . . [emphasis mine] had serious short and long-term health impacts, including skin infections, gastrointestinal and respiratory ailments, and increased [the] risk of cancers, and neurological and reproductive problems.[22]

Equally important in discussing the carbon emissions/human rights nexus is the opposite argument that is often made especially in the specific context of emissions from fossil fuels. Fossil fuels are still significant economic resources upon which large sections of the global economy depend. While renewable energy sources like hydro, wind, solar and battery are proving to be cleaner and more sustainable alternatives to fossil fuel both now and in the long term, getting to the level where they totally replace the latter, it has been argued, will obviously still require some time.[23] There are economic implications tied to massive withdrawal from fossil fuels. Some of those implications could in fact be of a human rights nature.

A sudden withdrawal from fossil fuel companies could lead to significant job losses unless it is also accompanied by other measures to prevent such a huge socio-economic cost. This is why even in a carbon-constrained world, governments still provide consumer and producer subsidies to fossil fuel corporations – producer subsidies to maintain jobs and economic competitiveness and consumer subsidies to keep energy prices at a level more people can afford.[24] The latter is a concern to developed economies[25] but is also particularly salient in the context of developing economies where such subsidies are used to achieve certain social, economic and environmental objectives.[26] This includes alleviating energy poverty, improving equity and access, redistributing national resource wealth and protecting jobs.[27] In Nigeria, for example, the subsidies on the consumer side of the fossil fuel business has over time become a sensitive issue over which popular protests and litigation have been waged.[28] As would be expected, the court cases are framed mostly in human rights terms.[29]

As demonstrated with the recent decline in the price of crude oil in the international marketplace, loss of jobs and reduced standards of living could result from disturbances and volatility within the fossil fuel economy. In Nigeria, the 2015 sharp drop in the international price for crude has set off an economic recession, with a vast majority of its population suffering reduced standards of living as a result.[30] Similarly, Saudi Arabia, which is the world's largest exporter of crude petroleum, has long argued that "its reliance on hydrocarbon revenues means it would suffer in a carbon-constrained world."[31] The country took this viewpoint to the UNFCCC intersession held May 2016 in Bonn, Germany, where it argued that "in addition to scientific assessments of climate change, Parties would benefit greatly from a socioeconomic framing of the issue at hand in terms of the means of implementation and support available as they devise their policies to combat climate change."[32]

Dealing with the carbon challenge

Considering the extent to which carbon emissions contribute to global warming and climate change, it would be fair to argue that dealing with these challenges at the source would require substantial curbs to current global emissions. In other words, efforts to address climate change would remain unsuccessful if current or higher levels of carbon emissions are maintained. In the 2015 edition of the report that tracks global trends in CO_2 emissions, the conclusions reached included that (1) cumulative emissions of carbon dioxide will largely determine global mean surface warming by the late 21st century and beyond; (2) it would be possible, using a wide array of technological measures and changes in behaviour, to limit the increase in global mean temperature to 2°C above pre-industrial levels; (3) substantial emission reductions over the next few decades can reduce climate risks in the 21st century and beyond; and (4) without additional mitigation efforts to those in place today, and even with adaptation, warming by the end of the 21st century will lead to high and very high risks of severe, widespread and irreversible impacts, on a global scale.[33]

A major tool for mitigating climate change is therefore sustained global action to reduce carbon emissions. The Paris Agreement as earlier stated makes the reduction in greenhouse gas emissions a major plank of all multilateral international action to address climate change.[34] This objective is further amplified in all national climate change strategies regardless of the geographical location or economic circumstances of the countries proposing those policies.[35] Corporate climate change policies as well often make the reduction in carbon footprints traceable to business operations the most significant portion of the objectives of corporations proposing those policies.[36]

However, only limited reliance could be placed on voluntary "facilitative ... non-adversarial and non-punitive"[37] standards whether proposed at multilateral levels or by nation-states or corporations. The reason for this is apparently self-evident. Given that climate change is a global "commons" problem,[38] expressions of corporate voluntary action could be deceptive. There are strong, entrenched and well-funded economic interests that are tied to denying climate change and continuing fossil fuel exploitation that would not be deterred by voluntary commitments. The intervention of such interests tends to muddy the water of public conversation and render the climate change debate an ideologically polarized exercise.[39]

For example, some corporations, especially carbon-intensive ones, often claim to respond to the dangers of fossil fuel use and its impacts (including as drivers of climate change) through voluntary actions and subscription to climate-friendly codes of behaviour. For a variety of reasons, these corporations indicate action at the highest levels of their operations with the aim of reducing their carbon footprints and other GHG emissions. They claim to do this by way of their approaches to doing business as well as by supporting government policies to reduce the impact of fossil fuel use, as for example combating climate change. However, while a good number of companies claim to be actively involved in activities that contribute towards reducing GHG emissions, there tends to be a disconnection between these claims and the actions that the corporations who make them take.[40] Some examples of this disconnection are provided below.

In dealing with the challenge of carbon emissions, four different trends in terms of response strategy have been observed. Depending on where an actor situates itself in the mix for solutions to the challenge, these trends have included the following: (1) public regulation through international and domestic rule making; (2) self-regulation, particularly at the corporate level through voluntary codes and principles; (3) citizen-led litigation to force government action in situations where the claimants suspect the government to be hesitant in imposing carbon-limiting regulation; and (4) divestment (on ethical/moral grounds) from corporations whose businesses produce high carbon emissions and therefore imperil the climate/environment. This chapter's concentration is on the last strategy, that is, divestment as a tool for curbing carbon emissions and therefore advancing human rights. However, before carrying out the analysis that it deserves, especially in relation to how this strategy could further human rights protection, it is essential to describe how the other three strategies contribute to the search for solutions to the carbon challenge.

Public policy and regulation are usually at the core of strategies deployed to deal with carbon emissions. They have both an international as well as domestic dimension where the objective primarily is to compel or facilitate reductions in carbon emissions through incentives, peer compliance through pledge and review (as in the Paris Agreement) or the imposition of legal sanctions. At the international level, the best example of such a regulation is the Paris Agreement adopted in 2015, whose major goal is to hold the global average temperature to a certain level through reduction in greenhouse gas emissions. At the domestic level, greenhouse gas emissions regulations could take the form of carbon taxation,[41] a cap and trade system,[42] financial penalty for, as an example, gas flaring,[43] the requirement of carbon emissions testing for vehicles,[44] environmental impact and air pollution assessment measures for a variety of projects, etc.

Corporations have also been responding to the climate change challenge through two other means: voluntary actions by individual corporations and subscription to climate-friendly codes of behaviour within and across industry sectors. Under the first means, corporations are acting at the highest levels of their governance with the aim of scaling back on their carbon footprints. They do this by way of their approaches to doing business as well as by supporting government policies to combat climate change.[45] However, while a good number of companies claim to be actively involved in activities that contribute towards addressing the climate challenge, this does not translate to emissions reductions *per se*.[46] Regarding industry-wide initiatives, examples had come from the aluminium sector under the auspices of the International Aluminium Institute made up of a group of major aluminium corporations, the Cement Sustainability Initiative within the World Business Council for Sustainable Development and the International Iron and Steel Institute representing over 200 steel-producing companies from almost all of the regions of the world.[47]

There is, however, a little bit of history regarding how businesses came to embrace voluntary carbon emissions policies. They were not always receptive to the very idea of reducing carbon emissions. Previously, in response to growing scientific consensus on the relationship between GHG emissions and global warming, several energy-intensive corporations ganged up to challenge this science under the Global Climate Coalition (GCC).[48] This group lobbied to convince policy makers that mandatory controls on emissions were unjustified.[49] By the late 1990s, however, some members of the GCC ate humble pie and began shifting their positions. These corporations more recently took public positions consistent with the belief that it is possible to achieve a transition to a low-carbon economy.

Just before the COP21 meeting in Paris, Bob Dudley, who was BP's chief executive officer, expressed the company's belief that "the best mechanism to drive a shift to a lower carbon future is to put a price on carbon. That can be done via taxes or by cap-and-trade systems. Either can be effective if well-constructed."[50] He had earlier joined nine other CEOs of the world's largest oil and gas companies to sign a declaration ahead of the Paris meeting, in which they described the global ambition of a 2°C future as "a challenge for the whole of society."[51]

On the other hand, these promises flew in the face of actual corporate behaviour. For example, recent statistics from the UK-based non-profit Influence Map ranked BP highest on the list of 100 global companies trying very hard to subvert climate policies.[52] As well, a report published in 2016 involving BP and seven other fossil fuel–based corporations further illustrates the difference between corporate claims of climate action and the result when those claims are subjected even to the most minimal scrutiny.[53] The study focused on five leading investor-owned oil and gas companies that were ranked in terms of their cumulative GHG emissions, including Chevron, ExxonMobil, BP, Royal Dutch Shell and ConocoPhillips. The rest were three leading investor-owned US-based coal companies that were also ranked in terms of their cumulative GHG emissions, namely Peabody Energy, CONSOL Energy and Arch Coal. These corporations were ranked on a five-band spectrum, with "advanced" having the highest rating of companies demonstrating best practices and "egregious" ranking lowest and rating companies behaving irresponsibly. The other rankings were "good" for companies meeting emerging societal expectations, "fair" for companies whose performances were neither positive nor negative, and "poor" for companies falling short of emerging societal expectations.[54]

The result was that none of the companies were rated "advanced" or "good." BP and ConocoPhillips were rated "fair," while Shell got stuck between "fair" and "poor." Chevron, CONSOL Energy and ExxonMobil were rated "poor," even as Arch Coal and Peabody Energy were rated between "poor" and "egregious."[55] Unsurprisingly, the last two corporations were undergoing United States Chapter 11 bankruptcy protection as the report was being prepared.[56]

In addition to public regulation and voluntary corporate actions to reduce carbon emissions, citizens and groups in different regions of the world are taking active interest in ratcheting up social action processes in this domain. They are doing so through litigation intended primarily to pressure governments and corporations to take concrete steps to drive the carbon emissions reduction agenda forward. Earlier in the chapter, it was shown how some of those suits presented emissions reduction as central to protecting human rights and public safety. Some examples have already been provided on how this is happening in various jurisdictions, including the Greenpeace complaint before the Human Rights Commission of Philippines and the case that SERAC filed on behalf of the Ogonis against the Nigerian government before the African Human Rights Commission.

A similar case was also filed in the Netherlands to compel more ambitious cuts to GHG emissions which otherwise exposed Dutch citizens and the international community to dangerous climate change that could result in irreversible damage to human health and the environment.[57] Pakistan has received such a case from a local farmer who claimed that the country's hesitance in tackling emissions and climate change posed a serious threat to water, food and energy security in Pakistan and therefore offends the fundamental right to life enshrined in Section 9 of the constitution of Pakistan.[58] More than these, in October 2016 major litigation was launched in Norway in which plaintiffs claimed that licensing new oil exploration in the Barents Sea after the country had ratified the Paris Climate Change Agreement amounted to significant contribution to major greenhouse gas emissions that would exacerbate global warming.[59] In a press statement accompanying the launching of the suit, Nature & Youth, Norway and Greenpeace Nordic who initiated it asserted, "It is impossible to reconcile the petroleum production permitted by the Licensing Decision with the reduction in [GHG] emissions that Norway must contribute in order to avoid devastating and irreversible climate change."[60]

Carbon divestment as human rights strategy

Another strategy that citizens and concerned institutions could use to generate pressure towards reducing GHG emissions is by divesting from carbon intensive businesses. This is defined broadly as the process of selling assets for financial or social goals carried out for financial reasons, as for example when a certain industry underperforms, or for ethical reasons, as when a corporation carries on a business that is misaligned to the investor's morals.[61] While divestment could be driven by cold business/profit calculations (as for example non-competitive fossil fuel assets that could lose future value and thus be "stranded"), it could also be for ethical/moral/reputational reasons like protecting the environment for future generations.[62]

Most of the literature on divestment from carbon-intensive businesses tend to analyze the issue as simply an investment decision or one that is based solely on corporate performance and profitability alone. Put in a different way, the argument is that if returns on an investment are going to be negatively impacted by fossil fuel corporations becoming less profitable because of their assets depreciating in value or becoming "stranded," investors might be tempted to move their funds to more competitive alternatives, such as renewable energy businesses. The motivation would, however, not be ethics or morals, but rather the potential loss or reduction in earnings. In fact, it seems that most divestments from fossil fuel corporations, for example, are driven more by profit and risk calculations than by social concerns such as human rights.[63]

If investors pulled out of carbon-intensive businesses, as is currently the case with the coal industry, this could force them out of operation and therefore drive down GHG emissions. If the argument that reduced emissions would have positive effects on climate conditions, this would also mean an improvement in human rights that are impacted by carbon emissions and therefore climate change.[64] However, to evaluate if divestments could be a significant human rights protection strategy, it might be necessary to study the reasons that individual and institutional investors might want to take their funds out of carbon-intensive businesses. We must, however, first dispose of the point that even where divestment from those businesses is informed by profit considerations alone, this could still have positive implications for reduced emissions and enhanced human rights conditions.

Rubin has argued, for example, that divestment campaigns

> were originally motivated by concerns that anthropogenically induced climate change threatens human civilization with potentially catastrophic consequences arising from a wide range of impacts, including the increased incidence of extreme weather events, threats to world food production, widespread island and coastal flooding from rising sea levels, and the migration of diseases.[65]

If this argument were accepted, it would mean that human rights concerns were central to divestment decisions. He, however, believes that this social activism–based concern is now giving way to an understanding that "climate change will severely limit future fossil fuel consumption [which] not only renders much of today's fossil fuel reserves unburnable, and hence of no value, but suggests that even current production levels are unlikely to remain economically viable."[66] This has implications for future loss or reduction in asset earnings.

To give an example, CALSTRS, which was mentioned earlier in this chapter, justified its commitment of $2.5 billion to a low-carbon index on grounds of lessons learned from "the decline of coal companies and in the value of their reserves"[67] as well as because of taxes that could be imposed on greenhouse gas emitters and fossil fuel reserve holders. It seems, therefore, that the group was persuaded more by earnings calculations in reaching its decision than by social or human rights considerations. In a 2014 article, Fossil Free offered seven reasons that holders of coal, oil and gas stocks should sell them off. None of those reasons addressed human rights or other environmental implications of holding on to such stocks.[68] Yet, the concern of this chapter is with divestments from carbon-intensive businesses that could be explained as resulting from "social" considerations of which human rights could be an element.

The questions therefore are, On what basis are decisions to divest from carbon-intensive businesses made? Are human rights and other social concerns factored into those decisions at all? My reading of the literature indicates that human rights does not appear to be a separate consideration for reaching divestment decisions, whether by individual shareholders or institutional investors. Rather, it is subsumed within corporate social responsibility (CSR) questions extending from environmental management to employee relations, etc.[69] Human rights and sustainable development are often included as components of CSR.[70] In fact, most corporate sustainability reporting guidelines have human rights as among the major issues.[71]

An article published in 2009 showed that when a consulting firm asked investors about the impact of environmental, social and governance factors in mainstream investment considerations, the result indicated that sustainability was rated 39%, employee relations 33%, human rights 26%, water 25%, environmental management 18% and climate change 7%.[72] On the contrary, corporate governance was viewed as very important by 64% of respondents.[73] For emphasis, this result is from a broad CSR analysis that covers all business types and not just those whose activities produce carbon emissions.

On its part, the citizen group Divest Waterloo, which initiates and supports fossil fuel divestment campaigns in Ontario, Canada's Kitchener-Waterloo region, emphasises both financial and ethical dimensions of divestment in which there is a win on both sides. It advises institutions, pension plans, individuals and businesses that divestment from fossil fuels could prevent "droughts, fires, flooding, global food shortages and the displacement of environmental refugees," all of which happen because of carbon emissions and climate change.[74] It is clear that this group's concerns mirror core human rights issues that could result from fossil fuel burning and carbon emissions, where investing in businesses that carry out these activities is tantamount to actively encouraging them and the impacts that they bring. In addition, the group says divesting makes as much financial sense as it does ethical sense.

In what follows, I will provide a few examples of recent investments decisions made by institutional investors to show what role, if any, social and human rights considerations played in making them reach their decisions. The Institutional Investors Group on Climate Change (IIGCC), which controls assets worth €22 trillion, in its report for 2016 called on the auto industry to shift to low-carbon products.[75] The group stated that the report is intended to enable investors to engage with boards of automotive companies regarding their efforts "to address climate change risk and place sustainability at the heart of the industry's future."[76] The report therefore concluded, "To achieve sustainable long-term returns on [our] investments for clients and beneficiaries, investors must both ensure that each investment is prepared for the challenges of climate change and ensure that robust policy action is taken to address the energy transition."[77] To the extent that IIGCC's goals encompass sustainability and the challenge of climate change, I would argue that human rights are clearly consequential as well.[78]

In September 2016, it was reported that Norway's $900 million wealth fund would discontinue investment in Duke Energy, which is the biggest United States power firm by generation capacity, because the company breached environmental laws at its coal-fired plants.[79] Norway's Central Bank, while announcing the decision to discontinue investment, said it was based on an assessment of the risk of severe environmental damage, and noted: "For many years, these companies have, among other things, repeatedly discharged environmentally harmful substances from a large number of ash basins at coal-fired power plants in North Carolina."[80] The nature of this allegation leaves no doubt about the negative human rights potential of the infraction, even if this was not directly stated.

Earlier in the year, the same fund dropped from its portfolio at least 52 companies linked to coal.[81] Among the companies barred were American Electric Power Inc., China Shenhua Energy Company Ltd, Whitehaven Coal Ltd, Tata Power Co. and Peabody Energy Corporation. The decision of the Norwegian government in this case was based on new investment criteria that excluded companies that derive at least 30% of their revenue or base their activities up to this percentage on coal.[82] It appeared, however, that the decision was made from an investment risk-based assessment rather than on social grounds, even though coal as a fossil fuel produces well-known negative human rights impacts.

Furthermore, Japan's Pension Fund announced in August 2016 that it would start investing in companies picked not only for their earnings potentials but also for their environmental, social and governance merits.[83] While the group used the phrase "social factors" to describe its major rationale for deciding to tailor its investments, it did not define what those factors might be. However, in October of the same year, the Fund replaced its "Stewardship Enhancement Group" with a new "Stewardship & ESG Division" committed to enhancing the Fund's fiduciaries by fostering "sustainable" growth.[84] While there is no specific mention of human rights as a consideration of high value, I will argue that they are accommodated in the "social" category as in the ESG factors which could also be an element in both the "environmental" (for example, avoidance of air pollution) and "governance" (for example, non-discrimination and access to information) categories.

Finally, among these examples, seven Catholic institutions on four continents announced in October 2016 a decision to divest from fossil fuel companies as their way of fighting global warming. The language of their announcement provided clear proof that the decision was informed by serious human rights concerns. Father Peter Bisson, the provincial of the Jesuits in English Canada, one of the groups that issued the statement, said, "Climate change is already affecting poor and marginalised communities globally, through drought, rising sea levels, famine and extreme weather. We are called to take a stand."[85] Bishop João Mamede Filho representing Catholic institutions in Brazil offered, "We cannot accommodate and continue allowing economic interests that seek exorbitant profits before the well being of people, to destroy biodiversity and ecosystems, nor continue dictating our energy model based on fossil fuels."[86] The divestment decision affected the coal industry while some of the groups redirected their investments to renewable energy corporations.[87]

The reasons offered by the various groups above for divestment from fossil fuel–based and other carbon-intensive businesses tend to signal a strong social concern. These examples indicate that a decision to divest could be based on concern for sustainability, the discharge of harmful materials into the environment, the destruction of biodiversity or the interests of marginalized groups in society. Each of these issues exemplifies an objective that could be pursued to catalyze one or another form of positive social change. When these concerns are subjected to further analysis, it should be clear that they embed fragments of human rights goals, which would position divestment in the climate context as a significant vehicle for justice and human rights protection.[88]

Conclusion

This chapter examined the possible human rights implications and effectiveness of divestment from carbon-intensive businesses. It began by highlighting the risk that climate change poses to humanity and the extent that carbon-intensive business activities are driving the phenomenon. The chapter followed up this initial introduction with a closer analysis of the relationship between carbon emissions and climate change, as well as an identification of the various ways that corporations could be exposed to the climate discourse. Continuing the analysis, the chapter then used the relationship between climate change and human rights to draw out a similar relationship between carbon emissions and human rights. It was shown that while carbon emissions have various human rights implications, this is hardly described directly as a human rights problem in the literature. Instead, the connection is indirectly made through the general discussion on climate change and its traceability to carbon emissions.

The chapter also indicated that though most literature reviewed during writing does not refer to carbon emissions as a human rights problem *per se*, this is not the case in domestic and transnational climate litigation, where the claims are generally built upon the impact of carbon pollution on a range of human rights guarantees. The inevitably led to a discussion on domestic and international strategies for responding to the climate challenge. Some of the prominent strategies identified in this regard are public policy and regulation at the domestic and international levels, voluntary corporate action, litigation and divestment. As divestment is the major concern of the chapter, it had a more expanded treatment. The main objective of the analysis was to show whether divestment from carbon intensive corporations could pass as a human rights protection strategy.

The first point that must be made is that when the goal in mind is reducing carbon emissions and thus combating climate change, all options should be on the table – regulation, policy options, corporate action, litigation and divestment. The big picture is that these strategies make sense not just from the economic and sustainability standpoints but also from the perspective of other social benefits that could be derived as well. One of those social benefits could be human rights protection, which flows from reducing carbon emissions with its net positive benefits to a less risky global climate.

Speaking of divestment specifically, if there is agreement on the fact that carbon intensity drives climate change which in turn imperils human rights guarantees, this understanding must be integrated in divestment decision-making. In other words, portfolio managers or investment funds that intend to divest from carbon-intensive businesses should state clearly whether they are doing so to secure long-term shareholder earnings interests or for social/moral/ethical reasons. If the latter, they must go further to unpack those social/moral/ethical reasons and if possible indicate whether human rights is one of them. This way, divestment could then be directly linked to improved human rights conditions that should accrue from reduced carbon emissions. This will also clear out some of the fog in the debate whether social considerations such as human rights are material to investment decision-making.

Notes

1 J. A. Richards and K. Bloom, *Making a Killing: Who Pays the Real Cost of Big Oil, Coal and Gas?* (2015), available online at www.boell.de/sites/default/files/making-a-killing.pdf, at 2.

2 R. Heede, 'Tracing Anthropogenic Carbon Dioxide and Methane Emissions to Fossil Fuel and Cement Producers, 1854–2010', 122 *Climatic Change* (2014) 122.

3 See M. Scott, Fossil Fuel Assets in Danger of Being Stranded, *Forbes* (11 December 2013), available online at www.forbes.com/sites/mikescott/2013/12/11/fossil-fuel-assets-in-danger-of-being-stranded/#4cfcff6b16fb reporting that Bloomberg had unveiled a product that allows "investors to calculate how much of their money is tied up in assets that are "stranded" because they are carbon-intensive and therefore likely to be hit by tighter regulations, carbon prices or other constraints"; See also A. Ansar, B. Caldecott and J. Tilbury, *Stranded Assets and the Fossil Fuel Divestment Campaign: What does Divestment Mean for the Valuation of Fossil Fuel Assets?* Smith School of Enterprise and the Environment, University of Oxford (2013), available online at www.fossilfuelsreview.ed.ac.uk/resources/Evidence%20-%20Investment,%20Financial,%20Behavioural/Smith%20School%20-%20Stranded%20Assets.pdf (last visited 4 August 2016); A. Park and C. Ravenel, 'Integrating Sustainability into Capital Markets: Bloomberg LP and ESG's Quantitative Legitimacy', 25 *Journal of Applied Corporate Finance* (2013) 62; M. L. Linnenluecke *et al.*, 'Divesting From Fossil Fuel Companies: Confluence Between Policy and Strategic Viewpoints', 40 *Australian Journal of Management* (2015) 478.

4 See D. Carrington, *Climate Change Will Wipe $2.5tn Off Global Financial Assets: Study* (4 April 2016), available online at www.theguardian.com/environment/2016/apr/04/climate-change-will-blow-a-25tn-hole-in-global-financial-assets-study-warns?CMP=share_btn_tw.

5 *Ibid.*

6 See Fossil Fuel Assets Taking Huge Hits, *Market Forces* (16 August 2016), available online at www.marketforces.org.au/fossil-fuel-assets-taking-huge-hits/.

7 A. Pfeiffer *et al.*, 'The "2°C Stock" for Electricity Generation: Committed Cumulative Carbon Emissions From the Electricity Generation Sector and the Transition to a Green Economy', 179 *Applied Energy* (2016) 1395.

8 C. Shearer *et al.*, *Boom and Burst 2016: Tracking the Global Coal Plant Pipeline* (2016), available online at https://sierraclub.org/sites/www.sierraclub.org/files/uploads-wysiwig/Final%20Boom%20and%20Bust%20report_0.pdf.

9 *Ibid.*, at 3.

10 See for example C. Okereke, 'An Exploration of Motivations, Drivers and Barriers to Carbon Management: The UK FTSE 100', 25 *European Management Journal* (2007) 475; B. Eberlein and D. Matten, 'Business Response to Climate Change Regulation in Canada and Germany: Lessons for MNCs From

Emerging Economies', 86 *Journal of Business Ethics* (2009) 241; H. K. Jeswani, W. Wehrmeyer and Y. Mulugetta, 'How Warm Is Corporate Response to Climate Change? Evidence From Pakistan and the UK', 17 *Business Strategy and the Environment* (2008) 46.

11 J. Galbreath, 'To What Extent Is Business Responding to Climate Change? Evidence From a Global Wine Producer', 104 *Journal of Business Ethics* (2011) 421; See also *BMO Financial Group: Environmental Policy* (2016), available online at www.bmo.com/cr/images/BMOEnvironmentalPolicy_April2016.pdf.

12 See P. Sethi, 'Investing in Socially Responsible Companies is a Must for Public Pension Funds – Because There Is No Better Alternative', 56 *Journal of Business Ethics* (2005) 99; S. Barker *et al.*, 'Climate Change and the Fiduciary Duties of Pension Fund Trustees – Lessons from the Australian Law', 6 *Journal of Sustainable Finance & Investment* (2016) 211; M. Gold and A. Scotchmer, *Climate Change and the Fiduciary Duties of Pension Fund Trustees in Canada, Koskie Minksy LLP* (1 September 2015), available online at www.turnbackthetide.ca/tools-and-resources/whatsnew/2015/KoskieMinskyLLP.pdf.

13 See *CalSTRS Commits $2.5 Billion to Low-Carbon Index* (14 July 2016), available online at www.calstrs.com/news-release/calstrs-commits-25-billion-low-carbon-index.

14 *Ibid.*

15 M. Carney, *Breaking the Tragedy of the Horizon – Climate Change and Financial Stability*, Speech to the Insurance Market Lloyd's of London (29 September 2015), available online at www.bankofengland.co.uk/publications/Pages/speeches/2015/844.aspx, at 3; See also C. Brown and S. Seck, 'Insurance Law Principles in an International Context: Compensating Losses Caused by Climate Change', 50 *Alberta Law Review* (2013) 541; E. Mills, 'Insurance in a Climate of Change', 309 *Science* (2005) 1040; W. Botzen and J. Van Den Bergh, 'Insurance Against Climate Change and Flooding in the Netherlands: Present, Future, and Comparison With Other Countries', 28 *Risk Analysis* (2008) 413; M. Hawker, 'Climate Change and the Global Insurance Industry', 32 *The Geneva Papers* (2007) 22; H. Kunreuther and E. Michel-Kerjan, 'Climate Change, Insurance of Large-Scale Disasters, and the Emerging Liability Challenge', 155 *University of Pennsylvania Law Review* (2007) 1795; D. Estrin and S. Vern Tan, *Thinking Outside the Boat about Climate Change Loss and Damage: Innovative Insurance, Financial and Institutional Mechanisms to Address Climate Harm Beyond the Limits of Adaptation* (2016), available online at www.cigionline.org/sites/default/files/workshop_washington_march2016.pdf.

16 See Ansar, Caldecott and Tilbury, *supra* note 3, at 9.

17 J. Grewal, G. Serafeim and A. Yoon, *Shareholder Activism on Sustainability Issues*, 17-003 Harvard Business School Working Paper (2016), available online at www.hbs.edu/faculty/Pages/item.aspx?num=51379, at 12.

18 See A. Johnston and P. Morrow, *Fiduciary Duties of European Investors: Legal Analysis and Policy Recommendations*, Legal Studies Research Paper Series No. 2016-04, University of Oslo Faculty of Law (2016), available online at http://papers.ssrn.com/sol3/papers.cfm?abstract_id=2783346##.

19 See Philippines prepares to Summon 47 Companies to Account for Climate Change, Common Dreams (27 May 2016), available online at www.commondreams.org/newswire/2016/05/27/philippines-prepares-summon-47-companies-account-climate-change.

20 See Greenpeace, *Philippines Launches World's First National Human Rights Investigation Into 50 Big Polluters*, Greenpeace International (4 December 2015), available online at www.greenpeace.org/international/en/press/releases/2015/Philippines-launches-worlds-first-national-human-rights-investigation-into-50-big-polluters/.

21 AFCommHR, *Social and Economic Rights Action Center & Another v. Nigeria*, Judgment (Admissibility, Merits and Holding), 27 October 2001; See generally, E. Grant, 'International Human Rights Courts and Environmental Human Rights: Re-Imagining Adjudicative Paradigms', 6 *Journal of Human Rights and the Environment* (2015) 156; A. Boyle, 'Human Rights and the Environment: Where Next?', 23 *The European Journal of International Law* (2012) 613; L. Stevens, 'The Illusion of Sustainable Development: How Nigeria's Environmental Laws Are Failing the Niger Delta', 36 *Vermont Law Review* (2011) 387; see in particular, R. Lord *et al.*, *Climate Change Liability: Transnational Law and Practice* (2012) 39; with regard to climate justice and broader environmental struggles S. Atapattu, *Human Rights Approaches to Climate Change: Challenges and Opportunities* (2016) 7.

22 *Social and Economic Rights Action Center & Another v. Nigeria*, *supra* note 21.

23 *Ibid.*

24 See for example, C. Beaton *et al.*, *A Guidebook to Fossil-Fuel Subsidy Reform for Policy-Makers in South East Asia* (2013) 16.

25 To justify expanded subsidies to fossil fuel corporations in the UK in 2015, a spokesperson for its Department of Energy and Climate Change stated, "We are committed to meeting our decarbonization targets – we've made record investments in renewables and are focusing on low-carbon secure energy sources, such as nuclear and shale gas. *However, this will not happen overnight – oil and gas will continue to play a role so we can ensure hardworking families and businesses have access to secure, affordable energy.*" (emphasis mine). See D. Carrington, UK Becomes Only G7 Country to Increase Fossil Fuel Subsidies, *The Guardian* (12 November 2015), available online at www.theguardian.com/environment/2015/nov/12/uk-breaks-pledge-to-become-only-g7-country-increase-fossil-fuel-subsidies.

26 M. Bazilian and I. Onyeji, 'Fossil Fuel Subsidy Removal and Inadequate Public Power Supply: Implications for Business', 45 *Energy Policy* (2012) 1–2.

27 *Ibid.*, at 2.

28 O. C. Okafor, 'The Precarious Place of Labour Rights and Movements in Nigeria's Dual Economic and Political Transition, 1999–2005', 51 *Journal of African Law* (2007) 68.

29 O. C. Okafor, 'Between Elite Interests and Pro-Poor Resistance: The Nigerian Courts and Labour-Led Anti-Fuel Price Hike Struggles (1999–2007)', 54 *Journal of African Law* (2010) 95.

30 See Nigeria's Economy Slips Into Recession, *BBC News* (31 August 2016), available online at www.bbc.com/news/business-37228741; K. Barnato, Nigeria's Government Hints at Asset Sales to Lift Economy out of Recession, *CNBC* (19 September 2016), available online at www.cnbc.com/2016/09/19/nigeria-government-hints-at-asset-sales-to-lift-economy-out-of-recession.html.

31 See E. King, *Saudi Arabia Warns UN Over 1.5C Climate Study* (12 September 2016), available online at www.climatechangenews.com/2016/09/12/saudi-arabia-warns-un-over-1-5c-climate-study/.

32 See Matters Relating to Science Review: Advice on how the Assessment of the Intergovernmental Panel on Climate Change Can Inform the Global Stocktake, Submission by Saudi Arabia, SBSTA 44 Agenda item 6(b), available online at http://www4.unfccc.int/Submissions/Lists/OSPSubmissionUpload/102_264_131179740143972731-Saudi%20Arabia%20Submission%20on%20Global%20Stocktake%20and%20IPCC_%20SBSTA.pdf.

33 PBL Netherlands Environmental Assessment Agency and European Commission, Trends in Global CO_2 Emissions 2015 Report: Background Studies (2015), available online at http://edgar.jrc.ec.europa.eu/news_docs/jrc-2015-trends-in-global-co2-emissions-2015-report-98184.pdf, at 14.

34 Article 4(1), Paris Agreement, (2015) I-54113.

35 See for example *Building Nigeria's Response to Climate Change: National Adaptation Strategy and Plan of Action on Climate Change in Nigeria* (November 2011), available online at http://citeseerx.ist.psu.edu/viewdoc/download;jsessionid=27C7F6A7FD5FCAAD495C69457F2C5F7D?doi=10.1.1.367.6707&rep=rep1&type=pdf (last visited 14 November 2016); *National Strategy on Climate Change and Low Carbon Development for Rwanda: Baseline Report* (2011), available online at http://cdkn.org/wp-content/uploads/2010/12/FINAL-Baseline-Report-Rwanda-CCLCD-Strategy-super-low-res.pdf.

36 See for example ConocoPhillips, *Taking Action on Climate Change*, available online at www.conocophillips.com/sustainable-development/environment/climate-change/climate-change-action-plan/Pages/default.aspx (last visited 14 November 2016); ExxonMobil, *Our Action on Climate Change*, available online at http://corporate.exxonmobil.com/en/current-issues/climate-policy/climate-perspectives/our-position.

37 Article 15(2), Paris Agreement, (2015) I-54113.

38 G. Hardin, 'The Tragedy of the Commons', 162 *Science* (1968) 1243; K. Engel and S. Saleska, 'Subglobal Regulation of the Global Commons: The Case of Climate Change', 32 *Ecology Law Quarterly* (2005) 183.

39 See for example J. Farrell, 'Corporate Funding and Ideological Polarization About Climate Change', 113 *PNAS* (2016) 92.

40 See for example I. A. Saeverud and J. B. Skjaerseth, 'Oil Companies and Climate Change: Inconsistencies Between Strategy Formulation and Implementation?', 7 *Global Environmental Politics* (2007) 42.

41 Q. M. Liang, Y. Fan and Y. M. Wei, 'Carbon Taxation Policy in China: How to Protect Energy and Trade-Intensive Sectors', 29 *Journal of Policy Modeling* (2007) 311; K. Harrison, 'The Comparative Politics of Carbon Taxation', 6 *Annual Review of Law and Social Science* (2010) 507; K. Harrison, 'A Tale of Two Taxes: The Fate of Environmental Tax Reform in Canada', 29 *Review of Policy Research* (2012) 383; D. G. Duff, 'Carbon Taxation in British Columbia', 10 *Vermont Journal of Environmental Law* (2008) 87; B. G. Rabe and C. P. Borick, 'Carbon Taxation and Policy Labeling: Experience From American States and Canadian Provinces', 29 *Review of Policy Research* (2012) 358.

42 B. C. Murray, R. G. Newell and W. A. Pizer, 'Balancing Cost and Emissions Certainty: An Allowance Reserve for Cap-and-Trade', 3 *Review of Environmental Economics and Policy* (2009) 1; C. Fischer,

'Combining Rate-Based and Cap-and-Trade Emissions Policies', 3 *Climate Policy* (2003) 89; G. Luderer *et al.*, 'On the Regional Distribution of Mitigation Costs in a Global Cap-and-Trade Regime', 114 *Climatic Change* (2012) 59.

43 See O. Amao, 'Corporate Social Responsibility, Multinational Corporations and the Law in Nigeria: Controlling Multinationals in Host States', 52 *Journal of African Law* (2008) 89; J. T. Ayoola, 'Gas Flaring and Its Implications for Environmental Accounting in Nigeria', 4 *Journal of Sustainable Development* (2011) 244.

44 See N. Lutsey and D. Sperling, 'America's Bottom-up Climate Change Mitigation Policy', 36 *Energy Policy* (2008) 673; D. Thornton, R. Kagan and N. Gunningham, 'Compliance Costs, Regulation, and Environmental Performance: Controlling Truck Emissions in the US', 2 *Regulation & Governance* (2008) 275; D. Bauner, 'Global Innovation vs. Local Regulation: Introduction of Automotive Emission Control in Sweden and Europe', 7 *International Journal of Environmental Technology and Management* (2007) 244.

45 For an illustration of how businesses are responding to climate change in developing countries, see UN Global Compact and United Nations Environmental Program, *Business and Climate Change: Toward Resilient Companies and Communities* (2012), available online at www.unglobalcompact.org/docs/issues_doc/Environment/climate/Business_and_Climate_Change_Adaptation.pdf; see also Carbon Disclosure Project, *Insights Into Climate Change Adaptation by UK Companies* (2012), available online at http://webarchive.nationalarchives.gov.uk/20130402151656/http://archive.defra.gov.uk/environment/climate/documents/cdp-adaptation-report.pdf.

46 See for example I. A. Saeverud and J. B. Skjaerseth, *supra* note 40.

47 See Centre for European Political Studies, *Global Sectoral Industry Approaches to Climate Change: CEPS Task Force Report* (2008), available online at http://aei.pitt.edu/9529/2/9529.pdf.

48 See Jeswani, Wehrmeyer and Mulugetta, *supra* note 10, at 48.

49 *Ibid.*

50 BP Global, *Why We Want to Act on Climate Change* (12 November 2015), available online at www.bp.com/en/global/corporate/bp-magazine/observations/why-we-want-to-act-on-climate-change.html (last visited on 12 October 2016); See also A. Kolk and D. Levy, 'Winds of Change: Corporate Strategy, Climate Change and Oil Multinationals', 19 *European Management Journal* (2001) 501; E. Lowe and R. Harris, 'Taking Climate Change Seriously: British Petroleum's Business Strategy', 5 *Corporate Environmental Strategy* (1998) 22.

51 BP Global, *Oil and Gas CEOs Jointly Declare Action on Climate Change* (16 October 2015), available online at www.bp.com/en/global/corporate/press/press-releases/oil-and-gas-ceos-jointly-declare-action-on-climate-change.html (last visited 12 October 2016).

52 A. Nelsen, BP Tops the List of Firms Obstructing Climate Action in Europe, *The Guardian* (21 September 2015), available online at www.theguardian.com/environment/2015/sep/21/bp-tops-the-list-of-firms-obstructing-climate-action-in-europe (last visited 12 October 2016).

53 See K. Mulvey *et al.*, *The Climate Accountability Scorecard: Ranking Major Fossil Fuel Companies on Climate Deception, Disclosure, and Action, Union of Concerned Scientists* (2016), available online at www.ucsusa.org/sites/default/files/attach/2016/10/climate-accountability-scorecard-full-report.pdf.

54 *Ibid.*, at 2.

55 *Ibid.*, at 3.

56 *Ibid.*, at 1.

57 R. Cox, 'A Climate Change Litigation Precedent: Urgenda Foundation v The State of the Netherlands', 34 *Journal of Energy & Natural Resources Law* (2016) 143; S. Roy and E. Woerdman, 'Situating Urgenda Versus The Netherlands With Comparative Climate Change Litigation', 34 *Journal of Energy & Natural Resources Law* (2016) 165.

58 *Asghar Leghari v. Federation of Pakistan*, Lahore WP No. 25501/2015 No.1, HC Green Bench, Pakistan (31 August 2015), available online at https://elaw.org/pk_Leghari (last visited 14 November 2016).

59 *The People v. Arctic Oil*, Oslo District Court, Norway (18 October 2016), available online at www.greenpeace.org/norway/Global/norway/Arktis/Dokumenter/2016/legal_writ_english_final_20161018.pdf.

60 *Ibid.*

61 See C. Hunt, O. Weber and T. Dordi, 'A Comparative Analysis of the anti-Apartheid and Fossil Fuel Divestment Campaigns', 6 *Journal of Sustainable Finance & Investment* (2016) 64, available online at www.tandfonline.com/doi/abs/10.1080/20430795.2016.1202641?journalCode=tsfi20.

62 See C. J. Cleveland and R. Reibstein, *The Path to Fossil Fuel Divestment for Universities: Climate Responsible Investment, Boston University* (12 February 2015), available online at http://energyincontext.com/

wp-content/uploads/2015/02/University-Divestment-Fossil-Fuels-Cleveland_Reibstein_02_13_15.pdf; Fossil Free Stanford, *The Case for Fossil Fuel Divestment at Stanford University* (2015), available online at www.fossilfreestanford.org/uploads/2/3/4/0/23400882/thecaseforfossilfueldivestmentatstanforduniversity2015.pdf.

63 J. Rubin, 'The Case for Divesting From Fossil Fuels in Canada', 112 *CIGI Papers* (2016), available online at www.cigionline.org/publications/case-divesting-fossil-fuels-canada, at 2 (last visited 9 November 2016); See also B. Longstreth, The Financial Case for Divestment of Fossil Fuel Companies by Endowment Fiduciaries, *Huffington Post Politics* (13 February 2014), available online at www.maine.edu/wp-content/uploads/2014/02/TAB-1.1-The-Financial-Case-for-Divestment.pdf; A. Gore and D. Blood, The Coming Carbon Asset Bubble, *The Wall Street Journal* (30 October 2013), available online at www.genfound.org/media/pdf-wsj-carbon-asset-bubble-30-10-13.pdf; F. Van der Ploeg, 'Fossil Fuel Producers Under Threat', 32 *Oxford Review of Economic Policy* (2016) 206; J. Balzac, *Corporate Responsibility: Promoting Climate Justice Through the Divestment of Fossil Fuels and Socially Responsible Investment*, available online at https://papers.ssrn.com/sol3/papers.cfm?abstract_id=2871803; G. Alexander and R. Buchholz, 'Corporate Social Responsibility and Stock Market Performance', 21 *Academy of Management Journal* (1978) 479; L. Becchetti and R. Ciciretti, 'Corporate Social Responsibility and Stock Market Performance', 19 *Applied Financial Economics* (2009) 1283.

64 For a description of some of these impacts, see N. Schneider, 'Revisiting Divestment', 66 *Hastings Law Journal* (2015) 589, at 600–601.

65 See Rubin, *supra* note 63, at 2.

66 *Ibid.*, at 3.

67 See *supra* note 13.

68 DC Divest, 7 *Reasons to Sell your Coal, Oil and Gas Stocks, Fossil Free* (3 April 2014), available online at http://gofossilfree.org/7-reasons-to-sell-coal-oil-and-gas-stocks/.

69 L. T. Starks, 'EFA Keynote Speech: Corporate Governance and Corporate Social Responsibility: What Do Investors Care About? What should Investors Care About?', 44 *The Financial Review* (2009) 461, at 465.

70 See E. Garriga and D. Melé, 'Corporate Social Responsibility Theories: Mapping the Territory', 53 *Journal of Business Ethics* (2004) 51, at 61.

71 See Boston College Center for Corporate Citizenship and Ernst & Young LLP, *Value of Sustainability Reporting* (2013), available online at www.confluencellc.com/uploads/3/7/9/6/37965831/value-ofsustainabilitysummary.pdf, at 12 (last visited on 9 November 2016); Global Reporting Initiative, *Sustainability Reporting Guidelines* (2011), available online at www.globalreporting.org/resourcelibrary/G3.1-Guidelines-Incl-Technical-Protocol.pdf, at 32; Letter of International Corporate Accountability Roundtable to the U.S. Securities and Exchange Commission on Business and Financial Disclosures (19 July 2016), available online at www.sec.gov/comments/s7-06-16/s70616-161.pdf, at 2.

72 See Starks, *supra* note 69, at 465.

73 *Ibid.*

74 See Divest Waterloo, *Divestment is In Ethical Win*, available online at http://divestwaterloo.ca/why-divest/divestment-is-an-ethical-win/.

75 IIGCC *et al.*, *Investor Expectations of Automotive Companies: Shifting Gears to Accelerate the Transition to Low Carbon Vehicles* (2016), available online at www.iigcc.org/files/publication-files/IIGCC_2016_Auto_report_v13_Web.pdf.

76 *Ibid.*

77 *Ibid.*, at 5.

78 It has been argued that sustainable development, for example, incorporates international human rights. See E. Dorsey *et al.*, 'Falling Short of Our Goals: Transforming the Millennium Development Goals Into Millennium Development Rights', 28 *Netherland Quarterly of Human Rights* (2010) 516; P. G. Ferreira, 'Did the Paris Agreement Fail to Incorporate Human Rights in Operative Provisions? Not If You Consider the 2016 SDGs', 113 *CIGI Papers* (2016), available online at www.cigionline.org/publications/did-paris-agreement-fail-incorporate-human-rights-operative-provisions, at 6.

79 G. Fouche, Norway's Fund Barred From Investing in U.S. Firm Duke Energy, *Reuters* (7 September 2016), available online at www.reuters.com/article/us-norway-swf-duke-energy-idUSKCN11D0WB.

80 *Ibid.*

81 M. Holter, Norway's $860 Billion Fund Drops 52 Companies Linked to Coal, *Bloomberg* (14 April 2016), available online here www.bloomberg.com/news/articles/2016-04-14/norway-s-860-billion-fund-drops-52-companies-linked-to-coal.

82 *Ibid.*
83 R. Kodaira, Japan Pension Fund to Pick Socially Responsible Stocks, *Nikkei Asian Review* (2 August 2016), available online at http://asia.nikkei.com/Politics-Economy/Economy/Japan-pension-fund-to-pick-socially-responsible-stocks?platform=hootsuite.
84 See *Establishment of Stewardship & ESG Division* (1 October 2016), available online at www.gpif.go.jp/en/topics/pdf/20161003_release_of_stewardship_enhancement_division_en.pdf.
85 M. Hood, Catholic Groups Divest from Fossil Fuels, *Agence France-Presse* (5 October 2016), available online at http://news.abs-cbn.com/overseas/10/04/16/catholic-groups-divest-from-fossil-fuels.
86 S. Page, Catholic Groups Announce Massive Divestment From Fossil Fuels: The Church Gets It, *Think Progress* (4 October 2016), available online at https://thinkprogress.org/catholic-groups-announce-massive-divestment-from-fossil-fuels-f774e4346bc9#.o1c9u1n9e.
87 Hood, *supra* note 85.
88 See E. Bratman *et al.*, 'Justice Is the Goal: Divestment as Climate Change Resistance', 6 *Journal of Environmental Studies and Sciences* (2016) 677, at 679.

Index

Page numbers in italics indicate figures on the corresponding page.